Mathematical Physics in Theoretical Chemistry

Developments in Physical & Theoretical Chemistry

Series Editor

James E. House

With the new series *Developments in Physical & Theoretical Chemistry*, Elsevier introduces a collection of volumes that highlight timely and important developments in this interdisciplinary field. The series aims to present useful and timely reference works dealing with significant areas of research in which there is rapid growth. Through the contributions of specialists, these volumes will provide essential background on appropriate and relevant topics and provide surveys of the literature at a level to be useful to advanced students and researchers. In this way, the volumes will address the underlying theoretical and experimental background on the topics for researchers entering the topic fields and function as useful reference works of lasting value. A primary goal for the volumes in the series is to provide a strong educational thrust for advanced study in particular fields. Each volume will have an editor who is intimately involved in work constituting the topic of the volume. Although contributions to volumes in the series will include those of established scholars, contributions from those who are rising in prominence will also be included.

2018 *Physical Chemistry of Gas–Liquid Interfaces*
Jennifer A. Faust and James E. House, Editors
2019 *Mathematical Physics in Theoretical Chemistry*
S.M. Blinder and J.E. House, Editors

Developments in Physical & Theoretical Chemistry

J. E. House, Series Editor

Mathematical Physics in Theoretical Chemistry

Edited by

S. M. Blinder

University of Michigan, Ann Arbor, MI and Wolfram Research, Champaign, IL, USA

J. E. House

Illinois Wesleyan University, Bloomington, IL; and Illinois State University, Normal, IL, USA

ELSEVIER

Elsevier
Radarweg 29, PO Box 211, 1000 AE Amsterdam, Netherlands
The Boulevard, Langford Lane, Kidlington, Oxford OX5 1GB, United Kingdom
50 Hampshire Street, 5th Floor, Cambridge, MA 02139, United States

Library of Congress Cataloging-in-Publication Data
A catalog record for this book is available from the Library of Congress

British Library Cataloguing-in-Publication Data
A catalogue record for this book is available from the British Library

ISBN 978-0-12-813651-5

For information on all Elsevier publications
visit our website at https://www.elsevier.com/books-and-journals

Publisher: Susan Dennis
Acquisition Editor: Anneka Hess
Editorial Project Manager: Amy M. Clark
Production Project Manager: Prem Kumar Kaliamoorthi
Cover Designer: Victoria Pearson

Typeset by SPi Global, India

Contents

Contributors

S.M. Blinder
University of Michigan, Ann Arbor, MI, United States

Caila Bruzzese
Department of Chemistry, Brock University, St. Catharines, Ontario, Canada

Kimberly Jordan Burch
Department of Mathematics, Indiana University of Pennsylvania, Indiana, PA, United States

Andrew L. Cooksy
Department of Chemistry and Biochemistry, San Diego State University, San Diego, CA, United States

Guido Fano
University of Bologna, Bologna, Italy

James W. Furness
Department of Physics and Engineering Physics, Tulane University, New Orleans, LA, United States

David Z. Goodson
Department of Chemistry and Biochemistry, University of Massachusetts Dartmouth, North Dartmouth, MA, United States

Justin K. Kirkland
Department of Chemistry, University of Tennessee, Knoxville, TN, United States

Errol Lewars
Department of Chemistry, Trent University, Peterborough, ON, Canada

Devin A. Matthews
Institute for Computational Engineering and Sciences, The University of Texas at Austin, Austin, TX, United States

Egor Ospadov
Department of Physics, Brock University, St. Catharines; Department of Chemistry, The University of Western Ontario, London, Ontario, Canada

Stuart M. Rothstein
Department of Physics; Department of Chemistry, Brock University, St. Catharines, Ontario, Canada

John F. Stanton
Department of Chemistry, University of Florida, Gainesville, FL, United States

Jianwei Sun
Department of Physics and Engineering Physics, Tulane University, New Orleans, LA, United States

Jacob Townsend
Department of Chemistry, University of Tennessee, Knoxville, TN, United States

Inga S. Ulusoy
Department of Chemistry, Michigan State University, East Lansing, MI, United States

Konstantinos D. Vogiatzis
Department of Chemistry, University of Tennessee, Knoxville, TN, United States

Angela K. Wilson
Department of Chemistry, Michigan State University, East Lansing, MI, United States

Yubo Zhang
Department of Physics and Engineering Physics, Tulane University, New Orleans, LA, United States

Mathematical physics in theoretical chemistry

CONTENTS

INTRODUCTION

Theoretical chemistry provides a systematic account of the laws governing chemical phenomena in matter. It applies physics and mathematics to describe the structure and interaction of atoms and molecules, the fundamental units of matter. Through the end of the 19th century, chemistry remained predominantly a descriptive and empirical science.[1] True, there had been developed by then a consistent quantitative foundation based on the notions of atomic and molecular weights, combining proportions, thermodynamic quantities, and the fundamental ideas of molecular stereochemistry. Chemistry was certainly far more rational than its ancient roots in alchemy but was still largely a collection of empirical facts about the behavior of matter. Immanuel Kant, in his Critique of Pure Reason, claimed that "in any special doctrine of nature there can be only as much proper science as there is mathematics therein."[2] This can serve as our philosophical rationalization for emphasizing mathematical methods (specifically the field designated *mathematical physics*) in theoretical chemistry.

[1] A very intriguing account of the historical development of modern chemistry is given by Mary Jo Nye [1].
[2] Quoted in the online Stanford Encyclopedia of Philosophy.

The developments of physics in the 20th century made all of chemistry explicable, in principle, by quantum mechanics. As summarized by Dirac: "The underlying physical laws necessary for the mathematical theory of a large part of physics and the whole of chemistry are thus completely known, and the difficulty is only that the exact application of these laws leads to equations much too complicated to be soluble" [2]. By its very nature, quantum mechanics *is* mathematical physics and thereby we establish the connection which is the theme of this volume. However, the loophole noted by Dirac, the existence of chemical problems too mathematically complex to be solved exactly, justifies the survival of parts of chemistry as an empirical science. In this category are semiempirical concepts of chemical bonding and reactivity. This has also led to computational models promoting rational drug design. These have also stimulated applications of other branches of mathematics, for example, information theory and graph theory applied to the definition of various chemical indices.

The primary objective of theoretical chemistry is to provide a coherent account for the structure and properties of atomic and molecular systems. Techniques adapted from mathematics and theoretical physics are applied in attempts to explain and correlate the structures and dynamics of chemical systems. In view of the immense complexity of chemical systems, theoretical chemistry, in contrast to theoretical physics, generally uses more approximate mathematical techniques, often supplemented by empirical or semiempirical methods.

This volume begins with an introduction to the quantum theory for atoms and small molecules, expanding upon the original applications of mathematical physics in chemistry. This field is now largely subsumed within a subdiscipline known as *computational chemistry*. Chapter 1 begins with an introduction to the Hartree-Fock method, which is the conceptual foundation for computational chemistry. Chapter 2 discusses the basis functions employed in these computations, now largely dominated by Gaussian functions. Chapter 3 describes some post-Hartree-Fock methods, which seek to attain "chemical accuracy" in atomic and molecular computations, in particular, configuration interaction, many-body perturbation theory, and coupled-cluster theory. Chapter 10 discusses diagrammatic techniques borrowed from theoretical physics, which can enhance the efficiency of computations. Chapter 7 is an account of the development of personal computers and their applications to computational chemistry.

For larger molecules and condensed matter, alternative approaches, including density functional theory (Chapter 4) and quantum Monte-Carlo (Chapter 6), are becoming popular computational methods. Some additional topics covered in this volume are vibrational partition functions (Chapter 5), singularity analysis of perturbation theories (Chapter 9), and chemical applications of graph theory (Chapter 8).

Finally, Chapter 11 introduces the principles of the quantum computer, which has the speculative possibility of exponential enhancement of computational power for theoretical chemistry, as well as many other applications.

REFERENCES

[1] Nye MJ. From chemical philosophy to theoretical chemistry. Berkeley: University of California Press; 1993.
[2] Dirac PAM. Quantum mechanics of many-electron systems. Proc R Soc A (Lond) 1929;123:714–33.

CHAPTER

Introduction to the Hartree-Fock method

1

S.M. Blinder
University of Michigan, Ann Arbor, MI, United States

A fundamental bottleneck in both classical and quantum mechanics is the *three-body problem*. That is, the motion of systems in which three or more masses interact cannot be solved analytically, so that approximation methods must be utilized. This chapter introduces the basic ideas of the self-consistent field (SCF) and Hartree-Fock (HF) methods, which provide the foundation for the vast majority of computational work on the electronic structure of atoms and molecules. More advanced generalizations of HF are discussed in Chapter 3. Conceptual developments beyond HF, including density-functional and Monte-Carlo methods, are introduced in subsequent chapters.

1 HARTREE SELF-CONSISTENT FIELD THEORY

A precursor of SCF methods might have been the attempts to study the motions of electrons in many electron atoms in the 1920s, on the basis of the Old Quantum Theory. The energy levels of a valence electron, such as the 3*s*-electron in sodium, could be reproduced quite closely if the Bohr orbits of the innerelectrons were smeared out into a continuous spherically symmetric charge distribution [1–3]. After the development of wave mechanics in 1926, it was recognized by Hartree [4] that Bohr orbits must be replaced by continuous charge clouds of electrons, such that the charge density of a single electron is given by $\rho(\mathbf{r}) = -e|\psi(\mathbf{r})|^2$. Here, e is the magnitude of the electron charge (1.602×10^{-19} coulomb) and the charge density $\rho(\mathbf{r})$ follows the Born interpretation of the atomic orbital $\psi(\mathbf{r})$.

The approaches to atomic and molecular structure that are to be described in this chapter are classified as ab initio ("from the beginning") methods, since no experimental or semiempirical parameters are used (other than the fundamental physical constants).

The simplest application of Hartree's SCF method is the helium atom, with two electrons. Electron 1, which occupies the atomic orbital $\psi_1(\mathbf{r}_1)$, moves in the field of the nucleus and electron 2. The potential energy of an electron with charge $-e$ a distance r from a nucleus of charge $+Ze$ follows directly from Coulomb's law, with

Mathematical Physics in Theoretical Chemistry. https://doi.org/10.1016/B978-0-12-813651-5.00001-2

$$V(r) = -\frac{Ze^2}{r}. \tag{1}$$

(We use Gaussian units to avoid the unnecessary factors $4\pi\epsilon_0$, and, in any event, we will soon be switching to atomic units.) To review, the Schrödinger equation for a hydrogen-like atom can be written

$$\left\{-\frac{\hbar^2}{2m}\nabla^2 - \frac{Ze^2}{r}\right\}\psi(\mathbf{r}) = \epsilon\psi(\mathbf{r}), \tag{2}$$

where the energy for principal quantum number n is given by $\epsilon_n = -Z^2e^2/2a_0n^2$, with a_0 equal to the Bohr radius $\hbar^2/me^2$. The one-electron functions $\psi(\mathbf{r})$, when used in the context of a multielectron system, are called *orbitals* [5], an adjective, used as a noun, to denote the quantum-mechanical analog of classical *orbits*. For an electron at point $\mathbf{r}$ interacting with the charge distribution of a second electron in an atomic orbital $\psi(\mathbf{r}')$, the potential energy is given by

$$V(\mathbf{r}) = e^2\int d^3\mathbf{r}'\frac{|\psi(\mathbf{r}')|^2}{|\mathbf{r}-\mathbf{r}'|}. \tag{3}$$

Thus, the total potential energy for electron 1 is given by

$$V_1(\mathbf{r}_1) = V_1[\psi_2] = -\frac{Ze^2}{r_1} + e^2\int d^3\mathbf{r}_2\frac{|\psi_2(\mathbf{r}_2)|^2}{|\mathbf{r}_1-\mathbf{r}_2|}, \tag{4}$$

where the notation $V_1[\psi_2]$ indicates that V_1 is a *functional* of ψ_2, emphasizing the dependance on the charge distribution of electron 2. In Hartree's method, electron 1 obeys the effective one-particle Schrödinger equation

$$\left\{-\frac{\hbar^2}{2m}\nabla^2 + V_1[\psi_2]\right\}\psi_1(\mathbf{r}_1) = \epsilon_1\psi_1(\mathbf{r}_1), \tag{5}$$

where ϵ_1 is the orbital energy of electron 1, negative for bound states. Analogously, interchanging the labels 1 and 2, the orbital function for electron 2 is the solution of

$$\left\{-\frac{\hbar^2}{2m}\nabla^2 + V_2[\psi_1]\right\}\psi_2(\mathbf{r}_2) = \epsilon_2\psi_2(\mathbf{r}_2). \tag{6}$$

The coupled integro-differential equations (5), (6), known as the *Hartree equations*, can be represented in symbolic form by

$$\mathcal{H}_i^{\text{eff}}\psi_i(\mathbf{r}_i) = \epsilon_i\psi_i(\mathbf{r}_i), \quad i = 1, 2. \tag{7}$$

These are coupled in the sense that the solution to the first equation enters the second equation (via the effective Hamiltonian operator $\mathcal{H}_2^{\text{eff}}$ containing $V_2[\psi_1]$), and vice versa. A solution to these equations can be found, in principle, by a successive approximation procedure. An initial "guess" of the functions ψ_1 and ψ_2 is used to compute the potential energies $V_1[\psi_2]$ and $V_2[\psi_1]$. Each Hartree equations can then be solved to give "first-improved" orbital functions $\psi_1^{(1)}$ and $\psi_2^{(1)}$. These, in turn, are used to recompute $V_1^{(1)}$ and $V_2^{(1)}$, and the new Hartree equations are solved to give second-improved orbital functions. The iterative procedure is continued until the input and output functions agree to within some desired accuracy. The orbital functions and potential fields are then said to be *self-consistent*. The usual quantum-mechanical restrictions on a bound state wavefunction—that it be everywhere single-valued, finite, and continuous—apply at each stage of the computation. Each Hartree equation is thus an eigenvalue problem, soluble only for certain discrete values of ϵ_i (in general, different in each stage). For the helium atom the orbital functions ψ_1 and ψ_2 turn out to be identical. This does not violate the Pauli principle since the two orbitals can have opposite spins. Note that the Hartree method does not itself take spin into account.

Extension of the Hartree method to an N-electron atom is straightforward. Each electron now moves in the potential field of the nucleus plus the overlapping charge clouds of $N-1$ other electrons. Now N coupled integro-differential equations are to be solved:

$$\mathcal{H}_i^{\text{eff}}\psi_i(\mathbf{r}_i) = \epsilon_i\psi_i(\mathbf{r}_i), \quad i = 1\ldots N, \tag{8}$$

where

$$\mathcal{H}_i^{\text{eff}} = -\frac{\hbar^2}{2m}\nabla^2 + V_i[\psi_1, \psi_2\ldots\psi_N], \tag{9}$$

and

$$V_i[\psi_1, \psi_2\ldots\psi_N] = -\frac{Ze^2}{r_i} + \sum_{j\neq i} e^2\int d^3\mathbf{r}_j\frac{|\psi_j(\mathbf{r}_j)|^2}{|\mathbf{r}_i - \mathbf{r}_j|}. \tag{10}$$

Each set of orbital functions $\psi_1\ldots\psi_N$ can be identified with an electronic configuration, for example, $1s^22s^22p^63s$ for the Na atom. It is left to the good sense of the user not to allow more than two of the orbitals $\psi_1\ldots\psi_N$ to be the same.[1] The different orbital pairs should also be constructed to be mutually orthogonal. The eigenvalues ϵ_i should be negative for bound orbitals. Their magnitudes are approximations to the ionization energies of the corresponding electrons.

At this point, it is convenient to introduce atomic units, which simplifies all of the previous formulas by removing the repetitive physical constants. We set

$$\hbar = e\left(\text{or } \frac{e}{\sqrt{4\pi\epsilon_0}}\right) = m = 1. \tag{11}$$

[1]Ignoring this restriction has been dubbed "inconsistent field theory."

The unit of length is the *Bohr* equal to the Bohr radius $a_0 = \hbar^2/me^2 = 0.529177 \times 10^{-10}$ m. The unit of energy is the *Hartree*, equal to e^2/a_0, corresponding to 27.2114 eV. Expressed in atomic units, the Schrödinger equation for a hydrogen-like atom (2) simplifies to

$$\left\{-\frac{1}{2}\nabla^2 - \frac{Z}{r}\right\}\psi(\mathbf{r}) = \epsilon\psi(\mathbf{r}), \tag{12}$$

with $\epsilon_n = -Z^2/2n^2$.

Hartree's SCF method, as described so far, followed entirely from intuitive considerations of atomic structure. We turn next to a more rigorous quantum-theoretical derivation of the method [6,7]. The first step is to write down the Hamiltonian operator for an N-electron atom. Now using atomic units, neglecting magnetic interactions and other higher-order effects:

$$H = \sum_{i=1}^{N}\left\{-\frac{1}{2}\nabla_i^2 - \frac{Z}{r_i}\right\} + \sum_{i>j=1}^{N}\frac{1}{r_{ij}}. \tag{13}$$

The one-electron parts of the Hamiltonian—the kinetic energy and nuclear attraction operators—are contained in the first summation. The second summation, over $N(N-1)/2$ distinct pairs i, j, represents the interelectronic repulsive interactions. The interelectronic distances are denoted $r_{ij} = |\mathbf{r}_i - \mathbf{r}_j|$. The N-electron wavefunction is approximated by a *Hartree product*:

$$\Psi(\mathbf{r}_1 \ldots \mathbf{r}_N) = \psi(\mathbf{r}_1)\psi(\mathbf{r}_2)\ldots\psi(\mathbf{r}_N), \tag{14}$$

where $\psi(\mathbf{r}_i)$ are the one-electron orbitals. These should consist of mutually orthonormal functions

$$\int d^3\mathbf{r}\,\psi_i^*(\mathbf{r})\psi_j(\mathbf{r}) = \langle\psi_i|\psi_j\rangle = \delta_{ij}, \tag{15}$$

with none repeated more than twice (maximum of two electrons per atomic orbital). Note that we have now introduced Dirac notation, for compactness. A fully separable wavefunction such as Eq. (14) would be exact only if the Hamiltonian were a sum of one-electron parts. This is not the case since the electron coordinates are inextricably mixed by the r_{ij}^{-1} terms, representing mutual electron repulsion. We therefore must consider approximate solutions of the N-particle Schrödinger equation, optimized in accordance with the variational principle. This means minimizing the ratio of integrals

$$\mathscr{E} = \frac{\int\ldots\int d^3\mathbf{r}_1\ldots d^3\mathbf{r}_N\,\Psi^* H\Psi}{\int\ldots\int d^3\mathbf{r}_1\ldots d^3\mathbf{r}_N\,|\Psi|^2} = \frac{\langle\Psi|H|\Psi\rangle}{\langle\Psi|\Psi\rangle}. \tag{16}$$

This gives an upper limit to the exact ground state energy E_0: $\mathscr{E} \geq E_0$.

We next give a derivation of the Hartree equations. Using the orthonormalized orbitals $\psi_i(\mathbf{r})$, satisfying Eq. (15), the total wavefunction is found to be normalized as well:

$$\langle \Psi | \Psi \rangle = \langle \psi_i \psi_2 \ldots \psi_N | \psi_1 \psi_2 \ldots \psi_N \rangle = \langle \psi_1 | \psi_1 \rangle \langle \psi_2 | \psi_2 \rangle \ldots \langle \psi_N | \psi_N \rangle = 1. \tag{17}$$

Thus the variational energy can be written, with detailed specification of Ψ and H,

$$\mathscr{E} = \sum_i \left\langle \psi_1 \psi_2 \ldots \psi_N \left| \left\{ -\frac{1}{2}\nabla_i^2 - \frac{Z}{r_i} \right\} \right| \psi_1 \psi_2 \ldots \psi_N \right\rangle \tag{18}$$
$$+ \sum_{i>j} \left\langle \psi_1 \psi_2 \ldots \psi_N | r_{ij}^{-1} | \psi_1 \psi_2 \ldots \psi_N \right\rangle,$$

where we have separately written the contributions from the one-electron and two-electron parts of the Hamiltonian. We now define the one-electron integrals

$$H_i = \int d^3\mathbf{r}\, \psi_i^*(\mathbf{r}) \left\{ -\frac{1}{2}\nabla^2 - \frac{Z}{r} \right\} \psi_i(\mathbf{r}), \tag{19}$$

and the two electron integrals

$$J_{ij} = \iint d^3\mathbf{r} d^3\mathbf{r}' \frac{|\psi_i(\mathbf{r})|^2 |\psi_j(\mathbf{r}')|^2}{|\mathbf{r} - \mathbf{r}'|}. \tag{20}$$

The H_i are known as *core integrals*, while the J_{ij} are called *Coulomb integrals* since they represent the electrostatic interactions of interpenetrating electron-charge clouds. After carrying out the integrations implicit in Eq. (18), we obtain

$$\mathscr{E} = \sum_i H_i + \sum_{i>j} J_{ij}, \tag{21}$$

as an approximation to the total energy of the N-electron atom.

We can now apply the variational principle to determine the "best possible" set of atomic orbitals $\psi_1 \ldots \psi_N$. Formally, a minimum of $\mathscr{E}$ is sought by variation of the functional forms of the ψ_i. The minimization is not unconditional; however, since the N normalization conditions (15) must be maintained. A conditional minimum problem becomes equivalent to an unconditional problem by application of Lagrange's method of undetermined multipliers. The ψ_i and ψ_i^* are formally treated as independent functional variables. The Lagrange multipliers are denoted ϵ_i in anticipation of their later emergence as energies in the Hartree equations. Accordingly, we seek the minimum of the functional

$$\mathscr{L}[\psi_1 \ldots \psi_N, \psi_1^* \ldots \psi_N^*] = \mathscr{E}[\psi_1 \ldots \psi_N, \psi_1^* \ldots \psi_N^*] - \sum_{i=1}^{N} \epsilon_i \langle \psi_i | \psi_i \rangle. \tag{22}$$

Expressing $\mathscr{L}$ in terms of the original integrals, using Eqs. (15), (19), (20), we obtain

$$\mathscr{L}[\psi, \psi^*] = \sum_i \int d^3\mathbf{r}\, \psi_i^*(\mathbf{r}) \left\{ -\frac{1}{2}\nabla^2 - \frac{Z}{r} + \sum_{j \neq i} \int d^3\mathbf{r}' \frac{|\psi_j(\mathbf{r}')|^2}{|\mathbf{r} - \mathbf{r}'|} - \epsilon_i \right\} \psi_i(\mathbf{r}). \tag{23}$$

The variation of $\mathscr{L}[\psi, \psi^*]$ in terms of variations in all the ψ_i and ψ_i^* is given by

$$\delta\mathscr{L} = \sum_i \frac{\partial \mathscr{L}}{\partial \psi_i}\delta\psi_i + \sum_i \frac{\partial \mathscr{L}}{\partial \psi_i^*}\delta\psi_i^* = 0. \tag{24}$$

Since the minimum in $\mathscr{L}$ is unconditional, this result must hold for arbitrary variations of all the $\delta\psi_i$ and $\delta\psi_i^*$. This is possible only if each of the coefficients of these variations vanish, that is,

$$\frac{\partial \mathscr{L}}{\partial \psi_i} = \frac{\partial \mathscr{L}}{\partial \psi_i^*} = 0, \quad i = 1 \ldots N. \tag{25}$$

Let us focus on one particular term in the variation $\delta\mathscr{L}$, namely the term linear in $\delta\psi_k^*$ for some $i = k$. From the condition $\frac{\partial \mathscr{L}}{\partial \psi_k^*} = 0$ applied to Eq. (23), we are led to the Hartree equations[2]

$$\left\{-\frac{1}{2}\nabla^2 - \frac{Z}{r} + \sum_{j \neq k} \int d^3\mathbf{r}' \frac{|\psi_j(\mathbf{r}')|^2}{|\mathbf{r} - \mathbf{r}'|}\right\} \psi_k(\mathbf{r}) = \epsilon_k \psi_k(\mathbf{r}), \quad k = 1 \ldots N, \tag{26}$$

in agreement with Eqs. (8)–(10). We have used the facts that the first summation $\sum_i$ reduces to a single term with $i = k$ and the vanishing of the integral $\int d^3\mathbf{r} \ldots$ for arbitrary values of $\delta\psi_k^*$ implies that the remaining integrand is identically equal to 0.

2 DETERMINANTAL WAVEFUNCTIONS

The electron in each orbital $\psi_i(\mathbf{r})$ is a spin $\frac{1}{2}$ particle and thus has two possible spin orientations w.r.t. an arbitrary spatial direction, $m_s = +\frac{1}{2}$ or $m_s = -\frac{1}{2}$. The spin function is designated σ, which can correspond to one of the two possible spin states $\sigma = \alpha$ or $\sigma = \beta$. We define a composite function, known as a *spin-orbital*

$$\phi(x) = \psi(\mathbf{r})\sigma, \quad \sigma = \begin{cases} \alpha \\ \beta, \end{cases} \tag{27}$$

denoting by x the four-dimensional manifold of space and spin coordinates. For example, a hydrogen-like spin-orbital is labeled by four quantum numbers, so $a = \{n, l, m, m_s\}$. We will abbreviate combined integration over space coordinates and summation over spin coordinates by

$$\sum_{\text{spin}} \int d^3\mathbf{r} = \int dx. \tag{28}$$

[2]The Hartree equations might appear today to have only historical significance, but their generalization leads to the Kohn-Sham equations of modern density-functional theory.

A Hartree product of spin-orbitals now takes the form

$$\Psi(1\dots N) = \phi_a(1)\phi_b(2)\dots\phi_n(N). \tag{29}$$

For further brevity, we have replaced the variables x_i simply by their labels i.

To be physically valid, a simple Hartree product must be generalized to conform to two quantum-mechanical requirements. First is the Pauli exclusion principle, which states that no two spin-orbitals in an atom can be the same. This allows an orbital to occur twice, but only with opposite spins. Second, the metaphysical perspective of the quantum theory implies that individual interacting electrons must be regarded as indistinguishable particles. One cannot uniquely label a specific particle with an ordinal number; the indices given must be interchangeable. Thus each of the N electrons must be equally associated with each of the N spin-orbitals. Since we have now undone the unique connection between electron number and spin-orbital label, we will henceforth designate the spin-orbital labels as lowercase letters $a, b, \dots, n$ while retaining the labels $1, 2, \dots, N$ for electron numbers. The simplest example is again the $1s^2$ ground state of helium atom. Let the two occupied spin-orbitals be $\phi_a(1) = \psi_{1s}(1)\alpha(1)$ and $\phi_b(2) = \psi_{1s}(2)\beta(2)$. To fulfill the necessary quantum requirements, we can construct the (approximate) ground state wavefunction in the form

$$\Psi_0(1,2) = \frac{1}{\sqrt{2}}\Big(\phi_a(1)\phi_b(2) - \phi_a(2)\phi_b(1)\Big). \tag{30}$$

Inclusion of the term with interchanged particle labels, $\phi_a(2)\phi_b(1)$, fulfills the indistinguishability requirement. The factor $\frac{1}{\sqrt{2}}$ preserves normalization for the linear combination (assuming that ϕ_a and ϕ_b are individually orthonormalized). The exclusion principle is also satisfied, since the function would vanish identically if spin-orbitals a and b were the same. A general consequence of the Pauli principle is the *antisymmetry principle* for identical fermions, whereby

$$\Psi(2,1) = -\Psi(1,2). \tag{31}$$

The function (30) has the form of a 2×2 determinant

$$\Psi_0(1,2) = \frac{1}{\sqrt{2}}\begin{vmatrix} \phi_a(1) & \phi_b(1) \\ \phi_a(2) & \phi_b(2) \end{vmatrix}. \tag{32}$$

The generalization for a function of N spin-orbitals, which is consistent with the Pauli and indistinguishability principles, is an $N \times N$ Slater determinant[3]

$$\Psi(1\dots N) = \frac{1}{\sqrt{N!}}\begin{vmatrix} \phi_a(1) & \phi_b(1) & \dots & \phi_n(1) \\ \phi_a(2) & \phi_b(2) & \dots & \phi_n(2) \\ \vdots & \vdots & \ddots & \vdots \\ \phi_a(N) & \phi_b(N) & \dots & \phi_n(N) \end{vmatrix}. \tag{33}$$

[3]The determinantal form was first proposed by Heisenberg [8,9] and Dirac [10]. Slater first used it in the application to a many-electron system [11].

There are $N!$ possible permutations of N electron among N spin-orbitals, which accounts for the normalization constant $1/\sqrt{N!}$. A general property of determinants is that they identically equal to 0 if any two columns (or rows) are equal; this conforms to the Pauli exclusion principle. A second property is that, if any two columns are interchanged, the determinant changes sign. This expresses the antisymmetry principle for an N-electron wavefunction:

$$\Psi(\ldots j \ldots i \ldots) = -\Psi(\ldots i \ldots j \ldots). \tag{34}$$

A closed-shell configuration of an atom or molecule contains $N/2$ pairs of orbitals, doubly occupied with α and β spins; this can be represented by a single Slater determinant. However, an open shell configuration must, in general, be represented by a sum of Slater determinants, so that $\Psi(1 \ldots N)$ will be an eigenfunction of total spin and orbital angular momenta. As a simple illustration, consider the $1s^2$ and $1s2s$ configurations of helium atom. The $1s^2$ ground state can be represented by a single determinant

$$\Psi_0(1,2) = \frac{1}{\sqrt{2}} \begin{vmatrix} \phi_{1s\alpha}(1) & \phi_{1s\beta}(1) \\ \phi_{1s\alpha}(2) & \phi_{1s\beta}(2) \end{vmatrix}, \tag{35}$$

which is an eigenfunction of the spin with eigenvalues $S = 0, M_S = 0$. The $1s2s$ states with $S = 1, M_S = \pm 1$ can likewise be represented by single determinants:

$$\Psi(1,2) = \frac{1}{\sqrt{2}} \begin{vmatrix} \phi_{1s\alpha}(1) & \phi_{2s\alpha}(1) \\ \phi_{1s\alpha}(2) & \phi_{2s\alpha}(2) \end{vmatrix}, \tag{36}$$

for $S = 1, M_S = +1$ and

$$\Psi(1,2) = \frac{1}{\sqrt{2}} \begin{vmatrix} \phi_{1s\beta}(1) & \phi_{2s\beta}(1) \\ \phi_{1s\beta}(2) & \phi_{2s\beta}(2) \end{vmatrix}, \tag{37}$$

for $S = 1, M_S = -1$. The states with the same configuration for $M_S = 0$ must, however, be written as a sum of two determinants:

$$\Psi(1,2) = \frac{1}{\sqrt{2}} \left[\frac{1}{\sqrt{2}} \begin{vmatrix} \phi_{1s\alpha}(1) & \phi_{2s\beta}(1) \\ \phi_{1s\alpha}(2) & \phi_{2s\beta}(2) \end{vmatrix} \pm \frac{1}{\sqrt{2}} \begin{vmatrix} \phi_{1s\beta}(1) & \phi_{2s\alpha}(1) \\ \phi_{1s\beta}(2) & \phi_{2s\alpha}(2) \end{vmatrix} \right]. \tag{38}$$

The (+) sign corresponds to the $S = 1, M_S = 0$ state, and is the third component of the $1s2s\ {}^3S$ term, while the (−) sign corresponds to $S = 0, M_S = 0$ and represents the $1s2s\ {}^1S$ state.

3 HARTREE-FOCK EQUATIONS

The HF method is most usefully applied to molecules. We must, therefore, generalize the Hamiltonian to include the interaction of the electrons with multiple nuclei, located at the points $\mathbf{R}_1, \mathbf{R}_2, \ldots$, with nuclear charges $Z_1, Z_2, \ldots$:

$$H = \sum_i \left\{ -\frac{1}{2}\nabla_i^2 - \sum_A \frac{Z_A}{r_{iA}} \right\} + \sum_{i>j} \frac{1}{r_{ij}}. \tag{39}$$

We use the abbreviation $r_{iA} = |\mathbf{r}_i - \mathbf{R}_A|$. In accordance with the Born-Oppenheimer approximation, we assume that the positions of the nuclei $\mathbf{R}_1, \mathbf{R}_2, \ldots$ are fixed. Thus there are no nuclear kinetic energy terms such as $-\frac{1}{2M_A}\nabla_A^2$. The internuclear potential energy $V_{\text{nucl}}(R) = \sum_{A,B} \frac{Z_A Z_B}{R_{AB}}$ is constant for a given nuclear conformation, which is added to the result after the electronic energy is computed. Note that the total energy $\mathscr{E}(R)$ as well as the one-electron energies $\epsilon_i(R)$ are dependent on the nuclear conformation, abbreviated simply as R. It is of major current theoretical interest to plot *energy surfaces*, which are the molecular energies as functions of the conformation parameters R.

We are now ready to calculate the approximate variational energy corresponding to HF wavefunctions [12,13]

$$\mathscr{E} = \langle \Psi_{\text{HF}} | H | \Psi_{\text{HF}} \rangle. \tag{40}$$

We will now refer to the one-electron functions making up a Slater determinant as *molecular orbitals*. To derive the energy formulas, it is useful to reexpress the determinantal functions in a more directly applicable form. Recall that an $N \times N$ determinant is a linear combination of $N!$ terms, obtained by permutation of the N electron labels $1, 2, \ldots, N$ among the N molecular orbitals. Whenever necessary, we will label the spin-orbitals by $r, s \ldots n$ to distinguish them from the particle labels $i, j \ldots N$. We can then write

$$\Psi_{\text{HF}} = \frac{1}{\sqrt{N!}} \sum_{p=1}^{N!} (-1)^p \mathscr{P}_p \big[\phi_r(1)\phi_s(2)\ldots\phi_n(N) \big], \tag{41}$$

where $\mathscr{P}_p$ is one of $N!$ permutations labeled by $p = 1 \ldots N!$. Permutations are classified as either *even* or *odd*, according to whether they can be composed of an even or an odd number of binary exchanges. The products resulting from an even permutation are *added*, in the linear combination, while those from an odd permutation are *subtracted*. Even permutations are labeled by even p, odd permutations by odd p. Thus each product in the sum is multiplied by $(-1)^p$. Let us first consider the normalization bra-ket of Ψ_{HF}

$$\langle \Psi_{\text{HF}} | \Psi_{\text{HF}} \rangle = \frac{1}{N!} \left\langle \sum_{p=1}^{N!} (-1)^p \mathscr{P}_p[\phi_r(x_1)\phi_s(x_2)\ldots] \middle| \sum_{p'=1}^{N!} (-1)^{p'} \mathscr{P}_{p'}[\phi_r(x_1')\phi_s(x_2')\ldots] \right\rangle. \tag{42}$$

Because of the orthonormality of the molecular orbitals $\phi_r, \phi_s, \ldots$, the only nonzero terms of this double summation will be those with $x_1' = x_1,\ x_2' = x_2, \ldots,\ x_N' = x_N$. There will be $N!$ such terms, thus the bra-ket reduces to

$$\langle \Psi_{HF}|\Psi_{HF}\rangle = \langle \phi_r(x_1)\phi_s(x_2)\ldots|\phi_r(x_1)\phi_s(x_2)\ldots\rangle = \langle \phi_r|\phi_r\rangle\langle \phi_s|\phi_s\rangle\ldots\langle \phi_n|\phi_n\rangle = 1. \quad (43)$$

The core contributions to the energy involves terms in the one-electron sum in Eq. (39). Defining the core operator

$$\mathscr{H}(x) = \left\{-\frac{1}{2}\nabla^2 - \sum_A \frac{Z_A}{|\mathbf{r}-\mathbf{R}_A|}\right\}, \quad (44)$$

the expression for the core integral H_r reduces to

$$H_r = \langle \phi_r(x)|\mathscr{H}(x)|\phi_r(x)\rangle, \quad r = 1, 2, \ldots, n. \quad (45)$$

In analogy with Eq. (43) for case of the normalization bra-ket, all the other factors $\langle \phi_b|\phi_s\rangle$, $s \neq r$ are equal to 1. This is analogous to Eq. (19), the definition of the core integral in the Hartree method, except that now spin-orbitals, rather than simple orbitals are now used. Actually, the scalar products of the spin functions σ_r give factors of 1, so that only the space-dependent orbital functions are involved in the computation, just as in the Hartree case.

We consider next the interelectronic repulsions r_{ij}^{-1}. Following an analogous calculation, all contributions except those containing particle numbers i or j give factors of 1. What remains is

$$\langle r_{ij}^{-1}\rangle = \langle \phi_r(x_i)\phi_s(x_j)|r_{ij}^{-1}|\phi_r(x_i)\phi_s(x_j)\rangle - \langle \phi_r(x_i)\phi_s(x_j)|r_{ij}^{-1}|\phi_r(x_j)\phi_s(x_i)\rangle. \quad (46)$$

The minus sign reflects the fact that interchanging two particle labels i, j multiplies the wavefunction by -1. The first term earlier corresponds to a Coulomb integral (20); again these are labeled by spin-orbitals, but the computation involves only space-dependent orbital functions:

$$J_{rs} = \langle \phi_r(x_i)\phi_s(x_j)|r_{ij}^{-1}|\phi_r(x_i)\phi_s(x_j)\rangle. \quad (47)$$

The second term in Eq. (46) gives rise to an *exchange integral*:

$$K_{rs} = \langle \phi_r(x_i)\phi_s(x_j)|r_{ij}^{-1}|\phi_r(x_j)\phi_s(x_i)\rangle. \quad (48)$$

This represents a purely quantum-mechanical effect, having no classical analog, and arising from the antisymmetry principle. In terms of the orbitals $\psi(\mathbf{r})$, after carrying out the formal integrations over the spin, we can write

$$J_{ij} = \iint d^3\mathbf{r}\, d^3\mathbf{r}'\, \frac{|\psi_i(\mathbf{r})|^2|\psi_j(\mathbf{r}')|^2}{|\mathbf{r}-\mathbf{r}'|} \quad (49)$$

and

$$K_{ij} = \iint d^3\mathbf{r}\, d^3\mathbf{r}'\, \psi_i(\mathbf{r})\psi_j(\mathbf{r}')\frac{1}{|\mathbf{r}-\mathbf{r}'|}\psi_i(\mathbf{r}')\psi_j(\mathbf{r})\langle \sigma_i|\sigma_j\rangle. \quad (50)$$

Unlike J_{ij}, K_{ij} involves the electron spin. Because of the scalar product of the spins associated with ϕ_i and ϕ_j, the exchange integral vanishes if $\sigma_i \neq \sigma_j$, in other words, if spin-orbitals i and j have opposite spins, α, β or β, α.

The expression for the approximate total energy can now be given by the summation

$$\mathscr{E} = \sum_i H_i + \sum_{i>j} \left(J_{ij} - K_{ij}\right). \tag{51}$$

Note that $K_{ii} = J_{ii}$, which would cancel any presumed electrostatic self-energy of a spin-orbital. The effective one-electron equations for the HF spin-orbitals can be derived by a procedure analogous to that of Eqs. (22)–(26). A new feature is that the Lagrange multipliers must now take account of N^2 orthonormalization conditions $\langle\phi_i|\phi_j\rangle = \delta_{ij}$, leading to N^2 multipliers λ_{ij}. Accordingly

$$\mathscr{L}[\phi, \phi^*] = \mathscr{E}[\phi, \phi^*] - \sum_{i>j} \lambda_{ij}\langle\phi_i|\phi_j\rangle. \tag{52}$$

The Lagrange multipliers λ_{ij} can be represented by a Hermitian matrix. It should therefore be possible to perform a unitary transformation to diagonalize the λ-matrix. Fortunately, we do not have to do this transformation explicitly; we can just assume that the set of spin-orbitals ϕ_i are the results after this unitary transformation has been carried out. The new diagonal matrix elements can be designated $\epsilon_i = \lambda_{ii}$. Again, we see that the ϵ_i will correspond to the one-electron energies in the solutions of the HF equations. As a generalization of Eq. (26), the contribution to the variation $\delta\mathscr{L}$ linear in $\delta\phi_i^*$ is given by

$$\begin{aligned}&\left\{-\frac{1}{2}\nabla^2 - \sum_A \frac{Z_A}{|\mathbf{r} - \mathbf{R_A}|} + \sum_{j\neq i} \int dx' \, \frac{|\phi_j(x')|^2}{|\mathbf{r} - \mathbf{r}'|}\right\} \phi_i(x) \\ &\quad - \sum_{j\neq i} \left\{\int dx' \, \frac{\phi_j^*(x')\phi_i(x')}{|\mathbf{r} - \mathbf{r}'|}\right\} \phi_j(x) = \epsilon_i \phi_i(x).\end{aligned} \tag{53}$$

The effective HF Hamiltonian $\mathscr{H}_{\mathrm{HF}}$ is known as the *Fock operator*, designated $\mathscr{F}$. Finally, the HF equations can be written

$$\mathscr{F}\phi_i(x) = \epsilon_i\phi_i(x) \quad i = 1, 2, \ldots, n. \tag{54}$$

In contrast to the Hartree equations (26), $\mathscr{F}\phi_i(x)$ also produces terms linear in the other spin-orbitals ϕ_j, $j \neq i$. Just as in the Hartree case, the coupled set of HF integro-differential equations can, in principle, be solved numerically, using the analogous self-consistency approach, with iteratively improved sets of spin-orbitals.

The significance of the one-electron eigenvalues ϵ_i can be found by premultiplying the HF equation (53) by $\phi_i^*(x)$ and integrating over x. Using the definitions of H_i, J_{ij}, and K_{ij}, we find

$$\epsilon_i = H_i + \sum_j (J_{ij} - K_{ij}). \tag{55}$$

Consider now the difference in energies of the N-electron system and the $(N-1)$-electron system with the spin-orbital ϕ_k removed

$$\mathscr{E}(N) - \mathscr{E}(N-1) = \sum_{i=1}^{N} H_i + \sum_{i>j=1}^{N} (J_{ij} - K_{ij}) - \left[\sum_{i=1}^{N-1} H_i + \sum_{i>j=1}^{N-1} (J_{ij} - K_{ij}) \right]_{i,j \neq k} = H_k + \sum_{j=1}^{N-1} (J_{kj} - K_{kj}) = \epsilon_k. \tag{56}$$

Therefore, the magnitudes of the eigenvalues ϵ_k are approximations for the ionization energies of the corresponding spin-orbitals ϕ_k. Since the ϵ_k are negative, $\mathrm{IP}_k = |\epsilon_k|$. This result is known as *Koopmans' theorem.* It is not exact since it assumes "frozen" spin-orbitals, when the N-electron system becomes an $(N-1)$-electron positive ion. In actual fact, the separately optimized orbitals for an atom or molecule and its positive ion will be different.

It can be shown that the magnitudes of the Coulomb and exchange integrals satisfy the inequalities

$$\frac{1}{2}(J_{ii} + J_{jj}) \geq J_{ij} \geq K_{ij} \geq 0. \tag{57}$$

In general, K_{ij} is an order of magnitude smaller than the corresponding Coulomb integral J_{ij}. HF expressions for the total energy can readily explain why the triplet state of, for example, the $1s2s\ ^3S$ configuration of helium atom is lower in energy than the singlet of the same configuration $1s2s\ ^1S$. Denoting the two-determinant functions in Eq. (38) as $\Psi(^{1,3}S)$ for the (+) and (−) signs, respectively, we compute the expectation value of the two-electron Hamiltonian for helium (with $Z = 2$). After some algebra, the following result is found:

$$\mathscr{E}(^{1,3}S) = \left\langle \Psi(^{1,3}S) \left| \sum_{i=1}^{2} \left\{ -\frac{1}{2}\nabla_i^2 - \frac{2}{r_i} \right\} + \frac{1}{r_{12}} \right| \Psi(^{1,3}S) \right\rangle = H_{1s} + H_{2s} + J_{1s,2s} \pm K_{1s,2s}. \tag{58}$$

Therefore, since $K > 0$, the triplet, with the (−) sign, has the lower energy. One caution, however, is again the fact that the singlet and triplet states have different optimized orbitals, so that the values of $K_{1s,2s}$ (as well as $J_{1s,2s}$, H_{1s}, and H_{2s}) are not equal. But even with separately optimized orbitals, the conclusion remains valid.

One can also give a simple explanation of Hund's first rule based on exchange integrals. For a given electron configuration, the term with maximum multiplicity has the lowest energy. The multiplicity $2S + 1$ is maximized when the number of parallel spins is as large as possible, while conforming to the Pauli principle. But more parallel spins give more contributions of the form $-K_{ij}$, thus lower energy.

4 HARTREE-FOCK EQUATIONS USING SECOND QUANTIZATION

In much of the recent literature on theoretical developments beyond the HF method ("post-HF"), it has become common to express operators and state vectors using *second quantization*, which is based on *creation* and *annihilation operators*. This formalism was originally introduced to represent physical processes that involved actual creation or destruction of elementary particles, photons, or excitations (such as phonons). In a majority of applications of second quantization to quantum chemistry, no electrons are actually created or annihilated. The operators just serve as a convenient and operationally useful device in terms of which quantum-mechanical states, operators, commutators, and expectation values can be represented. To make the notation more familiar to the reader, we will, in this section, reexpress the HF equations in the language of second quantization.

A common way to introduce creation and annihilation operators is via an alternative algebraic approach to the one-dimensional harmonic oscillator. The Schrödinger equation, in atomic units, can be written

$$\frac{1}{2}\left\{-\frac{d^2}{dq^2}+\omega^2q^2\right\}\psi_n(q)=\epsilon_n\psi_n(q),\quad \epsilon_n=\left(n+\frac{1}{2}\right)\omega,\quad n=1,2,\ldots. \tag{59}$$

Now define the operators

$$a=\frac{1}{\sqrt{2}}(q+ip)=\frac{1}{\sqrt{2}}\left(q+\frac{d}{dq}\right),\quad a^\dagger=\frac{1}{\sqrt{2}}(q-ip)=\frac{1}{\sqrt{2}}\left(q-\frac{d}{dq}\right), \tag{60}$$

where $p = -id/dq$ is the dimensionless momentum operator. The canonical commutation relation $[q,p]=i$ implies

$$[a,a^\dagger]=1, \tag{61}$$

and the Hamiltonian operator then simplifies to

$$H=\left(a^\dagger a+\frac{1}{2}\right)\omega. \tag{62}$$

With the wavefunction $\psi_n(x)$ written in Dirac notation as $|n\rangle$, the Schrödinger equation in Eq. (59) becomes

$$H|n\rangle=\left(a^\dagger a+\frac{1}{2}\right)\omega|n\rangle=\left(n+\frac{1}{2}\right)\omega|n\rangle. \tag{63}$$

This implies the relation

$$a^\dagger a|n\rangle=n|n\rangle. \tag{64}$$

The harmonic oscillator equations can be reinterpreted as representing an assembly of photons, or other Bose-Einstein particles, in which $|n\rangle$ is the state with n particles and $a a^\dagger$ is the *number operator*, which counts the number of particles in the state, called the *occupation number*.

Consider now the commutation relation

$$[a^\dagger a, a^\dagger] = a^\dagger [a, a^\dagger] + [a^\dagger, a^\dagger] a = a^\dagger. \tag{65}$$

Applying this to the state $|n\rangle$, we have

$$[a^\dagger a, a^\dagger]|n\rangle = a^\dagger a a^\dagger |n\rangle - a^\dagger a^\dagger a|n\rangle = a^\dagger |n\rangle, \tag{66}$$

which can be rearranged to

$$a^\dagger a(a^\dagger |n\rangle) = (n+1)(a^\dagger |n\rangle). \tag{67}$$

The interpretation of the last equation is that $a^\dagger |n\rangle$ is an eigenfunction of the number operator $a^\dagger a$ with the eigenvalue $n+1$. Thus $a^\dagger$ is a *creation operator*, which increases the number of bosons in the state $|n\rangle$ by 1. The norm of $a^\dagger |n\rangle$ is given by

$$\langle a^\dagger n | a^\dagger n\rangle = \langle n | a a^\dagger | n\rangle = (n+1)\langle n|n\rangle. \tag{68}$$

Thus, if both $|n\rangle$ and $|n+1\rangle$ are normalized, we have the precise relation for the creation operator

$$a^\dagger |n\rangle = \sqrt{n+1}|n+1\rangle. \tag{69}$$

By an analogous sequence of steps beginning with $[a^\dagger a, a] = -a$, we find

$$a|n\rangle = \sqrt{n}|n-1\rangle, \tag{70}$$

showing that a acts as the corresponding *annihilation operator* for the bosons.

The $n = 0$ ground state of the harmonic oscillator corresponds to a state containing no bosons, $|0\rangle$, called the *vacuum state*. A state $|n\rangle$ can be built from the vacuum state by applying $a^\dagger$ n times:

$$|n\rangle = \frac{1}{\sqrt{n!}}(a^\dagger)^n |0\rangle. \tag{71}$$

By contrast, the annihilation operator applied to the vacuum state gives zero. The vacuum state is said to be *quenched* by the action of the annihilation operator.[4]

$$a|0\rangle = 0. \tag{72}$$

In the event that the state contains several different variety of bosons, with occupation numbers $n_1, n_2, \ldots$, the corresponding states will be designated

$$|\Psi\rangle = |n_1, n_2 \ldots n_N\rangle. \tag{73}$$

[4]An interesting philosophical conundrum to ponder is the difference between the vacuum state and zero. One way to look at it: the vacuum state is like an empty box; zero means that the box is also gone.

This is called the *occupation-number representation*, the *n-representation*, or *Fock space*. In the language of second quantization, one does not ask "which particle is in which state," but rather, "how many particles are there in each state." The *vacuum state*, in which all $n_i = 0$, will be abbreviated by

$$|\mathrm{O}\rangle = |0_1, 0_2 \ldots 0_N\rangle. \tag{74}$$

(Another common notation is $|\mathrm{vac}\rangle$.)

There will exist creation and annihilation operators $a_1^\dagger, a_2^\dagger \ldots, a_1, a_2 \ldots$. Assuming that the bosons do not interact, a and $a^\dagger$ operators for different varieties will commute. The following generalized commutation relations are satisfied:

$$[a_i, a_j^\dagger] = \delta_{ij}, \quad [a_i, a_j] = [a_i^\dagger, a_j^\dagger] = 0. \tag{75}$$

For bosons, the occupation numbers n_i are not restricted and the wavefunction of a composite state is symmetric w.r.t. any permutation of indices. Things are, of course, quite different for the case of electrons, or other fermions. The exclusion principle limits the occupation numbers for fermions, n_i to either 0 or 1. Also, as we have seen, the wavefunction of the system is antisymmetric for any odd permutation of particle indices.

The behavior of fermions can be elegantly accounted for by replacing the boson commutation relations (75) by corresponding *anticommutation relations*. The anticommutator of two operators is defined by

$$\{A, B\} \equiv AB + BA, \tag{76}$$

and the basic anticommutation relations for fermion creation and annihilation operators are given by

$$\{a_i, a_j^\dagger\} = \delta_{ij}, \quad \{a_i, a_j\} = \{a_i^\dagger, a_j^\dagger\} = 0. \tag{77}$$

These relations are intuitively reasonable, since the relation $a_i^\dagger a_j^\dagger = -a_j^\dagger a_i^\dagger$ is an alternative expression of the antisymmetry principle (34), while $a_i^\dagger a_i^\dagger = 0$ accords with the exclusion principle.

The state (73) can be constructed by successive operations of creation operators on the N-particle vacuum state

$$|\Psi\rangle = a_1^\dagger a_2^\dagger \ldots a_N^\dagger |0_1, 0_2 \ldots 0_N\rangle = \prod_k a_k^\dagger |\mathrm{O}\rangle. \tag{78}$$

Let us next consider the representation of matrix elements in second-quantized notation. We wish to replace the expectation value of an operator $\mathscr{A}$ for the state $|\Psi\rangle$ with one evaluated for the vacuum state $|\mathrm{O}\rangle$:

$$\langle\Psi|\mathscr{A}|\Psi\rangle \Rightarrow \langle\mathrm{O}|\mathscr{A}_{\mathrm{SQ}}|\mathrm{O}\rangle. \tag{79}$$

We introduce the convention that an annihilation operator, say a_k, acting on the N-electron vacuum state, which we temporarily designate $|O_N\rangle$, produces the $(N-1)$-electron vacuum state $|O_{N-1}\rangle$, in which 0_k is deleted. For a one-electron operator, say $\mathscr{H}(x)$, such as the core term (45) in the HF equations, we can write

$$\sum_k \langle\Psi|\mathscr{H}(x_k)|\Psi\rangle = \sum_r H_{rr}, \quad H_{rr} = \langle\phi_r(x)|\mathscr{H}(x)|\phi_r(x)\rangle, \tag{80}$$

noting that the k summation is over particles, while the r summation is over spin-orbital labels. We can now show that the rule for transcribing a single-particle operator into second-quantized form is given by

$$\sum_k \mathscr{H}(x_k) \Rightarrow \sum_r H_{rr}\, a_r^\dagger a_r, \tag{81}$$

noting that

$$\left\langle O\left|\sum_r H_{rr}\, a_r^\dagger a_r\right|O\right\rangle = \sum_r H_{rr}\left\langle O_N|a_r^\dagger a_r|O_N\right\rangle = \sum_r H_{rr}\langle O_{N-1}|O_{N-1}\rangle = \sum_r H_{rr}, \tag{82}$$

since the $|O_{N-1}\rangle$ are normalized. This agrees with Eq. (80) and verifies the rule (81). For a two-particle operator, such as the electron repulsion r_{ij}^{-1}, we have

$$\sum_{i>j} r_{ij}^{-1} = \frac{1}{2}\sum_{i\neq j} r_{ij}^{-1} \Rightarrow \frac{1}{2}\sum_{r,s,r',s'} \langle\phi_r(i)\phi_s(j)|r_{ij}^{-1}|\phi_{r'}(i)\phi_{s'}(j)\rangle a_r^\dagger a_s^\dagger a_{r'} a_{s'}, \tag{83}$$

where $a_{r'}a_{s'} = a_r a_s$ or $-a_s a_r$. With use of the truncated vacuum state $|O_{N-2}\rangle$, in analogy with the above, we find

$$\begin{aligned}
\sum_{i>j}\langle\Psi|r_{ij}^{-1}|\Psi\rangle &= \frac{1}{2}\sum_{r,s,r',s'} \langle O|\langle\phi_r(i)\phi_s(j)|r_{ij}^{-1}|\phi_{r'}(i)\phi_{s'}(j)\rangle a_r^\dagger a_s^\dagger a_{r'} a_{s'}|O\rangle \\
&= \sum_{r>s}\langle\phi_r(i)\phi_s(j)|r_{ij}^{-1}(|\phi_r(i)\phi_s(j)\rangle - |\phi_s(i)\phi_r(j)\rangle)\langle O_{N-2}|O_{N-2}\rangle \\
&= \sum_{r>s}(J_{rs} - K_{rs}),
\end{aligned} \tag{84}$$

where we have introduced the notation for Coulomb and exchange integrals (47), (48).

5 ROOTHAAN EQUATIONS

A significant improvement in the practical solution of the HF equations was introduced by Roothaan [14]. Almost all current work on atomic or molecular electronic structure is based on this and related procedures. Essentially, the integro-differential equations for the $\phi_i(x)$ are transformed into linear algebraic equations for a set of coefficients $c_{i\alpha}$. The resulting matrix equations are particularly suitable computers based on von Neumann architecture. Accordingly, the spin-orbitals are represented as linear combinations of a set of n basis functions $\{\chi_1(x), \chi_2(x), \ldots, \chi_n(x)\}$:

$$\phi_i(x) = \sum_{\alpha=1}^{n} c_{i\alpha}\chi_\alpha(x), \quad i = 1\ldots N. \tag{85}$$

For $n = N$, we have what is called a *minimal basis set*. For more accurate computations, larger basis sets are used, with $n > N$, even $n \gg N$. Most often, the $\chi_\alpha(x)$ are atomic-like functions, centered about single nuclei in a molecule. Conceptually, this is a generalization of the simple LCAO MO method, in which molecular orbitals of small molecules are approximated by a linear combination of atomic orbitals.

We begin with the HF equations (54)

$$\mathscr{F}\phi_i(x) - \epsilon_i\phi_i(x) = 0, \quad i = 1\ldots N, \tag{86}$$

and substitute the expansion (85), to give

$$\sum_\alpha \Big\{\mathscr{F}\chi_\alpha(x) - \epsilon_i\chi_\alpha(x)\Big\}c_{i\alpha} = 0. \tag{87}$$

Next we multiply both sides by $\chi_\beta^*(x)$ and do the four-dimensional integration over x. The result can be written

$$\sum_\alpha \Big\{\langle\chi_\beta(x)|\mathscr{F}|\chi_\alpha(x)\rangle - \epsilon_i\langle\chi_\beta(x)|\chi_\alpha(x)\rangle\Big\}c_{i\alpha} = 0. \tag{88}$$

Introducing the abbreviations

$$F_{\beta\alpha} = \langle\chi_\beta(x)|\mathscr{F}|\chi_\alpha(x)\rangle, \quad S_{\beta\alpha} = \langle\chi_\beta(x)|\chi_\alpha(x)\rangle, \tag{89}$$

we can write

$$\sum_\alpha \Big\{F_{\beta\alpha} - \epsilon_i S_{\beta\alpha}\Big\}c_{i\alpha} = 0, \tag{90}$$

which can be further abbreviated as the matrix equation

$$\Big\{\mathbf{F} - \epsilon\mathbf{S}\Big\}\mathbf{c} = 0. \tag{91}$$

We need to show, in more detail, the structure of the Fock operator $\mathscr{F}$ and the corresponding matrix $F_{\beta\alpha}$. First, we derive the core, Coulomb, and exchange integrals in terms of the basis functions χ_α. For the core contribution, substitute Eq. (85) into Eq. (45) to give

$$H_i = \sum_{\alpha,\beta} c_{i\beta}^* c_{i\alpha}[\beta|\alpha], \tag{92}$$

having defined the *one-electron integrals* over the basis functions

$$\left[\beta|\alpha\right] = \langle\chi_\beta(x)|\mathcal{H}|\chi_\alpha(x)\rangle. \tag{93}$$

The *core operator* is defined by

$$\mathcal{H} = -\frac{1}{2}\nabla^2 - \sum_A \frac{Z_A}{|\mathbf{r}-\mathbf{R}_A|}. \tag{94}$$

The Coulomb and exchange integrals are given by

$$J_{ij} = \sum_{\alpha,\beta}\sum_{\alpha',\beta'} c^*_{i\beta}c^*_{j\alpha}c_{i\beta'}c_{j\alpha'}\left[\beta\beta'|\alpha,\alpha'\right], \tag{95}$$

and

$$K_{ij} = \sum_{\alpha,\beta}\sum_{\alpha',\beta'} c^*_{i\beta}c^*_{j\alpha}c_{i\alpha'}c_{j\beta'}\left[\beta\alpha'|\alpha,\beta'\right], \tag{96}$$

in terms of the *two-electron integrals*

$$\left[\alpha\beta|\gamma\delta\right] = \iint dx\,dx'\; \chi^*_\alpha(x)\chi_\beta(x)\frac{1}{|\mathbf{r}-\mathbf{r}'|}\chi^*_\gamma(x')\chi_\delta(x'). \tag{97}$$

In an alternative notation, more consistent with Dirac notation and favored by physicists,

$$\langle\alpha\beta|\gamma\delta\rangle = \left\langle\chi_\alpha(x)\chi_\beta(x')\right|\frac{1}{|\mathbf{r}-\mathbf{r}'|}\left|\chi_\gamma(x)\chi_\delta(x')\right\rangle = \left[\alpha\gamma|\beta\delta\right]. \tag{98}$$

We can define the *Coulomb operator*, a term in the Fock operator, by

$$\mathcal{J} = \sum_\beta \int dx' \frac{|\chi_\beta(x')|^2}{|\mathbf{r}-\mathbf{r}'|}, \tag{99}$$

and the *exchange operator* $\mathcal{K}$, such that

$$\mathcal{K}\chi_\alpha(x) = \sum_\beta \int dx' \frac{\chi^*_\beta(x')\chi_\alpha(x')}{|\mathbf{r}-\mathbf{r}'|}\chi_\beta(x). \tag{100}$$

Formally, $\mathcal{K}$ acting on $\chi_\alpha(x)$ gives a term linear in $\chi_\beta(x)$.

The Fock operator, acting on a function $\chi_\alpha(x)$, can now be written explicitly as

$$\mathcal{F} = \mathcal{H} + \mathcal{J} - \mathcal{K}, \tag{101}$$

and the matrix elements as

$$F_{\beta\alpha} = H_{\beta\alpha} + J_{\beta\alpha} - K_{\beta\alpha} = [\beta|\alpha] + [\beta\beta|\alpha\alpha] - [\beta\alpha|\alpha\beta]. \tag{102}$$

We also need an expression for the orthonormalization bra-kets:

$$\langle \phi_i(x)|\phi_j(x)\rangle = \sum_{\alpha,\beta} c^*_{i\beta} c_{j\alpha} S_{\beta\alpha}, \tag{103}$$

where

$$S_{\beta\alpha} = \langle \chi_\beta(x)|\chi_\alpha(x)\rangle, \tag{104}$$

known as *overlap integrals*. The basis functions need not belong to an orthonormal set. Once all the needed one- and two-electron and overlap integrals for a given basis set are calculated, no further integrations are necessary. Thus the problem reduces to linear algebra involving the coefficients $c_{i\alpha}$.

Writing the matrix equation (91) in explicit form, we have

$$\{\mathbf{F} - \epsilon\mathbf{S}\}\mathbf{c} = 0 \quad \Rightarrow \quad \begin{bmatrix} F_{11} - \epsilon S_{11} & F_{12} - \epsilon S_{12} & \dots & F_{1n} - \epsilon S_{1n} \\ F_{21} - \epsilon S_{21} & F_{22} - \epsilon S_{22} & \dots & F_{2n} - \epsilon S_{2n} \\ \vdots & \vdots & \ddots & \vdots \\ F_{n1} - \epsilon S_{n1} & F_{n2} - \epsilon S_{n2} & \dots & F_{nn} - \epsilon S_{nn} \end{bmatrix} \begin{bmatrix} c_1 \\ c_2 \\ \vdots \\ c_n \end{bmatrix} = 0. \tag{105}$$

This system of n simultaneous linear equations, usually called the *Roothaan equations*, has n nontrivial solutions for $\epsilon_i, c_{i1}, c_{i2} \dots c_{in}$ $(i = 1 \dots n)$. The condition for a nontrivial solution is the vanishing of the determinant of the coefficients, that is,

$$\begin{vmatrix} F_{11} - \epsilon S_{11} & F_{12} - \epsilon S_{12} & \dots & F_{1n} - \epsilon S_{1n} \\ F_{21} - \epsilon S_{21} & F_{22} - \epsilon S_{22} & \dots & F_{2n} - \epsilon S_{2n} \\ \vdots & \vdots & \ddots & \vdots \\ F_{n1} - \epsilon S_{n1} & F_{n2} - \epsilon S_{n2} & \dots & F_{nn} - \epsilon S_{nn} \end{vmatrix} = 0. \tag{106}$$

This is an nth degree polynomial equation in ϵ. The n roots correspond to the eigenvalues ϵ_i of the HF equations. The corresponding eigenvectors $\mathbf{c}_i$, whose elements are the coefficients $c_{i1} \dots c_{in}$, are then found by solution of the set of homogeneous linear equations:

$$\begin{aligned} (F_{11} - \epsilon_i S_{11})c_{i1} + (F_{12} - \epsilon_i S_{12})c_{i2} + \cdots + (F_{1n} - \epsilon_i S_{1n})c_{in} &= 0 \\ (F_{21} - \epsilon_i S_{21})c_{i1} + (F_{22} - \epsilon_i S_{22})c_{i2} + \cdots + (F_{2n} - \epsilon_i S_{2n})c_{in} &= 0 \\ &\vdots \\ (F_{n1} - \epsilon_i S_{n1})c_{i1} + (F_{n2} - \epsilon_i S_{n2})c_{i2} + \cdots + (F_{nn} - \epsilon_i S_{nn})c_{in} &= 0. \end{aligned} \tag{107}$$

This is type of generalized matrix eigenvalue problem, which is readily solved by efficient computer programs particularly suited for von Neumann-type computer architecture.

Note that the elements of the Fock matrix $F_{\alpha\beta}$ themselves depend on the coefficients $c_{i\alpha}$, in the Coulomb and exchange terms. But these coefficients are not obtained until the Roothaan equations are solved. So, paradoxically, it seems like we need to solve the equations before we can even write them down! Clearly, however, a recursive approach can be applied. A first "guess" of the coefficients, say $c^{(0)}_{i\alpha}$,

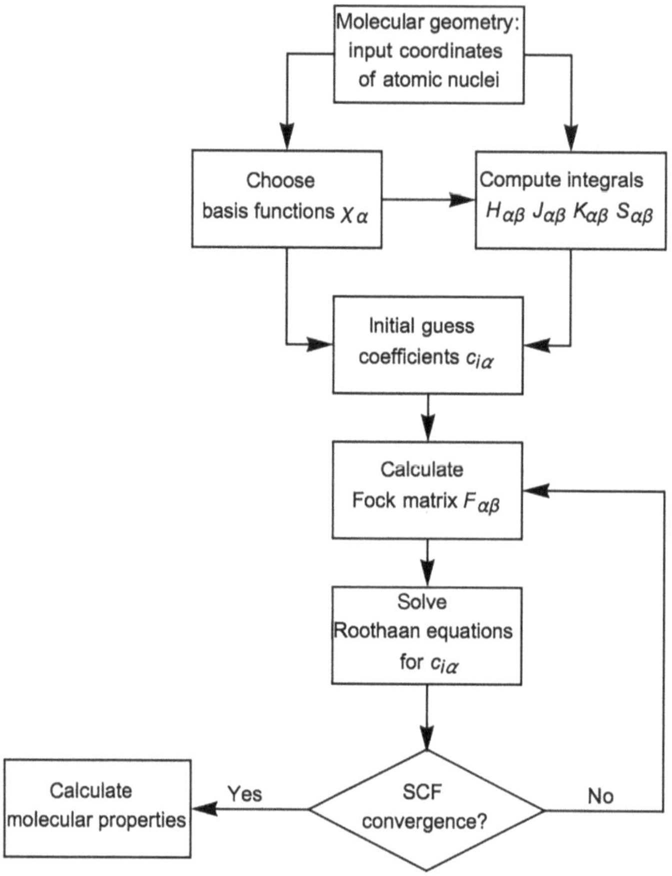

FIG. 1

Flowchart for molecular SCF computation.

is used to construct a Fock matrix. The solution then gives an "improved" set of coefficients, say $c_{i\alpha}^{(1)}$. These are then used, in turn, to construct an improved Fock matrix, which can be applied to obtain a second improved set of coefficients $c_{i\alpha}^{(2)}$. And this procedure is repeated until *self-consistency* is attained, to some desired level of accuracy. A algorithmic flowchart for a Hartree-Fock-Roothaan SCF computation is shown in Fig. 1.

The integrals involving the basis functions χ_α, namely $[\alpha|\beta], [\alpha\beta|\gamma\delta]$ and $S_{\alpha\beta}$ remain constant during the computations involving the Roothaan equations. But more accurate computations require ever-larger basis sets and increasingly difficult computations of the Coulomb and exchange integrals. In computations on molecules, the integrals $[\alpha\beta|\gamma\delta]$ can involve basis functions centered on as many as four different atoms. For many years, the computation of three- and four-center integrals was the most significant impediment to progress in molecular-structure computations. This barrier has since been largely surmounted by the introduction of Gaussian basis functions (to be discussed in Chapter 2), as well as the continuing improvement in computational speed and capacity.

The HF method, as described in this chapter, has been applied mainly using single Slater-determinant wavefunctions. This is an instance of a *mean-field approximation*, in which each electron is described as interacting with the averaged nonlocal potential produced by the other electrons. The neglected effects of instantaneous particle-like interactions are known as *electron correlation*. For example, London dispersion forces are due mainly to electron correlation and are thus not correctly accounted for by HF methods. The *total energy* computed by the HF method generally gives over 99% of the experimental value. This might appear impressive, but lots of important chemistry happens in the remaining 1%. The error due to correlation is generally of the order of 0.04 Hartree per electron pair. This is equivalent to about 1 eV or 100 kJ/mol, the same order of magnitude as electron excitations and molecular binding energies. These represent the *differences* between energies with comparable errors, thus HF computations are of only limited value in accounting for spectroscopic or chemical parameters.[5] For example, in the N_2 molecule, the correlation energy represents 0.5% of the total energy but about 50% of the binding energy. To be useful for chemistry, molecular energies must be computed to a precision of at least six decimal places, equivalent to approximately 0.1 kJ/mol.

In addition, HF computations generally neglect magnetic interactions (e.g., spin-orbit coupling) and relativistic effects. These become increasingly important for systems containing heavy atoms, beginning around $Z > 20$. Also, usually neglected are the effects of finite nuclear masses, leading to what is known as *mass polarization*.

HF computations are, on the other hand, quite good for representing the shapes of atomic and molecular orbitals and for describing electronic charge distributions, as, for example, radial distribution functions for atoms. Also, in biomolecules, reaction sites and docking geometries can often be successfully predicted.

6 ATOMIC HF RESULTS

The most frequently used basis functions in atomic HF computations are *Slater-type orbitals* (STOs), first introduced in 1930 [15]. These are exponential functions suggested by hydrogenic solutions of the Schrödinger equation, but lacking radial nodes, and thus are not mutually orthogonal. STOs have the general form (not normalized):

$$\chi_{n,l,m}(r,\theta,\phi) = r^{n-1} e^{-\zeta r}\, Y_{l,m}(\theta,\phi), \tag{108}$$

[5] An analogy due to C.A. Coulson: First, we weigh the ship with the captain aboard and then we weigh the ship with the captain ashore; the difference gives the captain's weight.

where $Y_{l,m}(\theta, \phi)$ is a spherical harmonic, n is an integer (usually) that plays the role of the principal quantum number, and ζ is an *effective nuclear charge*, chosen to account for the partial shielding by other electrons. STOs exhibit exponential decay at long range and appropriate behavior as $r \to 0$ (Kato's cusp condition[6]). Slater originally suggested a set of rules for the effective nuclear charge

$$\zeta = \frac{Z - \sigma}{n}, \tag{109}$$

where σ is a *screening constant*. Clementi and Raimondi [16] gave improved orbital exponents based on optimized SCF computations. For example, for argon, $Z = 18$:

$$\zeta_{1s} = 17.5075, \quad \zeta_{2s} = 6.1152, \quad \zeta_{2p} = 7.0041, \quad \zeta_{3s} = 2.5856, \quad \zeta_{3p} = 2.2547.$$

A *minimal basis set* is one in which the number of basis functions n is the same as the number of occupied orbitals N. The small number of variable exponential parameters ζ limits the flexibility of the wavefunction. A first step in improving upon a minimal basis set is a sum of two terms in each basis function, so that the radial part of the STO is generalized to

$$c_1\, r^{n-1} e^{-\zeta_1 r} + c_2\, r^{n-1} e^{-\zeta_2 r}. \tag{110}$$

This is known as a *double-zeta basis set*. The optimal orbital exponents ζ_1 and ζ_2 are generally larger and smaller than the original ζ, which allows for better representation of the "expanded" and "contracted" regions of the atomic orbital. There are obvious generalizations to multiple-zeta basis sets, with three or more terms.

STOs are most appropriate for computations on atomic systems. For molecular quantum-chemical computations, however, efficient evaluation of multicenter Coulomb and exchange integrals, involving basis functions centered on more than one atom, are extremely difficult using STOs (although recent theoretical developments and increasing computer power are improving the situation). For this reason, alternative *Gaussian-type orbitals* (GTOs) were first proposed by Boys [17]. These have the general form

$$\chi(x, y, z) = x^m y^n z^p e^{-\alpha r^2}, \tag{111}$$

with angular dependence usually represented by the Cartesian factors. The main advantage of Gaussian basis functions is the *Gaussian product theorem*, according to which product of two GTOs centered on two different atoms can be transformed into a finite sum of Gaussians centered on a point along the axis connecting them. For example, a two-center integral involving $1s$ functions centered about nuclei A and B can be reduced using

---

[6] In a many-electron system, $\frac{1}{\Psi}\frac{\partial \Psi}{\partial r_{iA}}\Big|_{r_{iA}\to 0} = -Z_A$ and $\frac{1}{\Psi}\frac{\partial \Psi}{\partial r_{ij}}\Big|_{r_{ij}\to 0} = \pm\frac{1}{2}$.

$$e^{-\alpha(\mathbf{r}-\mathbf{R}_A)^2} \times e^{-\beta(\mathbf{r}-\mathbf{R}_B)^2} = Ke^{-(\alpha+\beta)(\mathbf{r}-\mathbf{R}_0)^2}, \quad \mathbf{R}_0 = \frac{\alpha\mathbf{R}_A + \beta\mathbf{R}_B}{\alpha+\beta}. \tag{112}$$

As a consequence, four-center integrals can be reduced to finite sums of two-center integrals, with a speedup of several orders of magnitude compared with STOs.

Gaussian functions, with their $e^{-\alpha r^2}$ exponential dependence, compared to the more realistic $e^{-\zeta r}$ dependence, are obviously not very good physical representations of atomic orbitals. They are inaccurate in describing the cusp behavior, as $r \to 0$, and the exponential decay, as $r \to \infty$. In order to minimize these deficiencies but still take advantage of the computational properties of Gaussian functions, Pople [18] suggested the use of *contracted Gaussian type orbitals* (CGTOs). These are linear combination of Gaussian "primitives," with coefficients and exponents fixed, to simulate the behavior of STOs. Fig. 2 shows the optimal approximations of a 1*s* STO by up to three Gaussian primitives. The approximations of STOs by a sum of three Gaussians are called the STO-3G basis set, and are the simplest CGTO set, which can produce HF results of useful accuracy. Basis sets will be discussed in much greater detail in the following chapters.

HF computations have by now been carried out on all the atoms of the periodic table and most atomic ions. Table 1 shows the results for atomic numbers Z = 2–20, including the experimental values obtained from a sum of the Z ionization energies of each atom [19]. The energy values are expressed in Hartrees. As discussed earlier, even the best HF computations cannot reproduce the experimental energies, largely due to the exclusion of correlation energy.

The eigenvalues of the Roothaan matrix represent the energies of the individual spin-orbitals, which, by Koopmans' theorem, are approximations to ionization energies of the atom. The first ionization energy, which is the difference in ground-state energies of the atom and its positive ion, is of particular significance. The periodic properties of these ionization energies are well known and these trends are very well accounted for by high-quality HF computations. Fig. 3 shows a recent compendium [20] of the experimental and calculated ionization energies for elements H through Xe ($Z = 1-54$).

As mentioned earlier, HF computations are fairly reliable in determining the shapes of electronic distributions in atoms and molecules. In particular, the shell structure of many-electron atoms can be exhibited in HF results. An early success was the detection of the radial distribution function for the argon atom, as shown in Fig. 4. The SCF computation of Hartree [21] is compared with the electron-diffraction experimental result of Bartell and Brockway [22]. This is a very impressive result, both for theory and for experiment.

7 POST-HF METHODS

Several extensions of the HF method have been developed to enable more accurate computations on atoms and molecules, in particular, to account for the effects of

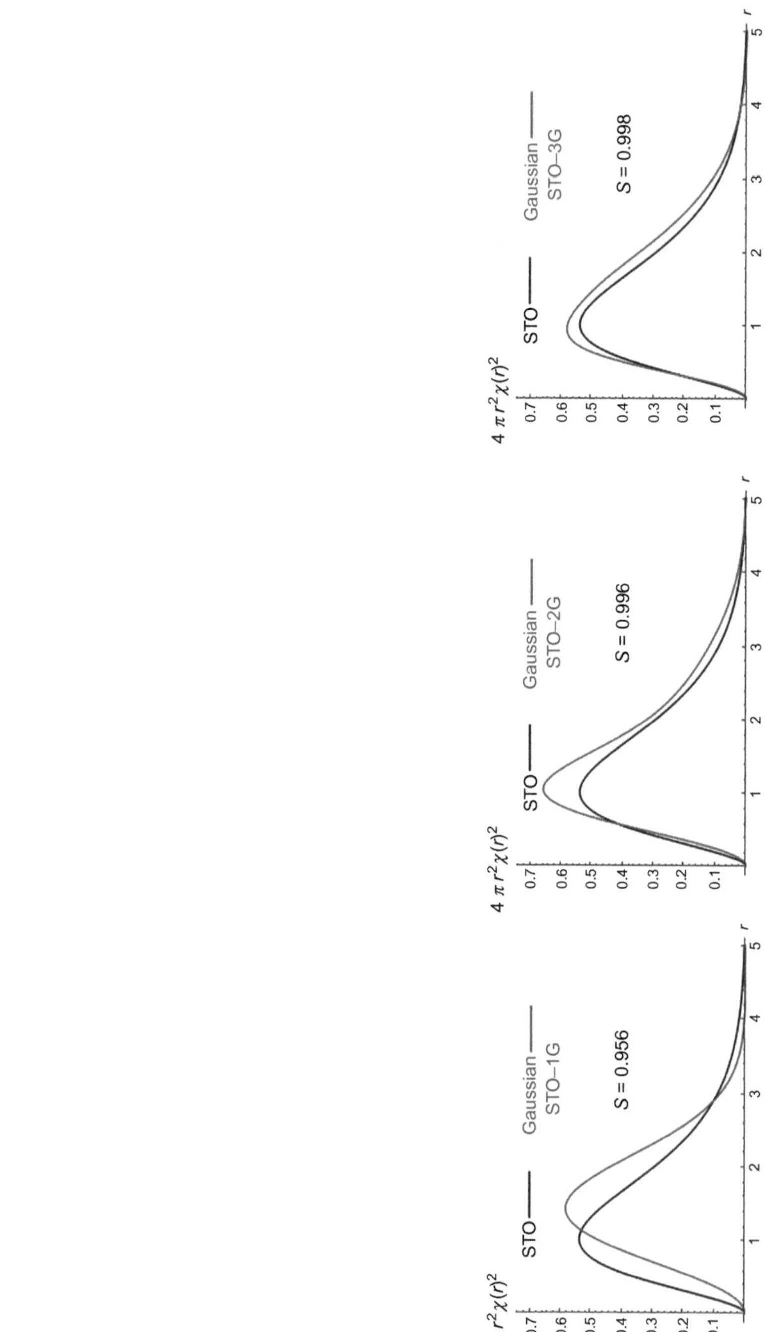

FIG. 2

Radial distribution functions for contracted Gaussians representing the 1*s* STO. The overlap integral *S* is shown for each of the STO-nG functions.

Table 1 HF Computations for Atoms

Z	Atom	Configuration	Exact	H-F
2	He	$1s^2\ ^1S$	−2.903385	−2.8616799
3	Li	$1s^22s\ ^2S$	−7.477976	−7.432726
4	Be	$1s^22s^2\ ^1S$	−14.668449	−14.573023
5	B	$1s^22s^22p\ ^2P$	−24.658211	−24.529060
6	C	$1s^22s^22p^2\ ^3P$	−37.855668	−37.688618
7	N	$1s^22s^22p^3\ ^4S$	−54.611893	−54.400934
8	O	$1s^22s^22p^4\ ^3P$	−75.109991	−74.809398
9	F	$1s^22s^22p^5\ ^2P$	−99.803888	−99.409349
10	Ne	$1s^22s^22p^6\ ^1S$	−128.830462	−128.547098
11	Na	$[Ne]3s\ ^2S$	−162.428221	−161.858911
12	Mg	$[Ne]3s^2\ ^1S$	−200.309935	−199.614636
13	Al	$[Ne]3s^23p\ ^2P$	−242.712031	−241.876707
14	Si	$[Ne]3s^23p^2\ ^3P$	−289.868255	−288.854362
15	P	$[Ne]3s^23p^3\ ^4S$	−341.946219	−340.718780
16	S	$[Ne]3s^23p^4\ ^3P$	−399.034923	−397.504895
17	Cl	$[Ne]3s^23p^5\ ^2P$	−461.381223	−459.482072
18	Ar	$[Ne]3s^23p^6\ ^1S$	−529.112009	−526.817512
19	K	$[Ar]4s\ ^2S$	−601.967492	−599.164786
20	Ca	$[Ar]4s^2\ ^1S$	−680.101971	−676.758185

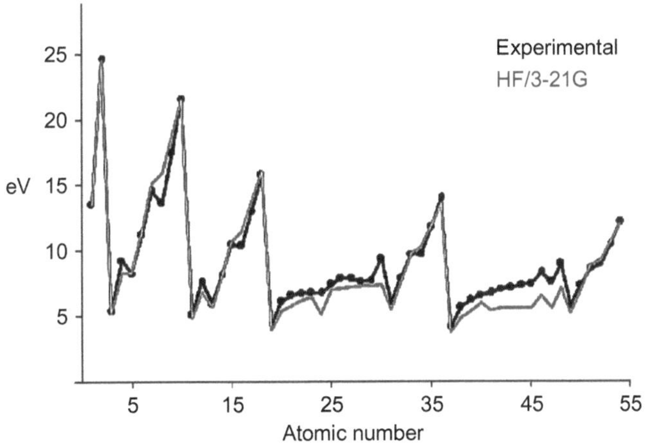

FIG. 3

Experimental and calculated IE value of atoms H-Xe.

(Krishnamohan GP, Mathew T, Saju S, Joseph JT. World J Chem Educ 2017;5:112–9; licensed under a Creative Commons Attribution 4.0 International License.)

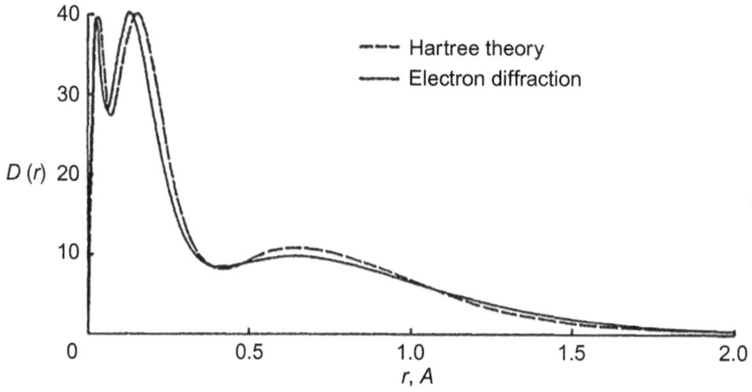

FIG. 4

Radial distribution function for argon atom [22].

electron correlation. Such improved accuracy is necessary to treat chemically significant properties of molecules, including bond energies and geometric parameters—bond lengths and angles. In addition, there are spectroscopic frequencies, dipole moments, magnetic properties, and NMR coupling constants. Post-HF methods will be discussed in detail in the following chapters, but we will give a short introduction here to the main ideas of configuration interaction (CI), Møller-Plesset perturbation theory and the coupled-cluster method.

A single Slater-determinant $|\Phi_0\rangle$, with occupied spin-orbitals $\phi_1, \phi_2 \dots \phi_N$ can serve as a *reference configuration* for an N-electron system. Basis sets of dimension $n > N$ will also produce higher-energy spin-orbitals $\phi_r, \phi_s \dots$, with $\epsilon_r, \epsilon_s > \epsilon_N$, known as *virtual spin-orbitals*, which are unoccupied in the ground state. These might have energies $\epsilon_r, \epsilon_s > 0$. Now an excited configuration can be constructed by promoting one of the electrons from an occupied spin-orbital, say ϕ_i, to one of the virtual orbitals, say ϕ_r. We write the corresponding Slater determinant as $|\Phi_i^r\rangle$. We can also construct *doubly excited* configurations by promoting *two* electrons from occupied spin-orbitals ϕ_i, ϕ_j to virtual spin-orbitals ϕ_r, ϕ_s. Such determinants are denoted $|\Phi_{ij}^{rs}\rangle$. In the jargon of quantum chemistry, the singly and doubly excited configurations are called *singles* and *doubles*, respectively. This can be extended to give *triples*, *quadruples*, etc. The improved ground state, enhanced by CI, can now be represented by a linear combinations of Slater determinants:

$$|\Psi\rangle = c_0|\Phi_0\rangle + \sum_{i,r} c_i^r |\Phi_i^r\rangle + \sum_{i,j,r,s} c_{ij}^{rs} |\Phi_{ij}^{rs}\rangle + \cdots \quad (i,j \dots \le N;\ r,s \dots > N). \tag{113}$$

The coefficients $c_0, c_i^r \dots$ are optimized using a linear variational method. This CI function can *in principle* approach the exact solution of the N-electron Schrödinger equation, as completeness of the one-electron basis set $\{\phi_i(x)\}$ is enhanced and the number of CI contributions $|\Phi\rangle$ is increased. But practical consideration and computational limitations require significant truncation of the CI space. A widely used approximation, called "CISD," truncates the CI expansion to just single and double excitations relative to the reference configuration. Since the Hamiltonian contains only one- and two-electron terms, only singly and doubly excited configurations can interact directly with $|\Phi_0\rangle$.[7] CISD computations can typically account for over 90% of the correlation energy in small molecules. Appropriate strategies for a variety of CI computations will be discussed in detail in succeeding chapters.

A second post-HF method we will describe is *Møller-Plesset perturbation theory* [23]. Recalling the molecular Hamiltonian (Eq. 39), we treat the core terms as an unperturbed Hamiltonian

$$\mathcal{H}_0 = \sum_i \left\{ -\frac{1}{2}\nabla_i^2 - \sum_A \frac{Z_A}{r_{iA}} \right\}, \tag{114}$$

and the electron repulsion terms as a perturbation

$$\mathcal{V} = \sum_{i>j} \frac{1}{r_{ij}} = \frac{1}{2}\sum_{i\neq j} \frac{1}{r_{ij}}. \tag{115}$$

The solution of the HF equations for the ground-state configuration wavefunction $|\Phi_0\rangle$ implies the relation

$$\langle \Phi_0|\mathcal{H}_0|\Phi_0\rangle + \langle \Phi_0|\mathcal{V}|\Phi_0\rangle = E_0^{\mathrm{HF}}. \tag{116}$$

Thus the HF energy formally represents the sum of the unperturbed and first-order perturbation energies

$$E_0^{\mathrm{HF}} = E_0^{(0)} + E_0^{(1)}. \tag{117}$$

In Rayleigh-Schrödinger perturbation theory, the second-order energy is given by

$$E_0^{(2)} = -\sum_{n\neq 0} \frac{\langle \Phi_0|\mathcal{V}|\Phi_n\rangle\langle \Phi_n|\mathcal{V}|\Phi_0\rangle}{E_n^{(0)} - E_0^{(0)}}, \tag{118}$$

where the $|\Phi_n\rangle$ are excited configurations, such as those encountered in CI. We limit the $|\Phi_n\rangle$ to double excitations, noting that, by Brillouin's theorem, single excitations do not contribute.

To evaluate the matrix elements of $\mathcal{V}$, we recall the result for a single Slater determinant

[7] If the same basis functions (not individually optimized) are used in building all configurations, then Brillouin's theorem states that all $\langle \Phi_0|\Phi_i^r\rangle = 0$, so that single excitations do not contribute to the ground state.

$$\langle \Phi_0 | \mathcal{V} | \Phi_0 \rangle = \frac{1}{2} \sum_{i,j} \Big(\langle ij|ij \rangle - \langle ij|ji \rangle \Big), \tag{119}$$

using the notation of Eq. (98). It can be shown analogously that

$$\langle \Phi_0 | \mathcal{V} | \Phi_{ij}^{rs} \rangle = \frac{1}{2} \sum_{i,j,r,s} \Big(\langle ij|rs \rangle - \langle ij|sr \rangle \Big). \tag{120}$$

Therefore, the second-order Møller-Plesset energy, often denoted E_{MP2}, is given by

$$E_0^{(2)} = -\frac{1}{4} \sum_{i,j}^{\mathrm{occ}} \sum_{r,s}^{\mathrm{virt}} \frac{|\langle ij|rs \rangle - \langle ij|sr \rangle|^2}{\epsilon_r + \epsilon_s - \epsilon_i - \epsilon_j}, \tag{121}$$

representing an approximate correlation energy correction to the HF energy E_0^{HF}.

Finally, we mention *coupled cluster* (CC) techniques. These were adapted from nuclear physics and applied to atoms and molecules, largely the work of Čížek [24] and Paldus. Formally, CC applies an exponential cluster operator to the HF wavefunction to obtain the exact solution:

$$|\Psi\rangle = e^T |\Phi_0\rangle. \tag{122}$$

The cluster operator is written in the form

$$T = T_1 + T_2 + T_3 + \cdots, \tag{123}$$

where T_1 is the operator for single excitations, T_2 for double excitations, and so forth. The excitation operators act as follows:

$$T_1 |\Phi_0\rangle = \sum_{i,r} t_i^r \, |\Phi_i^r\rangle, \quad T_2 |\Phi_0\rangle = \frac{1}{4} \sum_{i,j,r,s} t_{ij}^{rs} |\Phi_{ij}^{rs}\rangle, \ldots. \tag{124}$$

The exponential operator e^T is actually defined by its Taylor series expansion. Considering only the cluster operators T_1 and T_2, we can write:

$$e^T = 1 + T + \frac{1}{2!} T^2 + \cdots = 1 + T_1 + T_2 + \frac{1}{2} T_1^2 + T_1 T_2 + \frac{1}{2} T_2^2 + \cdots. \tag{125}$$

These formal relations must, of course, be converted to explicit formulas for actual computations on atoms and molecules. For example, the number of excitations must necessarily be truncated. Coupled-cluster theory, in its several variations, is the de facto standard of modern ab initio computational chemistry, able to accurately account for the chemical properties of a large variety of moderate-sized molecules.

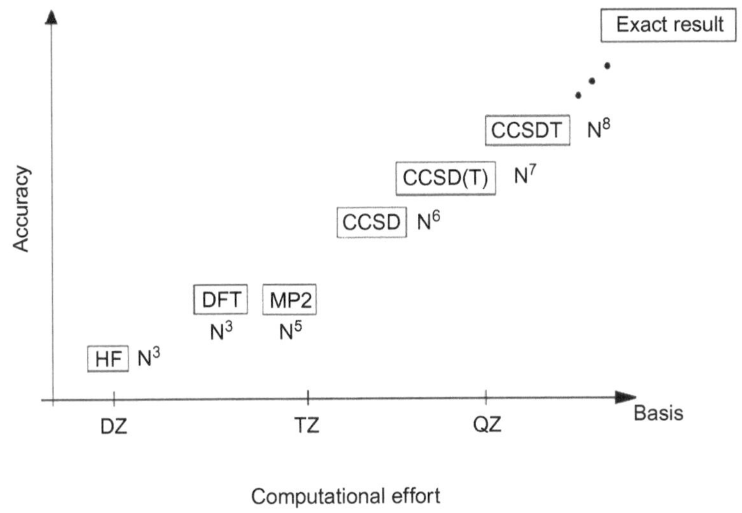

FIG. 5

Pople diagram for ab initio computations.

(Adapted, with permission, from lecture by Alexander A. Auer at MMER Summerschool; 2014, Available from: https://cec.mpg.de/fileadmin/media/Presse/Medien/Auer_Electron_Correlation.pdf.)

A *Pople diagram*[8] provides a general description of the computational effort needed to achieve a given level of accuracy in ab initio molecular computations, as shown in Fig. 5. Key to the diagram: Power of N shows the scaling of the number of computer operations for each method; double-, triple-, quadruple-zeta (DZ, TZ, QZ), Hartree-Fock (HF), density-functional theory (DFT), second-order Møller-Plesset (MP2), coupled-cluster singles and doubles, single, doubles and triples (CCSD, CCSDT), usually experimental result (exact result).

REFERENCES

[1] Fues E. Z Phys 1922;11:364.
[2] Hartree DR. Proc Camb Phil Soc 1923;21:625.
[3] Lindsay RB. J Math Phys 1924;3:191.
[4] Hartree DR. Proc Camb Phil Soc 1928;24(89):111, 246.
[5] Mulliken RS. Phys Rev 1932;41:49
[6] Slater JC. Phys Rev 1928;32:339.
[7] Gaunt JA. Proc Camb Phil Soc 1928;24:328.

[8]Introduced by Pople at the Symposium on Atomic and Molecular Quantum Theory, Sanibel Island, Florida, 1965.

[8] Heisenberg W. Z Phys 1926;38:411.
[9] Heisenberg W. Z Phys 1927;39:499.
[10] Dirac PAM. Proc R Soc (Lond) 1926;A112:661.
[11] Slater JC. Phys Rev 1929;34:193.
[12] Slater JC. Phys Rev 1930;35:210.
[13] Fock V. Z Phys 1930;61:126.
[14] Roothaan CCJ. Rev Mod Phys 1951;23:69.
[15] Slater JC. Atomic shielding constants. Phys Rev 1930;36:57.
[16] Clementi E, Raimondi DL. Atomic screening constants from SCF functions. J Chem Phys 1963;38:2686–9.
[17] Boys SF. Electronic wave functions. I. A general method of calculation for the stationary states of any molecular system. Proc R Soc Lond Ser A 1950;200:542.
[18] Pople JA, Hehre WJ. J Comput Phys 1978;27:161.
[19] Malykhanov YB, Gorshunov MV. Energies of atoms and ions calculated by the Hartree-Fock method. J Appl Spectrosc 2013;80:631–6.
[20] Krishnamohan GP, Mathew T, Saju S, Joseph JT. World J Chem Educ 2017;5:112–9.
[21] Hartree DR, Hartree W. Proc R Soc (Lond) 1938;166:450.
[22] Bartell LS, Brockway LO. Phys Rev 1953;90:833.
[23] Møller C, Plesset MS. Phys Rev 1934;46:618.
[24] Čížek J. J Chem Phys 1966;45:4256.

CHAPTER

Slater and Gaussian basis functions and computation of molecular integrals

2

Inga S. Ulusoy, Angela K. Wilson

Department of Chemistry, Michigan State University, East Lansing, MI, United States

1 INTRODUCTION

As introduced in the prior chapter, addressing the Schrödinger equation requires a number of approximations, and among them is the Hartree-Fock approximation. In applying this approximation to molecular systems, one of the objectives is to determine the best set of molecular orbitals for a molecular system. While the prior chapter addresses specifics of the Hartree-Fock approach, here, the focus is upon basis sets, the mathematical representations of molecular orbitals needed to begin such quantum mechanical calculations. These basis sets enable an initial guess at the molecular orbitals for use in the self-consistent field (SCF) procedure. While this chapter does delve into some of the mathematical concepts related to basis sets and molecular integrals, more in-depth concepts are covered elsewhere. As well, there are aspects of this chapter that will be useful to those who wish to gain insight about basis sets, without delving so much into the theory. Included in this chapter is an introduction to useful terminology.

2 GENERAL REPRESENTATION OF MOLECULAR ORBITALS

In most quantum mechanical calculations, it is important to begin a calculation with a good description of the wave function, as this can impact the overall quality of the calculation, which, in turn, impacts the quality of the properties predicted (i.e., bond length, bond angle, dissociation energies, reaction energies). With this in mind, there are numerous choices that can be made for choosing functions that can be used to express molecular orbitals in terms of a set of known functions, and among these are exponential, Gaussian, polynomial, cube, and plane wave functions. However, factors that play an important role in these choices include a desire to ensure that any functional form chosen enables calculations to be feasible, and, preferably, practical,

Mathematical Physics in Theoretical Chemistry. https://doi.org/10.1016/B978-0-12-813651-5.00002-4

and, that the functions be physically meaningful. In general, molecular orbitals ϕ_i can be expressed as a linear combination of K basis functions, denoted as χ_μ, where $c_{\mu i}$ are coefficients that result in the lowest energy wave function in the Hartree-Fock approach

$$\phi_i = \sum_{\mu=1}^{K} c_{\mu i}\chi_\mu \tag{1}$$

This is an exact relationship provided that $K = \infty$, resulting in an infinite, or complete, basis set expansion. (A set of basis functions is called a basis set.) This is entirely impractical, as the ab initio Hartree-Fock method scales (in terms of computer time) formally as $\approx K^4$ with the number of basis functions (scaling is also often discussed in terms of molecule size), and this scaling increases quite substantially for other electronic structure methods, which are discussed in a later chapter. (And, when the size of the calculation increases, the computer memory and disk space requirements also increase.)

In considering a basis set which is physically meaningful, ensuring that the set is well defined for any nuclear configuration, and, thus, useful for a theoretical model, a particular set of basis functions can be associated with each nucleus, with dependence upon only the charge of the nucleus. These functions can have the symmetry properties of atomic orbitals, with classifications based on their angular properties (e.g., s, p, d, f, etc. types). This results in a focus upon atomic basis functions, which are the typical form of basis functions used for ab initio calculations. The preceding equation, then, is often referred to as a linear combination of atomic orbitals approach to describe molecular orbitals.

3 SLATER- AND GAUSSIAN-TYPE ORBITALS: MATHEMATICS

Most common to quantum chemistry are basis functions that can take two primary forms, Slater-type atomic orbitals (STOs) or Gaussian-type atomic orbitals (GTOs), also referred to as Slater-type orbitals [1] and Gaussian-type orbitals [2].

In tying these concepts to undergraduate physical chemistry courses, orbitals for hydrogenic orbitals take the general form of:

$$\Psi_{nlm}(r,\theta,\phi) = R_{nl}(r)Y_{lm}(\theta,\phi) \tag{2}$$

where $R_{nl}(r)$ is a radial wave function and $Y_{lm}(\theta,\phi)$ is a spherical harmonic, and n, l, and m represent the quantum numbers for an atom. The exact radial wave function for hydrogen is shown in Fig. 1, with a cusp at the nucleus ($r = 0.0\ a_0$) and exponential decay for larger r. This cusp is described through Kato's cusp condition, a theorem which states the electron density has a cusp at the position of the nuclei.

The functional form for molecules becomes far more complicated. Slater introduced a simpler form of

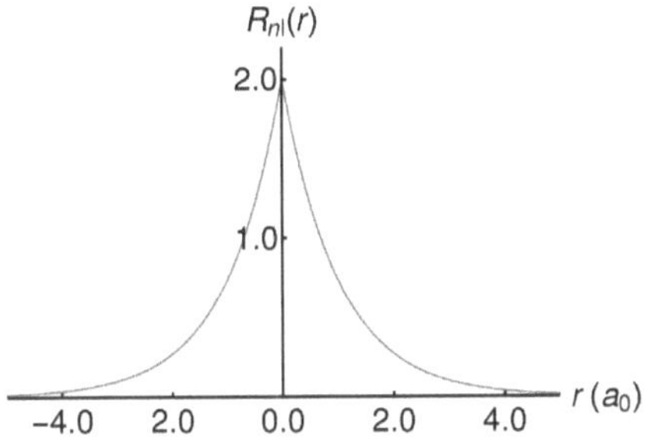

FIG. 1

The exact radial wave function for hydrogen, with the cusp at the nucleus ($r = 0.0\ a_0$) and exponential decay for larger r.

$$R_{nl}(\zeta, r) = (2\zeta)^{n+1/2}\,[(2n!)]^{-1/2}\,r^l \exp[-\zeta r] \tag{3}$$

where

$$\zeta = Z - \frac{\sigma}{n^*} \tag{4}$$

Z is the atomic number, σ is a shielding constant, and n^* is an effective principal quantum number. (As the intent here is to draw a connection to undergraduate physical chemistry course, and not delve fully into the theory, more details of this approach can be found in Refs. [3,4].) In Fig. 2, the radial part of the wave function is shown for a varying exponent ζ (on the left), leading to a "softening" of the cusp, and for different angular momentum quantum numbers l (on the right), leading to a shift of electron density toward larger r, and a node at the nucleus. This form represents the leading term in the Laguerre polynomials, which are typically discussed in calculus-based undergraduate physical chemistry texts. To obtain a description of the entire

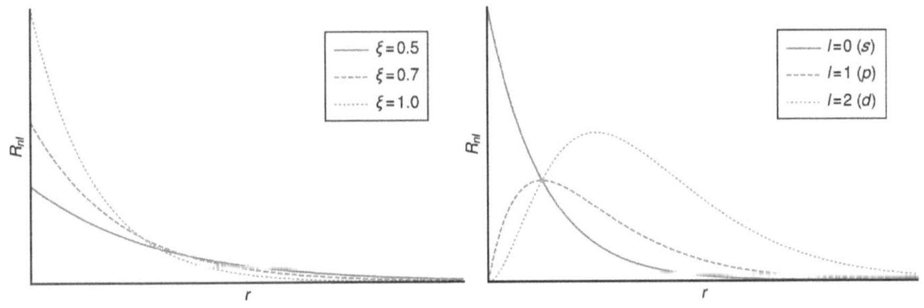

FIG. 2

Slater's approximation to the radial part of the wave function $R_{nl}(\zeta, r)$. On the *left* for different exponents ζ, providing a "softening" of the cusp for smaller ζ. On the *right* for different angular momentum quantum number l, for a principal quantum number $n = 4$ and a unit exponent.

orbital, $R_{nl}(r)$ is multiplied by the appropriate angular part, and, for a 1s orbital, for example, the orbital would be described as

$$\phi_{1s}(\zeta, r) = \left(\frac{\zeta}{\pi}\right)^{1/2} \exp[-\zeta r] \tag{5}$$

STOs seem to be quite useful, as they result in the correct description of the wave function, satisfying Kato's cusp condition and providing exponential behavior at extended distances.

However, quite unfortunately, they are not well suited for numerical work overall. In general, STOs work for atoms (one-center integrals), for diatomics (two-center integrals), which requires numerical integration, and for linear polyatomics, where the numerical integration becomes even more difficult. However, for nonlinear molecules, there is no algorithm to address the required mathematics. While STOs have been used in calculations, their use is quite limited.

GTOs are alternative functions of interest. They are useful, as they do have the angular symmetries of atomic orbitals. An example of the normalized form for a 1s orbital is

$$G_s(\alpha, r) = \left(\frac{2\alpha}{\pi}\right)^{3/4} \exp\left[-\alpha r^2\right] \tag{6}$$

Though this form presents a much more mathematically solvable form than STOs, the drawbacks of GTOs are that they do not have the appropriate behavior at the origin (they are not able to represent the cusp, "Kato's condition") and they decay too quickly (decaying like $\exp[-r^2]$), despite at least having a maximum and decaying, as shown in Fig. 3.

On the upside, a *combination* of Gaussian functions can be used to describe a hydrogenic orbital, enabling better behavior at both the nucleus, as well as at a distance. And, thus, a combination of GTOs is typically used in electronic structure calculations, and this combination is called a contracted GTO (cGTO) or contracted

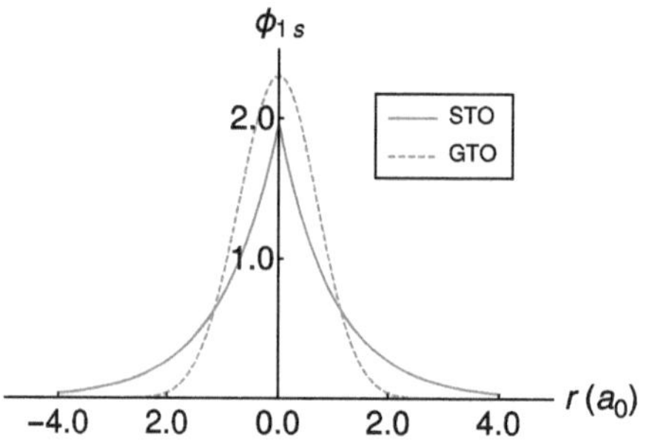

FIG. 3

Unit exponent normalized STO versus GTO. The cusp is not well represented by a single Gaussian function, and the GTO decays much more quickly than the STO.

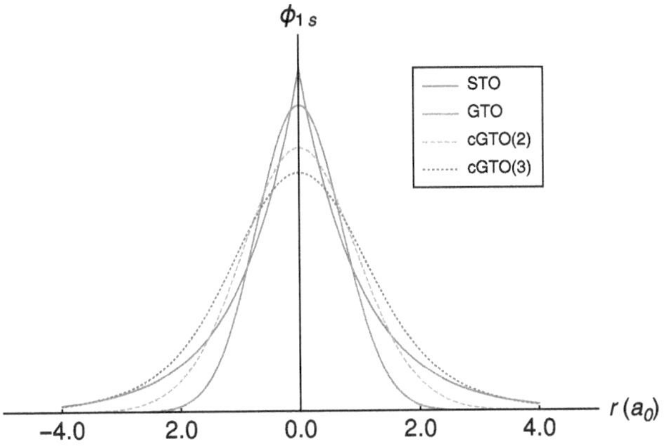

FIG. 4

A GTO and cGTOs to represent the hydrogenic 1*s* atomic orbital, overlayed with an STO for comparison. The GTO uses one Gaussian function to reproduce the shape of the 1*s* function; cGTO(2) and cGTO(3) are linear combinations of two and three Gaussian functions to better represent the cusp and exponential decay for longer distances.

function [5], where each Gaussian in the contraction is referred to as a primitive function

$$\phi_\mu = \sum_s d_{\mu s} G_s(\alpha, r) \tag{7}$$

The fitting coefficients $d_{\mu s}$ in the linear combination are obtained from a fitting procedure, as are the exponents α. Two different cGTO expansions are shown in Fig. 4, in comparison with an STO and a GTO. One expansion uses two Gaussian functions, and another three, to model the shape of the 1*s* hydrogenic orbital. The cGTOs improve upon the deficiencies of GTOs in representing the radial wave function, in particular, the failure to reproduce the cusp and the exponential decay for larger distances from the nucleus. Such fitted contractions are combined in sets, and constitute the basis sets used in electronic structure calculations.

4 BASIS SET TYPES FOR QUANTUM MECHANICAL CALCULATIONS

We will now begin to look at the format of basis functions and basis sets. Despite the details provided within this chapter on the mathematics behind basis sets and molecular integrals, there is a very wide variety of basis sets that are already available, and, for most practical quantum chemical calculations, all that must be done is to select an appropriate basis set, rather than delve into the mathematical complexities. With that being said, however, basis sets come in many different sizes and are of different quality, and the size and quality can have a very substantial impact upon the computational resources (i.e., computer time) required and the accuracy of the properties predicted.

Basis set selection should never be random, and, instead, selection should be made based on the utility of the basis set for past calculations using the same or similar ab initio or density functional method, predicting similar types of properties, and examining very similar molecules when possible. Comparing calculated results with experiment is another important gauge. (However, it is possible to get the "right results for the wrong reason," which is undesirable, as an important part of computational chemistry is the transferability of approaches (including basis sets) from one molecule to the next. This will be discussed in more detail shortly.) But, overall, the basic philosophy behind basis set selection, for most problems, is to use a basis set that is practical, yet reliable for the prediction that is being made. This is quite a generic description, but, a useful one, as questions to consider in basis set choice are, how accurate does the property prediction need to be? Are qualitative results sufficient? Or, are the very best calculations feasible and needed? And, how long might the calculation run? This insight comes from both experience and familiarity with prior research.

As mentioned, there are many types of basis sets. Here, it is not possible to cover all types, but, an overview of some of the most widely used basis sets is provided, as well as a discussion about some of the important features to consider in the selection of a basis set.

A first consideration is basis set size. The smallest basis set is called a minimal basis set. A minimal basis set is comprised of the number of basis functions required to accommodate all of the electrons while maintaining overall spherical symmetry. To illustrate, for hydrogen and helium, all that is needed is to describe a single $1s$ orbital, so this requires one basis function. For lithium and beryllium, what is needed are $1s$ and $2s$ orbitals; however, typically, the shells across a row of the periodic table are all included to maintain symmetry. So, for these elements, $1s$, $2s$, $2p_x$, $2p_y$, and $2p_z$ are all included, as they are for the remainder of the row, boron through neon, resulting in five basis functions for each element. Thus, for sodium through argon, basis functions to describe these five orbitals plus $3s$, $3p_x$, $3p_y$, and $3p_z$ are used, resulting in nine basis functions. To summarize,

Minimal basis set:

$$\begin{aligned} \text{H, He} &= 1s \\ \text{Li-Ne} &= 1s, 2s, 2p_x, 2p_y, 2p_z \\ \text{Na-Ar} &= 1s, 2s, 2p_x, 2p_y, 2p_z, 3s, 3p_x, 3p_y, 3p_z \\ &\vdots \end{aligned}$$

An example of a minimal basis set is an STO-KG set, where $K = 2$–6, and K represents the number of primitive Gaussian functions used to describe a single

STO [6–8]. STO-3G, which is the most common of the STO-KG basis sets, can be described as:

$$\phi_{1s}(r) = c_1 G_1(\alpha_1, r) + c_2 G_2(\alpha_2, r) + c_3 G_3(\alpha_3, r) \tag{8}$$

While these basis sets are computationally efficient, there are many drawbacks to minimal basis sets. First, there is a significant imbalance in the description of the atoms. For example, while boron, with its three electrons, is described by eight basis functions, chlorine, with its seven electrons, is also described by eight basis functions, so, it is likely that the descriptions of the elements at the end of the periodic row (e.g., O, Cl, Ne) will be poorer than the elements at the first part of a row, and, in fact, the basis sets show a significant level of nonuniformity in terms of their ability to make predictions. Another drawback is that the atoms are described by fixed Gaussians. What this means is that there is no flexibility in size—there is no ability to expand or contract an orbital based on the environment, such as polarity, charge, and other environments. These drawbacks make STO-KG (and other minimal basis sets) generally unacceptable approaches for modern computation.

To address these issues, there are a number of remedies. By expanding the numbers of basis functions in the basis set such as providing additional shells of valence basis functions helps to offset the imbalance in the description of the atoms. These additional basis functions enable flexibility in the radial size. As well, additional p and d functions are useful in describing anisotropic and polar molecules. One form of improved basis set is a double-ζ ("double-zeta") basis set. In these types of sets, the minimal basis set is doubled, so that there are two $1s$ functions representing hydrogen and helium, and 10 functions representing Li-Ne, as follows:

$$\begin{aligned}
\text{H, He} &= 1s', 1s'' \\
\text{Li-Ne} &= 1s', 1s'', 2s', 2s'', 2p_x', 2p_x'', 2p_y', 2p_y'', 2p_z', 2p_z'' \\
\text{Na-Ar} &= 1s', 1s'', 2s', 2s'', 2p_x', 2p_x'', 2p_y', 2p_y'', 2p_z', 2p_z'', 3s', 3s'', 3p_x', 3p_x'', 3p_y', 3p_y'', 3p_z', 3p_z''
\end{aligned}$$

In general, the effect of innershell electrons on molecular bonding is minimal, so a well-utilized approach is to double only the number of basis functions representing the valence region, so, for Li-Ne, only the $2s$, $2p_x$, $2p_y$, and $2p_z$ are doubled. Such a basis set is referred to as a split-valence basis set. For the double-ζ basis set, there is no change in the set for hydrogen and helium, as the electrons are valence electrons. So, now, the split-valence sets are of the following composition for double-ζ basis sets:

$$\begin{aligned}
\text{H, He} &= 1s', 1s'' \\
\text{Li-Ne} &= 1s', 2s', 2s'', 2p_x', 2p_x'', 2p_y', 2p_y'', 2p_z', 2p_z'' \\
\text{Na-Ar} &= 1s', 2s', 2p_x', 2p_y', 2p_z', 3s', 3s'', 3p_x', 3p_x'', 3p_y', 3p_y'', 3p_z', 3p_z''
\end{aligned}$$

Innershell electrons, however, can be important to the total energy, so, for calculations where a well-described total energy is needed, additional representation of the core orbitals may be necessary.

To tie this terminology to basis sets that will be found in computational chemistry software packages, two common examples of split-valence basis sets are 3-21G and 6-31G sets [9–11], which are part of one of the most popular series of basis sets, known as the Pople sets, developed by Nobel Laureate John Pople. Overall, sets of this form can be written as N-KLG, where N signifies the number of primitive Gaussian functions that comprise each core atomic orbital basis function. K and L provide insight about the valence orbitals. The fact that there are two numbers, K and L, indicates that the valence orbitals are composed of two basis functions each. The first one is a combination of K primitive Gaussian functions, and the second one is comprised of L primitive Gaussian functions. (If the form is N-KLMG, this just means that the valence orbitals are composed of three basis functions each.) Thus, the notation 3-21G represents two basis functions instead of one for each valence atomic orbital. Each innershell atomic orbital is represented by a single function, which is written in terms of three Gaussian primitive functions. For the valence atomic orbitals, the term is written as an expansion of two Gaussian primitives, and the second term is written as one Gaussian primitive.

For modern calculations, basis sets such as 3-21G are generally not of sufficient quality for the prediction of most properties of interest, and, therefore, are not recommended (unless they are used as a part of a hierarchy of basis sets, as will be discussed shortly). However, for very large molecules (i.e., hundreds of atoms), for example, a double-ζ level basis set may be used in light of the computational cost to try to gain some qualitative data, though caution is encouraged. Greater flexibility in the basis set is best. And, in light of continuous improvements in computer hardware and the ever-evolving development of more efficient computational chemistry methodologies, there will be a point where any calculations of such quality will become obsolete.

Basis sets can be increased further in size, and common notation is to increase the ζ-level ("zeta-level"). Whereas for the double-ζ basis sets, the number of basis functions of each orbital type was doubled from a minimal basis set (with a simpler form doubling only the number of functions for the valence-shell orbitals), the numbers of basis functions can be tripled (triple-ζ basis set), quadrupled (quadruple-ζ basis set), quintupled (quintuple-ζ basis set), or increased by even higher multiples of sets. With each increase, however, comes an increase in computational cost, which becomes very critical as molecule size increases, in terms of both numbers of atoms and numbers of electrons.

The basis sets discussed so far have all been composed of functions that are constrained to be centered at the nucleus. But, chemically, this does not account for polar molecules or strained rings, where the charge may be displaced from the atom centers. To account for this, extra basis functions of higher-angular momentum can be used to help account for this charge displacement. There are numerous examples of basis sets that incorporate polarization, and, among these are the 6-31G* (or 6-31G(d)) and 6-31G** (or 6-31G(d,p)) sets, where the 6-31G split-valence Pople set now includes an additional set of *d* basis functions for nonhydrogen atoms (noted by "*" or "(d)") and also extra *p* basis functions on hydrogen and helium (noted by "**" or "(d,p)") [12–15]. These additional functions can be important to appropriately describe bonding in many molecules.

While polarization functions are useful, in general, the basis sets considered so far are more suitable for molecules in which the electrons are tightly held or fairly tightly held to the nuclear centers. But, for molecules where there is significant electron density that is far removed from the centers, such as for anions, where calculations using basis sets without appropriate flexibility can incorrectly predict that the outermost valence electrons are unbound, additional functions are needed. Highly diffuse functions are incorporated within the basis set to describe the long-range behavior of molecular orbitals, which can have a dramatic impact upon calculations of properties such as proton affinities, electron affinities, Rydberg states, binding energies of van der Waals compounds, and inversion barriers. Within the Pople-style basis sets, the inclusion of highly diffuse functions is indicated by a "+," and examples of such sets include 3-21+G, 6-31+G, and 6-31+G* [16], where the latter set also represents the inclusion of polarization functions.

Another family of basis sets, which are broadly used in computational chemistry, is the correlation consistent basis sets [17–19]. These sets, which have their foundation strongly coupled to atomic natural orbitals [20], have unique, important features that are related to their design. They are constructed so that correlation energy is accounted for in a systematic way—thus the name—"correlation consistent." At this point in the book it is important to note that the Hartree-Fock approximation is simply that: an approximation. In order to improve upon the Hartree-Fock approximation, electron correlation must be introduced. Electron correlation can be defined as the improvement in energy beyond that of the energy obtained from Hartree-Fock. In most practical calculations, electron correlation is needed and thus basis sets designed to take correlation energy into account are vital to this end.

The correlation consistent basis sets are constructed in shells, for example, for B-Ne atoms, the structure of the primitive and contracted functions for these sets is illustrated below. The shell-like structure and correlation consistency become clearest when considering the contracted functions as in Table 1. The notation for the primitive functions and contracted functions for any basis set can be described in the format primitive functions/contracted functions. For example, for the cc-pVDZ basis set earlier, the notation can be expressed as $9s4p1d/3s2p1d$.

The change in size of the basis set from cc-pVDZ to cc-pVTZ comes from an enhancement of the set (beyond *s* and *p* orbitals) with a *d* and an *f* functions. For the next step from cc-pVTZ to cc-pVQZ, there is an enhancement by a *d*, an *f*, and a *g* functions. (Note that for each atom and cc-pV(X+d)Z basis set, all of the functions from the cc-pVXZ basis sets are reoptimized when an additional *d* function is included in the basis set. The "enhancement" is not simply adding a *d* function.)

Table 1 Numbers of Primitive Functions and Contracted Functions for Each Type of Angular Momentum in the Basis Set, for Each ζ-Level of the Correlation Consistent Basis Sets for B-Ne

	Primitive Functions	Contracted Functions
cc-pVDZ	9*s*4*p*1*d*	3*s*2*p*1*d*
cc-pVTZ	10*s*5*p*2*d*1*f*	4*s*3*p*2*d*1*f*
cc-pVQZ	12*s*6*p*3*d*2*f*1*g*	5*s*4*p*3*d*2*f*1*g*
cc-pV5Z	14*s*8*p*4*d*3*f*2*g*1*h*	6*s*5*p*4*d*3*f*2*g*1*h*
cc-pV6Z	16*s*10*p*5*d*4*f*3*g*2*h*1*i*	7*s*6*p*5*d*4*f*3*g*2*h*1*i*

So, for B-Ne, there is an increase in shell structure (beyond the *s* and *p* orbitals) for the contracted functions as follows:

cc-pVDZ	
cc-pVTZ	+1*d*1*f*
cc-pVQZ	+1*d*1*f*1*g*
cc-pV5Z	+1*d*1*f*1*g*1*h*
cc-pV6Z	+1*d*1*f*1*g*1*h*1*i*

The configurations provided for the atoms B-Ne become larger for higher elements of the periodic table, though follow similar types of patterns. Further information can be found in Refs. [17–19]. It should be noted that there is a different form of the sets, denoted cc-pV(X+d)Z (so, i.e., cc-pV(D+d)Z, cc-pV(T+d)Z, etc.) that should always be used as for the atoms Al-Ar in lieu of the cc-pVXZ sets, as the cc-pVXZ basis sets were modified, based upon noted deficiencies in the sets [21].

The shell-like structure of the correlation consistent basis sets is shown in Fig. 5. For each of the basis set levels in this basis set family, an increase in

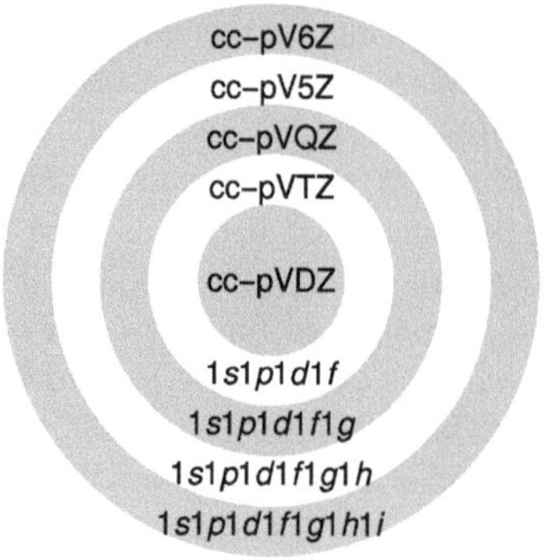

FIG. 5

This general picture illustrates the increasing shell structure for the correlation consistent basis sets for the atoms B-Ne.

basis set size leads to the addition of higher-angular momentum functions. The term "correlation consistent" is a key to the construction of the basis sets. Each basis function contributes to the total energy of the molecule. And, here, the focus is upon the correlation energy (how the Hartree-Fock energy is improved upon). For the correlation consistent basis sets, the higher-angular momentum shells are comprised of functions that contribute similar amounts of correlation energy. For example, for the cc-pVTZ basis set, both a *d* and an *f* functions are included within the basis set, as both contribute a similar amount of correlation energy. For the cc-pVQZ basis set, a *d*, an *f*, and a *g* functions are all included, as they contribute similar amounts of correlation energy. Fig. 6 is included to illustrate the concept of correlation consistency, demonstrating the correlation energy that can be attributed to each function of the basis set. For cc-pVTZ, the notations "+1*d*" and "+1*f*" represent the contribution of the second *d* function (beyond a first *d* function which occurs for the cc-pVDZ basis set) and an *f* function. In this notation, the "+" always refers to the addition of the selected function to the corresponding lower-level basis set. The need for functions that describe orbitals beyond 2*p* orbitals for B-Ne may seem, at first, to be a bit surprising. However, the general mathematical meaning of an orbital should be considered. Basically, an orbital describes the most probable location of an electron within a molecule. To allow more flexibility in describing an electron, higher-angular momentum functions can be utilized. These higher-angular

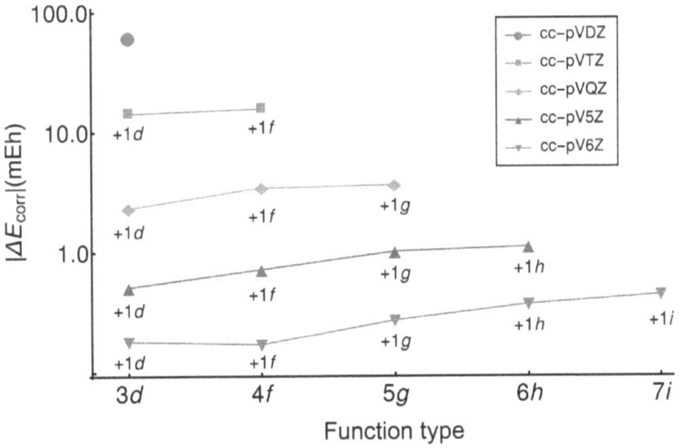

FIG. 6

Functions are added to the Hartree-Fock *sp* basis functions in shells, with each shell contributing similar amounts of correlation energy. Here is a representative example for the correlation consistent basis set family for oxygen. The *y*-axis represents the contribution of each function to the correlation energy in milli-Hartree (mEh), and the *x*-axis represents each angular momentum function. The number in front of the function type represents the angular momentum number, *n*, for each type of function (i.e., "3*d*," where the "*d*" orbital has principal quantum number $n = 3$).

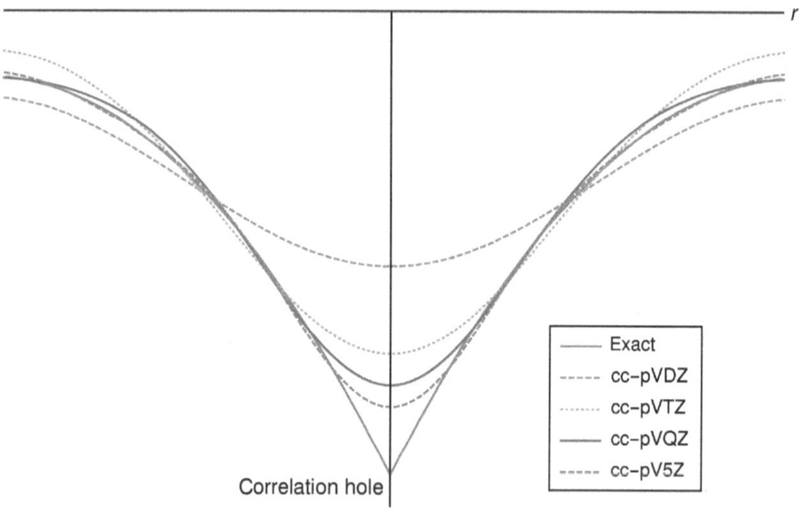

FIG. 7

The exact correlation hole for He and reproduced using different basis sets. With the addition of higher-angular momentum functions in the basis, the description of the cusp and long-range behavior is much improved.

momentum functions can become very important in describing the wave function, particularly in the region of the cusp. This is shown in Fig. 7, for the He atom. The correlation hole, basically the Coulomb integral with one-electron integrated out, exhibits a cusp at the position of the reference electron, similarly to the cusp of the radial wave function. This behavior of electron correlation with distance between pairs of electrons is not very well reproduced for basis sets with comparatively low-angular momentum functions such as cc-pVDZ. With increasing basis set size and especially higher momentum functions, the approximation to the cusp becomes much improved. Furthermore, the long-range behavior is also obtained in much better agreement. Overall, higher-angular momentum functions can significantly impact the prediction of molecular properties, particularly energetic properties.

A unique consequence of the construction of the correlation consistent basis sets is that for a number of properties (i.e., bond lengths, dissociation energies), there is a systematic change in the property description upon increasing the size in the basis set. In fact, the description converges toward the value that would be obtained, in principle, at the complete, or infinite, basis set (CBS) limit. Note that the CBS limit is the limit that would be reached for the approximate method (i.e., Hartree-Fock or some other electronic structure method) being used. This does not guarantee the exact answer and perfect description of properties to be in alignment with experiment. However, this unique convergence toward the CBS limit does enable a much greater understanding of the method (Hartree-Fock or other electronic structure method) that is being employed.

In any calculation, many approximations and choices are made. However, in general, two primary choices must be made—one is the choice of Hartree-Fock or other electronic structure method, and, the second is the choice of basis set. Both of these choices are approximations—necessary approximations—that lead to errors from an exact solution of the Schrödinger equation. By eliminating the error in the basis set by approaching the CBS limit, the error resulting from choice of method can be assessed. This ability to approach the CBS limit with the correlation consistent basis sets has enabled a general understanding of the utility of many different electronic structure methods for the prediction of molecular properties, enabling the understanding of what methods are needed to describe a property well.

The ability to reach the CBS limit also helps to eliminate false positives, commonly referred to in the field as "Pauling points," where the error in the basis set and error in the method choice result in what may look like the right answer, but is not (Ernest Davidson referred to this as obtaining the "right answer for the right reason" [22]), and, only by deeper examination of the basis set or method choices, can the error be noted and addressed. The danger with a Pauling point is that making such a combination of basis set and method likely will not result in a consistent method and basis set approach for the study of other molecules.

The behavior of a generic property Q with increasing basis set size is shown in Fig. 8. If the calculated property value does not cross the experimental property value for any basis set (left-hand side of the figure), the apparent error is always larger than the intrinsic error, where the apparent error is related to the difference between the calculated property value with the selected basis set, and the experimental value; while the intrinsic error of the method is related to the difference of calculated property value at CBS limit and the experimental value. Shown on the right-hand side is a case that exhibits a Pauling point at the TZ basis set: The calculated property value matches the experimental one, but only because of an error cancelation between incomplete basis set and selected electronic structure method.

For molecular properties such as bond lengths and energies, the series of energies or bond lengths calculated at each basis set level (i.e., cc-pVDZ, cc-pVTZ, cc-pVQZ) can be used within a number of well-established, utilized, and demonstrated formula for estimating the complete basis set limit. There are many formulas that have been utilized, some requiring only points from two basis set levels, and some requiring three or more basis set levels. These formulas are well described in the literature. Overall, for energies, the most effective strategies entail a separation of the Hartree-Fock energy from the correlation energy, as these two energies do converge with respect to increasing basis set size at different rates. As well, more effective strategies typically entail the use of three or more points, beginning with the cc-pVDZ basis set, as it can serve as an effective anchor point in the extrapolation scheme, though some strategies utilize basis sets beginning with the cc-pVTZ basis set.

The earliest CBS extrapolation formula for the correlation consistent basis sets was introduced by Xantheas and Feller [23–25], which can be expressed as

$$E(X) = E(\text{CBS}) + A\exp[-BX] \tag{9}$$

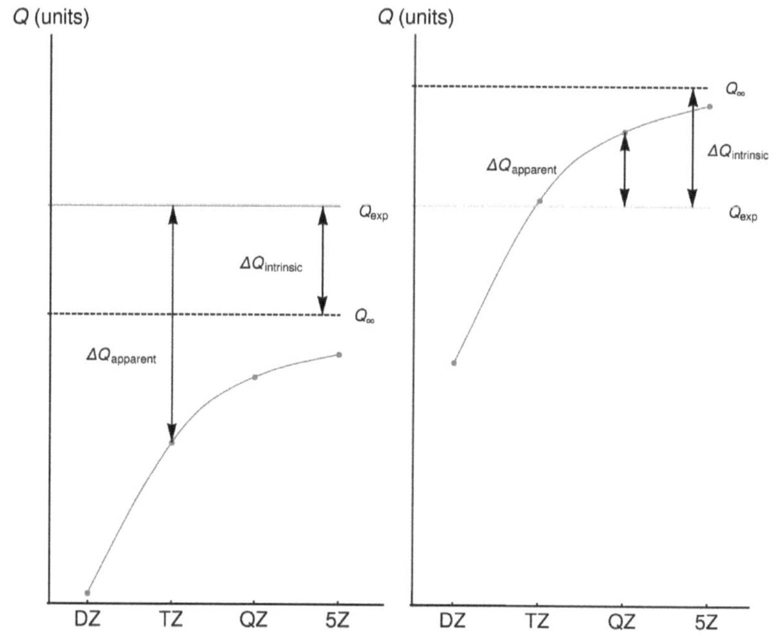

FIG. 8

The behavior of a generic property Q with increasing basis set size: On the *left graph*, the calculated property is continually underestimated. On the *right graph*, the calculated property value is close to the experimental value for the TZ basis set ("Pauling point").

$E(X)$ represents the energy obtained at each basis set level, X, $E(\mathrm{CBS})$ represents the energy at the CBS limit, which will be determined in the calculation, and A and B are parameters that will be determined in the calculation. While this formula is still extraordinarily useful, there are a number of other formulas which have also found common use. Among these are the following:

$$E(X) = E(\mathrm{CBS}) + A\exp[-1.63X] \tag{10}$$

is a formula that is widely used for a two-point extrapolation of the energies arising from the Hartree-Fock portion of a calculation (which are available in the calculation output). Often, the extrapolation is done based on energies arising from the cc-pVTZ and cc-pVQZ calculations.

Another useful scheme, a mixed exponential/Gaussian form by Peterson, Woon, and Dunning [26], which can be referred to as the "CBS-P" scheme is a useful approach for the extrapolation of correlation energies

$$E(X) = E(\mathrm{CBS}) + A\exp[-(X-1)] + B\exp[-(X-1)^2] \tag{11}$$

Two other widely used forms are the inverse cubic power scheme by Schwartz, Halkier et al., and Helgaker [27–30], which is denoted S3, or "CBS-S3"

$$E(X) = E(\text{CBS}) + A l_{\text{max}}^{-3} \tag{12}$$

where l_{max} is the higher-angular momentum used in the basis set, and the inverse quartic scheme by Kutzelnigg and Martin [31–33], denoted S4, or "CBS-S4"

$$E(X) = E(\text{CBS}) + A\left(l_{\text{max}} + \frac{1}{2}\right)^{-4} \tag{13}$$

These equations represent only a subset of possible extrapolation schemes to the CBS limit, though, in practice, these forms represent the most widely used schemes. Overall, the performance is quite similar, though there can be sensitivities based upon the method being used, properties considered, and molecules of interest, as each of these factors can impact the convergence behavior with respect to increasing basis set size. Therefore, it is always best to select an approach that is relevant to the problem and method of interest.

There are many forms of the correlation consistent basis sets. Just as the Pople-style basis sets can be modified for additional flexibility, so can the correlation-consistent basis sets. Among the most widely used modifications are the inclusion of diffuse functions in the sets, which is denoted with the prefix "aug," so the sets take the form aug-cc-pVDZ, aug-cc-pVTZ, aug-cc-pVQZ, aug-cc-pV5Z, and aug-cc-pV6Z [34,35]. For these sets, an additional diffuse function of each type of angular momentum is included in the basis set. For weakly bound systems such as He_2, Ne_2, and Ar_2, additional diffuse functions can be included, with the functions generated via an even-tempered expansion [36] to obtain multiply augmented basis sets such as doubly augmented (i.e., d-aug-cc-pVTZ) and triply augmented (i.e., t-aug-cc-pVTZ) basis sets [37]. An even-tempered expansion results in the generation of additional diffuse functions using

$$\alpha_i = \alpha\beta^{i-1} \tag{14}$$

where α_i represents the new diffuse function, α represents the most diffuse function, and β represents the separation between the most diffuse and second-most diffuse function in the aug-cc-pVXZ basis set. This can be done for each angular momentum function. Most practical basis sets are generated using an even-tempered expansion for spacing between at least the higher-angular momentum basis functions. However, these technical details are provided in other references, and, here, the formula is simply provided as a quick means to provide additional diffuseness to a basis set if needed.

Other widely used forms of the correlation consistent basis sets are the core-valence, cc-pCVXZ [38], and weighted core-valence, cc-pwCVXZ [39], basis sets. (Diffuse functions can also be incorporated in these sets, that is, aug-cc-pCVXZ and aug-cc-pwCVXZ.) In most electron correlation calculations, the frozen core approximation—which constrains the lowest-lying molecular orbitals to be doubly occupied in all configurations—is employed. This restriction can enable significant cost savings in electron correlation calculations by reducing the number of configurations that must be considered. This approximation is also useful as innershell electrons play much less of a role in describing the chemical properties of a molecule than the valence electrons, so the impact of this approximation is typically minimal. However, for highly accurate geometries (subpicometer accuracy) and thermochemistry ("chemical accuracy" of errors <1 kcal/mol), more explicit description of the core electrons can be important. Most basis sets used in quantum chemical calculations do not provide sufficient flexibility in the core region to describe the correlation of the core electrons accurately. However, the correlation-consistent core-valence and weighted core-valence basis sets are very useful all-electron basis sets (all-electron sets mean that all electrons of an atom are explicitly described by basis functions), when the core region needs to be better described. Both of these families of basis sets result in systematic convergence to the CBS limit for molecular energies and spectroscopic properties. The difference between the sets is that while the cc-pCVXZ basis sets have a noted bias toward excitation of pairs of electrons from the core orbitals over pairs that were mixed between the core and valence orbitals, for the cc-pwCVXZ basis sets, a weighting of core-valence excitations over core-core excitations were utilized in the optimization of the sets. Both sets of basis sets are effective, though the cc-pwCVXZ generally provides more rapid convergence of the core-valence correlation energy.

There are many additional forms of the correlation consistent basis sets, including a pseudopotential form and a relativistic form, which will be discussed in the following section.

5 PSEUDOPOTENTIALS OR ECPS AND RELATIVISTIC EFFECTS

Basis functions in the Gaussian-based form are quite useful, as they do correspond to the chemical picture of a molecule, and the description of the inner electrons is really not any more difficult than is the description of the valence electrons. As much of the chemical properties of a molecule are described by the valence electrons, core electrons can be described without as much representation (i.e., the core does not have to be described as well as the valence electrons) as needed for the valence electrons. However, as the number of electrons that should be included in the core grows substantially in the lower portion of the periodic table, so do the numbers of basis functions necessary in calculations. Because of the optimization and use of so many basis functions, calculations can become quite complicated and expensive.

In lieu of all-electron calculations, instead, pseudopotentials, or effective core potentials (ECPs), can be used to replace the core electrons with an effective potential. There are a number of important benefits to ECPs. They reduce the basis set size, and reduce the number of electrons that need to be addressed specifically in calculations. In the early implementation of ECPs, these were the primary reasons to utilize ECPs. However, now, another advantage of ECPs is that they can facilitate the incorporation of relativistic effects (scalar and/or spin-dependent effects; these terms are discussed in a later chapter) in calculations. Relativistic effects are important, even in the earlier portion of the periodic table, as they can have an impact on the energy and shape of valence orbitals, and, thus, on bonding, and many other properties.

ECPs are often classified as *large-core* or *small-core* ECPs. These categories are based on the numbers of electrons to include in the core. Large-core ECPs include all electrons with the exception of those in the valence shell. Small-core ECPs include all electrons with the exception of the valence shell and the next lowest shell (subshell). For heavy elements, electrons in the subshell may play a role in describing the chemistry of molecules correctly, and, as such, polarization of this subshell can be important, and, thus, the more costly route of using small-core ECPs is often best.

Among the most widely used ECPs are the Los Alamos National Laboratory ECPs [40–42], which originally, were available in nonrelativistic forms for atoms through the first row of the transition metals, and were available in relativistic forms for atoms beyond the first row of the transition metals. However, as the community has learned much more about elements across the period table, these ECPs have been expanded to both relativistic and nonrelativistic forms for metals, and polarization and diffuse functions for elements of the *p* block. For the heaviest elements, the Stuttgart-Dresden ECPs [43,44] are widely used. And, a hierarchical set, the Dyall sets [45], of ECPs for the heaviest elements, akin to the correlation consistent basis sets, but for ECPs, is also useful. There is a correlation consistent form of ECP-based sets, denoted cc-pVXZ-PP [46], built upon small-core multiconfigurational dirac Hartree-Fock (MCDHF) adjusted pseudopotentials for the Stuttgart-Dresden ECP family [47].

Relativistic effects can also be incorporated within all-electron calculations by recontracting (reoptimizing the contraction coefficients) the basis functions explicitly for the Douglass-Kroll-Hess (DKH) Hamiltonian [48], which is used in lieu of the Hartree-Fock Hamiltonian. For the correlation consistent basis sets, these basis sets are then called the cc-pVXZ-DK basis set family. Such basis sets are ideal for calculations employing the DKH Hamiltonian.

6 BASIS SETS FOR DENSITY FUNCTIONAL APPROACHES

The basis sets described earlier are used widely in the literature for both ab initio electronic structure calculations as well as density functional approaches (referred to as density functional theory [DFT]). However, the basis sets described so far are ones that were explicitly designed using ab initio electronic structure calculations. For example, Dunning's correlation consistent basis sets are developed, in part, utilizing a configuration interaction with single and double excitation method (discussed in a later chapter). While DFT can be used to predict the same types of properties as ab initio electronic structure calculations, the premise of both types of methods are quite different (see the respective chapters). And, quite different convergence is experienced for predicted molecular properties with respect to basis set size. For example, molecular structures are typically quite well converged and well described using a triple-ζ quality basis set. Energies may be nearly converged using a triple-ζ quality basis set, but often may require a quadruple-ζ quality basis set to reach convergence, though note, even at convergence, energetic predictions may be near or quite far from experiment, and, thus, when the calculations are feasible, ab initio correlation calculations are often preferred for energetic predictions that require quantitative, and not just qualitative accuracy.

Because of the quite different methodologies, basis sets have been developed explicitly for density functional approaches. Examples are the polarization consistent basis sets [49], which are similar in design as the correlation consistent basis sets, though are optimized using DFT, resulting in a hierarchical framework of basis sets, pc-0, pc-1, pc-2, pc-3, pc-4, corresponding to single-ζ, double-ζ, triple-ζ, quadruple-ζ, and quintuple-ζ basis sets. (Note that single-ζ basis sets should never be used, as they are akin to a minimum basis set, at best.) In comparing the basis set sizes with those of the correlation consistent basis sets, only the double-ζ pc-1 set is smaller in size than the corresponding cc-pVDZ basis set. All other polarization consistent basis sets are larger than their correlation consistent counterparts. In terms of applications, for the atoms C–F, there is slightly faster convergence of atomic properties for the polarization consistent sets. For studies of molecular properties, overall, the performance is similar for both families of basis sets. However, it must be noted that this is an active area of investigation, and there are multiple studies that advocate for the use of DFT-based basis sets. To date, however, in practice, there has been much success using ab initio-developed basis sets for DFT calculations, and, this is the most common practice for molecular calculations.

(Here, approaches for periodic systems are not addressed, as they are beyond the scope of this chapter.)

7 BASIS SET SUPERPOSITION ERROR

Throughout this chapter, insight has been provided about practical considerations that should be made in calculations. Another consideration should be basis set superposition error (BSSE). While it might be tempting to mix basis sets of many different types for the atoms in a molecule (and, note that it some cases, this does become necessary, particularly when working with the heaviest elements, where all families of basis sets may not be available across the periodic table), this is an exercise that should be approached with caution, as this can result in the use of unbalanced basis sets. As well, using basis sets that are too small can also result in an unbalanced description of atoms within a molecule. An earlier example was provided by a minimum basis set, where lithium, with only three electrons, was described by five basis functions, and oxygen, with eight electrons, was also described by just five basis functions. The problem with unbalanced descriptions and basis sets that are not sufficient is that as the atoms within a molecule interact or molecules interact, their basis functions overlap, and each of the monomers (atoms or molecules) "borrows" basis functions from other monomers in an attempt to improve its own basis set. In the calculation then of a property such as binding energy or interaction energy, where, for illustrative purposes, the energy is defined for a dimer AB, comprised of monomers A and B, as

$$D(\mathrm{AB}) = E(\mathrm{AB})_{\mathrm{AB}} - E(\mathrm{A})_{\mathrm{A}} - E(\mathrm{B})_{\mathrm{B}} \tag{15}$$

where $D(\mathrm{AB})$ represents the dissociation energy of dimer AB, $E(\mathrm{AB})_{\mathrm{AB}}$ represents the total energy of AB determined in the presence of basis sets for A and B (represented by the subscript AB), and $E(\mathrm{A})_{\mathrm{A}}$ and $E(\mathrm{B})_{\mathrm{B}}$ each represent the total energies of the atoms. In this case, the energy of the individual atoms or molecules will not benefit from the "borrowed" basis functions, yet the molecule or molecular system will benefit. This imbalance results in the BSSE, and can be addressed in a number of different ways. Among the most popular solutions are to use large basis sets that can eliminate, or nearly eliminate BSSE, or to utilize an approximation called a counterpoise correction [50]. The counterpoise correction utilizes calculations on each monomer in the presence of the basis sets for the entire system, rather than simply utilizing only the basis set on the monomer alone for each monomer. For example, the counterpoise correction for the dissociation energy, $\Delta D_{\mathrm{counterpoise}}$ can be expressed as

$$\Delta D_{\mathrm{counterpoise}} = E(\mathrm{A})_{\mathrm{AB}} + E(\mathrm{B})_{\mathrm{AB}} - E(\mathrm{A})_{\mathrm{A}} - E(\mathrm{B})_{\mathrm{B}} \tag{16}$$

This enables the energy to be adjusted (corrected) by providing a more balanced description of the basis sets.

BSSE can occur for any calculation if there are deficiencies (incompleteness) in the basis set. However, as basis sets increase in size, the prevalence of BSSE is greatly diminished, and is zero at the CBS limit. BSSE can impact many properties, but is particularly important for properties such as the energies of van der Waals complexes, hydrogen bonds, and interaction energies. The impact is generally far more dramatic upon electron correlation calculations than for Hartree-Fock calculations.

8 SOLUTION OF THE INTEGRALS OVER ATOMIC ORBITALS

The calculation of the electronic energy, such as in the Hartree-Fock approach, that was introduced in the prior chapter, requires the evaluation of integrals over molecular orbitals. This integral evaluation is central to any ab initio electronic structure calculation and is typically carried out only once in the beginning of the calculation. The integrals can be divided into two groups, depending on the number of electrons that are involved in the integration: the one-electron integrals require integration over two orbitals that are occupied by the same electron, with one-electron operators acting on the second spin orbital. The two-electron integrals require integration over four orbitals that are occupied by two different electrons, and entails the interaction between electron pairs.

Together with the basis set, the one- and two-electron integrals constitute the foundation for all electronic structure calculations. There are libraries available which enable an efficient evaluation of the integrals, for example, the Libint [51,52] library; Libcint [53] or as in BAGEL [54], to name a few, with many more available.

The computational demand of the integral evaluation can be reduced by local schemes and using density fitting (DF) techniques. The reasoning behind DF, and the auxiliary basis that is required, is also introduced here.

8.1 ONE-ELECTRON INTEGRALS

The one-center integrals, or one-electron integrals, involve integration over pairs of spin orbitals i and j (ϕ_i and ϕ_j), that is

$$\langle i|h|j\rangle = \int \phi_i^*(\mathbf{x}_1)h(\mathbf{r}_1)\phi_j(\mathbf{x}_1)d\mathbf{x}_1 \tag{17}$$

where $\mathbf{x}_1$ is a combined variable denoting space and spin coordinate for electron one, $\mathbf{x}_1 = \{\mathbf{r}_1, \omega_1\}$. A spin orbital is a molecular orbital multiplied with an α or β function for the electronic spin. As described earlier, the molecular orbitals are represented as a linear combination of Gaussian functions or atomic orbitals $G(\alpha, \mathbf{r})$ (see Eq. 1), that are denoted by Greek indices μ, ν. The preceding integrals form the matrix elements in the matrices required in the Roothaan [55] equation, that is, the representation of the differential Hartree-Fock equation in an algebraic form, and are constructed in the atomic orbital basis. Multiplication with the coefficients $c_{\mu i}$ of the molecular orbitals then yields the matrices in the molecular orbital basis. The one-electron matrices that need to be constructed in an electronic structure calculation are: the overlap matrix $\mathbf{S}$, the electronic kinetic energy matrix $\mathbf{T}$, and the electron-nuclear attraction or one-electron Coulomb matrix $\mathbf{V}$

$$S_{\mu\nu} = \langle \mu|\nu\rangle = \int \chi_\mu^*(\mathbf{r}_1)\chi_\nu(\mathbf{r}_1)d\mathbf{r}_1 \tag{18}$$

$$T_{\mu\nu} = \langle \mu|T_e|\nu\rangle = \int \chi_\mu^*(\mathbf{r}_1)\left(-\frac{1}{2}\frac{\partial^2}{\partial \mathbf{r}_1^2}\right)\chi_\nu(\mathbf{r}_1)d\mathbf{r}_1 \tag{19}$$

$$V_{\mu\nu} = \langle \mu|V_{en}|\nu\rangle = \int \chi_\mu^*(\mathbf{r}_1)\left(-\sum_A \frac{Z_A}{(\mathbf{r}_1 - \mathbf{R}_A)}\right)\chi_\nu(\mathbf{r}_1)d\mathbf{r}_1 \tag{20}$$

and, for molecular properties such as the dipole moments, the position integrals

$$X_{\mu\nu} = \langle \mu|\mathbf{r}_1|\nu\rangle = \int \chi_\mu^*(\mathbf{r}_1)\mathbf{r}_1\chi_\nu(\mathbf{r}_1)d\mathbf{r}_1 \tag{21}$$

For all these integrals, the spin part has been integrated out so that only the spatial part remains for orbitals with same spin function (as $\langle\alpha|\alpha\rangle = \langle\beta|\beta\rangle = 1$ and $\langle\alpha|\beta\rangle = \langle\beta|\alpha\rangle = 0$). For a set of K primitive basis functions, the overlap matrix $\mathbf{S}$ then contains $K \times K$ elements. Typically, the molecular orbitals will be normalized but not generally be orthogonal so that its entries lie between 0 and 1, with the magnitude signifying the extent of the overlap and the sign relating to the phase of the molecular orbitals and their relative orientation. The kinetic energy matrix $\mathbf{T}$ and the electron-nuclear attraction matrix $\mathbf{V}$ are often combined to form the Fock matrix $\mathbf{F} = \mathbf{T} + \mathbf{V}$, which is the one-electron term in the Hartree-Fock equations.

The Gaussian basis functions can either be represented in Cartesian or spherical harmonic coordinates. In the following, we mostly focus on Cartesian coordinates as the description is more intuitive; but the derivation for spherical harmonics follows the same scheme. As the primitive or atomic orbital basis functions require a term for the angular dependence of the orbital, for higher-angular momentum such as p or d, the Gaussian function is multiplied with integer powers of its Cartesian coordinate,

$$G(\mathbf{r}_A, \alpha_A) = x_A^i y_A^j z_A^k \exp\left[-\alpha_A \mathbf{r}_A^2\right] \tag{22}$$

with the sum of the powers i, j, k related to the angular momentum quantum number l as $l = i + j + k$, and the distance $\mathbf{r}_A = \mathbf{r} - \mathbf{R}_A$. Alternatively, the basis functions can be represented using spherical harmonics

$$G(\mathbf{r}_A, \alpha_A) = P_l^{|m|}(\theta_A, \phi_A) r_A^l \exp[-\alpha_A r_A] \tag{23}$$

for spherical coordinates θ, ϕ, and r, spherical- or solid-harmonic polynomial P and magnetic quantum number m.

At this point, the great advantage of using Gaussian functions instead of Slater functions or other representations of the primitive basis becomes apparent. The one-electron integrals involve integration over Gaussian functions that can either be centered both at $\mathbf{R}_A$ or one at another atomic center $\mathbf{R}_B$. It is easy to write this product as a function residing on the same center $\mathbf{R}_P$ that lies in between $\mathbf{R}_A$ and $\mathbf{R}_B$

$$\exp\left[-\alpha_A \mathbf{r}_{iA}^2\right] \cdot \exp\left[-\alpha_B \mathbf{r}_{jB}^2\right] = E_{ij} \exp[-\alpha_P \mathbf{r}_P] \tag{24}$$

with

$$\alpha_P = \alpha_A + \alpha_B \tag{25}$$

$$E_{ij} = \exp\left[-\frac{\alpha_A \alpha_B}{\alpha_A + \alpha_B} (\mathbf{R}_A - \mathbf{R}_B)^2\right] \tag{26}$$

The vector $\mathbf{r}_P$ denotes the position relative to center P.

Such a product Gaussian is depicted in Fig. 9. For small E_{ij} (long distances between centers A and B), the product Gaussian becomes very small, as the exponential in E_{ij} decays to zero with increasing $(\mathbf{R}_A - \mathbf{R}_B)$. A product Gaussian for two overlapping functions is shown on the left-hand side of the figure, where the product Gaussian is centered on $\mathbf{R}_P$. Shown on the right-hand side of the figure are two Gaussians that are well separated and their product Gaussian is zero.

As such, prescreening techniques can be utilized to discard small integrals resulting from spatially well-separated functions before the integration is done.

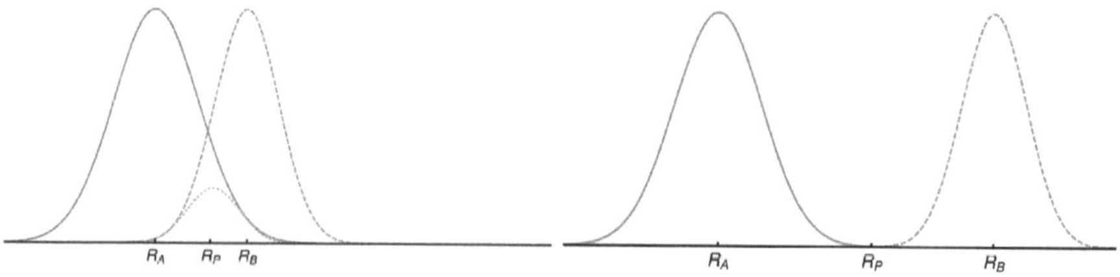

FIG. 9

Gaussian basis functions at different centers and their overlap. The functions on the *left* overlap and their product results in a new Gaussian centered at R_P. The functions on the *right* are well separated and result in no overlap.

This is part of local methods, but mostly used for the two-electron integrals. The Gaussian product rule is very advantageous as it reduces a two-center integral to an one-center integral (and a four-center integral to a two-center integral).

The product of two Gaussians is often called charge distribution or overlap distribution Ω_{ij}, and integration over it is still cumbersome because of the appearance of the Cartesian monomials x_A^i and x_B^j. For a Gaussian product in x-direction, the integral over the charge distribution reads

$$\int \Omega_{ij}(x)dx = E_{ij} \int x_A^i x_B^j \exp\left[-\alpha_P x_P\right] dx \tag{27}$$

However, the charge distribution can be expanded in Hermite Gaussians instead; here, the prefactors are obtained as sums of derivatives with respect to the position of the center along x-direction, x_P,

$$\Omega_{ij}(x) = \sum_{t=0}^{i+j} E_{ij} \left(\frac{\partial}{\partial X_P}\right)^t \exp\left[-\alpha_P x_P\right] = \sum_{t=0}^{i+j} E_{ij,t} \Lambda_t(x_P) \tag{28}$$

An integral over a Hermite Gaussian is simply [56]

$$\int \Lambda_t(x)dx = \delta_{t0}\sqrt{\frac{\pi}{\alpha_P}} \tag{29}$$

Furthermore, the recurrence relation [56–58] gives

$$E_{i+\lambda,j,t} = \mathbf{R}_{AP}E_{ij} + \frac{1}{2\alpha_p}E_{ij,t-1} + (t+\lambda)\, E_{ij,t+\lambda} \tag{30}$$

so that all E_{ij} can be obtained for all quantum numbers in all Cartesian directions.

The standard procedure to solve the one-electron integrals involves several steps: First, the GTOs are factored into directions x, y, and z as

$$\mathrm{G}_{ijk}(\alpha_A,\mathbf{r}_A) = x_A^i y_A^j z_A^k \exp\left[-\alpha_A \mathbf{r}_A^2\right] \tag{31}$$

$$= x_A^i \exp\left[-\alpha_A x_A^2\right] \cdot y_A^j \exp\left[-\alpha_A y_A^2\right] \cdot z_A^k \exp\left[-\alpha_A z_A^2\right] \tag{32}$$

$$= \mathrm{G}_i(\alpha_A, x_A)\mathrm{G}_j(\alpha_A, y_A)\mathrm{G}_k(\alpha_A, z_A) \tag{33}$$

For example, for the overlap integral $S_{\mu\nu}$ one obtains

$$S_{\mu\nu} = \mathrm{G}_{ijk}(\alpha_A,\mathbf{r}_A)\mathrm{G}_{lmn}(\alpha_B,\mathbf{r}_B) \tag{34}$$

$$= S_{il}S_{jm}S_{kn} \tag{35}$$

Second, making use of the product rule (Eq. 24) leads to a contraction of two Gaussians into one Gaussian—the charge distribution or overlap distribution $\Omega_{il}(x)$. And third, the charge distribution is represented as Hermite Gaussian. Through the use of the recurrence relation and Eq. (29), it then follows that for the preceding example

$$S_{il} = \int \Omega_{il}(x)dx = \sum_{t=0}^{i+l} E_{il,t} \int \Lambda_t(x_P)dx = E_{il,0}\sqrt{\frac{\pi}{\alpha_P}} \tag{36}$$

The full overlap integral $S_{\mu\nu}$ over all three Cartesian directions thus reads

$$S_{\mu\nu} = E_{il,0}E_{jm,0}E_{kn,0}\left(\frac{\pi}{\alpha_P}\right)^{3/2} \tag{37}$$

For the dipole integrals, the procedure is very similar to the overlap integrals. The matrix elements that need to be evaluated are of the type

$$X_{il} = \int x_C \Omega_{il}(x)dx \tag{38}$$

Using the earlier recurrence relation and making use of $x_C = x_P + (x_P - x_C)$, the matrix elements are obtained as

$$X_{il} = \sum_{t=0}^{i+l} E_{il,t} \int x_C \Lambda_t(x_P)dx = E_{il,0}\sqrt{\frac{\pi}{\alpha_P}}(x_P - x_C) + E_{il,1}\sqrt{\frac{\pi}{\alpha_P}} \tag{39}$$

The evaluation of the kinetic energy integral involves the computation of derivatives of Gaussian functions. Separation of the kinetic energy operator in its Cartesian components, as well as the Gaussian basis functions in the Cartesian directions, results in integrals over derivatives multiplied with the integrals for the perpendicular directions (for a derivation, see Ref. [59]). As the derivative of a Gaussian function is again a Gaussian, the kinetic energy integral can be rewritten as a product of overlap integrals.

The electron-nuclear attraction integral, or one-electron Coulomb integral, is not separable in the Cartesian directions due to the $1/\mathbf{r}$-dependence. Instead, an auxiliary function is introduced, the Boys function [60], by which the 3D integral is reduced to a 1D integral for a specified range. The Boys function is given by

$$F_n(x) = \int_0^1 \exp\left[-xt^2\right] t^{2n} dt \tag{40}$$

The value of x depends on the exponents α_A and centers $\mathbf{R}_A$; but not on the angular momentum quantum number and its constituent coefficients i, j, and k. In the evaluation of the one-electron Coulomb integrals, different numerical techniques are used depending on the values of the parameters of the atomic orbital functions. The derivation of the expression is cumbersome and can be found elsewhere [61].

8.2 TWO-ELECTRON INTEGRALS

The two-electron or four-center integrals involve integration over four spin orbitals i, j, k, l and two electronic coordinates $\mathbf{x}_1$ and $\mathbf{x}_2$

$$\langle ij|kl\rangle = \int \phi_i^*(\mathbf{x}_1)\phi_j^*(\mathbf{x}_2)\frac{1}{r_{12}}\phi_k(\mathbf{x}_1)\phi_l(\mathbf{x}_2)d\mathbf{x}_1 d\mathbf{x}_2 \tag{41}$$

resulting in matrix elements between atomic orbital basis functions $\mu, \nu, \sigma, \lambda$ (after integration over the spin coordinate)

$$V_{\mu\nu\sigma\lambda} = \langle\mu\nu|V_{ee}|\sigma\lambda\rangle = \int \chi_\mu^*(\mathbf{r}_1)\chi_\nu^*(\mathbf{r}_2)\frac{1}{r_{12}}\chi_\kappa(\mathbf{r}_1)\chi_\lambda(\mathbf{r}_2)d\mathbf{r}_1 d\mathbf{r}_2 \tag{42}$$

Within the Hartree-Fock approach, the two-electron integrals are typically classified into Coulomb integrals J and exchange integrals K

$$J_{ij} = \langle ij|ij\rangle \tag{43}$$

$$K_{ij} = \langle ij|ji\rangle \tag{44}$$

with J_{ij} corresponding to the repulsion between two molecular orbitals because of the electrostatic interaction of two electrons, and K_{ij} representing the exchange interaction. While both integrals have positive values, J_{ij} can be understood from a simple classical interpretation, whereas K_{ij} results from the antisymmetry of the one-determinantal Hartree-Fock wave function and basically leads to a correlation between electrons with parallel spin.

While the spin part can be integrated out, the spatial part needs to be evaluated for all possible combinations of the molecular orbitals. However, if the molecular orbitals are real-valued and not complex, which is the case in standard electronic structure theory, the four-center integrals do not need to be evaluated for all possible combinations of i, j, k, and l, but obey eightfold permutational symmetry where

$$\begin{aligned}\langle ij|kl\rangle &= \langle kl|ij\rangle = \langle ji|lk\rangle = \langle lk|ji\rangle \\ &= \langle kj|il\rangle = \langle il|kj\rangle = \langle li|jk\rangle = \langle jk|li\rangle\end{aligned} \tag{45}$$

Still, the computation of the two-electron integrals is quite involved and expensive in terms of computing time and memory demands. There are $K^4/8$ matrix elements that need to be evaluated for a total of K basis functions. Even more so, the four-index transformation from atomic to molecular orbital basis is computationally expensive. The four-index transformation relates the atomic orbital to the molecular orbital basis and needs to be carried out for each set of updated molecular orbitals in the SCF cycle

$$\langle ij|kl\rangle = \sum_{\mu}\sum_{\nu}\sum_{\sigma}\sum_{\lambda} C^*_{\mu i} C^*_{\nu j} C_{\sigma k} C_{\lambda l} \langle \mu\nu|\sigma\lambda\rangle \tag{46}$$

Also, this is required in MP2 (second-order Møller-Plesset perturbation theory) calculations to obtain the perturbative energy correction, which requires multiplication of $K \times K$ matrices and thus scales as K^5 with the number of basis functions.

The actual evaluation of the two-electron integrals is only carried out once in the atomic orbital basis and the integrals are then stored and transformed into the respective molecular orbital basis when required. The evaluation of the integrals follows the one of the one-electron Coulomb integrals, and also makes use of Boys functions. Different computational schemes are available and an overview is given in [62].

8.3 RESOLUTION OF THE IDENTITY AND DENSITY FITTING

The resolution of the identity (RI) or DF approach is used to reduce the computational cost of the evaluation of the two-center integrals, in particular for medium to large molecules with many basis functions. In this approach, the four-center (or two-electron) integrals are reduced to products of two- and three-center integrals (see, e.g., Refs. [63,64]). As such, the computational demand of the four-center integral evaluation is much reduced, and a significant speed-up of the electronic structure calculation is achieved.

DF methods mostly focus on the Coulomb interaction between pairs of electrons, as these are the computational more demanding part in the two-electron integrals; and as in many DFT calculations, the exchange integrals do not appear (only for hybrid functionals, which include a portion of Hartree-Fock exchange K).

We start by rewriting the two-electron Coulomb integral J in chemist's notation [4] in the atomic orbital basis

$$(\mu\nu|\sigma\lambda) = \int \chi^*_\mu(\mathbf{r}_1)\chi_\nu(\mathbf{r}_1)\frac{1}{r_{12}}\chi_\sigma(\mathbf{r}_2)^*\chi_\lambda(\mathbf{r}_2)d\mathbf{r}_1 d\mathbf{r}_2 \tag{47}$$

For real basis functions, this is then expressed as a repulsion integral between two-electron densities $\rho(\mathbf{r})$

$$(\mu\nu|\sigma\lambda) = \int \rho_{\mu\nu}(\mathbf{r_1})\frac{1}{r_{12}}\rho_{\sigma\lambda}(\mathbf{r_2})d\mathbf{r_1}d\mathbf{r_2} \tag{48}$$

The exact densities $\rho_{\mu\nu}$ are then approximated using an auxiliary basis set ξ

$$\tilde{\rho}_{\mu\nu}(\mathbf{r}) = \sum_{a=1}^{N_{\text{fit}}} d_{\mu\nu,a}\xi_a(\mathbf{r}) \tag{49}$$

with fitting coefficients $d_{\mu\nu,a}$. The auxiliary basis is discussed in some more detail herein. As the densities are products of two Gaussians, they can be described by a single Gaussian function located in between the original functions as described earlier (see Fig. 9). These product Gaussians are then described by a new, auxiliary basis function that resides on a center in between the original functions. The coefficients are obtained by minimizing the integral over the difference for the exact and approximate one-electron density, weighted by the Coulomb operator $1/r_{12}$

$$\Delta_{ij} = \int \Big[\rho_{\mu\nu}(\mathbf{r}_1) - \tilde{\rho}_{\mu\nu}(\mathbf{r}_1)\Big]\frac{1}{r_{12}}\Big[\rho_{\mu\nu}(\mathbf{r}_2) - \tilde{\rho}_{\mu\nu}(\mathbf{r}_2)\Big]d\mathbf{r}_1 d\mathbf{r}_2 \tag{50}$$

Representing the one-electron density through an approximate one resembles inserting the resolution of identity in the two-electron integrals

$$(\mu\nu|\sigma\lambda) \approx \sum_{a=1}^{N_{\text{fit}}} (\mu\nu|a)\,(a|\sigma\lambda) \tag{51}$$

Because the auxiliary basis is incomplete, the DF scheme is not exact, but very nearly so for well-fitted auxiliary basis sets.

It is also possible to fit the exchange term K using an auxiliary basis [64]. This requires the fitting of products of basis functions, instead of one-electron densities, and a larger number of auxiliary basis functions are needed for a good fit. The J term requires more computational time than the K term for medium-sized and large molecules (i.e., for 500 basis functions, K took 10% of the time that it took to evaluate J [65]), while for small molecules, K is more demanding than J. Most applications therefore target primarily the Coulomb integrals as the DF technique is typically applied to larger molecules.

Not only the evaluation and storage of the DF integrals is advantageous in terms of computing times and memory requirements, also the effort for the four-index transformation is much reduced, even though the auxiliary basis is typically much larger than the standard basis. As a result, DF is often used in MP2 calculations.

8.4 AUXILIARY BASIS SETS

The numerical quality of the DF approach largely depends on the auxiliary basis set. These sets can be obtained through different fitting procedures, either using the Coulomb operator to weight the difference between exact and approximate density or using an attenuated Coulomb operator [66]. Fits that make use of the locality of the basis functions are also computationally advantageous, such as in atomic RI [67]. Other approaches include a Cholesky decomposition [68]. A fit of the Coulomb integrals is typically referred to as "J-fit" or "RI-J"; if the auxiliary basis also contains terms for the exchange integrals, then it is typically termed as "JK-fit" or "RI-JK." As the fitting for the exchange operator is typically more extensive, basis sets that are fitted for K can also be used for J [64].

The auxiliary basis sets for a J-fit are obtained through fitting of the coefficients to

$$d_{\mu\nu,b} = \sum_{a=1}^{N_{\mathrm{fit}}} (\mu\nu|a)\,[J^{-1}]_{ab} \tag{52}$$

which is a linear combination of three-center $(ij|a)$ and two-center J_{ab} integrals

$$(\mu\nu|a) = \int \chi_\mu(\mathbf{r}_1)\chi_\nu(\mathbf{r}_1)\frac{1}{r_{12}}\xi_a(\mathbf{r}_2)d\mathbf{r}_1 d\mathbf{r}_2 \tag{53}$$

$$J_{ab} = \int \xi_a(\mathbf{r}_1)\frac{1}{r_{12}}\xi_b(\mathbf{r}_2)d\mathbf{r}_1 d\mathbf{r}_2 \tag{54}$$

For a K-fit, the difference in the exact K and the approximate one is evaluated. The auxiliary basis sets are typically about three to four times larger than the original basis sets. Especially the $l = 0$ (s-)part is more extended, but also higher-angular functions need to be included. For example, for hydrogen, additional basis functions up to an angular momentum of $l = 4$ (f functions) are required, and for atoms with higher-angular momentum functions even up to i functions, especially for correlated methods [69]. However, for large originating basis sets (increasing quality of the basis set), the size of the auxiliary basis decreases in comparison [70].

Many auxiliary basis sets are available, starting with the sets from Karlsruhe [65,70–73]. Further, sets for Dunning's correlation consistent basis [64,69], atomic natural orbital (ANO) basis sets [68], and even sets for lanthanides [74], to name a few. The DF approach has become a standard tool in many quantum chemistry codes and is widely available in most implementations.

9 SUMMARY

In this chapter, an overview of many of the concepts behind basis sets and molecular integrals for electronic structure (ab initio and DFT) calculations have been provided. Important to any electronic structure calculation is the choice of the means to describe the orbitals (or wave function), as this selection can have a very significant impact upon the prediction of atomic and molecular properties. Bigger basis sets do not necessarily mean better basis sets. So, important to calculations is the selection of a quality basis set that can enable calculations to be feasible, while taking into account the chemical properties that need to be predicted, and the level of prediction (accuracy) that is desired. New approaches continue to become available, and the utility of these approaches should always be gauged before embarking upon applications with new approaches.

Basis sets and molecular integrals are very active research areas in computational chemistry, with very rich histories. Additional insight can be gained about the field by reading the representative papers included within the chapter as well as reading many of the corresponding historical papers that were not included. A particularly useful resource is the EMSL Gaussian Basis Set Library, which is a repository of many of the available basis sets. This source includes literature references for many basis sets, and provides basis sets that available for each atom in formats suitable for a wide range of computational chemistry software, which is useful if a specific basis set is not already implemented within a computational chemistry software package.

ACKNOWLEDGMENTS

The authors thank Lucas Aebersold, Prajay Patel, and Zachary Stewart for their useful suggestions to this chapter.

REFERENCES

[1] Slater JC. Atomic shielding constants. Phys Rev 1930;36:57–64.
[2] Huzinaga S. Gaussian-type functions for polyatomic systems. I. J Chem Phys 1965;42: 1293.
[3] Helgaker T, Jørgensen P, Olsen J. Molecular electronic-structure theory. 1st ed. Chichester: John Wiley & Sons, Ltd; 2000.
[4] Szabo A, Ostlund NS. Modern quantum chemistry: introduction to advanced electronic structure theory. 1st ed. Mineola (NY): Dover Publications, Inc.; 1996.
[5] Raffenetti RC. General contraction of Gaussian atomic orbitals: core, valence, polarization, and diffuse basis sets; molecular integral evaluation. J Chem Phys 1973;58:4452–8.
[6] Hehre WJ, Steward RF, Pople JA. Self-consistent molecular-orbital methods. I. Use of Gaussian expansions of Slater-type atomic orbitals. J Phys Chem 1969;51:2657–64.
[7] Hehre WJ, Ditchfield R, Steward RF, Pople JA. Self-consistent molecular orbital methods. IV. Use of Gaussian expansions of Slater-type orbitals. Extension to second-row molecules. J Phys Chem 1970;52:2769–73.
[8] Pietro WJ, Levi BA, Hehre WJ, Stewart RF. Molecular orbital theory of the properties of inorganic and organometallic compounds. 1. STO-VG basis sets for third-row main-group elements. Inorg Chem 1980;19:2225–9.
[9] Binkley JS, Pople JA, Hehre WJ. Self-consistent molecular orbital methods. 21. Small split-valence basis sets for first-row elements. J Am Chem Soc 1980;102:939–47.
[10] Gordon MS, Binkley JS, Pople JA, Pietro WJ, Hehre WJ. Self-consistent molecular-orbital methods. 22. Small split-valence basis sets for second-row elements. J Am Chem Soc 1983;104:2797–803.
[11] Dobbs KD, Hehre WJ. Molecular orbital theory of the properties of inorganic and organometallic compounds. 4. Extended basis sets for third- and fourth-row, main-group elements. J Comput Chem 1986;7:359–78.

[12] Hariharan PC, Pople JA. The effect of d-functions on molecular orbital energies for hydrocarbons. Chem Phys Lett 1972;66:217.
[13] Krishnan R, Frisch MJ, Pople JA. Contribution of triple substitutions to the electron correlation energy in fourth order perturbation theory. J Chem Phys 1980;72:4244–5.
[14] Francl MM, Pietro WJ, Hehre WJ, Binkley JS, Gordon MS, DeFrees DJ, et al. Self-consistent molecular orbital methods. XXIII. A polarization-type basis set for second-row elements. J Chem Phys 1982;77:3654–65.
[15] Rassolov V, Pople JA, Ratner M, Windus TL. 6-31G* basis set for atoms K through Zn. J Chem Phys 1998;109:1223–9.
[16] Clark T, Chandrasekhar J, Spitznagel GW, Schleyer PvR. Efficient diffuse function-augmented basis sets for anion calculations. III.* The 3-21+G basis set for first-row elements, Li-F. J Comp Chem 1983;4:294–301.
[17] Dunning Jr TJ. Gaussian basis sets for use in correlated molecular calculations. I. The atoms boron through neon and hydrogen. J Chem Phys 1989;90:1007–23.
[18] Woon DE, Dunning Jr TH. Calculation of the electron affinities of the second row atoms: Al–Cl. J Chem Phys 1993;98:1358–71.
[19] Wilson AK, Woon DE, Peterson KA, Dunning Jr TH. Gaussian basis sets for use in correlated molecular calculations. IX. The atoms gallium through krypton. J Chem Phys 1999;110:7667–76.
[20] Almlöf J, Helgaker T, Taylor PR. Gaussian basis sets for high-quality ab initio calculations. J Phys Chem 1988;92:3029–33.
[21] Dunning Jr TH, Peterson KA, Wilson AK. Gaussian basis sets for use in correlated molecular calculations. X. The atoms aluminum through argon revisited. J Chem Phys 2001;114:9244–53.
[22] Davidson E. General introduction. Computational quantum chemistry—1984. Faraday Symp Chem Soc 1984;19:7–15.
[23] Xantheas SS, Dunning Jr TH. Theoretical estimate of the enthalpy of formation of sulfhydryl radical (HSO) and HSO-SOH isomerization energy. J Phys Chem 1993;97:18–9.
[24] Feller D. Application of systematic sequences of wave functions to the water dimer. J Chem Phys 1992;96:6104–14.
[25] Feller D. The use of systematic sequences of wave functions for estimating the complete basis set, full configuration interaction limit in water. J Chem Phys 1993;98:7059–71.
[26] Peterson KA, Woon DE, Dunning Jr TH. Benchmark calculations with correlated molecular wave functions. IV. The classical barrier height of the $H + H_2 \rightarrow H_2 + H$ reaction. J Chem Phys 1994;100:7410–5.
[27] Halkier T, Helgaker T, Jørgensen P, Klopper W, Koch H, Olsen J, et al. Basis-set convergence in correlated calculations on Ne, N_2, and H_2O. Chem Phys Lett 1998;286:243–52.
[28] Helgaker T, Klopper W, Koch H, Noga J. Basis-set convergence of correlated calculations on water. J Chem Phys 1997;106:9639–46.
[29] Schwartz C. Importance of angular correlations between atomic electrons. Phys Rev A 1962;126:1015–9.
[30] Schwartz C. Methods in computational physics, vol. 2. New York: Academic; 1963. p. 262–5.
[31] Kutzelnigg W, Morgan JD. Rates of convergence of the partial-wave expansions of atomic correlation energies. J Chem Phys 1992;96:4484–508.
[32] Martin JML. Ab initio total atomization energies of small molecules—towards the basis set limit. Chem Phys Lett 1996;259:669–78.

[33] Martin JML, Lee TJ. The atomization energy and proton affinity of NH_3. An ab initio calibration study. Chem Phys Lett 1996;258:136–43.

[34] Kendall RA, Dunning Jr TH, Harrison RJ. Electron affinities of the first-row atoms revisited. Systematic basis sets and wave functions. J Chem Phys 1992;96:6796–806.

[35] Woon DE, Dunning Jr TH. Gaussian basis sets for use in correlated molecular calculations. IV. Calculation of static electrical response properties. J Chem Phys 1994;100:2975–88.

[36] Raffenetti RC, Ruedenberg K. Even-tempered representations of atomic self-consistent-field wave functions. Ames (IA): AMES Laboratory; 1973.

[37] van Mourik T, Wilson AK, Dunning Jr TH. Benchmark calculations with correlated molecular wavefunctions. XIII. Potential energy curves for He_2, Ne_2 and Ar_2 using correlation consistent basis sets through augmented sextuple zeta. Mol Phys 1999;96: 529–47.

[38] Woon DE, Dunning Jr TH. Gaussian basis sets for use in correlated molecular calculations. V. Core-valence basis sets for boron through neon. J Chem Phys 1995;103:4572–85.

[39] Peterson KA, Dunning Jr TH. Accurate correlation consistent basis sets for molecular core–valence correlation effects: the second row atoms Al–Ar, and the first row atoms B–Ne revisited. J Chem Phys 2002;117:10548–60.

[40] Wadt WR, Hay PJ. Ab initio effective core potentials for molecular calculations. Potentials for the transition metal atoms Sc to Hg. J Chem Phys 1985;82:270–83.

[41] Wadt WR, Hay PJ. Ab initio effective core potentials for molecular calculations. Potentials for main group elements Na to Bi. J Chem Phys 1985;82:284–98.

[42] Wadt WR, Hay PJ. Ab initio effective core potentials for molecular calculations. Potentials for K to Au including the outermost core orbitals. J Chem Phys 1985;82:299–310.

[43] Dolg M, Wedig U, Stoll H, Preuss H. Energy-adjusted ab initio pseudopotentials for the first row transition elements. J Chem Phys 1987;86:866–72.

[44] Andrae D, Haussermann U, Dolg M, Stoll H, Preuss H. Energy-adjusted *ab initio* pseudopotentials for the second and third row transition elements. Theor Chim Acta 1990;77:123–41.

[45] Dyall KG, Fægri Jr K. Optimization of Gaussian basis sets for Dirac–Hartree–Fock calculations. Theor Chim Acta 1996;94:39–51.

[46] Peterson KA. Systematically convergent basis sets with relativistic pseudopotentials. I. Correlation consistent basis sets for the post-d group 13–15 elements. J Chem Phys 2003;119:11099.

[47] Metz B, Stoll H. Small-core multiconfiguration-Dirac–Hartree–Fock-adjusted pseudopotentials for post-d main group elements: application to PbH and PbO. J Chem Phys 2000;113:2563–9.

[48] Douglas M, Kroll NM. Quantum electrodynamical corrections to the fine structure of helium. Ann Phys 1974;82:89–155.

[49] Jensen F. Polarization consistent basis sets: principles. J Chem Phys 2001;115:9113–25.

[50] Boys SF, Bernardi F. The calculation of small molecular interactions by the differences of separate total energies. Some procedures with reduced errors. Mol Phys 1970;19:553–66.

[51] Valeev E. Libint library. Available from: http://www.chem.vt.edu/chem-dept/valeev/libint/ [Accessed August 2018].

[52] Valeev EF, Janssen CL. Second-order Møller-Plesset theory with linear R12 terms (MP2-R12) revisited: auxiliary basis set method and massively parallel implementation. J Chem Phys 2004;121(3):1214–27.

[53] Sun Q. Libcint: an efficient general integral library for Gaussian basis functions. J Comput Chem 2015;36:1664–71.
[54] Shiozaki T. BAGEL: Brilliantly Advanced General Electronic-structure Library. Wiley Interdiscip Rev Comput Mol Sci 2018;8(1):e1331.
[55] Roothaan CCJ. New developments in molecular orbital theory. Rev Mod Phys 1951;23:69–89.
[56] Mcmurchie LE, Davidson ER. One- and two-electron integrals over Cartesian Gaussian functions. J Comput Phys 1978;26:218–31.
[57] Obara S, Saika A. Efficient recursive computation of molecular integrals over Cartesian Gaussian functions. J Chem Phys 1986;84:3963–74.
[58] Obara S, Saika A. General recurrence formulas for molecular integrals over Cartesian Gaussian functions. Coulomb Integr J Chem Phys 1988;89:2067.
[59] Hô M, Hernández-Pérez JM. Evaluation of Gaussian molecular integrals II. Kinetic-energy integrals. Math J 2013;15:1–10.
[60] Rys J, Dupuis M, King HF. Computation of electron repulsion integrals using the Rys quadrature method. J Comput Chem 1983;4:154–7.
[61] Hô M, Hernández-Pérez JM. Evaluation of Gaussian molecular integrals III. Nuclear–electron attraction integrals. Math J 2014;16:1–13.
[62] Reine S, Helgaker T, Lindh R. Multi-electron integrals. Wiley Interdiscip Rev Comput Mol Sci 2012;2(2):290–303.
[63] Kendall RA, Früchtl HA. The impact of the resolution of the identity approximate integral method on modern ab initio algorithm development. Theor Chem Acc 1997;97:158–63.
[64] Weigend F. A fully direct RI-HF algorithm: implementation, optimised auxiliary basis sets, demonstration of accuracy and efficiency. Phys Chem Chem Phys 2002;4:4285–91.
[65] Eichkorn K, Weigend F, Treutler O, Ahlrichs R. Auxiliary basis sets for main row atoms and transition metals and their use to approximate Coulomb potentials. Theor Chem Acc 1997;97:119–24.
[66] Jung Y, Sodt A, Gill PMW, Head-Gordon M. Auxiliary basis expansions for large-scale electronic structure calculations. Proc Natl Acad Sci 2005;102(19):6692–7.
[67] Sodt A, Head-Gordon M. Hartree-Fock exchange computed using the atomic resolution of the identity approximation. J Chem Phys 2008;128:104106.
[68] Pedersen TB, Aquilante F, Lindh R. Density fitting with auxiliary basis sets from Cholesky decompositions. Theor Chem Acc 2009;124:1–10.
[69] Kritikou S, Hill JG. Auxiliary basis sets for density fitting in explicitly correlated calculations: the atoms H-Ar. J Chem Theor Comput 2015;11:5269–76.
[70] Weigend F. Accurate Coulomb-fitting basis sets for H to Rn. Phys Chem Chem Phys 2006;8:1057.
[71] Eichkorn K, Treutler O, Öhm H, Häser M, Ahlrichs R. Auxiliary basis sets to approximate Coulomb potentials. Chem Phys Lett 1995;240:283–90.
[72] Weigend F, Häser M, Patzelt H, Ahlrichs R. RI-MP2: optimized auxiliary basis sets and demonstration of efficiency. Chem Phys Lett 1998;294:143–52.
[73] Hellweg A, Hättig C, Höfener S, Klopper W. Optimized accurate auxiliary basis sets for RI-MP2 and RI-CC2 calculations for the atoms Rb to Rn. Theor Chem Acc 2007;117:587–97.
[74] Chmela J, Harding ME. Optimized auxiliary basis sets for density fitted post-Hartree–Fock calculations of lanthanide containing molecules. Mol Phys 2018;116:1523–38.

CHAPTER

Post-Hartree-Fock methods: configuration interaction, many-body perturbation theory, coupled-cluster theory

3

Jacob Townsend[a], Justin K. Kirkland[a], Konstantinos D. Vogiatzis

Department of Chemistry, University of Tennessee, Knoxville, TN, United States

1 INTRODUCTION

1.1 THE MANY-BODY PROBLEM OF ELECTRON CORRELATION

At a fundamental level, the chemistry of matter can be interpreted by the physical interactions of electrons and nuclei. Quantum mechanics describes the physics at small scales and low energies. Therefore, it is not an exaggeration to say that chemistry is fully described by the solutions of the cornerstone equation of quantum mechanics: the Schrödinger equation. The first theoretical physicists to apply quantum mechanics in a chemical system (the hydrogen molecule) were Walter Heitler and Fritz London. The two researchers moved to Zurich with the intention to work with Erwin Schrödinger [1]. Schrödinger was not fond of collaborating with others, so the two found themselves working on their own. After some time, they showed their results to Schrödinger, who "had been very favorably surprised since he did not expect his equation to describe the whole of chemistry" [2].

Even as the scientific world approaches the centennial of quantum mechanics, it is remarkable that the only chemical systems for which exact solutions can be obtained are single-electron systems, such as the hydrogen atom and the hydrogen molecule cation (H_2^+). This issue arises from the correlated motion of the electrons: how the motion of an electron is affected by the motion of the other electrons present in the atom or molecule. Since no analytical solutions are known for systems with more than two electrons, Heitler and London used an approximate wave function for

[a]Both authors contributed equally to this work.

Mathematical Physics in Theoretical Chemistry. https://doi.org/10.1016/B978-0-12-813651-5.00003-6

solving the two-electron problem of H_2, which marked the beginning of what is now known as quantum chemistry. Since the seminal publication of Heitler and London in 1930, many approximate quantum chemical models have been developed. These models aim to describe the many-body problem of the electron correlation.

There are two sources of the electron correlation: symmetry and repulsion [3]. Correlation that originates from the spin and permutation symmetry conditions of the wave function is called Fermi correlation. The Coulombic repulsion between the electrons is a physically different source of correlation, and it is referred to as Coulomb correlation. For practical purposes, the Coulomb correlation is divided further into dynamic and strong (or nondynamic or static) correlation. As it will be discussed in the following paragraphs, these two types are not physically different, but this distinction can help to categorize the standard quantum chemical models.

1.2 FERMI CORRELATION

Throughout this chapter, the Born-Oppenheimer approximation will be assumed, in which the nuclear part of the Hamiltonian is decoupled from the electronic part. Let us consider a system with N electrons and R nuclei. A simple approximation for the solution of the Schrödinger equation can be obtained in terms of an one-electron Hamiltonian

$$\hat{h}_I = -\frac{1}{2}\nabla_I^2 - \sum_{k=1}^{R} \frac{Z_k}{r_{Ik}}. \tag{1}$$

The one-electron Hamiltonian operator $\hat{h}_I$ includes a kinetic energy term for the Ith electron and an attractive potential from all nuclei, without considering any electron-electron interactions. The total Hamiltonian operator of the system can be written as a sum of separable one-electron Hamiltonians

$$\hat{H} = \sum_{I=1}^{N} \hat{h}_I. \tag{2}$$

The eigenfunctions χ_I of the one-electron operator (1) must satisfy the corresponding one-electron Schrödinger equation

$$\hat{h}_I \chi_I = \epsilon_I \chi_I. \tag{3}$$

We shall refer to χ_I as a spin-orbital, which is a product of a spatial function ϕ_i and a spin function σ_i. Notice that for spin-orbitals capital letters are used as indices, while lowercase letters are kept for spin-free functions. The nomenclature that is followed in this chapter is shown in Table 1. Since each one-electron operator will act on only one eigenfunction χ_I, the total wave function can be expressed as a product of one-electron eigenfunctions χ_I (Hartree product)

$$\Psi_{\mathrm{HP}} = \chi_1(1)\chi_2(2)\ldots\chi_N(N). \tag{4}$$

The Hartree product can be considered as a primitive, uncorrelated, many-electron wave function. As discussed in a previous chapter of this book, the Hartree product is fundamentally incomplete since it distinguishes between electrons: electron 1 can only be found in the spatial part of the spin-orbital χ_1, electron 2 in spin-orbital χ_2, etc. However, electrons are identical, indistinguishable particles, and we should be able to interchange two electrons without affecting the properties of the system (Pauli symmetry). Since electrons are fermions, a permutation between two electrons should result in a change in the sign of the wave function. An antisymmetrized function that fulfills this requirement is a Slater determinant (SD), which has the general form:

$$\Psi_{\mathrm{SD}} = \frac{1}{\sqrt{N!}}\begin{vmatrix} \chi_1(1) & \chi_2(1) & \ldots & \chi_N(1) \\ \chi_1(2) & \chi_2(2) & \ldots & \chi_N(2) \\ \vdots & \vdots & \ddots & \vdots \\ \chi_1(N) & \chi_2(N) & \ldots & \chi_N(N) \end{vmatrix}. \tag{5}$$

Since each electron is described by a spin-orbital χ_I, either the spatial or the spin part of the total wave function of the system should be antisymmetric.

In the SD, the motion of one electron depends on the position of the other electrons. Therefore, representation of the electronic wave function with an SD considers interelectronic interactions and the motion of the electrons is now correlated. This type of correlation that originates from the correct symmetry of the wave function represents the Fermi correlation. Fermi correlation is not related to the electron repulsion, but is a direct consequence of the quantum nature of matter.

1.3 COULOMB CORRELATION AND LIMITATIONS OF HARTREE-FOCK THEORY

By expressing the wave function as an SD, we introduce antisymmetry in a very elegant manner. The Hartree-Fock (HF) method uses a single SD as the electronic, zeroth-order wave function, and the orbitals are determined by minimizing the total energy of the system. Although HF includes Fermi correlation in an exact sense and electron repulsion in an average, mean-field manner, the method fails to describe the instantaneous Coulombic repulsion between electron pairs. This discrepancy results from HF considering only the average field of each electron.

The missing Coulomb correlation is added by expanding the spatial part of the wave function in orbital products. The basis of this approach goes back to matrix mechanics, where a vector space is chosen for the expansion of the problem. For the case of an N-electron molecule, such a (complete) expansion is constructed by forming N-particle functions. The level of theory that includes a complete N-particle expansion, that is, includes all possible orbital configurations within a given orbital

Table 1 A Summary of Functions and Index Conventions Followed in This Chapter

Functions	
$\chi_i, \chi_j, \ldots$	Spin-orbitals
$\phi_i, \phi_j, \ldots$	Spatial functions
$\sigma_i, \tau_j, \ldots$	Spin functions
Indices	
$P, Q, R, S, \ldots$	Arbitrary spin-orbitals
$p, q, r, s, \ldots$	Arbitrary spin-free orbitals
$I, J, K, L, \ldots$	Occupied spin-orbitals
$i, j, k, l, \ldots$	Occupied (inactive) spin-free orbitals
$t, u, v, w, \ldots$	Occupied (active) spin-free orbitals
$A, B, C, D, \ldots$	Unoccupied (virtual) spin-orbitals
$a, b, c, d, \ldots$	Unoccupied (virtual) spin-free orbitals
$\sigma, \tau, \ldots$	Spin (α or β)
$m, n, \ldots$	Slater determinants

basis is called full configuration interaction (FCI). The FCI wave function can be written as

$$|\Psi_{\mathrm{FCI}}\rangle = c_0|\Psi_0\rangle + \sum_{IA} c_I^A|\Psi_I^A\rangle + \sum_{I<J,A<B} c_{IJ}^{AB}|\Psi_{IJ}^{AB}\rangle + \sum_{I<J<K,A<B<C} c_{IJK}^{ABC}|\Psi_{IJK}^{ABC}\rangle + \cdots, \quad (6)$$

where $|\Psi_0\rangle$ is the reference wave function (e.g., the HF SD). Occupied spin-orbitals are termed as $I, J, K, \ldots$, unoccupied spin-orbitals are termed as $A, B, C, \ldots$. The $|\Psi_I^A\rangle$ function is an SD formed after replacing spin-orbital I with spin-orbital A, the $|\Psi_{IJ}^{AB}\rangle$ function is an SD formed after replacing spin-orbitals IJ with spin-orbitals AB, etc. In other words, we can perform one-electron, two-electron, three-electron, etc., excitations (or promotions) from the reference, zeroth-order wave function to the unoccupied (virtual or secondary) space in order to calculate the Coulomb electron correlation. For a complete basis set (CBS) that contains an infinite number of one-electron basis functions, the FCI wave function will be expanded in an infinite number of determinants, and FCI at the CBS limit will provide the exact solution to the electronic, nonrelativistic, time-independent, Born-Oppenheimer Schrödinger equation. Practically, even by using a relatively small basis set, the FCI expansion is still extremely large, and the FCI problem becomes intractable, so the method is applicable only on small systems with few electrons. Therefore, we must rely on approximate solutions to FCI in order to go beyond HF to determine the missing correlation energy. It should be noted that the common definition of the correlation energy

$$E_{\mathrm{corr}} = E_{\mathrm{FCI}} - E_{\mathrm{HF}} \quad (7)$$

implies that the term E_{corr} excludes the Fermi correlation, which is treated at the HF level.

From a practical standpoint, there are two approaches that can be applied for including more determinants in the wave function to recover the missing Coulomb correlation. The first approach is to use a zeroth-order wave function that goes beyond a single SD, and can be considered as a truncated CI. By forming a linear combination of SDs (or configuration state functions [CSFs], see Section 3), the description of the electronic state under consideration can be greatly improved. This pertains especially for cases where a system has degenerate or near-degenerate orbitals, such as transition metal complexes, biradicals, or excited states. The type of electron correlation that is included by a superposition of configurations is known as strong or static or nondynamic correlation. The second approach is to account for instantaneous correlation of the electron motion due to their mutual repulsion by introducing a set of determinants formed by excitations from the reference zeroth-order wave function to the virtual space. This type of correlation is called dynamic correlation and the choice of the zeroth-order wave function can be either a single SD (single-reference) or a linear combination of determinants (multireference). Typically, the strong correlation is important to correct the zeroth-order description of specific systems (vide infra), while the dynamic correlation is mandatory for achieving high accuracy.

In the rest of this chapter, we will discuss solutions to the exact electronic, nonrelativistic, time-independent Born-Oppenheimer Schrödinger equation in the absence of external fields. Since the FCI/CBS level of theory cannot be applied for practical calculations, we have to rely on approximate solutions. The standard models of quantum chemistry provide such approximate solutions, which can be grouped in three different theories: the configuration interaction (CI) theory, the coupled-cluster (CC) theory, and the many-body perturbation theory (MBPT). These three wave function-based quantum chemical methods can be grouped under the terminology: post-HF methods. These quantum chemical models are constructed in such a manner that their accuracy is systematically improved toward the FCI/CBS limit either by expanding the one-electron basis or the N-electron basis (Fig. 1). This enables us to estimate the error that is introduced by a truncated configuration expansion or a non-CBS and allows the derivation of extrapolation schemes to the full one- and N-electron functions. A short introduction to the second quantization, the mathematical formulation that is usually applied in post-HF methods, will be presented in Section 2. The three quantum chemical models or combinations of them (e.g., multireference perturbation theory) are discussed in Sections 3–5. Our aim is to analyze the general theory of these methods, discuss methodologies that can provide solutions closer to the exact solution and present emerging, state-of-the-art methodologies that are promising for moving the field of electronic structure theory beyond the current limitations of the standard models (Section 6). We refer the interested reader to the excellent textbooks and review articles that are listed at the end of this chapter for more details on the derivation, implementation, and application of post-HF methods.

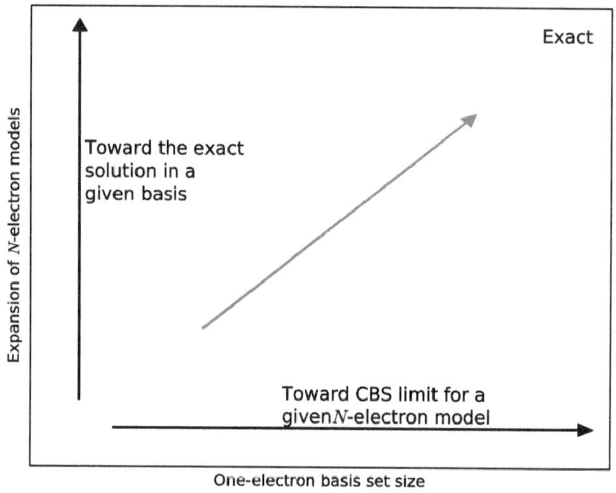

FIG. 1

Systematic approach to the exact solution of the electronic, nonrelativistic, time-independent Born-Oppenheimer Schrödinger equation. This approach includes a systematic expansion of the one-electron basis (x-axis) and of the N-electron models (y-axis).

2 BASICS OF SECOND QUANTIZATION

In the most common notation that is followed in quantum mechanics, an observable is represented by an operator and a state by a (wave) function. Here, we will follow a different formulation where both observables and states are represented by operators. This formulation is called second quantization and provides an elegant toolset for the development and implementation of standard and novel quantum chemical methods. The main advantages of second quantization are:

- its applicability to problems in which the total number of particles is neither fixed nor known;
- to provide a convenient and concise way to express an SD for the derivation of theories with a large number of determinants;
- the antisymmetric nature of the wave function is introduced from the algebra of the operators;
- the unified notation for both operators and states, which are expressed in terms of elementary operators; and
- the simplification of the final equations for direct implementation in a computer program.

The only drawback of second quantization is that it requires learning a new mathematical formulation. The same notation as in Helgaker et al. [4] is followed in this chapter. In this section, we will provide an introduction to this formulation and explain the most important relations that will help us to develop the basic models of molecular electronic structure theory.

2.1 FOCK SPACE

Let us firstly introduce an alternative expression of the SD of Eq. (5). We will represent an SD as an occupation-number (ON) vector

$$|\Psi_m\rangle = |\mathbf{m}\rangle = |m_1, m_2, \ldots, m_M\rangle, \quad m_I = \begin{cases} 1 & \chi_I \text{ is occupied} \\ 0 & \chi_I \text{ is unoccupied} \end{cases}, \tag{8}$$

where M is the number of all spin-orbitals and the occupation number m_I equals 1 if spin-orbital I is occupied and 0 if it is unoccupied. We have already seen in Section 1 that interchange of two columns in an SD changes the sign of the wave function. The same sign change holds for the interchange of two ONs that differ by one spin-orbital

$$|\ldots, m_I, m_J, \ldots\rangle = -|\ldots, m_J, m_I, \ldots\rangle. \tag{9}$$

As it can be understood, the order of the spin-orbitals in an ON vector plays a very important role in the second-quantization formalism.

We can express the overlap of two SDs that contain the same number of electrons as the innerproduct of two ON vectors $|\mathbf{m}\rangle$ and $|\mathbf{n}\rangle$

$$\langle \mathbf{m}|\mathbf{n}\rangle = \delta_{\mathbf{mn}} = \prod_{i=1}^{M} \delta_{m_i n_i}, \tag{10}$$

considering an orthonormal set of spin-orbitals. This definition allows the examination of states with variable number of electrons but same number of spin-orbitals.

The full vector space spanned by all possible ON vectors on a given number of M spin-orbitals is called *Fock space* and contains 2^M ON vectors. For a system with N electrons, we will work on a subspace $F(M,N)$ of the Fock space that spans over $\binom{M}{N}$ ON vectors. The special case of an ON vector with zero electrons is called vacuum state

$$|\text{vac}\rangle = |0_1, 0_2, \ldots, 0_M\rangle, \quad \langle \text{vac}|\text{vac}\rangle = 1. \tag{11}$$

2.2 ELEMENTARY OPERATORS

One of the main advantages of second quantization is that the expression of all operators and states in terms of two elementary *creation* and *annihilation* operators. The *creation operator* $\hat{a}_P^\dagger$, as its name implies, is an operator that when acting upon an empty spin-orbital P with an occupation number $k_P = 0_P$, it creates a particle (electron) in position P. (For sake of simplicity, the hat of the creation operator $\hat{a}_P^\dagger$

is omitted in our nomenclature.) If spin-orbital P is occupied (occupation number $k_P = 1_P$), then it returns a zero value

$$a_P^\dagger|m_1, m_2, \ldots, 0_P, \ldots, m_M\rangle = \Gamma_P^{\mathbf{m}}|m_1, m_2, \ldots, 1_P, \ldots, m_M\rangle, \tag{12}$$

$$a_P^\dagger|m_1, m_2, \ldots, 1_P, \ldots, m_M\rangle = 0. \tag{13}$$

The phase factor $\Gamma_P^{\mathbf{m}}$ ensures agreement between first- and second-quantization formalisms

$$\Gamma_P^{\mathbf{m}} = \prod_{Q=1}^{P-1} (-1)^{m_Q} \tag{14}$$

and it is equal to +1 for an even number of spin-orbitals prior to spin-orbital P ($Q < P$) and -1 for odd number of spin-orbitals. Eq. (13) properly maintains the Pauli Exclusion Principle: two electrons cannot occupy the same spin-orbital. Combination of Eqs. (12), (13) gives a more general definition of the creation operator

$$a_P^\dagger|\mathbf{m}\rangle = \delta_{m_P 0} \Gamma_P^{\mathbf{m}}|m_1, m_2, \ldots, 1_P, \ldots, m_M\rangle. \tag{15}$$

We can write the Hermitian adjoint of $a_P^\dagger$ as a_P, for which it holds

$$\langle 1_P|a_P^\dagger|0_P\rangle = 1 = \langle 0_P|a_P|1_P\rangle, \tag{16}$$

which implies that the Hermitian adjoint of the creation operator, known as the *annihilation operator*, is an operator that annihilates an electron and creates a hole in a vector $|\mathbf{m}\rangle$

$$a_P|\mathbf{m}\rangle = \delta_{m_P 1} \Gamma_P^{\mathbf{m}}|m_1, m_2, \ldots, 0_P, \ldots, m_M\rangle. \tag{17}$$

Operating twice with the creation operator of the same spin-orbital implies that we apply Eq. (15) twice, so

$$a_P^\dagger a_P^\dagger|\mathbf{m}\rangle = a_P^\dagger \delta_{m_P 0} \Gamma_P^{\mathbf{m}}|m_1, m_2, \ldots, 1_P, \ldots, m_M\rangle = 0. \tag{18}$$

The creation operators of two different spin-orbitals with $P > Q$ may act on an ON vector in two different ways

$$a_P^\dagger a_Q^\dagger|\mathbf{m}\rangle = \delta_{m_P 0}\delta_{m_Q 0} \Gamma_P^{\mathbf{m}} \Gamma_Q^{\mathbf{m}}|\ldots, 1_P, \ldots, 1_Q, \ldots\rangle, \tag{19}$$

$$a_Q^\dagger a_P^\dagger|\mathbf{m}\rangle = \delta_{m_Q 0}\delta_{m_P 0} \Gamma_P^{\mathbf{m}} \left(-\Gamma_Q^{\mathbf{m}}\right)|\ldots, 1_P, \ldots, 1_Q, \ldots\rangle. \tag{20}$$

The difference between Eqs. (19), (20) is in the number of spin-orbitals prior to spin-orbital P. In the second case, there is an additional spin-orbital prior to Q and this will change the sign of the phase factor $\Gamma_Q^{\mathbf{m}}$. In conclusion,

$$(a_P^\dagger a_Q^\dagger + a_Q^\dagger a_P^\dagger)|\mathbf{m}\rangle = 0. \tag{21}$$

For two arbitrary operators $\hat{A}$ and $\hat{B}$, the commutator $[\hat{A},\hat{B}]$ and anticommutator $[\hat{A},\hat{B}]_+$ relations are defined as

$$[\hat{A},\hat{B}] = \hat{A}\hat{B} - \hat{B}\hat{A}, \tag{22}$$

$$[\hat{A},\hat{B}]_+ = \hat{A}\hat{B} + \hat{B}\hat{A}. \tag{23}$$

Finally, the important anticommutator relation is deduced from Eq. (21)

$$a_P^\dagger a_Q^\dagger + a_Q^\dagger a_P^\dagger = [a_P^\dagger, a_Q^\dagger]_+ = 0. \tag{24}$$

Similarly, we can prove the following anticommutator relations between two annihilation and a creation/annihilation operators

$$[a_P, a_Q]_+ = 0, \tag{25}$$

$$[a_P^\dagger, a_Q]_+ = \delta_{PQ}. \tag{26}$$

Let us now return to the representation of the SD with an ON vector. The ON vector can be expressed as a product (string) of creation operators acting on the vacuum state without explicitly referring to the unoccupied spin-orbitals

$$|\mathbf{m}\rangle = \left[\prod_{P=1}^{M}\left(a_P^\dagger\right)^{m_P}\right]|\mathrm{vac}\rangle. \tag{27}$$

2.3 REPRESENTATION OF ONE- AND TWO-ELECTRON OPERATORS

Consider a general *one-electron operator* in first-quantization formalism (e.g., kinetic energy operator, dipole moment operator)

$$\hat{f}^{\mathrm{c}} = \sum_{I=1}^{N} \hat{f}^{\mathrm{c}}(\mathbf{x}_I), \tag{28}$$

where the superscript c is used to indicate the coordinate representation of first quantization. Let us examine the matrix elements $\langle\Psi_A|\hat{f}^{\mathrm{c}}|\Psi_B\rangle$ by applying the Slater-Condon rules, where Ψ_A and Ψ_B are two arbitrary SDs:

$$\langle\Psi_A|\hat{f}^{\mathrm{c}}|\Psi_B\rangle = \sum_{I=1}^{N}\langle\chi_I|\hat{f}^{\mathrm{c}}|\chi_I\rangle, \quad \text{if } \Psi_A \text{ and } \Psi_B \text{ are identical,} \tag{29}$$

$$\langle\Psi_A|\hat{f}^{\mathrm{c}}|\Psi_B\rangle = \langle\chi_P|\hat{f}^{\mathrm{c}}|\chi_Q\rangle, \quad \text{if } \Psi_A \text{ and } \Psi_B \text{ differ by one spin-orbital pair } PQ, \tag{30}$$

$$\langle\Psi_A|\hat{f}^{\mathrm{c}}|\Psi_B\rangle = 0, \quad \text{if } \Psi_A \text{ and } \Psi_B \text{ differ by more than one spin-orbital pair.} \tag{31}$$

In the second-quantization formalism, the one-electron operator is expressed as

$$\hat{f} = \sum_{PQ} f_{PQ} a_P^\dagger a_Q, \tag{32}$$

where the operators $a_P^\dagger a_Q$ shift one electron in an ON vector. The numerical parameters f_{PQ} are identical to the first-quantization matrix elements

$$f_{PQ} = \int \chi_P^*(\mathbf{x}) \hat{f}^c(\mathbf{x}) \chi_Q(\mathbf{x}) d\mathbf{x} \tag{33}$$

but they can be easily determined by evaluating the matrix elements of one-electron operator between two ON vectors, similarly to the Slater-Condon rules of first quantization. Let us consider two ON vectors $|\mathbf{m}\rangle$ and $|\mathbf{n}\rangle$ with

$$|\mathbf{m}\rangle = |m_1, \ldots, m_I, \ldots, m_J, \ldots, m_M\rangle, \tag{34}$$
$$|\mathbf{n}\rangle = |n_1, \ldots, n_I, \ldots, n_J, \ldots, n_M\rangle. \tag{35}$$

Three different cases can be considered:

1. ON vectors are identical ($m_I = n_I, I = 1, 2, \ldots, M$)

$$\langle \mathbf{m}|\hat{f}|\mathbf{n}\rangle = \sum_P f_{PP} \langle \mathbf{m}|a_P^\dagger a_P|\mathbf{n}\rangle = \sum_P m_P f_{PP}. \tag{36}$$

2. ON vectors differ in one occupation-number pair:

$$|\mathbf{m}\rangle = |m_1, \ldots, 1_I, \ldots, 0_J, \ldots, m_M\rangle, \tag{37}$$
$$|\mathbf{n}\rangle = |n_1, \ldots, 0_I, \ldots, 1_J, \ldots, n_M\rangle, \tag{38}$$
$$\langle \mathbf{m}|\hat{f}|\mathbf{n}\rangle = \Gamma_I^{\mathbf{m}} \Gamma_J^{\mathbf{n}} f_{IJ}. \tag{39}$$

3. ON vectors differ in more than one occupation-number pair:

$$\langle \mathbf{m}|\hat{f}|\mathbf{n}\rangle = 0. \tag{40}$$

The *two-electron operator* (e.g., electronic-repulsion operator, the two-electron part of the spin-orbit operator) is expressed in the first-quantization formalism as

$$\hat{g}^c = \frac{1}{2} \sum_{i \neq j} \hat{g}^c(\mathbf{x}_i, \mathbf{x}_j). \tag{41}$$

Analogous to the one-electron operators, the two-electron operator will give nonzero matrix elements between determinants only if they differ in two or less electron pairs. The two-electron operator expression in second quantization is

$$\hat{g} = \frac{1}{2} \sum_{PQRS} g_{PQRS} a_P^\dagger a_R^\dagger a_S a_Q. \tag{42}$$

Similar expressions to Eqs. (36)–(40) hold for the two-electron operator. The first- and second-quantization formalisms become equivalent once we substitute the parameter g_{PQRS} with the double integral

$$g_{PQRS} = \iint \chi_P^*(\mathbf{x}_1)\chi_R^*(\mathbf{x}_2)\hat{g}^c(\mathbf{x}_1,\mathbf{x}_2)\chi_Q(\mathbf{x}_1)\chi_S(\mathbf{x}_2)\mathrm{d}\mathbf{x}_1\mathrm{d}\mathbf{x}_2. \tag{43}$$

Notice the order of the creation and annihilation operators in Eq. (42), where the indices P and Q correspond to electron 1 with spin and spatial coordinates $\mathbf{x}_1$, and indices R and S to electron 2 with spin and spatial coordinates $\mathbf{x}_2$.

We are now ready to express the molecular electronic, Born-Oppenheimer, nonrelativistic Hamiltonian operator in the second-quantization formalism, which is given as:

$$\hat{H} = \sum_{PQ} h_{PQ} a_P^\dagger a_Q + \frac{1}{2}\sum_{PQRS} g_{PQRS} a_P^\dagger a_R^\dagger a_S a_Q + h_{\text{nuc}}. \tag{44}$$

The three individual terms of Eq. (44) are (in atomic units)

$$h_{PQ} = \int \chi_P^*(\mathbf{x})\left(-\frac{1}{2}\nabla^2 - \sum_X \frac{Z_X}{r_X}\right)\chi_Q(\mathbf{x})\mathrm{d}\mathbf{x}, \tag{45}$$

$$g_{PQRS} = \iint \chi_P^*(\mathbf{x}_1)\chi_R^*(\mathbf{x}_2)\frac{1}{r_{12}}\chi_Q(\mathbf{x}_1)\chi_S(\mathbf{x}_2)\mathrm{d}\mathbf{x}_1\mathrm{d}\mathbf{x}_2, \tag{46}$$

$$h_{\text{nuc}} = \frac{1}{2}\sum_{X \neq Y} \frac{Z_X Z_Y}{R_{XY}}. \tag{47}$$

The X and Y indices are running over all nuclei, Z_X, Z_Y are the nuclear charges, r_X is the electron-nucleus distance, r_{12} is the interelectronic distance, and R_{XY} is the internuclear distance. The h_{nuc} term corresponds to the nuclear repulsion energy and it is a scalar that is added in the total electronic energy.

2.4 ONE- AND TWO-ELECTRON DENSITY MATRICES

Let us consider the general expression of the normalized reference wave function $|\Psi_0\rangle$ as a linear combination of ON vectors

$$|\Psi_0\rangle = \sum_{\mathbf{m}} c_{\mathbf{m}}|\mathbf{m}\rangle, \quad \langle\Psi_0|\Psi_0\rangle = 1. \tag{48}$$

Eq. (48) is a general formula that covers all possible quantum chemical methods, since for a single-reference, single-determinant method, the expansion will contain only one ON vector. We will define the one- and two-electron density matrices as

$$\bar{D}_{PQ} = \langle\Psi_0|a_P^\dagger a_Q|\Psi_0\rangle, \tag{49}$$

$$\bar{d}_{PQRS} = \langle\Psi_0|a_P^\dagger a_R^\dagger a_S a_Q|\Psi_0\rangle. \tag{50}$$

The overbar denotes that the density matrices are expressed in the spin-orbital basis. As we will see in the following chapters, these two expressions are important for the formalism used in the electronic structure theory models.

2.5 SPIN-FREE OPERATORS

The electronic Hamiltonian discussed in Section 2.3 was expressed in terms of spin-orbitals, since they simplify the notation and derivation of equations of many quantum chemical methods. However, since we need to integrate separately the spatial part from the spin part of the wave function, we need then to provide also the form of the spin-free one- and two-electron operators:

$$\hat{f} = \sum_{pq} f_{pq} \sum_{\sigma} a^{\dagger}_{p\sigma} a_{q\sigma} = \sum_{pq} f_{pq} \hat{E}_{pq}, \tag{51}$$

$$\hat{g} = \frac{1}{2} \sum_{pqrs} g_{pqrs} \sum_{\sigma\tau} a^{\dagger}_{p\sigma} a^{\dagger}_{r\tau} a_{s\tau} a_{q\sigma} = \frac{1}{2} \sum_{pqrs} g_{pqrs} (\hat{E}_{pq}\hat{E}_{rs} - \delta_{qr}\hat{E}_{ps}), \tag{52}$$

where the products (strings) of creation and annihilation operators of Eqs. (32), (42) were substituted with summations over the spin functions σ and τ. For that purpose, the excitation operator $\hat{E}_{pq}$ was introduced

$$\hat{E}_{pq} = a^{\dagger}_{p\alpha} a_{q\alpha} + a^{\dagger}_{p\beta} a_{q\beta}. \tag{53}$$

The electronic Hamiltonian of Eq. (44) can be expressed in a spin-free formalism as:

$$\hat{H} = \sum_{pq} h_{pq} \hat{E}_{pq} + \frac{1}{2} \sum_{pqrs} g_{pqrs} (\hat{E}_{pq}\hat{E}_{rs} - \delta_{qr}\hat{E}_{ps}) + h_{\text{nuc}}. \tag{54}$$

Similarly, the spin-free one- and two-electron density matrices are given as

$$D_{pq} = \langle \Psi_0 | \hat{E}_{pq} | \Psi_0 \rangle, \tag{55}$$

$$d_{pqrs} = \langle \Psi_0 | \hat{E}_{pq}\hat{E}_{rs} - \delta_{qr}\hat{E}_{ps} | \Psi_0 \rangle. \tag{56}$$

3 CONFIGURATION INTERACTION THEORY

The first post-HF method that will be discussed is CI, due to its significance to the most important chemical concept: the chemical bond. In addition, many-body perturbation and CC theories are derivatives of CI. In the following paragraphs, the power of the CI method will be thoroughly discussed and become evident to the reader. We will analyze why restricted HF (RHF) fails to describe the dissociation of bonds, the shortcomings of unrestricted HF (UHF), and how CI theory provides the correct electronic structure, not only at the equilibrium and the dissociation limit, but for the full potential energy surface (PES), from the simplest single-bond of the H_2 molecule to the multiple bonds of the chromium and uranium dimers. CI theory has also been applied successfully to many chemical applications, such as spectroscopy, photochemistry and excited states, molecular magnetism, and more. The different variants of CI theory, such as the single-reference CI (SRCI), the multireference CI (MRCI), and the multiconfigurational self-consistent field (MCSCF) methods, will be analyzed in this section.

3.1 HF THEORY AND THE DISSOCIATION OF H_2

Let us consider the simplest chemical bond formed by two electrons: the single bond of the hydrogen molecule. With this definition we exclude the even simpler bond of H_2^+ since we are interested here in a system with two correlated electrons. The limitations of the HF theory are demonstrated for the dissociation of molecular hydrogen to two hydrogen atoms ($H_2 \rightarrow 2H$). We will use for our discussion a minimal basis for each hydrogen atom (just a 1s atomic orbital per hydrogen). By considering a linear combination of atomic orbitals and RHF theory, we can write the wave function of the H_2 molecule as a single SD:

$$\Psi_{\mathrm{RHF}}(1,2) = \frac{1}{\sqrt{2}} \begin{vmatrix} \sigma_g(1)\alpha(1) & \sigma_g(1)\beta(1) \\ \sigma_g(2)\alpha(2) & \sigma_g(2)\beta(2) \end{vmatrix}, \tag{57}$$

where σ_g is the bonding molecular orbital (MO) formed as an in-phase combination of the two 1s atomic orbitals of H_A and H_B

$$\sigma_g = N_g(1s_A + 1s_B) \tag{58}$$

and N_g is the normalization constant. Similarly, the out-of-phase combination of the two 1s orbitals provides the antibonding orbital σ_u

$$\sigma_u = N_u(1s_A - 1s_B). \tag{59}$$

While the SD of Eq. (57) can qualitatively describe the H_2 molecule at equilibrium, that is, the chemical bond at the interatomic distance $R_{\mathrm{H-H}} = 0.746$ Å, it deviates significantly at the dissociation limit (Fig. 2). At the equilibrium geometry, the σ_g bonding orbital will be doubly occupied, while the σ_u antibonding orbital will be unoccupied. This electronic configuration can be written as $(\sigma_g)^2(\sigma_u)^0$. If we expand the SD of Eq. (57), we obtain the wave function

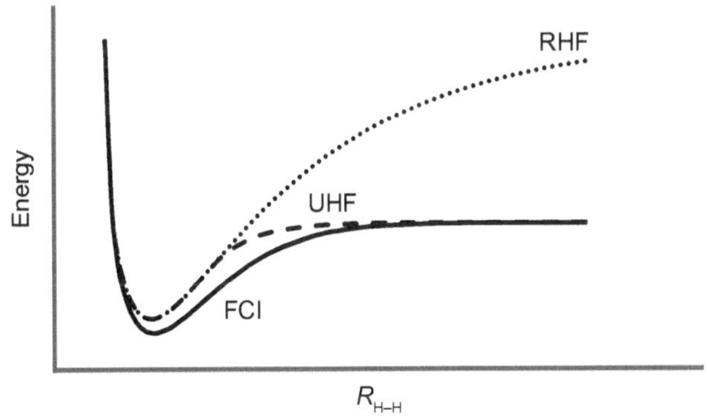

FIG. 2

Dissociation energy curve of the H_2 molecule calculated by RHF (*dotted line*) and UHF (*dashed line*). The FCI potential energy is shown as a *solid black line*.

$$\Psi_{\text{RHF}}(1,2) = \sigma_g(1)\sigma_g(2)\left\{\frac{1}{\sqrt{2}}[\alpha(1)\beta(2) - \beta(1)\alpha(2)]\right\}. \tag{60}$$

Eq. (60) can also expressed in a second-quantization formalism by considering four spin-orbitals formed by the minimal basis as

$$|\mathbf{m}\rangle = a^{\dagger}_{\sigma_g\alpha}a^{\dagger}_{\sigma_g\beta}|\text{vac}\rangle = a^{\dagger}_{\sigma_g\alpha}a^{\dagger}_{\sigma_g\beta}|0,0,0,0\rangle = |1,1,0,0\rangle. \tag{61}$$

The vacuum state $|\text{vac}\rangle$ consists of four unoccupied spin-orbitals, the first two correspond to the $\sigma_g\alpha$ and $\sigma_g\beta$ spin-orbitals, respectively; the last two to the $\sigma_u\alpha$ and $\sigma_u\beta$ spin-orbitals. The single-determinant wave function of RHF will keep the bonding MO as doubly occupied across the potential energy curve of Fig. 2, which gives the wrong dissociation limit for H_2. We can better understand this erroneous behavior if we expand the spatial part of the wave function of Eq. (60) as products of the atomic orbitals $1s_A$ and $1s_B$

$$\Psi_{RHF}(1,2) = N_g^2[1s_A(1)1s_A(2) + 1s_A(1)1s_B(2) + 1s_B(1)1s_A(2) + 1s_B(1)1s_B(2)]. \tag{62}$$

The first and fourth terms place both electrons on either hydrogen A or hydrogen B. These electronic configurations correspond to unnatural ionic terms ($H_2 \rightarrow H^- + H^+$), which cause the RHF wave function to deviate from the correct dissociation limit. By contrast, the second and third terms correctly describe the homolytic dissociation of the H_2 molecule ($H_2 \rightarrow 2H$).

Another approach is to use an UHF formalism. The UHF wave function is not a pure singlet as we would expect, but it is contaminated by a triplet configuration. Although the UHF provides the correct dissociation limit for H_2 and the correct MOs (Fig. 2), the erroneous wave function limits the applicability of the UHF method near the infinite separation. This error, which is known as *spin contamination error*, arises from the missing exchange term and disfavors the unrestricted formalism for the description of bond formation and dissociation.

An alternative solution is to use a wave function, which is a linear combination of two configurations (TC):

$$\Psi_{\text{TC}} = c_1\Psi_1 + c_2\Psi_2. \tag{63}$$

The coefficients c_1 and c_2 can be determined by using the variational principle to minimize the electronic energy. Such a flexible wave function can properly describe the electronic configuration of H_2 at both equilibrium and dissociation limit, since the coefficients will determine its properties at each point in the PES.

In the valence bond (VB) formalism for the H_2 molecule, we can apply an ionic and a covalent functions for Ψ_1 and Ψ_2, respectively:

$$\Psi_{\text{ion}} = N_{\text{ion}}[1s_A(1)1s_A(2) + 1s_B(1)1s_B(2)]\left\{\frac{1}{\sqrt{2}}[\alpha(1)\beta(2) - \beta(1)\alpha(2)]\right\}, \tag{64}$$

$$\Psi_{\text{cov}} = N_{\text{cov}}[1s_A(1)1s_B(2) + 1s_B(1)1s_A(2)]\left\{\frac{1}{\sqrt{2}}[\alpha(1)\beta(2) - \beta(1)\alpha(2)]\right\}, \tag{65}$$

where N_{ion} and N_{cov} are the normalization constants of Ψ_{ion} and Ψ_{cov}, respectively. The VB wave function can be expressed as:

$$\Psi_{\text{VB}} = c_{\text{ion}}\Psi_{\text{ion}} + c_{\text{cov}}\Psi_{\text{cov}}. \tag{66}$$

By varying the c_{ion} and c_{cov} coefficients, the correct wave function is obtained at the limit of infinite separation ($c_{\text{ion}} = 0, c_{\text{cov}} = 1$), as well as at the equilibrium ($c_{\text{ion}} = c_{\text{cov}} = 1/\sqrt{2}$). Do not forget that the square sum of the squared coefficients should remain equal to 1. By inspection of the VB wave function at the equilibrium, all terms of the RHF determinant are recovered Eq. (62). However, the VB theory is based on nonorthogonal atomic orbitals (basis functions), which leads to complicated equations that are difficult to be evaluated efficiently in quantum chemical programs for large-scale applications. For that reason, we will focus our attention on methods that include terms of orthogonal MOs.

Let us go back to the two-configuration scheme of Eq. (63) and examine the properties of the two SDs. The first SD corresponds to the RHF wave function of Eq. (60), where both electrons occupy the bonding σ_g orbital. The $(\sigma_g)^2$ electronic configuration is mixed with an excited state determinant Ψ_2 formed by occupying the antibonding MO (σ_u) with both electrons (or $(\sigma_u)^2$). As it was already mentioned, the c_1 and c_2 coefficients can be variationally defined and as in the VB case, the Ψ_{TC} can correctly describe the entire PES. At the equilibrium, the single determinant is recovered ($c_1 \approx 1, c_2 \approx 0$), while at the dissociation limit both configurations will have equal probabilities ($c_1 \approx -c_2$). In order to understand the importance of the latter, consider the H_2 molecule at large interatomic separation. The two noninteracting hydrogen atoms will form two degenerate MOs composed of in-phase and out-of-phase combinations of two nonoverlapping 1s atomic orbitals. These MOs should have the same electron occupancy and energy at the dissociation limit. Thus, the wave function formed by the two equally weighted configurations Ψ_1 and Ψ_2 can correctly describe the dissociation of the hydrogen molecule.

The methodology of adding configurations (SDs) for constructing more flexible wave functions is called CI. The Ψ_2 configuration of Eq. (63) can be considered as a two-electron excitation from the occupied bonding σ_g orbital to the unoccupied antibonding σ_u. The unoccupied HF orbitals might not have direct physical meaning since they are not variationally optimized during the SCF procedure, but can be exploited in post-HF methods for the improvement of the wave function, the electronic energy and molecular properties. Introduction of additional excited electronic configurations in the TC scheme of Eq. (63) will systematically increase the flexibility and the accuracy of the wave function. This approach is the basis of multiconfigurational (MC) or multireference quantum chemical models, which can correctly describe atoms and molecules with degenerate or near-degenerate orbitals (e.g., excited states, transition metals, lanthanides and actinides, biradicals, multiple bonds and more). The method that includes all possible configurations in the MC wave function is called full CI (FCI). The FCI theory will be discussed first since it is an exact theory and all other post-HF variants can be considered as approximations to it.

3.2 FCI THEORY

FCI is probably the most straightforward quantum chemical model, but the most complicated to efficiently develop in a computer program. In FCI, the electronic wave function is expressed as an expansion of all possible configurations formed by considering electronic excitations from a reference zeroth-order wave function $|\Psi_0\rangle$. For example, for a closed-shell system, the RHF wave function can be effectively used as the reference wave function, and all single-electron, double-electron, triple-electron, etc., excitations will form the FCI wave function:

$$|\Psi_{\text{FCI}}\rangle = c_0|\Psi_0\rangle + \sum_{I,A} c_I^A|\Psi_I^A\rangle + \sum_{\substack{I<J\\A<B}} c_{IJ}^{AB}|\Psi_{IJ}^{AB}\rangle + \sum_{\substack{I<J<K\\A<B<C}} c_{IJK}^{ABC}|\Psi_{IJK}^{ABC}\rangle + \sum_{\substack{I<J<K<L\\A<B<C<D}} c_{IJKL}^{ABCD}|\Psi_{IJKL}^{ABCD}\rangle + \cdots, \tag{67}$$

where the indices $I, J, K, \ldots$ correspond to occupied spin-orbitals, and $A, B, C, \ldots$ to unoccupied (virtual) spin-orbitals (Table 1). Therefore, the function Ψ_I^A describes a single excitation from spin-orbital I to spin-orbital A, the function Ψ_{IJ}^{AB} a double-excitation from spin-orbitals IJ to spin-orbitals AB and so on. In general, SDs are not always eigenfunctions of the spin operators and SD expansions might form wave functions with incorrect spin. For that reason, it is preferential to use CSFs, which are spin-adapted linear combinations of a few SDs. A second advantage of CSFs over SDs has to do with the size of the CI problem; CI expansions based on CSFs are shorter than the one formed by SDs.

In order to obtain the parameters $c_0, c_I^A, c_{IJ}^{AB}, \ldots$, we have to solve the FCI eigenvalue problem. Let us express Eq. (67) in the more general

$$|\Psi_0\rangle = \sum_m c_m|\Psi_m\rangle, \tag{68}$$

where m runs over all possible CSFs or SDs. The c_m coefficients are determined by variational minimization of the expectation value of the electronic energy

$$E = \min_c \frac{\langle\Psi_0|\hat{H}|\Psi_0\rangle}{\langle\Psi_0|\Psi_0\rangle}, \tag{69}$$

which in matrix notation is expressed as

$$\mathbf{Hc} = \mathbf{ESc}. \tag{70}$$

The expansion parameters are collected in the vector **c**, while the diagonal matrix **E** contains all the eigenvalues of the problem. If the CSFs are orthonormal, then the overlap matrix **S** becomes the identity matrix **I**. This equation holds also for SDs. The **H** matrix is the Hermitian Hamiltonian matrix with elements

$$H_{mn} = \langle \Psi_m | \hat{H} | \Psi_n \rangle, \tag{71}$$

where m and n are indices for two different CSFs (or SDs), and the Hamiltonian is given by Eq. (54).

The FCI Hamiltonian matrix is not a dense matrix. First, Brillouin's theorem states that the single-excitation configurations do not interact with the ground state Ψ_0. This means that all matrix elements of the form $\langle \Psi_I^A | \hat{H} | \Psi_0 \rangle$ are zero. Second, the ground state Ψ_0 does not couple with CSFs formed by triple or higher excitations. Similarly, single excitations Ψ_I^A do not couple with quadruple or higher excitations. The general rule is that a Hamiltonian matrix element will be nonzero if the CSFs differ by two or fewer electrons.

FCI forms a complete expansion so it retains all the properties of a well-behaved wave function, such as its size-extensivity. The FCI energy does not differentiate between the different types of correlation energy; it includes both strong (static) and dynamic correlations. As it was mentioned in Section 1, a FCI energy calculated with a CBS is the exact nonrelativistic, Born-Oppenheimer electronic energy for a given system.

The total number of CSFs ($K(n, N, S)$) in the FCI wave function for a system with N electrons is calculated using Weyl's formula:

$$K(n, N, S) = \frac{2S+1}{n+1} \binom{n+1}{\frac{1}{2}N - S} \binom{n+1}{\frac{1}{2}N + S + 1}, \tag{72}$$

where n is the number of orbitals ($2n$ spin-orbitals for a closed-shell system), and S the total spin of the N electron system. Eq. (72) does not consider the symmetry of the system, which can reduce significantly the size of the FCI space.

Therefore, since the FCI/CBS level provides the most accurate nonrelativistic electronic energy that we can obtain from electronic structure theory, why is it not generally applied for chemical applications? The first problem has to do with the infinite number of basis functions in a complete set. For practical applications, we need a finite set, and the larger the set is, the closer to the exact energy we are. But when the number of basis functions increases, the number of CSFs in the CI expansion of Eq. (72) becomes larger. Therefore, the second problem has to do with the size of the Hamiltonian matrix of Eq. (70). Since the number of CSFs increases *exponentially* with respect to the basis functions, the FCI method can be applied only to atoms or small diatomic or triatomic molecules, and usually with a small basis set. This exacerbates the second problem; the Hamiltonian matrix becomes intractably large.

3.2.1 Determinant-based CI

Even if spin-adapted CSFs provide the correct spin quantum number for the approximate wave function, there are specific computational advantages from an application of a determinant-based CI. Those advantages are related to the bookkeeping of the SD expansion, and the consequent efficient evaluation and storage of integrals and CI coefficients [5].

In order to properly generate and store all the possible SDs within a certain reference space, a systematic and efficient methodology for indexing all possible determinants should be applied. One method for performing this task is the graphical representation of alpha and beta electron strings, which allows a straightforward implementation in a computer code. This approach is based on the fact that an SD can be expressed as a product of alpha and beta string operators. The alpha string operator $\hat{\alpha}_{I_\alpha}$ is defined as a subsequent product of alpha creation operators, while the beta string operator $\hat{\beta}_{I_\beta}$ is a product of beta creation operators

$$|\Psi_{\text{SD}}\rangle = \hat{\alpha}_{I_\alpha}\hat{\beta}_{I_\beta}|\text{vac}\rangle, \tag{73}$$

where

$$\hat{\alpha}_{I_\alpha} = a^{\dagger}_{1\alpha}a^{\dagger}_{2\alpha}\ldots a^{\dagger}_{N_\alpha\alpha}, \quad \hat{\beta}_{I_\beta} = a^{\dagger}_{1\beta}a^{\dagger}_{2\beta}\ldots a^{\dagger}_{N_\beta\beta}. \tag{74}$$

$I_\alpha, I_\beta, J_\alpha, J_\beta, \ldots$ are indices for a particular string. We can construct all SDs formed by a CI expansion (truncated or full) by considering separately all possible ways to occupy n orbitals with alpha and beta electrons by also keeping in mind the proper spin. Thus, we need to consider how many alpha string operators we can construct by an N_α number of alpha electrons placed among n orbitals. The general expression for the calculation of all possible alpha strings is given by:

$$N^{\alpha}_{str} = \binom{n}{N_\alpha}. \tag{75}$$

A similar approach is followed for the beta strings, and the total number of SDs is given by the product of the number of alpha and beta strings.

$$N_{\text{SD}} = \binom{n}{N_\alpha}\binom{n}{N_\beta}. \tag{76}$$

Given the large number of determinants that can be generated even for systems of moderate size, application of a systematic method for indexing and storing the occupation strings is mandatory. Graphical representations organize all possible strings as a path on a diagram as seen in Fig. 3 [6]. This particular representation shows four electrons placed in six orbitals, which gives a total number of 15 strings Eq. (75). Each string is represented as a path on the graph; each path is a set of steps from the upper left node (with the label 15) to the bottom right node (label 1). Each node on the graph holds a label, which corresponds to the available paths from that node till the bottom right node (vide infra). Each node is also defined by two coordinates (e, o) corresponding to its position in the graph. The (e, o) coordinates are taken from the value of electrons on the x-axis and orbitals on the y-axis, respectively.

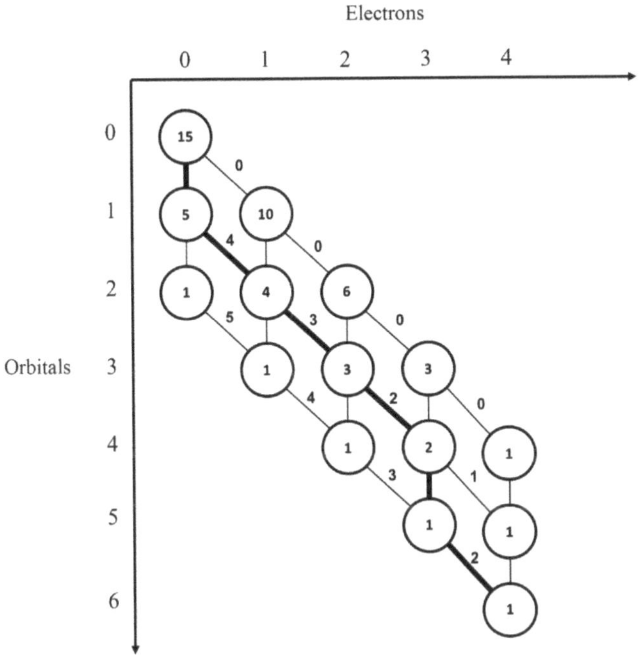

FIG. 3

Graphical representation of a string diagram of four electrons in six orbitals. The path of the string $a_2^\dagger a_3^\dagger a_4^\dagger a_6^\dagger$ is shown with a *bold line*.

For example, the upper left node has $(0, 0)$ coordinates. Two different types of steps are considered, a vertical step from node (e, o) to $(e, o + 1)$, which leaves orbital $o + 1$ unoccupied, and a diagonal step from node (e, o) to $(e + 1, o + 1)$, which places an electron on orbital $o + 1$. By considering the *lexical ordering* scheme, we can add labels on each node. By assigning a label equal to 1 ($\bar{W} = 1$) on the bottom right node, we can calculate the labels of all other nodes by applying the recursive formula

$$\bar{W}(e, o) = \bar{W}(e, o + 1) + \bar{W}(e + 1, o + 1). \tag{77}$$

A weight $Y_i(e, o)$ is assigned also to each step from node (e, o) to $(e + 1, o + 1)$, where the index i for vertical steps has a value of $i = 0$, and $i = 1$ for diagonal steps. Note that the weight $\bar{Y}_i(e, o)$ corresponds to the leaving node with coordinated (e, o). In lexical ordering, all vertical steps are defined to have a weight of zero ($\bar{Y}_0 = 0$), while the values $\bar{Y}_1$ for the diagonal steps are computed from the formula

$$\bar{Y}_1(e, o) = \sum_{i=1}^{o-e} \bar{W}(e + 1, o + 2 - i). \tag{78}$$

The index or address of a string is defined as the sum of weights of each step followed through the path

$$I_\sigma = 1 + \sum_{i=0}^{n} \bar{Y}_{L_i}(e_i, i), \tag{79}$$

where $\sigma = \alpha$ or β, L_i is the electron occupation For instance, the string $a_2^\dagger a_3^\dagger a_4^\dagger a_6^\dagger$ would have an index of $1 + (0 + 4 + 3 + 2 + 0 + 2) = 12$, which is unique to this particular string (bold line in Fig. 3).

The graphical representation of strings that described here does not take into consideration spatial symmetry, spin, or restrictions to the CI space. Those can be added, but the final format of the graphs becomes more complicated. However, such organized lexical order of strings, and consequently of SDs provide a straightforward approach to include large CI expansions in computer programs. Similar sophisticated approaches have been developed for CSF expansions, such as the graphical unitary group approach, which generates spin-adapted configurations from an SD basis [7].

3.2.2 CI Eigensolver

Eq. (70) is a standard linear algebra eigenvalue problem, which can be solved by diagonalization of the Hamiltonian matrix. However, for large CI expansions that contain millions of SDs/CSFs, it is impossible to calculate and store the full matrix. In addition, it would be impossible to diagonalize the large Hamiltonian matrix with standard numerical methods. An alternative approach for solving the CI problem is the application of iterative methods, such as the Davidson method, if only a small number of eigenvalues and eigenvectors are needed. Those involve the construction of a $\boldsymbol{\sigma}$ linear transformed trial vector as

$$\boldsymbol{\sigma} = \mathbf{Hv}, \tag{80}$$

where $\mathbf{v}$ is a trial CI vector. Therefore, instead of constructing the full Hamiltonian matrix $\mathbf{H}$, a small number of $\mathbf{Hv}$ contractions are performed. Those can be further vectorized by considering SDs in string format

$$\sigma(I_\alpha, I_\beta) = \sum_{J_\alpha, J_\beta} \langle \hat{\beta}_{J_\beta} \hat{\alpha}_{J_\alpha} | \hat{H} | \hat{\alpha}_{I_\alpha} \hat{\beta}_{I_\beta} \rangle \quad \mathrm{v}(J_\alpha, J_\beta), \tag{81}$$

where the Hamiltonian operator is given in Eq. (54). The product of Eq. (80) is *directly* calculated from one- and two-electron integrals and considers coupling coefficients for excluding zero contributions to the matrix elements. This methodology, known as *direct CI*, was developed by Roos and Siegbahn [8,9], and revolutionized CI theory since it contributed to the performance of large CI calculations.

3.3 TRUNCATED CI

3.3.1 Single-reference CI

One possible method for reducing the computational cost of the CI method is to limit the number of excitations considered in the CI expansion. Such truncated models are based on a single-reference determinant (thus the term "single-reference CI" or SRCI) and include only a specified number of excitations. For instance, compare the FCI wave function with that of the SRCI wave function truncated at single and double excitations (SR-CISD):

$$|\Psi_{\mathrm{FCI}}\rangle = \left(c_0 + \sum_{I,A} c_I^A a_A^\dagger a_I + \sum_{\substack{I<J\\A<B}} c_{IJ}^{AB} a_A^\dagger a_B^\dagger a_J a_I \right. \\ \left. + \sum_{\substack{I<J<K\\A<B<C}} c_{IJK}^{ABC} a_A^\dagger a_B^\dagger a_C^\dagger a_K a_J a_I + \cdots \right) |\Psi_0\rangle, \tag{82}$$

$$|\Psi_{\mathrm{CISD}}\rangle = \left(c_0 + \sum_{I,A} c_I^A a_A^\dagger a_I + \sum_{\substack{I<J\\A<B}} c_{IJ}^{AB} a_A^\dagger a_B^\dagger a_J a_I \right) |\Psi_0\rangle. \tag{83}$$

Such methods of truncation are an example of an a priori selection of the most important determinants in the expansion, which can capture a large and significant portion of the missing dynamic correlation energy from the reference single-determinant wave function. However, such truncated excitation expansions are not to be made without caution. For instance, determinants can play a varying role on different areas of the PES, and while a particular approximation may be adequate for some locations of the PES, it might not hold for other locations. In a similar vein, single-reference approaches in general will have difficulty describing multiple states as the dominant configuration for one state will rarely be the dominant reference function for another state.

An illustrative example on the convergence of a truncated CI expansion to the FCI limit is shown in Table 2 [10]. The number of CSF and the correlation energy of the ring H_6 model system are shown for the truncated CISD, CISDT, and CISDTQ expansions, and the corresponding percentage of the correlation energy that they capture with respect to the FCI expansion. All calculations were performed using a DZP basis set. Most commonly, CI methods are truncated at the CISD level since they provide quantitative agreement with the FCI energy (as up to 94% of the correlation energy is captured), triples increase the percentage to 96.73%, while almost 100% of the correlation energy can be recovered with the addition of quadruple excitations.

An additional, very important concern involves the concept of size-extensivity. When a method is size-extensive, the energy corresponding to the wave function will scale with the size of the system. Truncated CI methods lack size-extensivity, which

Table 2 Total Electronic Energies for the Ring H_6 Model System (in E_h) at the HF, Truncated CI, and FCI Levels

Method	Number of CSFs	Energy (E_h)	E_{corr} (%)
HF	1	−3.2519928	0.00
CISD	596	−3.3829989	94.54
CISDT	8680	−3.3860295	96.73
CISDTQ	70,263	−3.3904620	99.93
FCI	594,712	−3.3905630	100.00

From Fermann JT, Sherrill CD, Crawford TD, Schaefer III HF. Benchmark studies of electron correlation in six-electron systems. J Chem Phys 1994;100(11):8132.

can be seen by considering two noninteracting systems using the CID scheme (only double excitations). When the supersystem consisting of the two joined subsystems is considered, the expansion includes only double excitations. However, the wave function of the two separate systems forms a product that includes quadruple excitations. Therefore, the energy of the supersystem will not equal the sum of the energy of the two subsystems, which violates the size-extensivity of the quantum chemical model. One should note that when higher-order excitations are included in the CI scheme, the errors will correspond to the higher excitations, which decreases the overall size-extensivity error. As such, this source of error is lower for MRCI expansions (vide infra). A direct and inexpensive correction was suggested by Davidson, known as the "Davidson correction" (DC), who introduced a factor to scale the CI correlation energy [11]. The correction has the following form and is often added to the variationally obtained CI energy

$$\delta E_{DC} = (1 - c_0^2) \quad \delta E_{CI}, \tag{84}$$

where c_0 is the coefficient of the reference function. The notation δE corresponds to the correlation energy part of the total electronic energy of a method. While this correction does not make the energy strictly size-extensive, it does help to correct for a certain degree of the error introduced by the truncated models.

3.3.2 Multireference CI

A direct extension of the SRCI is to substitute the single-determinant reference wave function with an MC expansion

$$|\Psi_0\rangle = \sum_m c_m |\Psi_m\rangle. \tag{85}$$

In MRCI theory, all excitations up to a chosen level are performed from each reference configuration $|\Psi_m\rangle$, and the method is able to account for the relative importance of particular configurations at different points of a PES. The reference wave function of Eq. (85) is constructed by generating a CI vector representing the full configuration space, a task well-suited for MCSCF calculations (described in the next section). For the most part, MRCI with single and double excitations from the reference configurations (MRCISD) provides a highly accurate description of molecular systems, with accuracy comparable to SRCI with single, double, and triple excitations (SRCISDT), while consistently providing accurate descriptions of systems at the dissociation limit; MRCI is a highly accurate model when the most important configurations are present in the reference wave function.

3.4 MULTICONFIGURATIONAL SELF-CONSISTENT FIELD

The MCSCF method involves the simultaneous optimization of both the CI vector and the MO coefficients. It can be considered as a generalization of HF, which uses a multireference CI expansion rather than a single SD. The method provides a more flexible ansatz that can accurately describe the fundamental basis of chemical processes: the dissociation and formation of bonds. More generally, the MCSCF model can describe atomic and molecular systems that show degeneracies or near-degeneracies, which are usually present in excited states, biradicals, transition metals, and actinide chemistry. Since the MOs are optimized based on a linear combination of electronic configurations, useful chemical insights can be obtained for many applications, such as photochemistry, catalysis, magnetism, and so on.

The MCSCF wave function has two sets of variational parameters, and it can be represented as

$$|\Psi_0(\boldsymbol{\kappa},\boldsymbol{c})\rangle = \exp(-\hat{\kappa})\sum_m c_m|\Psi_m\rangle. \tag{86}$$

c_m are the expansion coefficients of the CI vector while the $\exp(-\hat{\kappa})$ operator performs the unitary transformations of the spin-orbitals. The wave function is built by variationally minimizing the expectation value of the energy with respect to the two sets of coefficients

$$E_{\text{MCSCF}} = \min_{\boldsymbol{\kappa},\boldsymbol{c}} \frac{\langle\Psi_0(\boldsymbol{\kappa},\boldsymbol{c})|\hat{H}|\Psi_0(\boldsymbol{\kappa},\boldsymbol{c})\rangle}{\langle\Psi_0(\boldsymbol{\kappa},\boldsymbol{c})|\Psi_0(\boldsymbol{\kappa},\boldsymbol{c})\rangle}. \tag{87}$$

Both CI and MO coefficients are optimized simultaneously, so an efficient approach that decouples them should be considered. This allows to solve separately the two problems: the CI problem and the orbital-optimization problem. For the CI step, the eigensolvers of standard CI theory can be applied, such as the Davidson algorithm, while the orbital-optimization step involves a set of orbital rotation coefficients $\boldsymbol{\kappa}$ expressed as unitary transformations. If the CI expansion forms a complete set (i.e., the FCI expansion), then the orbital rotations are redundant.

If we consider the nonrelativistic electronic Hamiltonian

$$\hat{H} = \sum_{pq} h_{pq}E_{pq} + \frac{1}{2}\sum_{pqrs} g_{pqrs}(E_{pq}E_{rs} - \delta_{qr}E_{ps}) \tag{88}$$

together with a normalized CI expansion, then the expectation value of the energy is given as

$$E_{\text{MCSCF}} = \langle\Psi_0|\hat{H}|\Psi_0\rangle = \sum_{pq} h_{pq}D_{pq} + \sum_{pqrs} g_{pqrs}d_{pqrs}. \tag{89}$$

The MCSCF wave function optimization is derived based on Eq. (89). The contributions from the one- and two-electron operators are included in the density matrices

$$D_{pq} = \langle \Psi_0 | \hat{E}_{pq} | \Psi_0 \rangle, \tag{90}$$

$$d_{pqrs} = \langle \Psi_0 | \hat{E}_{pq}\hat{E}_{rs} - \delta_{qr}\hat{E}_{ps} | \Psi_0 \rangle. \tag{91}$$

The energy expression of Eq. (89) contains all information on the variational parameters. The MO rotational parameters are included in the one- and two-electron integrals (h_{pq} and g_{pqrs}, respectively), while one- and two-electron density matrices (D_{pq} and d_{pqrs}, respectively), contain the information related to the CI coefficients.

In order to variationally minimize the MCSCF expression, a Taylor expansion of the energy expression is formed as a function of a set of parameters, λ, to the second order

$$E(\lambda) = E(0) + \sum_i \left(\frac{\partial E}{\partial \lambda_i}\right)_0 \lambda_i + \frac{1}{2}\sum_{i,j} \lambda_i \left(\frac{\partial^2 E}{\partial \lambda_i \partial \lambda_j}\right)_0 \lambda_j \tag{92}$$

or in vector notation:

$$E(\boldsymbol{\lambda}) = E(0) + \boldsymbol{g}^\dagger \boldsymbol{\lambda} + \frac{1}{2}\boldsymbol{\lambda}^\dagger \boldsymbol{H} \boldsymbol{\lambda}, \tag{93}$$

where $\boldsymbol{g}$ and $\boldsymbol{H}$ are the electronic gradient and Hessian, respectively. The problem then becomes finding the appropriate set of parameters $\boldsymbol{\lambda}$ such that the energy becomes stationary (i.e., $\partial E/\partial \lambda_i = 0$). This condition gives the following linear equations that can be solved iteratively:

$$\boldsymbol{g} + \boldsymbol{H}\boldsymbol{\lambda} = 0 \quad \boldsymbol{\lambda} = -\boldsymbol{H}^{-1}\boldsymbol{g}. \tag{94}$$

In the case of MCSCF calculations, the set of parameters $\boldsymbol{\lambda}$ include the orbital and CI expansion coefficients (κ and c, respectively). This allows us to write the gradient vector and Hessian matrix in the following block format:

$$\boldsymbol{g} = \begin{bmatrix} \boldsymbol{g}^c \\ \boldsymbol{g}^o \end{bmatrix} \quad \boldsymbol{H} = \begin{bmatrix} \boldsymbol{H}^{cc}\boldsymbol{H}^{co} \\ \boldsymbol{H}^{oc}\boldsymbol{H}^{oo} \end{bmatrix}. \tag{95}$$

The superscripts c and o refer to the differentiation with respect to the configuration expansion parameters and the orbital parameters, respectively, with $\boldsymbol{H}^{co} = \boldsymbol{H}^{(oc)\dagger}$. Due to this block format and the sparsity of the co and oc blocks of the Hessian matrix, we can effectively decouple the configuration and orbital parts of the calculation and solve for them separately in macro- and microiterative processes, respectively (also known as outer and inner iterations).

3.4.1 Complete active space self-consistent field

The complete active space self-consistent field (CASSCF) is a specific type of MCSCF method in which the number of determinants or CSFs used in the expansion of the CI vector are defined by dividing the orbitals into three subspaces. In the first

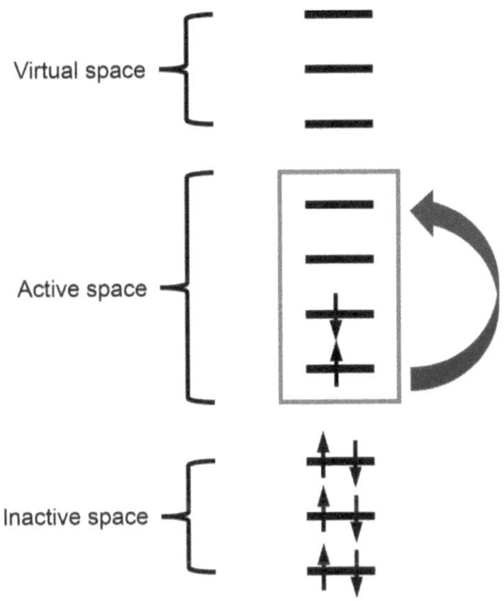

FIG. 4

Schematic representation of the three orbital subspaces of a CASSCF ansatz. The *arrow* represents electron excitations (promotions) in the active space for generating the FCI expansion.

subspace, termed as *inactive space*, all orbitals are doubly occupied. The second subspace is known as the *active space*, and within this orbital space, a full-CI expansion is considered. The electrons and orbitals included in the active space are those that contribute most to the multireference character of the particular system of study. The occupation number of the orbitals in the active space will be a noninteger number between 0 and 2. Finally, the third subspace, known as the *virtual space*, consists of orbitals that are kept unoccupied. The organization of the molecular space into three subspaces is shown schematically in Fig. 4. This method decreases the number of determinants or CSFs in the CI expansion since it limits the MC wave function within a specific subset of electrons and orbitals rather than on all electrons in all orbitals as one would do in traditional SRCI or MRCI method. As such, the CASSCF method cannot be considered merely as a "black box" method, as a priori knowledge of the chemical system under study is needed for the proper choice of the active orbitals. Another limitation of CASSCF is that the number of electrons and orbitals that can be included in the active space, since the CI expansion increases exponentially with respect to the number of electrons and orbitals. The current limitations of the conventional CASSCF implementations are 18 active electrons in 18 orbitals. Finally, it should be noted that the CAS concept is not the only approach for constructing the configuration space, but it is by far the most successful and widely applicable scheme for many chemical applications.

4 MANY-BODY PERTURBATION THEORY

MBPT, as it is presented here, is an alternative systematic technique for the approximate solution of the electron correlation. The term "perturbation" is used since a small correction "perturbs" the mathematical solution of a known solvable problem to approximate the exact unknown solution. It was initially discussed in the context of electronic structure by Møller and Plesset (abbreviated as MP) who corrected the HF method by applying second-order perturbation corrections, a method that is known nowadays as MP2 [12]. MP theory can be considered as a special case of the Rayleigh-Schrödinger perturbation theory (RSPT), and since it is routinely applied by quantum chemists, it will be discussed in the following section in detail.

In the early days of quantum chemistry, perturbation theory was little used since the main focus was the development of variational methods, such as HF and CI (as it will be explained in this section, perturbation theory is not variational). Brueckner, Goldstone, and others applied perturbative methods in the 1960s for the calculation of interactions in nuclear matter. During that period, Brueckner demonstrated that terms that depend to the N^2, N^3, etc., where N is the number of fermions, cancel each other, and proved that the RSPT series are size-consistent at every level. Those advances made quantum chemists to reexamine these methods for the description of electron correlation. The groups of Bartlett and Pople developed the first MPn methods for the low orders of the perturbation series ($n = 2{-}5$). However, it was proven that higher-order MPn methods do not converge to the FCI limit, demonstrating an unnatural behavior. With the development of CC and density functional theory, it became evident that MP theories were inaccurate and should be used with caution. Nowadays, MP2 calculations are still routinely performed for a large variety of applications and, in many cases, with modest computational effort (linear scaling with respect to the number of basis functions), while MP2 theory itself is usually the ideal workhorse to test new approximations before applying them to more sophisticated methods. In addition, MP2-type approaches have been successfully introduced for the estimation of the missing correlation energy from coupled-cluster due to a truncated basis set or for the computation of the missing dynamic correlation of MC theories.

4.1 RAYLEIGH-SCHRÖDINGER PERTURBATION THEORY

Consider that we want to solve the general eigenvalue problem for the nth state:

$$\hat{H}|\Psi_n\rangle = E_n|\Psi_n\rangle. \tag{96}$$

The exact Hamiltonian $\hat{H}$ can be expressed as the sum between two parts, a zeroth-order Hamiltonian $\hat{H}^{(0)}$, which is the unperturbed Hamiltonian operator, and a first-order perturbation correction $\hat{H}^{(1)}$

$$\hat{H} = \hat{H}^{(0)} + \hat{H}^{(1)}. \tag{97}$$

The choice of the zeroth-order Hamiltonian is based on a Schrödinger equation that we know how to solve (approximately or exactly)

$$\hat{H}^{(0)}|\Psi_n^{(0)}\rangle = E_n^{(0)}|\Psi_n^{(0)}\rangle, \tag{98}$$

where $|\Psi^{(0)}\rangle$ and $E^{(0)}$ are the known zeroth-order eigenfunctions and eigenvalues of $\hat{H}^{(0)}$. To organize the different levels of the perturbation expansion, we will introduce a parameter λ into the Hamiltonian operator

$$\hat{H} = \hat{H}^{(0)} + \lambda\hat{H}^{(1)}. \tag{99}$$

The parameter λ is used only for bookkeeping and will help us derive equations for the perturbation equations of the different orders. For example, first-order corrections depend on λ^1 (linear terms), second-order have a quadratic dependence (λ^2), and so on. Later, we will set λ to unity, while $\lambda = 0$ corresponds to the unperturbed problem. By expressing the Hamiltonian as in Eq. (99) implies that the wave function and the energy terms will also depend on λ. We can expand $|\Psi_n\rangle$ and E of Eq. (96) as a power series in λ:

$$|\Psi_n\rangle = |\Psi_n^{(0)}\rangle + \lambda|\Psi_n^{(1)}\rangle + \lambda^2|\Psi_n^{(2)}\rangle + \cdots \tag{100}$$

and

$$E_n = E_n^{(0)} + \lambda E_n^{(1)} + \lambda^2 E_n^{(2)} + \cdots, \tag{101}$$

where $|\Psi_n^{(0)}\rangle$ and $E_n^{(0)}$ are the zeroth-order wave function and energy, respectively; $|\Psi_n^{(1)}\rangle$ and $E_n^{(1)}$ are the first-order wave function and energy, respectively; and so on. The objective here is to choose the proper zeroth-order wave function so that the perturbation expansion will reach fast convergence and provide an accurate estimation of the exact $|\Psi_n\rangle$ and E by considering only the first few terms.

By substitution of Eqs. (99), (101) into the general Schrödinger equation (96), we get

$$\begin{aligned}(\hat{H}^{(0)} + \lambda\hat{H}^{(1)})(|\Psi_n^{(0)}\rangle + \lambda|\Psi_n^{(1)}\rangle + \lambda^2|\Psi_n^{(2)}\rangle + \cdots)\\ = (E_n^{(0)} + \lambda E_n^{(1)} + \lambda^2 E_n^{(2)} + \cdots)(|\Psi_n^{(0)}\rangle + \lambda|\Psi_n^{(1)}\rangle + \lambda^2|\Psi_n^{(2)}\rangle + \cdots).\end{aligned} \tag{102}$$

Eq. (102) might seem confusing, but it can be reorganized such as all terms that depend on λ^0, λ^1, λ^2, etc., are grouped together

$$\begin{aligned}&(\hat{H}^{(0)}|\Psi_n^{(0)}\rangle - E_n^{(0)}|\Psi_n^{(0)}\rangle)\lambda^0\\ &+ (\hat{H}^{(0)}|\Psi_n^{(1)}\rangle + \hat{H}^{(1)}|\Psi_n^{(0)}\rangle - E_n^{(0)}|\Psi_n^{(1)}\rangle - E_n^{(1)}|\Psi_n^{(0)}\rangle)\lambda^1\\ &+ (\hat{H}^{(0)}|\Psi_n^{(2)}\rangle + \hat{H}^{(1)}|\Psi_n^{(1)}\rangle - E_n^{(0)}|\Psi_n^{(2)}\rangle - E_n^{(1)}|\Psi_n^{(1)}\rangle - E_n^{(2)}|\Psi_n^{(0)}\rangle)\lambda^2\\ &+ \mathcal{O}(\lambda^3) = 0,\end{aligned} \tag{103}$$

where $\mathcal{O}(\lambda^3)$ stands for third- and higher-order terms. The first line of Eq. (103) includes all the coefficients of λ^0, the second of λ^1, and so on. We can easily see that for each line of Eq. (103), the exponent of λ is equal to the sum of the order of each product term that belongs to its coefficient, that is, between Hamiltonian/energy terms and wave function. For example, the sum of the orders of the $E_n^{(1)}|\Psi_n^{(1)}\rangle$ product is two, similar to the exponent of λ^2 (third line of Eq. 103). This rule helps to organize the various terms of the coefficients for each power of λ. If $\lambda = 0$, then Eq. (103) becomes identical to the zeroth-order problem of Eq. (98) that has known solutions.

Since λ is an arbitrary parameter, Eq. (103) holds only when the coefficient of each power of λ is equal to zero. Therefore, we can formulate now the Rayleigh-Schrödinger equations for each perturbative order (exponent of λ)

$$(\hat{H}^{(0)} - E_n^{(0)})|\Psi_n^{(1)}\rangle = -\hat{H}^{(1)}|\Psi_n^{(0)}\rangle + E_n^{(1)}|\Psi_n^{(0)}\rangle, \tag{104}$$

$$(\hat{H}^{(0)} - E_n^{(0)})|\Psi_n^{(2)}\rangle = -\hat{H}^{(1)}|\Psi_n^{(1)}\rangle + E_n^{(2)}|\Psi_n^{(0)}\rangle + E_n^{(1)}|\Psi_n^{(1)}\rangle, \tag{105}$$

$$(\hat{H}^{(0)} - E_n^{(0)})|\Psi_n^{(3)}\rangle = -\hat{H}^{(1)}|\Psi_n^{(2)}\rangle + E_n^{(3)}|\Psi_n^{(0)}\rangle + E_n^{(2)}|\Psi_n^{(1)}\rangle + E_n^{(1)}|\Psi_n^{(2)}\rangle. \tag{106}$$

Only the first three orders are given in Eqs. (104), (106). Let us now focus on the first-order correction by simplifying Eq. (104). If we multiply both sides from the left by $\langle\Psi_n^{(0)}|$, we obtain:

$$\langle\Psi_n^{(0)}|(\hat{H}^{(0)} - E_n^{(0)})|\Psi_n^{(1)}\rangle + \langle\Psi_n^{(0)}|\hat{H}^{(1)}|\Psi_n^{(0)}\rangle = E_n^{(1)}\langle\Psi_n^{(0)}|\Psi_n^{(0)}\rangle. \tag{107}$$

We consider a normalized zeroth-order wave function, so $\langle\Psi_n^{(0)}|\Psi_n^{(0)}\rangle = 1$ and we also choose that

$$\langle\Psi_n|\Psi_n^{(0)}\rangle = \langle\Psi_n^{(0)}|\Psi_n^{(0)}\rangle + \lambda\langle\Psi_n^{(1)}|\Psi_n^{(0)}\rangle + \lambda^2\langle\Psi_n^{(2)}|\Psi_n^{(0)}\rangle + \cdots = 1. \tag{108}$$

The condition of Eq. (108) is called *intermediate normalization* and allows us to further simplify the first term of Eq. (107)

$$\langle\Psi_n^{(0)}|(\hat{H}^{(0)} - E_n^{(0)})|\Psi_n^{(1)}\rangle = \langle(\hat{H}^{(0)} - E_n^{(0)})\Psi_n^{(0)}|\Psi_n^{(1)}\rangle = 0, \tag{109}$$

where we have also used the fact that $\hat{H}^{(0)} - E_n^{(0)}$ is Hermitian. The first-order equation (107) can now be rewritten as

$$E_n^{(1)} = \langle\Psi_n^{(0)}|\hat{H}^{(1)}|\Psi_n^{(0)}\rangle, \tag{110}$$

which is the expression for the first-order correction to $E_n^{(0)}$. Similarly, the second- and higher-order energy corrections can be evaluated, after solving separately for each power of λ, multiplying both sides from the left by $\langle\Psi_n^{(0)}|$. To summarize, the zeroth-order energy expression and the first three corrections from perturbation theory are:

$$E_n^{(0)} = \langle\Psi_n^{(0)}|\hat{H}^{(0)}|\Psi_n^{(0)}\rangle, \tag{111}$$

$$E_n^{(1)} = \langle\Psi_n^{(0)}|\hat{H}^{(1)}|\Psi_n^{(0)}\rangle, \tag{112}$$

$$E_n^{(2)} = \langle \Psi_n^{(0)} | \hat{H}^{(1)} | \Psi_n^{(1)} \rangle, \tag{113}$$
$$E_n^{(3)} = \langle \Psi_n^{(0)} | \hat{H}^{(1)} | \Psi_n^{(2)} \rangle. \tag{114}$$

As it was discussed before for the individual terms of Eq. (103), the sum of the perturbation orders of the right-hand side of Eqs. (111)–(114) is equal to the order of the energy correction at the left-hand side of those equations. It is also evident that the first-order energy correction depends on the wave function of the unperturbed system. Therefore, the only information that we need for the calculation of the first-order term is the definition of the first-order Hamiltonian and the known (approximate) zeroth-order wave function. For the evaluation of the second term, in addition to the first-order Hamiltonian, knowledge of the first-order wave function is needed. Similarly, for the third-order energy term, the second-order wave function is needed.

Let us discuss now the first-order correction to the wave function, which will allow us to evaluate the second-order perturbation theory energy correction. A usual approach that can be followed is to expand the first-order wave function $|\Psi_n^{(1)}\rangle$ in terms of the eigenfunctions of the zeroth-order Hamiltonian, which forms a complete set:

$$|\Psi_n^{(1)}\rangle = \sum_i c_{ni}^{(1)} |\Psi_i^{(0)}\rangle, \tag{115}$$

where the unknown first-order coefficients $c_{ni}^{(1)}$ have to be determined. For simplicity, the first-order notation of the coefficients is removed, so $c_{ni}^{(1)} = c_{ni}$. Substituting Eq. (115) in Eq. (104) gives

$$\sum_i c_{ni} (\hat{H}^{(0)} - E_n^{(0)}) |\Psi_i^{(0)}\rangle = -\hat{H}^{(1)} |\Psi_n^{(0)}\rangle + E_n^{(1)} |\Psi_n^{(0)}\rangle, \tag{116}$$
$$\sum_i c_{ni} (E_i^{(0)} - E_n^{(0)}) |\Psi_i^{(0)}\rangle = -\hat{H}^{(1)} |\Psi_n^{(0)}\rangle + E_n^{(1)} |\Psi_n^{(0)}\rangle, \tag{117}$$

since the zeroth-order Hamiltonian operating on $|\Psi_i^{(0)}\rangle$ will result in the zeroth-order energy of the ith state. By multiplying both sides from the left by $\langle \Psi_m^{(0)}|$, we get:

$$\sum_i c_{ni} \langle \Psi_m^{(0)} | (E_i^{(0)} - E_n^{(0)}) | \Psi_i^{(0)} \rangle = -\langle \Psi_m^{(0)} | \hat{H}^{(1)} | \Psi_n^{(0)} \rangle + E_n^{(1)} \langle \Psi_m^{(0)} | \Psi_n^{(0)} \rangle, \tag{118}$$
$$\sum_i c_{ni} (E_i^{(0)} - E_n^{(0)}) \delta_{mi} = -\langle \Psi_m^{(0)} | \hat{H}^{(1)} | \Psi_n^{(0)} \rangle + E_n^{(1)} \langle \Psi_m^{(0)} | \Psi_n^{(0)} \rangle. \tag{119}$$

All terms of the summation at the left-hand side of the previous equation will be zero except for $i = m$, while the last term is equal to 0. Therefore, we can rewrite:

$$c_{nm} (E_m^{(0)} - E_n^{(0)}) = -\langle \Psi_m^{(0)} | \hat{H}^{(1)} | \Psi_n^{(0)} \rangle = -H_{mn}^{(1)} \tag{120}$$

and for $m \neq n$

$$c_{nm} = \frac{H_{mn}^{(1)}}{E_n^{(0)} - E_m^{(0)}}, \tag{121}$$

where the matrix element $H_{mn}^{(1)} = H_{nm}^{(1)}$ on the numerator can be easily evaluated since the first-order Hamiltonian and zeroth-order wave functions are known. Now that we have expressed the first-order coefficients in terms of known quantities, we can substitute them in Eq. (115)

$$|\Psi_n^{(1)}\rangle = \sum_m \frac{H_{nm}^{(1)}}{E_n^{(0)} - E_m^{(0)}} |\Psi_m^{(0)}\rangle \tag{122}$$

since $i = m$. Substitution of the first-order wave function to Eq. (113) provides the desired expression for the second-order energy correction

$$\begin{aligned} E_n^{(2)} &= \langle \Psi_n^{(0)} | \hat{H}^{(1)} | \Psi_n^{(1)} \rangle \\ &= \sum_m \frac{H_{nm}^{(1)}}{E_n^{(0)} - E_m^{(0)}} \langle \Psi_n^{(0)} | \hat{H}^{(1)} | \Psi_m^{(0)} \rangle \\ &= \sum_m \frac{|H_{nm}^{(1)}|^2}{E_n^{(0)} - E_m^{(0)}}. \end{aligned} \tag{123}$$

Similar steps can be followed for the evaluation of the third-order energy Eq. (114). First, the second-order wave function should be expressed in terms of first-order Hamiltonian matrix elements (numerator) and known zeroth-order energy levels (denominator), and then inserted in Eq. (114).

RSPT is not the only form of perturbation theory that has been suggested for electronic structure theory calculations. Another example is the Brillouin-Wigner perturbation theory (BWPT), which was competing with theory RSPT in the early days of quantum chemistry. However, BWPT has two important limitations that make it less useful than RSPT: first, the BWPT equations are solved self-consistently since it requires knowledge of the exact energy, and second, it is not size-extensive.

4.1.1 Møller-Plesset perturbation theory

Let us now introduce Møller-Plesset perturbation theory (MPn, where n is the order of the expansion), a widely used perturbation scheme for electronic structure theory calculations. In MPn theory, which is a special case of RSPT, the zeroth-order problem is formulated by applying HF theory. First, the HF wave function is used as the zeroth-order wave function $\Psi_{\mathrm{HF}} = \Psi_0$. Eq. (98) can be then written as

$$\hat{H}^{(0)} |\Psi_0^{(0)}\rangle = E_0^{(0)} |\Psi_0^{(0)}\rangle, \tag{124}$$

where the subscript 0 denotes the solution for the ground state. The zeroth-order Hamiltonian is thus expressed as the sum of Fock operators (one-electron Hamiltonians for noninteracting particles)

$$\hat{H}^{(0)} = \sum_I \hat{f}(I) = \sum_I \hat{h}(I) + \sum_{IJ} [\hat{J}_J(I) - \hat{K}_J(I)]. \tag{125}$$

$\hat{h}(i)$ is the one-electron operator that includes the kinetic and the attractive nucleus-electron term for electron i, and $\hat{J}_J(i)$ and $\hat{K}_J(i)$ are the Coulomb and exchange operators, respectively.

Therefore, the zeroth-order energy (MP0) is simply the sum of all orbital energies

$$E^{(0)} = \sum_I \epsilon_I = \sum_I h_I + \sum_{IJ} \langle IJ||JI\rangle, \tag{126}$$

where the double bar notation in the integral corresponds to an antisymmetrized two-electron integral of the general type:

$$\langle PQ||RS\rangle = \iint \chi_P^*(\mathbf{x}_1)\chi_Q^*(\mathbf{x}_2)\frac{1}{r_{12}}[\chi_R(\mathbf{x}_1)\chi_S(\mathbf{x}_2) - \chi_S(\mathbf{x}_1)\chi_R(\mathbf{x}_2)]d\mathbf{x}_1 d\mathbf{x}_2. \tag{127}$$

We should remind the reader that spin-orbital $\chi_P^*(\mathbf{x}_1)$ is a function of the coordinates of electron 1, spin-orbital $\chi_Q^*(\mathbf{x}_2)$ of the coordinates of electron 2, while the antisymmetrized format of Eq. (127) includes both the Coulomb and exchange integral.

Notice that the MP0 energy *is not* the HF energy. Each orbital energy ϵ_I accounts for Coulomb and exchange interactions between electron I and all other electrons. Similarly, the orbital energy ϵ_J for electron J accounts also for the interelectronic, mean-field interaction between electrons I and J. The correct HF energy excludes this double counting since

$$E_{HF} = \sum_I h_I + \frac{1}{2}\sum_{IJ} \langle IJ||JI\rangle. \tag{128}$$

The HF wave function covers the occupied orbital space $\mathcal{P}$ (also known as primary space). The virtual (or secondary) space $\mathcal{Q}$ is accessed by functions that include single, double, triple, etc., electron excitations from occupied spin-orbitals $\chi_I, \chi_J, \ldots$ to virtual spin-orbitals $\chi_A, \chi_B, \ldots$. Such general functions $\Psi_I^A, \Psi_{IJ}^{AB}, \Psi_{IJK}^{ABC}, \ldots$ are orthogonal to the Ψ_{HF}, which simplifies the evaluation of the perturbative corrections.

The first-order perturbation operator should introduce electron correlation as excitations from the HF wave function to the virtual space, since the zeroth-order Hamiltonian includes electron correlation only in a mean-field approach. In MPn theory, the perturbation is defined as:

$$\hat{H}^{(1)} = \hat{H} - \hat{H}^{(0)} = \sum_{I \geq J} \frac{1}{r_{IJ}} - \sum_{I,J} [\hat{J}_J(I) - \hat{K}_J(I)]. \tag{129}$$

This expression can be interpreted as such: the second term will correct the double counting of the mean-field terms that were included in the first-order energy of Eq. (126) while the first term will be responsible for the post-HF correlation energy. This is achieved by creating *fluctuations*, that is, temporary changes in the energy with respect to the HF mean-field potential (some authors call the first-order perturbation as *fluctuation potential*). Substituting this operator in Eq. (112), where $|\Psi_n^{(0)}\rangle$ is the single-determinant HF wave function, we obtain the following expression for the first-order energy correction

$$E^{(1)} = \langle \Psi^{(0)} | \hat{H}^{(1)} | \Psi^{(0)} \rangle = -\frac{1}{2} \sum_{IJ} \langle IJ || JI \rangle. \tag{130}$$

The HF energy is recovered by addition of the energy correction to the zeroth-order energy of Eq. (126)

$$E_{\text{MP1}} = E^{(0)} + E^{(1)} = \sum_{i}^{\text{occ}} \epsilon_i - \frac{1}{2} \sum_{IJ}^{\text{occ}} \langle IJ || JI \rangle. \tag{131}$$

Therefore, the HF theory is correct up to the first order of the MPn expansion.

The first term that introduces corrections to the HF energy (and other properties) is the second-order perturbation correction. The general equation for the RSPT at the second order was given in Eq. (123), where the indices n and m correspond to different states. Since the n-state in MPn is the single-determinant HF ground state, all remaining m states will be considered as excited determinants. The determinants that involve single-electron excitations (Ψ_I^A) are zero (Brillouin's theorem) so they are excluded. Similarly, triply or higher excitations also vanish due to the two-electron nature of the perturbation. Therefore, the summation in Eq. (123) runs all over double-excited determinants Ψ_{IJ}^{AB}. We can also easily prove that the energy difference in the denominator between the zeroth-order and the double-excited determinant is simply the orbital energy difference between two formally virtual orbitals and the two occupied orbitals of the HF wave function since

$$\hat{H}^{(0)} |\Psi_{IJ}^{AB}\rangle = (E^{(0)} - (\epsilon_I + \epsilon_J - \epsilon_A - \epsilon_B)) |\Psi_{IJ}^{AB}\rangle. \tag{132}$$

The previous equation can be easily verified from the Slater-Condon rules. After introducing in Eq. (123) the corresponding terms from MP2 theory, we obtain the following important general expression:

$$\begin{aligned} E^{(2)} &= \frac{1}{4} \sum_{IJAB} \frac{|\langle \Psi^{(0)} | \hat{H}^{(1)} | \Psi_{IJ}^{AB} \rangle|^2}{\epsilon_I + \epsilon_J - \epsilon_A - \epsilon_B} \\ &= \frac{1}{4} \sum_{IJAB} \frac{|\langle IJ || AB \rangle|^2}{\epsilon_I + \epsilon_J - \epsilon_A - \epsilon_B} \\ &= \frac{1}{4} \sum_{IJAB} \frac{|g_{IAJB} - g_{IBJA}|^2}{\epsilon_I + \epsilon_J - \epsilon_A - \epsilon_B} \end{aligned} \tag{133}$$

(see also Eq. 43). From the previous expression we see that the only terms that we need in order to compute the second-order correction to the HF energy are the orbital energies and the four-index integrals of the numerator. The transformation of these integrals from atomic orbital basis functions to spin-orbital representation used to be the most time consuming step of an MP2 calculation (N^5, where N is the number of basis functions). Most recent implementations of MP2 involve the factorization of the four-index integrals into two- and three-index integrals, which are faster to evaluate. Such approaches are known as *resolution-of-identity*, *density fitting*, or *Cholesky decomposition*, and lead to significant speedups of the corresponding calculations. Alternative methodologies that lead to additional time savings are explicitly correlated methods, the Laplace-transformed MP2, the use of atomic orbitals (AO-MP2), fragmentation methods, exploration of the locality of correlation energy, or parallel implementations of the MP2 theory [13]. Such techniques, usually applied in combination, have allowed the execution of MP2 calculations for molecular systems with thousands of atoms. However, since the MOs usually are not corrected at the second order, the MP2 level of theory inherits all the deficiencies of the single-reference HF wave function, such as symmetry breaking or spin contamination, as in the case of unrestricted MP2.

The complexity and the computational effort increase as we go to higher orders of the MPn series. For example, by starting from Eq. (114) and following a similar procedure as for MP2, the third-order energy correction is given as:

$$\begin{aligned} E^{(3)} &= \frac{1}{8}\sum_{IJKLAB} \frac{\langle IJ||AB\rangle\langle KL||IJ\rangle\langle AB||KL\rangle}{(\epsilon_I+\epsilon_J-\epsilon_A-\epsilon_B)(\epsilon_K+\epsilon_L-\epsilon_A-\epsilon_B)} \\ &+ \frac{1}{8}\sum_{IJABCD} \frac{\langle IJ||AB\rangle\langle AB||CD\rangle\langle CD||IJ\rangle}{(\epsilon_I+\epsilon_J-\epsilon_A-\epsilon_B)(\epsilon_K+\epsilon_L-\epsilon_C-\epsilon_D)} \\ &+ \sum_{IJKABC} \frac{\langle IJ||AB\rangle\langle KB||CJ\rangle\langle AC||IK\rangle}{(\epsilon_I+\epsilon_J-\epsilon_A-\epsilon_B)(\epsilon_I+\epsilon_K-\epsilon_A-\epsilon_C)}. \end{aligned} \tag{134}$$

As in MP2 theory, only double excitations contribute at third order, and so on, the same terms as in MP2 are needed for the evaluation of $E^{(3)}$ (integrals in spin-orbital representation and orbital energies). Addition of triple excitations, as well as disconnected quadruple excitations, are introduced at the MP4 level. However, when we move to higher orders, the MPn series show an unnatural divergent behavior. For example, the energy corrections from the first 25 MPn orders for Ne are shown in Table 3, as they were calculated with two different basis sets [14]. It is evident that no convergence to the expected FCI limit can be achieved as the perturbation order increases, while increasing the size of the basis set does not correct this oscillating behavior.

4.1.2 Epstein-Nesbet perturbation theory

Epstein-Nesbet (EN) partitioning is an alternative to MP, but it leads to more complicated expressions than the MP variant. In the EN partitioning, the full Hamiltonian matrix is partitioned and includes projectors to SDs. For example, EN at the second-order introduces in the denominator contributions of the interactions

Table 3 MP*n* Energy Corrections (in m E_h) for Neon Using the aug-cc-pVDZ and the aug-cc-pVTZ′ Basis Sets (aug-cc-pVTZ′: aug-cc-pVTZ Without *f* Functions)

Order *n*	$E^{(n)}_{\text{aug-cc-pVDZ}}$	$E^{(n)}_{\text{aug-cc-pVTZ}'}$
2	−206.874	−244.690
3	−1.547	1.719
4	−5.686	−7.469
5	2.014	2.000
6	−1.582	−1.495
7	0.959	0.899
8	−0.707	−0.674
9	0.538	0.523
10	−0.440	−0.417
11	0.376	0.352
12	−0.334	−0.309
⋮	⋮	⋮
21	0.346	0.354
22	−0.415	−0.400
23	0.462	0.460
24	−0.519	−0.536
25	0.589	0.632

From Olsen J, Jørgensen P, Helgaker T, Christiansen O. Divergence in Møller-Plesset theory: a simple explanation based on a two-state model. J Chem Phys 2000;112(22):9736.

of two holes and two particles, in addition to the orbital energy differences of MP2 Eq. (133).

4.2 MULTIREFERENCE PERTURBATION THEORY

In Section 3, the advantages of an MC expansion over a single-reference wave function were presented. We also saw that even if MCSCF captures a percentage of the missing correlation energy from HF, it still lacks an important fraction of correlation energy necessary for quantitative results. We referred to this type of correlation as dynamic, and we discussed how it can be captured by means of CI theory (MRCI method). In MRCI, determinants that involve single and double excitations (MRCISD) are constructed from an MC zeroth-order wave function, which serves as the starting point of the method. However, though a powerful method, MRCI is not size-extensive, a problem that has been addressed with empirical corrections.

An alternative and computationally less demanding approach, which shares many similarities with MRCI, will be discussed here. By again applying an MC zeroth-order wave function, excitations to the virtual (or secondary) space can be accounted by means of perturbation theory, in a similar fashion as it was explained for a single-determinant wave function. A multireference perturbation expansion (MRPTn) can be constructed with the proper choice of the zeroth-order wave function. This will lead to second- and higher-order methodologies that account for both strong and dynamic correlation. If the initial choice of the MC wave function is a "good" choice that captures the strong correlation energy, then the perturbation corrections should be small.

The zeroth-order MCSCF wave function that approximates the ground state wave function $|\Psi_0\rangle$ is given as an expansion of all CSFs $|m\rangle$, as it was described in the previous section Eq. (68). The zeroth-order energy is variationally minimized with respect to the wave function parameters, the orbital rotation and CI coefficients c_m. In general, the expansion of Eq. (68) can be constructed from any CI scheme, but for sake of simplicity, we will assume here the CAS scheme, since CAS-type methodologies have been extensively applied in a plethora of chemical problems. In particular, we will focus our attention on two different CAS-based MRPT2 flavors, the complete active space second-order perturbation theory (CASPT2) [15,16] and the n-electron valence second-order perturbation theory (NEVPT2) [17].

The aim is to approximate the FCI, that is, the full manifold of all possible electronic configurations (configuration space). The configuration space is then divided into four subspaces:

(i) the V_0 subspace that contains all configurations describing the state under consideration and included in the zeroth-order wave function of Eq. (68);
(ii) the V_K subspace, which is orthogonal to $|\Psi_0^{(0)}\rangle$;
(iii) the V_{SD} subspace generated by all single- and double-electron excitations from $|\Psi_0^{(0)}\rangle$ to the virtual space; and
(iv) the V_X subspace that includes all higher excitations up to FCI from $|\Psi_0^{(0)}\rangle$ not included in the previous subspaces.

Triple or higher excitations do not contribute to the second order, therefore, only the V_{SD} subspace couples with the reference space V_0. The straightforward approach to construct the secondary space is by including all possible SDs or CSFs obtained by exciting one or two electrons from each configuration of V_0, which leads to a set of linear equations to be solved. However, this approach has an important drawback related to a growing number of first order wave function parameters as the number of reference functions increases. Alternatively, the first-order wave function can be generated by considering excitation operators acting on $|\Psi_0^{(0)}\rangle$:

$$|\Psi_0^{(1)}\rangle = \sum_{pqrs} C_{pqrs}\hat{E}_{pq}\hat{E}_{rs}|\Psi_0^{(0)}\rangle, \tag{135}$$

Table 4 The Eight Excitation Classes Involving Electrons Promoted From the Inactive (I) or Active (A) Orbitals to the Virtual (V) Orbitals

Internal	$\hat{E}_{ti}\hat{E}_{uv}\|\Psi_0^{(0)}\rangle$ $\hat{E}_{ti}\hat{E}_{uj}\|\Psi_0^{(0)}\rangle$	IA → AA II → AA
Semiinternal	$\hat{E}_{at}\hat{E}_{uv}\|\Psi_0^{(0)}\rangle$ $\hat{E}_{ai}\hat{E}_{tu}\|\Psi_0^{(0)}\rangle$ / $\hat{E}_{ti}\hat{E}_{au}\|\Psi_0^{(0)}\rangle$ $\hat{E}_{ti}\hat{E}_{aj}\|\Psi_0^{(0)}\rangle$	AA → VA IA → VA / IA → AV II → AV
External	$\hat{E}_{at}\hat{E}_{bu}\|\Psi_0^{(0)}\rangle$ $\hat{E}_{ai}\hat{E}_{bt}\|\Psi_0^{(0)}\rangle$ $\hat{E}_{ai}\hat{E}_{bj}\|\Psi_0^{(0)}\rangle$	AA → VV IA → VV II → VV

where, in spin-free notation, the excitation operators are given in Eq. (53). The subspace V_{SD} can then be generated by eight different types of excitations (excitation classes), as they are shown in Table 4, where i, j correspond to inactive, t, u, v to active, and a, b to virtual (secondary) orbital indices. Some authors separate those functions into internal, semiinternal, and external, when zero, one or two electrons are excited to the virtual space, respectively.

This ansatz is known as *internally contracted* and, compared to the CI-based expansion, it reduces dramatically the number of C_{pqrs} amplitudes when it is applied to large active spaces. The CASPT2 zeroth-order Hamiltonian is defined by the projection of the Fock operator onto the V_0 reference space and its compliment.

$$\hat{H}^{(0)} = \hat{P}\hat{F}\hat{P} + \hat{Q}\hat{F}\hat{Q}. \tag{136}$$

The projection operators are defined as $\hat{P} = |\Psi_0^{(0)}\rangle\langle\Psi_0^{(0)}|$ and $\hat{Q} = 1 - \hat{P}$. The reference (CASSCF) Fock operator $\hat{F}$ has the following format:

$$\hat{F} = \sum_{pq} f_{pq}\hat{E}_{pq}, \tag{137}$$

$$f_{pq} = f_{pq}^I + f_{pq}^A, \tag{138}$$

$$f_{pq}^I = h_{pq} + \sum_i (2g_{pqii} - g_{piiq}), \tag{139}$$

$$f_{pq}^A = \frac{1}{2}\sum_{tu} D_{tu}(2g_{pqtu} - g_{putq}), \tag{140}$$

where f_{pq}^I and f_{pq}^A are the inactive and active Fock matrices, respectively, and D_{tu} is the one-electron density matrix element Eq. (55).

There are important drawbacks associated with CASPT2, which should be discussed. The first is related to the V_{SD} subspace. The secondary space ($\hat{Q}\hat{F}\hat{Q}$) is not diagonal since the generated functions from Eq. (135) are no longer orthogonal to each other, which implies an iterative solver for the CASPT2 equations. In addition, the specific formulation of the Fock operator might lead to singularities, which is the reason behind the intruder state problem of CASPT2. This issue can be addressed either by expanding the active space with additional orbitals or by introducing a level shift in the zeroth-order Hamiltonian. An alternative MRPT2 method that solves the intruder state problem is the NEVPT2 scheme, which uses the Dyall's Hamiltonian with a partially contracted formalism, and provides a secondary space with a diagonal structure. Another issue that most MRPT2 implementations still have to confront has to do with the evaluation of fourth-order matrix elements (four-particle reduced density matrix).

As it is evident in this section, the straightforward nature of the single-determinant MP2 theory is lost once second-order perturbation theory is combined with an MC zeroth-order Hamiltonian. However, CASPT2/NEVPT2 calculations can solve chemical problems that single-reference methods cannot address, such as the dissociation of single or multiple bonds, excited states, intersection of PESs, transition metal and actinide chemistry, electronic structure of noninnocent ligands, and in general, molecular species with multiple spin centers. For example, in a benchmark study that included 225 electronically excited states, both CASPT2 and NEVPT2 provide comparable accuracy [18]. However, CASPT2 accuracy depends on the choice of the ionization potential-electron affinity shift of the zeroth-order Hamiltonian, a parameter that corrects the systematic underestimation of excitation energies.

5 COUPLED-CLUSTER THEORY

The basic formalism of CI and perturbation theory were presented in the previous two sections, and their strengths and weaknesses were discussed. CI is conceptually more straightforward to explain, but has one important limitation: the lack of size-extensivity. Additionally, implementing CI in efficient computer codes is also a very challenging task. In this section, CC theory, an alternative correlation scheme that does not suffer from either of these issues will be presented. The fundamental ideas of CC were firstly introduced by Coester and Kümmel in 1960 who applied them on the correlation problem present in nuclear physics [19]. Coester and Kümmel were also the first to express the cluster operator as an exponential ansatz. The same principles were applied for a first time on electron systems in 1960 by Čižek [20], and about 10 years later, the first computer programs and applications of CC theory were reported [21,22].

CC theory has been successfully applied in a large variety of chemical applications since it provides a systematic convergence to the one-electron limit (basis set) and the N-electron basis (correlation level). However, the complexity of the CC equations makes its efficient implementation in a computer program a challenging task. Here, the general formulation of CC theory will be presented and the equations of the simplest and most meaningful CC scheme, the CC with double excitations (CCD), will be discussed in more detail. In our discussion, we will also include alternative, hybrid schemes, such as the popular CC singles and doubles with perturbative triple excitations (CCSD(T)).

5.1 GENERAL CONSIDERATIONS

As we discussed in the previous sections, the CI ansatz captures the correlation missing from HF by applying electron excitations from a zeroth-order wave function to the N-electron space. The CI wave function includes an expansion of electronic configurations and it provides the exact nonrelativistic electronic energy at the FCI limit, that is, all possible electronic excitations are included in the CI expansion. However, the CI expansion needs to be truncated for practical computations, which introduces an important limitation in CI theory: the lack of size-extensivity. This important drawback arises from its linear parametrization of the wave function. To address this limitation, CC theory uses a nonlinear but separable product parametrization of the correlation energy and similarly to CI theory, the CC wave function is formed as a sum of determinants given by a series of operators acting on a reference wave function $|\Psi_0\rangle$.

$$|\Psi_{\mathrm{CC}}\rangle = \left[\prod_{I,A}(1 + t_I^A \hat{\tau}_I^A)\right]\left[\prod_{I>J,A>B}(1 + t_{IJ}^{AB}\hat{\tau}_{IJ}^{AB})\right]\ldots|\Psi_0\rangle, \tag{141}$$

where the t_I^A are the single-excitation cluster amplitudes, the t_{IJ}^{AB} are the double-excitation amplitudes, etc., which can be considered as the weight (or probability) of the specific excitation, and

$$\hat{\tau}_I^A = a_A^{\dagger} a_I, \tag{142}$$

$$\hat{\tau}_{IJ}^{AB} = a_A^{\dagger} a_B^{\dagger} a_J a_I, \tag{143}$$

$$\hat{\tau}_{IJK}^{ABC} = a_A^{\dagger} a_B^{\dagger} a_C^{\dagger} a_K a_J a_I \tag{144}$$

are the excitation operators. Here, we will focus our attention on the single-reference CC theory and thus, the HF wave function can be chosen as reference function. Each operator returns the reference state but also generates a set of excited configurations. If we consider only the two-electron excitation operator, we obtain

$$(1 + t_{IJ}^{AB}\hat{\tau}_{IJ}^{AB})|\Psi_0\rangle = |\Psi_0\rangle + t_{IJ}^{AB}|\Psi_{IJ}^{AB}\rangle. \tag{145}$$

One of the strengths of CC theory is that a specific configuration can be generated in multiple ways. For example, the quadruply excited configuration $|\Psi_{IJKL}^{ABCD}\rangle$ can be generated by the direct application of a four-electron operator

$$(1 + t_{IJKL}^{ABCD}\hat{\tau}_{IJKL}^{ABCD})|\Psi_0\rangle = |\Psi_0\rangle + t_{IJKL}^{ABCD}|\Psi_{IJKL}^{ABCD}\rangle \tag{146}$$

or from a product of two two-electron operators

$$(1 + t_{IJ}^{AB}\hat{\tau}_{IJ}^{AB})(1 + t_{KL}^{CD}\hat{\tau}_{KL}^{CD})|\Psi_0\rangle = |\Psi_0\rangle + t_{IJ}^{AB}|\Psi_{IJ}^{AB}\rangle + t_{KL}^{CD}|\Psi_{KL}^{CD}\rangle + t_{IJ}^{AB}t_{KL}^{CD}|\Psi_{IJKL}^{ABCD}\rangle. \tag{147}$$

5.2 EXPONENTIAL PARAMETRIZATION

Since the product of two operators with the same indices will return a zero value, for example,

$$\hat{\tau}_I^A \hat{\tau}_I^A = a_A^\dagger a_I a_A^\dagger a_I = 0, \tag{148}$$

we can express any operator in the product parametrization of Eq. (141) in an exponential format

$$1 + t_I^A \hat{\tau}_I^A + \frac{1}{2}(t_I^A)^2 \hat{\tau}_I^A \hat{\tau}_I^A + \cdots = \exp(t_I^A \hat{\tau}_I^A). \tag{149}$$

The nonlinear ansatz of Eq. (141) is recovered since the third and higher terms of Eq. (149) are zero. This expression holds for any level of excitation. The CC wave function can be expressed as a product of exponential operators, which is equivalent to

$$|\Psi_{\text{CC}}\rangle = \exp\left(\sum_{I,A} t_I^A \hat{\tau}_I^A + \sum_{I>J,A>B} t_{IJ}^{AB} \hat{\tau}_{IJ}^{AB} + \cdots\right)|\Psi_0\rangle. \tag{150}$$

This is the CC exponential ansatz, which can be summarized by

$$|\Psi_{\text{CC}}\rangle = \exp(\hat{T})|\Psi_0\rangle, \tag{151}$$

where $\hat{T}$ is the *cluster operator*:

$$\hat{T} = \hat{T}_1 + \hat{T}_2 + \hat{T}_3 + \cdots. \tag{152}$$

The subscript denotes the excitation operator for each cluster. The cluster operators are defined as:

$$\hat{T}_1 = \sum_{I,A} t_I^A a_A^\dagger a_I = \sum_{I,A} t_I^A \hat{\tau}_I^A, \tag{153}$$

$$\hat{T}_2 = \sum_{\substack{I<J\\A<B}} t_{IJ}^{AB} a_A^\dagger a_B^\dagger a_J a_I = \sum_{\substack{I<J\\A<B}} t_{IJ}^{AB} \hat{\tau}_{IJ}^{AB} = \frac{1}{4}\sum_{IJAB} t_{IJ}^{AB} \hat{\tau}_{IJ}^{AB}. \tag{154}$$

The last term of Eq. (154) implies that four equivalent configurations are generated. Since the double-excited configurations are antisymmetric, it should hold that $t_{IJ}^{AB} = -t_{IJ}^{BA} = -t_{JI}^{AB} = t_{JI}^{BA}$ (see also Section 2.2 and Eqs. 12, 20). Any N-fold excitation cluster operator can be written as the single- and double-excitation operators of Eqs. (153), (154). It is worth noting that given an infinite N-expansion, the CC equations provide the full-CI wave function. However, much like in CI theory, the CC equations can be truncated at a given excitation level, and in the case of CC theory, at a given excitation cluster operator. To better understand how the CC equations act on the reference wave function $|\Psi_0\rangle$, we will expand the exponential operator via a Taylor series:

$$\exp(\hat{T}) = 1 + \hat{T} + \frac{\hat{T}^2}{2!} + \frac{\hat{T}^3}{3!} + \cdots . \tag{155}$$

Each of the cluster operators can be grouped for a given electronic excitation level and compared to the equivalent CI operators $\hat{C}_N$, corresponding to the operator generating N number of electronic excitations from the reference. Collecting the terms for the first four excitation terms

$$\hat{C}_1 = \hat{T}_1, \tag{156}$$

$$\hat{C}_2 = \hat{T}_2 + \frac{1}{2}\hat{T}_1^2, \tag{157}$$

$$\hat{C}_3 = \hat{T}_3 + \hat{T}_1\hat{T}_2 + \frac{1}{6}\hat{T}_1^3, \tag{158}$$

$$\hat{C}_4 = \hat{T}_4 + \hat{T}_1\hat{T}_3 + \frac{1}{2}\hat{T}_2^2 + \frac{1}{2}\hat{T}_1^2\hat{T}_2 + \frac{1}{24}\hat{T}_1^4. \tag{159}$$

Configurations in the CI expansion formed by one-electron excitations ($\hat{C}_1$) can only be generated in CC via the $\hat{T}_1$ operator. By contrast, double excitations can be generated via the simultaneous two-electron excitation given by $\hat{T}_2$ (*connected clusters*), or through two independent electron excitations generated from $\hat{T}_1^2$ (*disconnected clusters*). Therefore, the CC formalism can be represented in an expansion synonymous to CI

$$|\Psi_{\text{CC}}\rangle = c_0|\Psi_0\rangle + \sum_{I,A} c_I^A|\Psi_I^A\rangle + \sum_{\substack{I<J\\A<B}} c_{IJ}^{AB}|\Psi_{IJ}^{AB}\rangle + \sum_{\substack{I<J<K\\A<B<C}} c_{IJK}^{ABC}|\Psi_{IJK}^{ABC}\rangle$$
$$+ \sum_{\substack{I<J<K<L\\A<B<C<D}} c_{IJKL}^{ABCD}|\Psi_{IJKL}^{ABCD}\rangle + \cdots , \tag{160}$$

where the CI coefficients correspond to the sum of connected and disconnected cluster amplitudes

$$c_I^A = t_I^A, \tag{161}$$

$$c_{IJ}^{AB} = t_{IJ}^{AB} + t_I^A t_J^B, \tag{162}$$

$$c_{IJK}^{ABC} = t_{IJK}^{ABC} + t_{IJ}^{AB} t_K^C + t_I^A t_J^B t_K^C. \tag{163}$$

Eqs. (160), (163) show again that a full-CC expansion would produce a nonlinear parametrization of the full-CI ansatz. However, for practical applications, the full-CC expansion cannot be evaluated (as FCI), and the true interest is in the behavior of the coupled-cluster operator when it is truncated at a given excitation level. Let us examine the behavior of truncating the cluster operator at the doubles level, by including only singles and doubles excitations. This level of theory is called *coupled-cluster singles and doubles (CCSD)* and the cluster operator from Eq. (153) takes the form

$$\hat{T}_{\text{CCSD}} = \hat{T}_1 + \hat{T}_2. \tag{164}$$

A Taylor expansion of the $\exp(\hat{T}_{\text{CCSD}})$ operator (as in Eq. 155) gives

$$\exp(\hat{T}_1 + \hat{T}_2) = 1 + (\hat{T}_1 + \hat{T}_2) + \frac{(\hat{T}_1 + \hat{T}_2)^2}{2!} + \frac{(\hat{T}_1 + \hat{T}_2)^3}{3!} + \cdots \tag{165}$$

$$= 1 + (\hat{T}_1) + \left(\hat{T}_2 + \frac{1}{2}\hat{T}_1^2\right) + \left(\hat{T}_1\hat{T}_2 + \frac{1}{6}\hat{T}_1^3\right) + \left(\frac{1}{2}\hat{T}_2^2 + \frac{1}{2}\hat{T}_1^2\hat{T}_2 + \frac{1}{24}\hat{T}_1^4\right) + \cdots. \tag{166}$$

In the last expression, terms that provide the same level of excitations are grouped together. Electronic configurations generated by one- and two-electron excitations are identical to the complete, untruncated CC scheme. However, configurations including three or more excitations will be represented solely by disconnected clusters, even if at the definition of the CCSD operator those higher terms were not included. The triples coefficient of CCSD can be compared to Eq. (163)

$$c_{IJK}^{ABC} = t_{IJ}^{AB} t_K^C + t_I^A t_J^B t_K^C. \tag{167}$$

Notice that the connected triples amplitudes t_{IJK}^{ABC} are absent from Eq. (167) since the CCSD wave function does not include the $\hat{T}_3$ operator. Instead, in the truncated CC scheme, weights of higher-order configurations are approximated by products of singles and doubles amplitudes, and higher-order correlation effects are indirectly captured. More concretely, if we compare the CISD and CCSD wave functions, the CC scheme contains more electron configurations (determinants), but retains the same level of coefficients to optimize.

5.3 SIZE-EXTENSIVITY AND COUPLED-CLUSTER THEORY

Another important consequence of the truncated CC scheme is that the method is *size-extensive*. As it was mentioned previously, this property is maintained due to the exponential parametrization of the wave function. Size-extensivity will ensure that the energies of two noninteracting fragments A and B should be equal to the energy as the supersystem AB, or simply $E_{AB} = E_A + E_B$. Suppose that wave functions are represented individually by two fragments such that

$$|\Psi_{\text{CC}}^A\rangle = \exp(\hat{T}_A)|\Psi_0^A\rangle, \tag{168}$$

$$|\Psi_{\text{CC}}^B\rangle = \exp(\hat{T}_B)|\Psi_0^B\rangle, \tag{169}$$

where $|\Psi_0^X\rangle$ is the reference determinant for fragment $X = A$ or B, composed of only localized orbitals, and $\hat{T}_X$ generates excited determinants on X. If there are no orbital interactions (e.g., infinite separation distance between A and B), excitations from fragment A to B will not occur. This can be mathematically expressed with the commutator relation $[\hat{T}_A, \hat{T}_B] = 0$. Given a size-extensive noninteracting reference wave function which is multiplicatively separable $|\Psi_0^{AB}\rangle = |\Psi_0^A\rangle|\Psi_0^B\rangle$, it follows

$$\begin{aligned}|\Psi_{\rm CC}^{AB}\rangle &= \exp(\hat{T})|\Psi_0^{AB}\rangle \\ &= \exp(\hat{T}_A+\hat{T}_B)|\Psi_0^{AB}\rangle \\ &= \exp(\hat{T}_A+\hat{T}_B)|\Psi_0^{A}\rangle|\Psi_0^{B}\rangle \\ &= \exp(\hat{T}_A)|\Psi_0^{A}\rangle\exp(\hat{T}_B)|\Psi_0^{B}\rangle \\ &= |\Psi_{\rm CC}^{A}\rangle|\Psi_{\rm CC}^{B}\rangle,\end{aligned} \tag{170}$$

which results a multiplicative CC wave function for two noninteracting fragments A and B, and thus, the energies are additive, showing that CC is indeed size-extensive.

5.4 DERIVATION OF THE COUPLED-CLUSTER EQUATIONS

Up to this point, the general CC ansatz and its main properties have been discussed. The CC wave function was parameterized based on a set of unknown cluster amplitudes that need to be computed in order to obtain the CC energy expression. In a given one-electron basis, the Schrödinger equation is given as

$$\begin{aligned}\hat{H}|\Psi_{\rm CC}\rangle &= E|\Psi_{\rm CC}\rangle, \\ \hat{H}\exp(\hat{T})|\Psi_0\rangle &= E\exp(\hat{T})|\Psi_0\rangle,\end{aligned} \tag{171}$$

where the CC subscript from the energy expression is omitted. This expression holds only for the full-CC operator, while for an arbitrary truncated cluster operator $\hat{T}_N$, the Schrödinger equation is not satisfied. A straightforward approach to obtain the cluster amplitudes would be to invoke the variational principle and solve for $E_{\rm CC}$, similarly to CI theory

$$E = \min_{t_\mu} \frac{\langle\Psi_0|\exp(\hat{T}^\dagger)\hat{H}\exp(\hat{T})|\Psi_0\rangle}{\langle\Psi_0|\exp(\hat{T}^\dagger)\exp(\hat{T})|\Psi_0\rangle}. \tag{172}$$

However, the nonlinear dependence of the CC wave function on the amplitudes t_μ, where μ is an arbitrary excited determinant, forms an intractable set of nonlinear equations to solve. Instead, a *projection method* is usually followed for the calculation of the t_μ amplitudes. This methodology, which is followed in the majority of the CC implementations, provides a nonvariational energy, which does not represent and upper bound for the FCI limit, but the deviations are typically small and of little consequence. Variational CC has been suggested by applying the Lagrangian formulation [23].

For obtaining an expression for the evaluation of the cluster amplitudes, we need to project Eq. (171) onto all determinants generated by the cluster operator $\hat{T}$

$$\langle\mu_n|\hat{H}\exp(\hat{T})|\Psi_0\rangle = E\langle\mu_n|\exp(\hat{T})|\Psi_0\rangle, \tag{173}$$

where $\langle\mu_n|$ is the determinantal manifold. For example, in CCSD, the CC wave function is projected onto all single- and double-excited determinants, plus the

HF reference determinant. The set of Eq. (173) is called *unlinked CC amplitude equations*.

Alternatively, the Schrödinger equation (171) is multiplied from the left by the exponential operator $\exp(-\hat{T})$

$$\exp(-\hat{T})\hat{H}\exp(\hat{T})|\Psi_0\rangle = E|\Psi_0\rangle, \tag{174}$$

where the non-Hermitian operator

$$\hat{H}^T = \exp(-\hat{T})\hat{H}\exp(\hat{T}) \tag{175}$$

is known as the *similarity-transformed Hamiltonian*. The similarity-transformed Schrödinger equation (174) is projected onto all determinants and the following CC energy and amplitude equations are formed

$$\langle\Psi_0|\exp(-\hat{T})\hat{H}\exp(\hat{T})|\Psi_0\rangle = E, \tag{176}$$

$$\langle\mu_n|\exp(-\hat{T})\hat{H}\exp(\hat{T})|\Psi_0\rangle = 0. \tag{177}$$

Eq. (177) corresponds to the *linked CC equations*, which are equivalent to the unlinked equations (173). The number of the linked CC equations is the same with the number of amplitudes. The projected equations are nonlinear with respect to the amplitudes, and they are solved iteratively, usually after applying the Baker-Campbell-Hausdorff (BCH) expansion, a very powerful mathematical expression for the evaluation of the similarity-transformed Hamiltonian of Eq. (175)

$$\exp(-\hat{T})\hat{H}\exp(\hat{T}) = H + [H,T] + \frac{1}{2!}[[H,T],T] + \frac{1}{3!}[[[H,T],T],T] + \frac{1}{4!}[[[[H,T],T],T],T]. \tag{178}$$

The expansion terminates after the fourfold commutators because the Hamiltonian contains only one- and two-particle operators, and the maximum number of $\hat{T}$ operators that can provide contributions to the CC amplitudes is four. The closed form of the CC equations, even if they depend on an unterminated exponential expansion, is one of the strengths of the coupled-scheme. The similarity-transformed Hamiltonian can be now expressed via the BCH expansion

$$\begin{aligned}
\exp(-\hat{T})\hat{H}\exp(\hat{T}) &= \left(1 - \hat{T} + \frac{\hat{T}^2}{2!}\cdots\right)\hat{H}\left(1 + \hat{T} + \frac{\hat{T}^2}{2!}\cdots\right) \\
&= H + HT - TH + \frac{1}{2!}HT^2 - THT + \frac{1}{2!}\hat{T}^2\hat{H} + \cdots \\
&= H + [H,T] + \frac{1}{2!}([H,T]T - T[H,T]) + \cdots \\
&= H + [H,T] + \frac{1}{2!}[[H,T],T] + \cdots .
\end{aligned} \tag{179}$$

Let us turn now our attention on the CC energy expression. Eq. (176) can be further simplified since $\langle\Psi_0|\exp(-\hat{T}) = \langle\Psi_0|$, thus

$$\langle\Psi_0|\hat{H}\exp(\hat{T})|\Psi_0\rangle = E. \tag{180}$$

After expanding the cluster operator, we get

$$\langle\Psi_0|\hat{H}\left(1+\hat{T}+\frac{\hat{T}^2}{2!}+\frac{\hat{T}^3}{3!}+\cdots\right)|\Psi_0\rangle = E. \tag{181}$$

However, the Hamiltonian has one- and two-particle operators, therefore cluster operators of order $\hat{T}_3$ and greater do not contribute to the energy. In addition, the one particle operators do not directly contribute to the energy due to Brillouin's theorem:

$$\langle\Psi_0|\hat{H}\hat{T}_1|\Psi_0\rangle = 0. \tag{182}$$

The simplified CC energy expression is now rewritten as

$$\langle\Psi_0|\hat{H}\left(1+\hat{T}_2+\frac{1}{2}\hat{T}_1^2\right)|\Psi_0\rangle = E. \tag{183}$$

By inspection of Eq. (183), we see that only singles and doubles amplitudes contribute directly to the CC energy, regardless of the truncation level of the cluster operator $\hat{T}$. However, the higher-order amplitudes indirectly contribute to the energy since they are used to determine the singles and doubles amplitudes via the projected equations (177).

5.5 COUPLED-CLUSTER DOUBLES

Let us now examine in more detail the case of truncating the cluster operator to only contain doubles, a scheme known as *coupled-cluster doubles* (CCD) that considers the cluster operator

$$\hat{T} = \hat{T}_2. \tag{184}$$

For sake of simplicity, a closed-shell RHF reference is considered. Applying the doubles cluster operator to the exponential ansatz and expanding we obtain

$$|\Psi_{\text{CCD}}\rangle = \exp(\hat{T}_2)|\Psi_0\rangle = \left(1+\hat{T}_2+\frac{\hat{T}_2^2}{2!}+\frac{\hat{T}_2^3}{3!}+\cdots\right)|\Psi_0\rangle. \tag{185}$$

Since the CCD wave function contains a two-particle operator, only even electron-excited determinants can be formed. Therefore, the CCD wave function becomes

$$|\Psi_{\text{CCD}}\rangle = |\Psi_0\rangle + \sum_{\substack{I<J\\A<B}} t_{IJ}^{AB}|\Psi_{IJ}^{AB}\rangle + \sum_{\substack{I<J<K<L\\A<B<C<D}} t_{IJ}^{AB}t_{KL}^{CD}|\Psi_{IJKL}^{ABCD}\rangle + \cdots, \tag{186}$$

where Ψ_0 is the reference wave function (e.g., HF), Ψ_{IJ}^{AB} and Ψ_{IJKL}^{ABCD} are the double and quadruple-excited determinants, respectively, with the corresponding connected doubles t_{IJ}^{AB} and disconnected quadruples $t_{IJ}^{AB}t_{KL}^{CD}$ cluster amplitudes. Now let us first derive an expression for the CCD energy. We multiply both sides from the left by $\langle\Psi_0|$ to the CCD Schrödinger equation

$$\langle\Psi_0|\hat{H}|\Psi_{\mathrm{CCD}}\rangle = E_{\mathrm{CCD}}\langle\Psi_0|\Psi_{\mathrm{CCD}}\rangle. \tag{187}$$

Noting that $\langle\Psi_0|\Psi_{\mathrm{CC}}\rangle = 1$ and expanding the CCD wave function

$$E_{\mathrm{CCD}} = \langle\Psi_0|\hat{H}|\Psi_0\rangle + \sum_{\substack{I<J\\A<B}} t_{IJ}^{AB}\langle\Psi_0|\hat{H}|\Psi_{IJ}^{AB}\rangle, \tag{188}$$

$$= E_{HF} + \sum_{\substack{I<J\\A<B}} t_{IJ}^{AB}\langle\Psi_0|\hat{H}|\Psi_{IJ}^{AB}\rangle, \tag{189}$$

$$= E_{HF} + \sum_{\substack{I<J\\A<B}} t_{IJ}^{AB}\langle IJ||AB\rangle, \tag{190}$$

where the double bar notation in the integral corresponds to the Eq. (127). Note that once again, the quadruple and higher determinants vanish due to the nature of the Hamiltonian operator, which contains only one- and two-particle terms.

5.6 COUPLED-CLUSTER MODELS AND CONVERGENCE TO THE FCI/CBS LIMIT

One of the strengths of CC theory is its systematic convergence to the FCI limit as the number of excitations is increased (size of the N-electron basis). This holds for a given one-electron basis set, and if both bases are approaching their limit, then the FCI/CBS is obtained (Fig. 1). Truncated CC hierarchies are expected to give results superior to equivalently truncated CI expansions since higher-order excitations are included in the CC formalism through disconnected clusters. For example, the CCD level contains determinants with quadruple excitations generated by the $\hat{T}_2^2$ operator, which are not found in CI until the CISDTQ level. As a result, the energies rapidly converge with respect to the expansion of the cluster operator. The most common truncation of the operator is at the doubles level, giving the CCSD model. The disconnected clusters are able to approximate the quadruples and higher-order terms, but severely underestimate contributions from triples excitations. Addition of the $\hat{T}_3$ operator gives the *coupled-cluster singles-doubles-and-triples* (CCSDT) level, which is a very computational demanding method due to the increased number of amplitudes to be optimized. Each iteration of CCSDT scales as N^8, where N is the number of basis functions, while CCSD has an N^6 dependence. An alternative and very successful approach is to include a noniterative triples correction to the CCSD model, resulting in the CCSD(T) level of theory. The general energy expression of CCSD(T) is given as

$$E_{\text{CCSD(T)}} = E_{\text{HF}} + E_{\text{CCSD}} + E_{\text{(T)}}, \tag{191}$$
$$E_{\text{(T)}} = E^{[4]} + E^{[5]}. \tag{192}$$

Two terms contribute to the perturbative triples, the fourth-order term $E^{[4]}$, which involves the CCSD doubles amplitudes, and the fifth-term $E^{[5]}$ that involves the CCSD singles amplitudes. The triples amplitudes are calculated perturbatively, in one step that scales as N^7. CCSD(T) has become a very successful electronic structure theory model for single-reference systems because it provides accurate results for energies and properties. Due to its accuracy and computational efficiency, it is usually termed as the "gold standard" of quantum chemistry.

To investigate the behavior of CCSD, CCSD(T), and CCSDT methods relative to full-CI limit, we discuss correlation energies for the water molecule at its equilibrium geometry, and for different symmetric O–H bond stretch values (Table 5) [24]. Results are shown as energy differences between the FCI limit and the truncated CC schemes. All calculations were performed with the cc-pVDZ basis set. In addition, the weight of the HF configuration in the FCI wave function is reported, which can be used to assess the choice of a single SD as a reference state. At equilibrium geometry of the water molecule, where $R_{\text{O-H}} = R_e$, the HF weight in the FCI expansion is 0.941, which means that all the remaining configurations will have infinitesimal weights and will have very small contribution to the energy and the electronic wave function of the molecule. This is an evidence that its electronic structure can be adequately described by a single determinant. Indeed, addition of dynamical correlation from CC systematically provides energies closer to the FCI limit. Lack of triples corrections at the CCSD level has a 3.744 m E_{h} deviation from FCI, which corresponds to recovery of more than 98% of the correlation energy. However, addition of the triples reduces the errors to <1 m E_{h}. Notice also that the difference between the full triples of CCSDT and the perturbative triples of CCSD(T) is only 0.165 m E_{h}. Along the symmetric stretch of the O–H bonds of the H_2O molecule, the energy deviations from the FCI limit for all models increase.

Table 5 Energy Differences (in E_{h}) From FCI for Three Difference Coupled-Cluster Models for the Symmetric O–H Bond Stretch of the Water Molecule

	R_e	$1.5R_e$	$2R_e$	$2.5R_e$	$3R_e$
HF Weight	**0.941**	**0.849**	**0.589**	**0.337**	**0.230**
HF	0.217821	0.269951	0.363954	0.476747	0.567554
CCSD	0.003744	0.010043	0.022032	0.020307	0.010849
CCSD(T)	0.000658	0.001631	−0.003820	−0.042564	−0.090512
CCSDT	0.000493	0.001423	0.001405	0.024752	0.040126

Notes: R_e is the equilibrium geometry of H_2O. The HF weight from the FCI wave function is also reported.

From Koch H, Christiansen O, Jørgensen P, de Merás AMS, Helgaker T. The CC3 model: an iterative coupled cluster approach including connected triples. J Chem Phys 1997;106(5):1808.

The reason is that a single-reference wave function cannot adequately describe the dissociation of two bonds, which becomes evident as the HF weight decreases and the HF energy deviation from the FCI limit increases. The near-degeneracies that formed at longer O–H bond distances imply that more than one electronic configurations are becoming important, and an MC reference wave function is needed, especially at O–H distances larger than $2R_e$. Application of an optimized CI expansion as reference wave function will include those near-degeneracies and will capture the so-called static correlation. Those effects are partially captured by the cluster operator and the large determinantal space of CCSDT, where deviations are not larger than 40.126 m E_{h}. Also notice that the CCSD(T) level overestimates the correlation energy of the stretched geometries, a result from the nonvariational nature of the perturbative triples.

Apart from size of the N-electron model, the second variable that affects the percentage of the missing correlation energy is the size of the one-electron basis. Up to this point, the discussion has been centered on using different electronic structure methods to approximate the FCI regime by selecting an appropriate N-electron description of the Fock space given a one-electron basis. Recall that the exact nonrelativistic electronic energy of an atomic or molecular system is obtained by FCI with an infinitely large basis set (or CBS limit), as shown in Fig. 1. However, the consideration of basis set, or the number and type of functions to describe the spin-orbitals that the wave function is built on, is also a crucial aspect in obtaining accurate energies. For variational wave function methods, expanding the basis set provides a systematic improvement toward the exact energy. An infinitely large basis set cannot be applied for practical calculations, but the goal is to include a sufficiently large number of basis functions. However, basis set convergence is slow in post-HF methods. The HF energy is minimized with respect to the MO coefficients, but the HF virtual orbitals responsible for electron correlation in post-HF methods are not efficiently described, leading to slow basis set convergence. For most post-HF methods, the asymptotic convergence to the basis set limit is known, so correlation energies calculated from truncated basis expansions can be extrapolated for the estimation of the CBS limit [25]. The most applied extrapolation scheme was suggested by Helgaker and co-workers [26] and is based on the two-point equation

$$E_{XY}^{CBS} = \frac{E_X^{corr} X^3 - E_Y^{corr} Y^3}{X^3 - Y^3}, \tag{193}$$

where E_X^{corr} and E_Y^{corr} are the correlation energies in the given basis with cardinal numbers X and Y, that is, two for double-zeta, three for triple-zeta, and so on. This extrapolation has shown to effectively estimate the CBS energy when the basis sets are of at least triple-zeta and higher quality.

The main reason for the slow convergence of the correlated schemes is the inefficiency of small basis expansions to describe the electron-electron repulsion. For example, consider two electrons both confined to an axis, where electron 1 is fixed at a particular point. The probability density of electron 2 should substantially decrease as it approaches electron 1, reaching a singularity when both occupy the same position, and the corresponding wave function should form a cusp, since the probability of finding two electrons in the same place is exactly zero. Since electron-electron coalescence is poorly described in the traditional one-electron basis, addition of auxiliary functions that depend *explicitly* on the interelectronic distance r_{12} can properly describe these electronic interactions. A family of methods known as *explicitly correlated* or *R12/F12 methods* is based on the correct physics provided by the augmented basis set, and accelerates the convergence to the CBS limit [27]. The was first suggested by Hylleraas in 1929, who introduced in the wave function of the helium atom terms that depended explicitly on the r_{12} term after observing exceptionally slow convergence on the calculation of the electronic energy with a traditional CI expansion [28]. Nowadays, a systematic addition of two-electron basis functions (*geminals*) complimentary to the one-electron basis increases the percentage of the correlation energies with a relatively small computational overhead. For example, an explicitly correlated CCSD(T) calculation with a triple-zeta quality basis set provides accuracy that is obtained from a conventional CCSD(T) calculation with a quadruple-zeta basis [29].

5.7 COUPLED-CLUSTER THEORY FOR EXCITED STATES

So far, we have considered a ground state HF wave function as reference for the CC exponential ansatz, which provides accurate energies and properties for molecular systems in their ground electronic state. CC variants have also been developed for excited states, which are based on the generation of excited state wave functions from the CC wave function. One approach known as *equation-of-motion coupled-cluster* (EOM-CC) considers a linear expansion of excited electronic configurations by applying an excitation operator $\hat{R}_k$ on $|\Psi_{\text{CC}}\rangle$

$$|\Psi_k\rangle = \hat{R}_k|\Psi_{\text{CC}}\rangle = \hat{R}_k \exp(\hat{T})|\Psi_0\rangle. \tag{194}$$

The excitation operator is defined as

$$\hat{R} = \hat{R}_0 + \hat{R}_1 + \hat{R}_2 + \cdots + \hat{R}_N, \tag{195}$$

where the terms of Eq. (195) are synonymous to the CI excitation operators. For example, the first three terms are given as

$$\hat{R}_0 = r_0, \tag{196}$$

$$\hat{R}_1 = \sum_{I,A} r_I^A a_A^\dagger a_I, \tag{197}$$

$$\hat{R}_2 = \sum_{\substack{I<J\\A<B}} r_{IJ}^{AB} a_A^\dagger a_B^\dagger a_J a_I. \tag{198}$$

If the excitation operator is limited to singles and doubles excitations, then the EOM-CCSD level is defined, addition of triples defines EOM-CCSDT, and so on.

Usually, the $\hat{T}$ operator is truncated at the same level as the excitation operator $\hat{R}$. Similar to a CI problem, the EOM energies and $r_o, r_I^A, r_{IJ}^{AB}, \ldots$ coefficients can be calculated by construction and diagonalization of the corresponding Hamiltonian matrix

$$\hat{H}|\Psi_k\rangle = \hat{H}\hat{R}_k \exp(\hat{T})|\Psi_0\rangle = E_k\hat{R}_k \exp(\hat{T})|\Psi_0\rangle. \tag{199}$$

Multiplying with $\exp(-\hat{T})$ from the left and considering that the operators $\hat{T}$ and $\hat{R}_k$ commute

$$\exp(-\hat{T})\hat{H}\exp(\hat{T})|\Psi_0\rangle = E_k \exp(-\hat{T})\exp(\hat{T})\hat{R}_k|\Psi_0\rangle = E_k\hat{R}_k|\Psi_0\rangle. \tag{200}$$

Practically, iterative methods for the calculation of a few eigenvalues are applied, such as the Lanczos-type methods. EOM-CC has several advantages over a truncated CI approach (e.g., CISD) for excited states. Using the CC ground state as a basis for generating excited states in the EOM-CC method provides a fairly compact and size-extensive initial wave function. With such an expansion and the fact that the EOM-CC model constructs excited determinants on an already correlated wave function, the distinction of electron correlation and an excited state is not evident. The EOM-CC model is a method closely mirroring CI, which treats excited states in a general sense rather than a single explicit excitation in a determinant, and can provide powerful methodologies for accurate calculations.

6 CONCLUSIONS AND OUTLOOK

Since the emergence of quantum chemistry, wave function theory methods have been applied in a multitude of applications and have contributed on the elucidation of many chemical phenomena. Almost every quantum chemical method uses HF theory as its initial point. In this chapter, the shortcomings of HF were presented and the fundamental equations of three post-HF theories were discussed. The dissociation of the H_2 bond was used in order to explain the different types of the missing electron correlation from HF. This helped differentiate between the single-reference post-HF methods, which are able to recover the short-range dynamic correlation, and the multireference (or MC) post-HF methods, that capture the strong (or static) correlation by applying a linear combination of SDs or CSFs as the reference wave function. The dynamic correlation can be calculated by CI, MBPT, or CC theory. These approaches offer a systematic expansion from HF to the FCI limit at a given one-electron basis, by considering a number of electron excitations from the reference wave function to the virtual (or secondary space). These excitations can involve a single-, double-, triple-, etc., electron excitation and form the N-electron expansion. A single-determinantal HF or an MC expansion can be used as reference wave function for these three families of methods. The second expansion that leads to the FCI limit is the size of the basis set, which is related with one of the most important strengths of post-HF methods: the systematic convergence to the CBS limit of the method by increase of the basis set size. The FCI scheme at the CBS limit gives the exact, nonrelativistic, Born-Oppenheimer electronic energy. Thus, a computational chemist has nowadays an arsenal with a large variety of accurate methods to use for solving chemical problems relevant to reactivity, spectroscopy, noncovalent interactions, and more.

With the development of CC theory, a systematic approach for the accurate calculation of the electron correlation was established. Once the exact theory was formulated, and efficient approximate schemes were developed (e.g., the popular CCSD(T)), the only challenge that remained was the transition from applications on small molecules to larger chemical species, without loss of the accuracy and reliability of the method. We will attempt to group these approaches in five categories: The first targets the faster convergence to the CBS limit, which can be achieved by explicitly correlated CC [27,30], or extrapolation schemes [26,31–33], as it was discussed in Section 5.6. A second approach explores the locality of electron excitations that contribute the most to the correlation energy [34–37]. Dynamic correlation is by definition a short-range effect, thus electron pairs which occupy orbitals that are located in close proximity will contribute more than a delocalized electron pair. Fragmentation schemes can also be added in this category [38–42]. Third, mathematical techniques such as Laplace-transformed expressions [43–46] or resolution-of-identity/density fitting techniques [47,48] for efficient integral evaluation provide significant speedups for correlated calculations. Fourth, modern hardware and high-performance computing [49] allows the performance of demanding computations on thousands of processors and with significantly larger memory and disk space. Finally, the truncation level of the cluster operator should also be considered for highly accurate results. Since correlation energy from higher-level clusters (e.g., full triples, quadruples) with a large basis set would be very computationally demanding, their contribution can be estimated by composite schemes, where high-excitation energies are calculated with smaller basis sets and they are added to the CBS limit of the lower clusters (e.g., CCSD(T)/CBS) [50–52]. These five categories, usually applied in combinations, aim on new implementations that reduce disk and memory requirements and allow the quantum chemical methods to scale linearly with respect to the size of the system.

CCSD(T), the gold standard of quantum chemistry, combined with the aforementioned approaches, can be applied routinely and reliably on hundreds of atoms. Therefore, the computation of dynamic correlation for molecules with a single-reference ground state should be considered as a tamed problem. The calculation of higher orders of CC theory (full triples, perturbative quadruples) is also feasible in common supercomputing facilities, which aids to achieve higher accuracy for many applications of interest (e.g., atomization energies, noncovalent interactions). A challenge for CC still remains the incorporation of strong (static) correlation. Combination of the CC expansion with an MC reference wave function captures both strong and dynamic correlations, and significant progress has been made toward implementations that will allow such a multireference CC scheme applicable to chemical problems [53]. Such approaches will contribute on the performance of accurate computations of excited states [54] and species with strong MC character.

We have discussed in detail that strongly correlated systems can be addressed by multireference methods (MCSCF, MRCI, or MRPT). The main drawback of those methods is the accurate, systematic, and black-box-type calculation of static or strong correlation. Traditional methods, such as the popular CASSCF, suffer from the size of the active space and the prior knowledge that the user should have on the system of interest. Usually, chemical intuition and/or many exploratory calculations are needed in order to perform such types of computations. Another issue is related to the missing dynamic correlation. MRCI is still a very computationally demanding method, while MRPT2 theories offer a cheaper alternative, but they also suffer from the size of the active space and the expensive terms that need to be evaluated. Advances on CI eigensolvers that include the density matrix renormalization group [55,56], selective CI [57], stochastic methods [58], or high-performance computing [59,60], allowed us to perform calculations beyond the limits of current CAS-type implementations.

Finally, it should be mentioned that the applicability of any post-HF method can be extended to larger systems by dividing the molecular species into two or more subsystems, and treat each subsystem at a different level of theory. A higher level of theory (such as post-HF methods) is applied on the subsystem of interest, while the environment is treated in a more approximate, lower level of theory (e.g., molecular mechanics or density functional theory). The interactions between the subsystems are introduced either via simple hybrid methods such as the quantum mechanics/molecular mechanics (QM/MM) method, which use mechanical or electrostatic embedding [61,62], or exact (self-consistent) quantum embedding [63].

As mentioned in Section 1, the scope of the chapter is to introduce the three families of post-HF theories, CI, MBPT, and CC, to a nonexpert who would like to obtain a good background on electronic structure theory. We hope that this chapter will serve as a starting point for interested readers to discover the power and applicability of wave function theory-based methods, either as developers or practitioners for chemical applications. For that reason, at the end of the chapter, we recommend some excellent textbooks and review articles that cover in more detail the theory, the implementation and the algorithmic techniques of the post-HF methods.

REFERENCES

[1] Heitler W, London F. Wechselwirkung neutraler atome und homöopolare bindung nach der quantenmechanik. Z Phys 1927;44(6):455–72.

[2] Gavroglu K, Simoes A. Neither physics nor chemistry. Cambridge, MA: The MIT Press; 2012.

[3] Tew DP, Klopper W, Helgaker T. Electron correlation; the main-body problem at the heart of chemistry. J Comput Chem 2007;28(8):1307–20.

[4] Helgaker T, Jørgensen P, Olsen J. Molecular electronic structure theory. Chichester: Wiley; 2000.

[5] Handy NC. Multi-root configuration interaction calculations. Chem Phys Lett 1980; 74(2):280–3.
[6] Duch W. GRMS or graphical representation of model spaces. Berlin: Springer; 1986.
[7] Shavitt I. The graphical unitary group approach and its application to direct configuration interaction calculations. In: Hinze J, editor. The unitary group for the evaluation of electronic energy matrix elements. Lecture notes in chemistry, vol. 22. Berlin, Heidelberg: Springer; 1981.
[8] Roos B. A new method for large-scale Cl calculations. Chem Phys Lett 1972;15(2): 153–9.
[9] Roos B, Siegbahn PEM. The direct configuration interaction method from molecular integrals. In: Schaefer HF, editor. Methods of electronic structure theory. New York: Plenum, Springer; 1977. p. 277–318.
[10] Fermann JT, Sherrill CD, Crawford TD, Schaefer III HF. Benchmark studies of electron correlation in six-electron systems. J Chem Phys 1994;100(11):8132.
[11] Langhoff SR, Davidson ER. Configuration interaction calculations on the nitrogen molecule. Int J Quant Chem 1974;8(1):61–72.
[12] Møller C, Plesset MS. Note on an approximation treatment for many-electron systems. Phys Rev 1934;46(7):618–22.
[13] Cremer D. Møller-Plesset perturbation theory: from small molecule methods to methods for thousands of atoms. WIREs Comput Mol Sci 2011;1(4):509–30.
[14] Olsen J, Jørgensen P, Helgaker T, Christiansen O. Divergence in Møller-Plesset theory: a simple explanation based on a two-state model. J Chem Phys 2000;112(22): 9736.
[15] Andersson K, Malmqvist PÅ, Roos BO, Sadlej AJ, Wolinsk K. Second-order perturbation theory with a CASSCF reference function. J Phys Chem 1990;94(14):5483–8.
[16] Andersson K, Malmqvist PÅ, Roos BO. Second-order perturbation theory with a complete active space self-consistent field reference function. J Chem Phys 1992;96(2): 1218.
[17] Angeli C, Cimiraglia R, Evangelisti S, Leininger T, Malrieu JP. Introduction of *n*-electron valence states for multireference perturbation theory. J Chem Phys 2001; 114(23):10252.
[18] Schapiro I, Sivalingam K, Neese F. Assessment of *n*-electron valence state perturbation theory for vertical excitation energies. J Chem Theory Comput 2013;9(8):3567–80.
[19] Coester F, Kümmel H. Short-range correlations in nuclear wave functions. Nucl Phys 1960;17:477–85.
[20] Čižek J. On the correlation problem in atomic and molecular systems. Calculation of wavefunction components in Ursell-type expansion using quantum-field theoretical methods. J Chem Phys 1966;45(11):4256.
[21] Pople JA, Krishnan R, Schlegel HB, Binkley JS. Electron correlation theories and their application to the study of simple reaction potential surfaces. Int J Quant Chem 1978;14(5):545–60.
[22] Bartlett RJ, Purvis GD. Many-body perturbation theory, coupled-pair many-electron theory, and the importance of quadruple excitations for the correlation problem. Int J Quant Chem 1978;14(5):561–81.
[23] Helgaker T, Jørgensen P. Configuration-interaction energy derivatives in a fully variational formulation. Theor Chem Acta 1989;75(2):111–27.
[24] Koch H, Christiansen O, Jørgensen P, de Merás AMS, Helgaker T. The CC3 model: an iterative coupled cluster approach including connected triples. J Chem Phys 1997; 106(5):1808.

[25] Helgaker T, Klopper W, Tew D. Quantitative quantum chemistry. Mol Phys 2008; 106(16–18):2107–43.
[26] Halkier A, Helgaker T, Jørgensen P, Klopper W, Koch H, Olsen J, et al. Basis-set convergence in correlated calculations on Ne, N_2, and H_2O. Chem Phys Lett 1998; 286(3–4):243–52.
[27] Hättig C, Klopper W, Köhn A, Tew DP. Explicitly correlated electrons in molecules. Chem Rev 2012;112(1):4–74.
[28] Hylleraas EA. Neue berechnung der energie des heliums im grundzustande, sowie des tiefsten terms von ortho-helium. Z Phys 1929;54(5–6):347–66.
[29] Tew DP, Klopper W, Neiss C, Hättig C. Quintuple-zeta quality coupled-cluster correlation energies with triple-zeta basis sets. Phys Chem Chem Phys 2007;9(16):1921–30.
[30] Kong L, Bischoff FA, Valeev EF. Explicitly correlated R12/F12 methods for electronic structure. Chem Rev 2012;112(1):75–107.
[31] Kutzelnigg W. The principle-quantum number (and the radial-quantum number) expansion of the correlation energy of two-electron atoms. Phys Chem Chem Phys 2008; 10(23):3460–8.
[32] Vogiatzis KD, Barnes E, Klopper W. Interference-corrected explicitly-correlated second-order perturbation theory. Chem Phys Lett 2011;503(1–3):157–61.
[33] Feller D, Peterson KA, Hill JG. On the effectiveness of CCSD(T) complete basis set extrapolations for atomization energies. J Chem Phys 2011;135(4):44102–19.
[34] Schütz M, Werner HJ. Local perturbative triples correction (T) with linear cost scaling. Chem Phys Lett 2000;318(4–5):370–8.
[35] Schütz M, Yang J, Chan GKL, Manby FR, Werner HJ. The orbital-specific virtual local triples correction: OSV-L(T). J Chem Phys 2013;138(5):054109.
[36] Riplinger C, Sandhoefer B, Hansen A, Neese F. Natural triple excitations in local coupled cluster calculations with pair natural orbitals. J Chem Phys 2013;139(13):134101.
[37] Guo Y, Riplinger C, Becker U, Liakos DG, Minenkov Y, Cavallo L, et al. An improved linear scaling perturbative triples correction for the domain based local pair-natural orbital based singles and doubles coupled cluster method [DLPNO-CCSD(T)]. J Chem Phys 2018;148(1):011101.
[38] Kobayashi M, Nakai H. Divide-and-conquer-based linear-scaling approach for traditional and renormalized coupled cluster methods with single, double, and noniterative triple excitations. J Chem Phys 2009;131(11):114108.
[39] Li W, Piecuch P, Gour JR, Li S. Local correlation calculations using standard and renormalized coupled-cluster approaches. J Chem Phys 2009;131(11):114109.
[40] Friedrich J, Dolg M. Fully automated incremental evaluation of MP2 and CCSD(T) energies: application to water clusters. J Chem Theory Comput 2009;5(2):287–94.
[41] Vogiatzis KD, Klopper W, Friedrich J. Non-covalent interactions of CO_2 with functional groups of metal-organic frameworks from a CCSD(T) scheme applicable to large systems. J Chem Theory Comput 2015;11(4):1574–84.
[42] Eriksen JJ, Baudin P, Ettenhuber P, Kristensen K, Kjærgaard T, Jørgensen P. Linear-scaling coupled cluster with perturbative triple excitations: the divide-expand consolidate CCSD(T) model. J Chem Theory Comput 2015;11(7):2984–93.
[43] Constans P, Ayala PY, Scuseria GE. Scaling reduction of the perturbative triples correction (T) to coupled cluster theory via Laplace transform formalism. J Chem Phys 2000;113(23):10451.
[44] Doser B, Lambrecht DS, Kussmann J, Ochsenfeld C. Linear-scaling atomic orbital-based second-order Møller-Plesset perturbation theory by rigorous integral screening criteria. J Chem Phys 2009;130(6):064107.

[45] Schmitz G, Hättig C. Perturbative triples correction for local pair natural orbital based explicitly correlated CCSD(F12*) using Laplace transformation techniques. J Chem Phys 2016;145(23):234107.
[46] Nagy PR, Kállay M. Optimization of the linear-scaling local natural orbital CCSD(T) method: redundancy-free triples correction using Laplace transform. J Chem Phys 2017;146(21):214106.
[47] Weigend F, Häser M, Patzelt H, Ahlrichs R. RI-MP2: optimized auxiliary basis sets and demonstration of efficiency. Chem Phys Lett 1998;294(1–3):143–52.
[48] Valeev EF. Improving on the resolution of the identity in linear R12 ab initio theories. Chem Phys Lett 2004;395(4–6):190–5.
[49] de Jong WA, Bylaska E, Govind N, Janssen CL, Kowalski K, Müller T, et al. Utilizing high performance computing for chemistry: parallel computational chemistry. Phys Chem Chem Phys 2010;12(26):6896–920.
[50] Tajti A, Szalay PG, Császár AG, Kállay M, Gauss J, Valeev EF, et al. Heat: high accuracy extrapolated ab initio thermochemistry. J Chem Phys 2004;121(23):11599–613.
[51] DeYonker NJ, Cundari TR, Wilson AK. The correlation consistent composite approach (ccCA): an alternative to the Gaussian-*n* methods. J Chem Phys 2006;124(11):114104.
[52] Vogiatzis KD, Haunschild R, Klopper W. Accurate atomization energies from combining coupled-cluster computations with interference-corrected explicitly correlated second-order perturbation theory. Theor Chem Acc 2014;133(3):1446.
[53] Lyakh DI, Musial M, Lotrich VF, Bartlett RJ. Multireference nature of chemistry: the coupled-cluster view. Chem Rev 2012;112(1):182–243.
[54] Sneskov K, Christiansen O. Excited state coupled cluster methods. WIREs Comput Mol Sci 2012;2(4):566–84.
[55] Chan GKL, Sharma S. The density matrix renormalization group in quantum chemistry. Annu Rev Phys Chem 2011;62:465–81.
[56] Knecht S, Hedegård ED, Keller S, Kovyrshin A, Ma Y, Muolo A, et al. New approaches for ab initio calculations of molecules with strong electron correlation. Chimia 2016;70(4):244–51.
[57] Huron B, Malrieu JP, Rancurel P. Iterative perturbation calculations of ground and excited state energies from multiconfigurational zeroth-order wavefunctions. J Chem Phys 1973;58(12):5745.
[58] Booth GH, Thom AJW, Alavi A. Fermion Monte Carlo without fixed nodes: a game of life, death, and annihilation in Slater determinant space. J Chem Phys 2009;131(5):054106.
[59] Müller T. Large-scale parallel uncontracted multireference-averaged quadratic coupled cluster: the ground state of the chromium dimer revisited. J Phys Chem A 2009;113(45):12729–40.
[60] Vogiatzis KD, Ma D, Olsen J, Gagliardi L, de Jong WA. Pushing configuration-interaction to the limit: towards massively parallel MCSCF calculations. J Chem Phys 2017;147(18):184111.
[61] Lin H, Truhlar DG. QM/MM: what have we learned, where are we, and where do we go from here? Theor Chem Acc 2007;117(2):185–99.
[62] Chung LW, Sameera WMC, Ramozzi R, Page AJ, Hatanaka M, Petrova GP, et al. The ONIOM method and its applications. Chem Rev 2015;115(12):5678–796.
[63] Sun Q, Chan GKL. Quantum embedding theories. Acc Chem Res 2016;49(12):2705–12.

FURTHER READING

[64] Szabo A, Ostlund NS. Modern quantum chemistry. New York: Dover; 1996.
[65] Helgaker T, Jørgensen P, Olsen J. Molecular electronic-structure theory. West Sussex: Wiley; 2000.
[66] Shavitt I, Bartlett RJ. Many-body methods in chemistry and physics. Cambridge: Cambridge University Press; 2009.
[67] Szalay PG, Müller T, Gidofalvi G, Lischka H, Shepard R. Multiconfiguration self-consistent field and multireference configuration interaction methods and applications. Chem Rev 2012;112:108–81.
[68] Roos BO, Lindh R, Malmqvist PA, Veryazov V, Widmark P-O. Multiconfigurational quantum chemistry. Hoboken, NJ: Wiley; 2016.
[69] Lewars EG. Computational chemistry. New York, NY: Springer; 2016.

CHAPTER

Density functional theory

4

Jianwei Sun, James W. Furness, Yubo Zhang
Department of Physics and Engineering Physics, Tulane University, New Orleans, LA, United States

1 INTRODUCTION

We used to think that if we know one, we knew two, because one and one are two. We are finding that we must learn a great deal more about 'and'.

Sir Arthur Eddington, from The Harvest of a Quiet Eye, by A. Mackay

Atoms, named by Democritus ~2500 years ago after the Greek word *atomos* meaning uncuttable or indivisible, were thought to be the smallest components from which our world was built until the early 20th century when it became known that the atoms are themselves divisible, composed of electrons and a nucleus. Nevertheless, the idea that all matter is composed at some level of elementary particles, the final unsplittable bricks, remains intriguing to physicists and underlies the field of particle physics. When we try to understand properties of matter built from those elementary particles or at the higher level of atoms, however, we find it is not a jig-saw like puzzle where small and simple things lead to large and complicated assemblies in a simple, seamless, and direct way. Instead we find putting one and one together does not simply lead us to two; extra attention must be paid to the "and" that combines them.

To illustrate this complexity we can consider the one electron hydrogenic system in which a single electron is moving around a fixed nucleus, a system for which the physics of the electron can be solved exactly to give the quantized electronic states. When one more electron is put into the system; however, the newly established states cannot be described simply as a combination of two states from two isolated identical electrons. The correlation in motions between them renders even this simple system unsolvable. In quantum chemistry and condensed matter physics, where usually the former studies molecules formed of a finite number of atoms while the latter covers systems comprised of an extremely large number of atoms in extended structures, these correlations between the motions of electrons play an essential role

Mathematical Physics in Theoretical Chemistry. https://doi.org/10.1016/B978-0-12-813651-5.00004-8

in forming useful properties we are interested in and approximating them remains a core challenge across these fields.

The systems of interest in quantum chemistry and condensed matter physics are usually in the low-energy domain where the internal structures of the nuclei and the relativistic effects can be neglected and thus can be described by the Hamiltonian,

$$\hat{H}_h = -\frac{\hbar^2}{2m_e}\sum_i \nabla_i^2 + \sum_{i,I}\frac{Z_I e^2}{|\mathbf{r}_i - \mathbf{R}_I|} + \frac{1}{2}\sum_{i \neq j}\frac{e^2}{|\mathbf{r}_i - \mathbf{r}_j|} - \sum_I \frac{\hbar^2}{2M_I}\nabla_I^2 + \frac{1}{2}\sum_{I \neq J}\frac{Z_I Z_J e^2}{|\mathbf{R}_I - \mathbf{R}_J|}, \quad (1)$$

where $\hbar$ is Planck's constant divided by 2π, e, m_e, and $\mathbf{r}_i$ are the charge, mass, and position of electron i, respectively. Electrons in Eq. (1) are denoted by lower case subscripts and nuclei, with charge Z_I, mass M_I, and position $\mathbf{R}_I$, are denoted by upper case subscripts. Because the nuclei are relatively massive they can often be regarded as classical particles, with their kinetic energy neglected in the Hamiltonian, while the much smaller electrons must be handled as fully quantum objects. This difference in size and mass allows the motion of the electrons and the nuclei to be decoupled, and the electronic energies may be computed for fixed nuclear positions, assuming the instantaneous electronic relaxation with respect to nuclear motions. This separation is termed the Born-Oppenheimer or *adiabatic* approximation [1], and by employing it we arrive at the fundamental Hamiltonian for the theory of electronic structure,

$$\hat{H} = \hat{T} + \hat{V}_{\text{ext}} + \hat{V}_{\text{ee}}, \quad (2)$$

where $\hat{T} = -\frac{1}{2}\sum_i \nabla_i^2$, $\hat{V}_{\text{ext}} = \sum_{i,I}\frac{Z_I}{|\mathbf{r}_i - \mathbf{R}_I|}$, and $\hat{V}_{\text{ee}} = \frac{1}{2}\sum_{i \neq j}\frac{1}{|\mathbf{r}_i - \mathbf{r}_j|}$; the Hartree atomic units of $\hbar = m_e = e = 4\pi/\epsilon_0 = 1$ are adopted throughout the text. The classical interaction between the nuclei is omitted as it is constant for a given nuclear configuration and can be absorbed into $\hat{V}_{\text{ext}}$. For most systems the Born-Oppenheimer is a good model; however, the approximation ceases to be adequate whenever the motion of nuclei becomes important, for example, for superconductivity where the phonon-electron coupling (a phonon can be understood as the vibration of ions in solids) leads to the formation of Cooper electron pairs.

The eigenstates of $\hat{H}$ satisfy the Schrödinger equations,

$$\hat{H}\Psi_k(\mathbf{r}_1, \sigma_1, \ldots, \mathbf{r}_N, \sigma_N) = E_k\Psi_k(\mathbf{r}_1, \sigma_1, \ldots, \mathbf{r}_N, \sigma_N), \quad (3)$$

where k is a complete set of many-electron quantum numbers, N is the number of electrons in the system, $\mathbf{r}_i$ and σ_i represent the space and spin coordinates of electron i, respectively. Because electrons are fermions, the wavefunction $\Psi_k(\mathbf{r}_1, \sigma_1, \ldots, \mathbf{r}_N, \sigma_N)$ must be antisymmetric with respect to permutation of the coordinates of the electrons.

In quantum chemistry, the Hartree-Fock (HF) approximation is usually taken as a starting point for solving Eq. (3) for the ground-state wavefunction and energy, which approximates the exact many-electron wavefunction by a single Slater determinant of one electron spin-orbitals, ignoring the electron-electron correlation. Many post-HF methods, such as configuration interaction and coupled cluster, have been proposed to approximate the electron-electron correlations neglected in the HF approximation which for the great majority of systems under study are essential for accurate predictions. Such post-HF methods can be expected to give improved results over basic HF calculations, although this added accuracy comes at the price of much increased computational effort.

An alternative approach for treating the many-electron problem is provided by density functional theory (DFT), which uses the electron density, $n(\mathbf{r})$, rather than $\Psi_k(\mathbf{r}_1, \sigma_1, \ldots, \mathbf{r}_N, \sigma_N)$ as its basic variational object. While the Hohenberg-Kohn (HK) theorems [2] give solid foundations for modern DFT, the Kohn-Sham (KS) approach [3] provides a practical way to solve for the ground-state electron density, and therefore the ground-state energy, by replacing the complicated many-electron problem with an auxiliary noninteracting system that leads to the one-electron Schrödinger-like KS equations. The genius of the KS approach is in explicitly separating the noninteracting kinetic energy and the long-range Hartree energy, which describes the classical electrostatic repulsion between electrons, from the remaining exchange-correlation energy that must be approximated by the unknown functional $E_{\text{xc}}[n_\uparrow(\mathbf{r}), n_\downarrow(\mathbf{r})]$. Because the exchange-correlation energy is usually only a small fraction of the ground-state total energy impressive accuracy can be obtained with relatively simple approximate functionals and without the large computational costs associated with the many-electron wave function.

In this chapter, some recent advances in DFT will be discussed along with the fundamentals. The fundamental aspects of DFT will be presented in Section 2 though it will not be possible to cover all fundamental aspects of DFT in the space of this text, and we refer readers to references [4–8] for a more comprehensive description. In Section 3, approaches for developing density functional approximations and the Jacob's ladder of DFT, which categorizes these approximations, will be discussed. The focus will be on the nonempirical approach to functional construction preferred by the authors, and we refer readers interested in other approaches to references [8,9]. A promising nonempirical density functional developed recently will be highlighted in Section 4, with emphasis on the ideas behind the construction. Finally, the outlook for new developments in DFT will be given. This chapter discusses only the ground-state DFT, and we refer readers interested in excited states to references for time-dependent DFT (TDDFT) [10–12].

2 FUNDAMENTALS OF DENSITY FUNCTIONAL THEORY

2.1 WAVEFUNCTION THEORY

In the electronic structure theory of many-electron systems we are interested in the eigenstates of Eq. (2), which are usually represented as many-electron wavefunctions $\Psi(\mathbf{r}_1\sigma_1, \ldots, \mathbf{r}_N, \sigma_N)$ in real space and completely determine all the properties of the system. Physical wavefunctions for electrons satisfy the fermion constraint that they are antisymmetric under exchange of two electron labels i and j,

$$\Psi(\mathbf{r}_1,\sigma_1,\ldots,\mathbf{r}_i,\sigma_i,\ldots,\mathbf{r}_j,\sigma_j,\ldots,\mathbf{r}_N,\sigma_N) = -\Psi(\mathbf{r}_1,\sigma_1,\ldots,\mathbf{r}_j,\sigma_j,\ldots,\mathbf{r}_i,\sigma_i,\ldots,\mathbf{r}_N,\sigma_N). \tag{4}$$

There are $N!$ distinct permutations of the labels $1, 2, \ldots, N$, which by Eq. (4) all give the same $|\Psi|^2$. Thus $N!|\Psi(\mathbf{r}_1,\sigma_1,\ldots,\mathbf{r}_N,\sigma_N)|^2 d^3\mathbf{r}_1 \ldots d^3\mathbf{r}_N$ is the probability to find any electrons with spin $\sigma_1,\ldots,\sigma_N$ in volume element $d^3\mathbf{r}_1 \ldots d^3\mathbf{r}_N$, and,

$$\frac{1}{N!}\sum_{\sigma_1,\ldots,\sigma_N}\int d^3\mathbf{r}_1\ldots\int d^3\mathbf{r}_N N!|\Psi(\mathbf{r}_1,\sigma_1,\ldots,\mathbf{r}_N,\sigma_N)|^2 = \int |\Psi|^2 = \langle\Psi|\Psi\rangle = 1, \tag{5}$$

where the $N!$ in the denominator takes care of the double counting in the multidimensional integral.

The electron spin density $n_\sigma(\mathbf{r})$ is defined so that $n_\sigma(\mathbf{r})d^3\mathbf{r}$ is the probability to find an electron with spin σ in volume element $d^3\mathbf{r}$ at $\mathbf{r}$. We find $n_\sigma(\mathbf{r})$ by integrating over the coordinates and spins of the $(N-1)$ other electrons,

$$\begin{aligned} n_\sigma(\mathbf{r}) &= \frac{1}{(N-1)!}\sum_{\sigma_2,\ldots,\sigma_N}\int d^3\mathbf{r}_2 \ldots\int d^3\mathbf{r}_N N!|\Psi(\mathbf{r},\sigma,\mathbf{r}_2,\sigma_2,\ldots,\mathbf{r}_N,\sigma_N)|^2 \\ &= N\sum_{\sigma_2,\ldots,\sigma_N}\int d^3\mathbf{r}_2 \ldots\int d^3\mathbf{r}_N|\Psi(\mathbf{r},\sigma,\mathbf{r}_2,\sigma_2,\ldots,\mathbf{r}_N,\sigma_N)|^2. \end{aligned} \tag{6}$$

Eqs. (5), (6) yield

$$\sum_\sigma \int d^3\mathbf{r}\, n_\sigma(\mathbf{r}) = N, \tag{7}$$

which shows that $n_\sigma(\mathbf{r})d^3\mathbf{r}$ is the average number of electrons of spin σ in volume element $d^3\mathbf{r}$.

The many-electron wavefunction $\Psi(\mathbf{r}_1\sigma_1,\ldots,\mathbf{r}_N,\sigma_N)$ has a great deal of information, which, however, is a function of $4N$ variables and not easy to calculate, store, apply, or even think about. Often we want no more than the total energy E (and possibly its derivatives), or perhaps also the spin densities $n_\uparrow(\mathbf{r})$ and $n_\downarrow(\mathbf{r})$ for the ground state. DFT assures that Ψ can be formally replaced by the observables $n_\uparrow$ and $n_\downarrow$ as the basic variational objects.

2.2 WAVEFUNCTION VARIATIONAL PRINCIPLE

A functional is a rule that assigns a number $F[f]$ to a function f, analogous to a function that is a rule assigning a number $f(x)$ to a number x. For example, $h[\Psi] = \langle\Psi|\hat{H}|\Psi\rangle$ is a functional of the trial wavefunction Ψ, given the Hamiltonian $\hat{H}$.

The wavefunction variational principle extremizes $\langle\Psi_k|\hat{H}|\Psi_k\rangle$ subject to the constraint $\langle\Psi_k|\Psi_k\rangle = 1$, that is, set the following first variation to zero:

$$\delta\{\langle\Psi|\hat{H}|\Psi\rangle/\langle\Psi|\Psi\rangle\} = 0, \tag{8}$$

which finds the ground- and excited-state wavefunctions, and hence energies. The Rayleigh-Ritz method finds the extrema or the minima in a restricted space of wavefunctions. For example, the HF approximation of the ground-state wavefunction is the Slater determinant Φ that minimizes $\langle\Phi|\hat{H}|\Phi\rangle/\langle\Phi|\Phi\rangle$. The configuration-interaction ground-state wavefunction generalizes the single Slater determinant in the HF approximation to an energy-minimizing linear combination of Slater determinants, often restricted to a subset of possible excitations from a reference HF determinant.

The constrained solution of Eq. (8) is equivalent by the method of Lagrange multipliers to the unconstrained solution of

$$\delta\{\langle\Psi|\hat{H}|\Psi\rangle - E\langle\Psi|\Psi\rangle\} = 0, \tag{9}$$

$$\langle\delta\Psi|\hat{H} - E|\Psi\rangle = 0. \tag{10}$$

As $\delta\Psi$ is an arbitrary variation, we recover the Schrödinger equation (3). Every eigenstate of $\hat{H}$ is an extremum of $\langle\Psi|\hat{H}|\Psi\rangle/\langle\Psi|\Psi\rangle$, and vice versa.

The wavefunction variational principle implies the Hellmann-Feynman theorem here and also implies the HK [2] density functional variational principle presented later.

2.3 HELLMANN-FEYNMAN THEOREM

The Hamiltonian $\hat{H}$ may be dependent on some parameter λ and we want to know the dependence of the energy E_λ on this parameter. For any normalized variational solution Ψ_λ extremizing $\langle\Psi_\lambda|\hat{H}_\lambda|\Psi_\lambda\rangle$ over a normalized class of trial functions (including any eigenstate of $\hat{H}_\lambda$), we define

$$E_\lambda = \langle\Psi_\lambda|\hat{H}_\lambda|\Psi_\lambda\rangle, \tag{11}$$

then,

$$\frac{dE_\lambda}{d\lambda} = \frac{d}{d\lambda'}\left\langle\Psi_{\lambda'}\middle|\hat{H}_\lambda\middle|\Psi_{\lambda'}\right\rangle\Bigg|_{\lambda'=\lambda} + \left\langle\Psi_\lambda\middle|\frac{\partial\hat{H}_{\lambda'}}{\partial\lambda'}\middle|\Psi_\lambda\right\rangle\Bigg|_{\lambda'=\lambda}. \tag{12}$$

The first term of Eq. (12) vanishes by the extremum property of Ψ_λ from the variational principle and the Hellmann-Feynman theorem is found as,

$$\frac{dE_\lambda}{d\lambda} = \left\langle\Psi_\lambda\middle|\frac{\partial\hat{H}_\lambda}{\partial\lambda}\middle|\Psi_\lambda\right\rangle. \tag{13}$$

2.4 HARTRE-FOCK APPROXIMATION

Due to the electron-electron interaction, it is extremely difficult to solve for the exact wavefunctions $\Psi(\mathbf{r}_1\sigma_1,\ldots,\mathbf{r}_N,\sigma_N)$. The HF approximation to the exact wavefunctions will be briefly covered here due to its importance and influence on DFT.

The HF approximation assumes that the electrons are independent [13] and move in the electrostatic field of the nuclei and the average field of the other $N-1$ electrons. We shall start with the case of an electron with spin σ ($\sigma = \uparrow, \downarrow$) moving in an external field $v(\mathbf{r})$. The Hamiltonian, or energy operator, for the system is given by,

$$\hat{h} = -\frac{1}{2}\nabla^2 + v(\mathbf{r}). \tag{14}$$

The energy eigenstates $\psi_i(\mathbf{r}, \sigma)$ and eigenvalues ϵ_i are solutions of the Schrödinger equation,

$$\hat{h}\psi_i(\mathbf{r}, \sigma) = \epsilon_i \psi_i(\mathbf{r}, \sigma), \tag{15}$$

and $|\psi_i(\mathbf{r}, \sigma)|^2 d^3\mathbf{r}$ is the probability of finding the electron in energy eigenstate $\psi_i(\mathbf{r}, \sigma)$ with spin σ in volume element $d^3\mathbf{r}$ at $\mathbf{r}$. Thus,

$$\sum_{\sigma} \int d^3\mathbf{r} |\psi_i(\mathbf{r}, \sigma)|^2 = \langle \psi | \psi \rangle = 1, \tag{16}$$

defining the normalization of the one-electron wavefunction. Because $\hat{h}$ commutes with $\hat{S}_z$, the quantum spin operator around z-axis, we can choose the ψ_i to be eigenstates of $\hat{S}_z$ and choose $\sigma \in \{\uparrow, \downarrow\}$ as a one-electron quantum number.

Now, consider the Hamiltonian for N noninteracting electrons,

$$\hat{H}_{\text{non}} = \sum_{i=1}^{N} \left[-\frac{1}{2}\nabla_i^2 + v(\mathbf{r}_i) \right]. \tag{17}$$

The eigenfunctions of the one-electron problem of Eqs. (14), (15) are spin-orbitals, which can be used to construct the antisymmetric eigenfunctions Φ of $\hat{H}_{\text{non}}$,

$$\hat{H}_{\text{non}} \Phi = E_{\text{non}} \Phi. \tag{18}$$

Let i denote the composite spin-space coordinate for $\{r_i, \sigma_i\}$ and construct the Slater determinant, or antisymmetrized product,

$$\Phi = \frac{1}{\sqrt{N!}} \sum_{P} (-1)^P \psi_{\alpha_1}(P1) \psi_{\alpha_2}(P2) \ldots \psi_{\alpha_N}(PN), \tag{19}$$

where the quantum label α_i now includes the spin quantum number σ. Here P is any permutation of the labels $1, 2, \ldots, N$, and $(-1)^P$ is $+1$ for an even permutation and -1 for an odd permutation. The total energy is then,

$$E_{\text{non}} = \epsilon_{\alpha_1} + \epsilon_{\alpha_2} + \cdots + \epsilon_{\alpha_N}, \tag{20}$$

and the density is given by $\sum_i^N |\psi_{\alpha_i}(\mathbf{r})|^2$. If any α_i equals any α_j in Eq. (19), we find the determinant $\Phi = 0$, which is not a normalizable wavefunction. This result is equivalent to the Pauli exclusion principle whereby two or more noninteracting electrons may not occupy the same spin orbital.

Suppose we use a single Slater determinant, Φ_{HF} of Eq. (19), as a trial noninteracting wavefunction. By varying the occupied spin-orbitals $\psi_i(\mathbf{r},\sigma)$ to find the extremum of $\langle\Phi_{\mathrm{HF}}|\hat{H}|\Phi_{\mathrm{HF}}\rangle$ subject to the normalization $\langle\psi_i|\psi_j\rangle = \delta_{ij}$ leads to the HF equations,

$$\left[-\frac{1}{2}\nabla^2 + v_{\mathrm{ext}}(\mathbf{r}) + \sum_j \int d^3\mathbf{r}'\psi_j^*(\mathbf{r}',\sigma')\psi_j(\mathbf{r}',\sigma')\frac{1}{|\mathbf{r}-\mathbf{r}'|}\right]\psi_i(\mathbf{r},\sigma)$$
$$-\sum_j \int d^3\mathbf{r}'\psi_j^*(\mathbf{r}',\sigma')\psi_i(\mathbf{r}',\sigma')\frac{1}{|\mathbf{r}-\mathbf{r}'|}\psi_j(\mathbf{r},\sigma) = \epsilon_i\psi_i(\mathbf{r},\sigma), \quad (21)$$

where the Lagrange multipliers $\{\epsilon_i\}$ are the HF orbital energies. The optimized orbitals are the solutions of the one-particle HF equations (21). This is a fairly tractable and very useful method, but it "neglects correlation" which can be important when the interactions between electrons are strong.

2.5 DENSITY VARIATIONAL PRINCIPLE

The density variational principle, rooted in the wavefunction variational principle, creates a firm foundation for DFT. It was first proposed by Hohenberg and Kohn [2] and then built into a practical computational scheme—the KS equations—by Kohn and Sham [3]. An alternative and elegant proof of this theory was provided by Levy [14]. As it is simpler and more constructive, we follow Levy's constrained search approach here.

Theorem 1 (Hohenberg-Kohn Theorem I)**.** *Let $n_g(\mathbf{r})$ be the, possibly degenerate, ground-state density for an N-electron system. Then $n_g(\mathbf{r})$ determines not only the electron number,*

$$N = \int d^3\mathbf{r}\, n_g(\mathbf{r}), \quad (22)$$

but also the external potential $v(\mathbf{r})$ and thus the Hamiltonian $\hat{H}$ and thereby everything about this system (e.g., the ground- and excited-state wavefunctions).

Proof. Let Ψ_{n_g} be any normalized antisymmetric wavefunction that yields the ground-state density $n_g(\mathbf{r})$. Then, by the wavefunction variational principle,

$$\left\langle\Psi_{n_g}\middle|\hat{H}\middle|\Psi_{n_g}\right\rangle = \left\langle\Psi_{n_g}\middle|\hat{T}+\hat{V}_{\mathrm{ee}}\middle|\Psi_{n_g}\right\rangle + \int d^3\mathbf{r}\, n_g(\mathbf{r})v(\mathbf{r}) \geq E_g, \quad (23)$$

where E_g is the lowest eigenvalue of $\hat{H}$, that is, the ground-state energy. Following the Levy constrained search, let $\Psi_{n_g}^{\min}$ be a normalized antisymmetric wavefunction, which yields the ground-state density $n_g(\mathbf{r})$ and minimizes $\langle\Psi_{n_g}|\hat{T}+\hat{V}_{\mathrm{ee}}|\Psi_{n_g}\rangle$. Then, $\Psi_{n_g}^{\min}$ is the, possibly degenerate, ground-state wavefunction Ψ_g and the result of the Levy constrained search over all wavefunctions that yield a given ground-state density $n_g(\mathbf{r})$. Thus $n_g(\mathbf{r})$ determines $\Psi_{n_g}^{\min}$ and so in principle Ψ_g. □

Theorem 2 (Hohenberg-Kohn Theorem II: Density Variational Principle)**.** *There exists a universal functional of the density, $F[n]$, such that for any N-representable density (i.e., any density that comes from some wavefunction for an N-electron system) $n(\mathbf{r})$, which yields a given number of electrons N, the energy functional is,*

$$E_v[n] = F[n] + \int d^3\mathbf{r} n(\mathbf{r}) v(\mathbf{r}) \geq E_g. \tag{24}$$

The equality holds when the density $n(\mathbf{r})$ is the, possibly degenerate, ground-state density for the external potential $v(\mathbf{r})$.

Proof. Define the universal functional

$$F[n] = \left\langle \Psi_n^{\min} \middle| \hat{T} + \hat{V}_{\text{ee}} \middle| \Psi_n^{\min} \right\rangle, \tag{25}$$

the minimum expectation value of the wavefunction yielding a density $n(\mathbf{r})$, which is not necessarily a ground-state density. Then,

$$E_v[n] = \left\langle \Psi_n^{\min} \middle| \hat{H} \middle| \Psi_n^{\min} \right\rangle = F[n] + \int d^3\mathbf{r} n(\mathbf{r}) v(\mathbf{r}) \geq E_g, \tag{26}$$

by the Rayleigh-Ritz variational principle [15]. The equality, $\Psi_n^{\min} = \Psi_g$ and $\langle \Psi_g | \hat{H} | \Psi_g \rangle = E_g$, stands if and only if density $n(\mathbf{r})$ is the ground-state density $n_g(\mathbf{r})$. □

The earlier Levy constrained search can, in principle, deliver the ground-state densities; however, it is even harder to implement in practice than the original wavefunction variational principle.

To find the ground-state density $n(\mathbf{r})$ and energy $E_g[n]$ of a system of N-electrons in an external potential $v(\mathbf{r})$, there are two problems that must be solved:

(i) Find a practical way to construct or approximate the universal density functional $F[n]$, which is independent of $v(\mathbf{r})$ and the same for all systems.

(ii) Given $F[n]$, which is defined and differentiable for all N-representable densities $n(\mathbf{r})$, that is, densities coming from an antisymmetric N-electron wavefunction, find the minimization of $E_v[n] = F[n] + \int d^3\mathbf{r} n(\mathbf{r}) v(\mathbf{r})$.

Problem (ii) leads to the Euler equation for the density and the KS equations.

Before we move to the derivation of the KS equations, let us introduce the functional derivative $\delta G/\delta n(\mathbf{r})$ that tells us how a functional $G[n]$ changes under a small variation $\delta n(\mathbf{r})$:

$$\delta G = \int d^3\mathbf{r} \left(\frac{\delta G}{\delta n(\mathbf{r})} \right) \delta n(\mathbf{r}). \tag{27}$$

For example, the local density approximation (LDA) exchange energy can be written as,

$$E_x^{\text{LDA}}[n] = A_x \int d^3\mathbf{r} n(\mathbf{r})^{4/3}, \tag{28}$$

where A_x is a constant. Then,

$$\delta E_x^{\text{LDA}} = A_x \int d^3\mathbf{r} \left\{ [n(\mathbf{r}) + \delta n(\mathbf{r})]^{4/3} - n(\mathbf{r})^{4/3} \right\}$$
$$= A_x \int d^3\mathbf{r} \frac{4}{3} n(\mathbf{r})^{1/3} \delta n(\mathbf{r}), \tag{29}$$

which gives the functional derivative of E_x^{LDA},

$$\frac{\delta E_x^{\text{LDA}}}{\delta n(\mathbf{r})} = A_x \frac{4}{3} n(\mathbf{r})^{1/3}. \tag{30}$$

In DFT, functionals like $F[n] = \int d^3\mathbf{r} f(\mathbf{r}, n, \nabla n, \nabla^2 n, \ldots, \nabla^l n)$ are often encountered. For such functionals, their functional derivatives [4,16,17] can be written as,

$$\delta F[n]/\delta n = \frac{\partial f}{\partial n} - \nabla \cdot \frac{\partial f}{\partial \nabla n} + \nabla^2 \frac{\partial f}{\partial \nabla^2 n} + \cdots + (-1)^l \nabla^l \frac{\partial f}{\partial \nabla^l n}. \tag{31}$$

2.6 KOHN-SHAM NONINTERACTING SYSTEM

Minimizing $E_v[n]$, for a fixed external potential $v(\mathbf{r})$, subject to the constraint $\int d^3\mathbf{r} n(\mathbf{r}) = N$ (N is fixed) leads to

$$\delta \left\{ E_v[n] - \mu \int d^3\mathbf{r} n(\mathbf{r}) \right\} = 0, \tag{32}$$

by the method of Lagrange multipliers in which μ is the Lagrange multiplier. Since

$$\delta \left\{ E_v[n] - \mu \int d^3\mathbf{r} n(\mathbf{r}) \right\} = \int d^3\mathbf{r} \left[\frac{\delta F[n]}{\delta n(\mathbf{r})} + v(\mathbf{r}) - \mu \right] \delta n(\mathbf{r}), \tag{33}$$

we obtain the Euler equation,

$$\frac{\delta F[n]}{\delta n(\mathbf{r})} + v(\mathbf{r}) = \mu. \tag{34}$$

Eq. (33) shows that the external potential $v(\mathbf{r})$ is uniquely determined by the ground-state density (or by any one of them, if the ground state is degenerate). The functional derivative $\delta F/\delta n(\mathbf{r})$ is defined via Eq. (33) for all densities which are *v-representable*, that is, come from antisymmetric N-electron ground-state wavefunctions for some choice of external potential $v(\mathbf{r})$.

Now, let us consider an auxiliary system of noninteracting electrons moving in some external potential $v_s(\mathbf{r})$ (Kohn-Sham noninteracting system). For this system, $\hat{V}_{\text{ee}}$ of Eq. (25) vanishes so $F[n]$ of Eq. (25) reduces to,

$$T_s[n] = \min_{\Psi \to n} \left\langle \Psi \middle| \hat{T} \middle| \Psi \right\rangle = \left\langle \Phi_n^{\min} \middle| \hat{T} \middle| \Phi_n^{\min} \right\rangle, \tag{35}$$

where $\Phi_n^{\min}$ is a single Slater determinant (or a combination of) noninteracting wavefunction, delivering the minimum of $\langle \hat{T} \rangle$ for a given density $n(\mathbf{r})$. Thus from Eq. (33),

$$\frac{\delta T_s}{\delta n(\mathbf{r})} + v_s(\mathbf{r}) = \mu. \tag{36}$$

Here, two assumptions have been made:

(i) $n(\mathbf{r})$ is both interacting and noninteracting v-representable, that is, the Kohn-Sham system of noninteracting electrons moving in an external potential $v_s(\mathbf{r})$ yields the ground-state density of the corresponding physical system, an interacting system of N-electrons moving in an external potential $v(\mathbf{r})$.

(ii) $\mu_s = \mu$. If there is any difference between μ_s and μ, the chemical potentials for noninteracting and interacting systems of the same density, it can be absorbed into $v_s(\mathbf{r})$.

Now let us turn to the definition of the exchange-correlation energy $E_{\mathrm{xc}}[n]$,

$$F[n] = T_s[n] + U[n] + E_{\mathrm{xc}}[n], \tag{37}$$

where $U[n]$ is the Hartree electrostatic energy given by,

$$U[n] = \frac{1}{2}\int d^3\mathbf{r} \int d^3\mathbf{r}' \frac{n(\mathbf{r})n(\mathbf{r}')}{|\mathbf{r}-\mathbf{r}'|}. \tag{38}$$

From the previous two assumptions we must have

$$v_s(\mathbf{r}) = v(\mathbf{r}) + u([n];\mathbf{r}) + v_{\mathrm{xc}}([n];\mathbf{r}), \tag{39}$$

where

$$u([n];\mathbf{r}) = \frac{\delta U[n]}{\delta n(\mathbf{r})} \tag{40}$$

is the Hartree potential, and

$$v_{\mathrm{xc}}([n];\mathbf{r}) = \frac{\delta E_{\mathrm{xc}}[n]}{\delta n(\mathbf{r})} \tag{41}$$

is the exchange-correlation potential. We see that the external potential $v_s(\mathbf{r})$ (the Kohn-Sham potential or effective potential) is determined by $n(\mathbf{r})$ to within an additive constant.

Thus for an auxiliary system of noninteracting N-electrons moving in the Kohn-Sham potential $v_s(\mathbf{r})$, the Hamiltonian is

$$\hat{H}_s = \sum_{i=1}^{N} -\frac{1}{2}\nabla_i^2 + v_s(\mathbf{r}_i), \tag{42}$$

and the ground-state wavefunction Ψ_s is the solution of the auxiliary many-electron Schrödinger equation,

$$\hat{H}_s\Psi_s = E_s\Psi_s. \tag{43}$$

Since the Kohn-Sham system is noninteracting, the Kohn-Sham wavefunction Ψ_s is a Slater determinant

$$\Psi_s = \frac{1}{\sqrt{N!}} \det[\psi_1, \ldots, \psi_N], \tag{44}$$

where ψ_i is the Kohn-Sham orbital satisfying the simplified one-electron Kohn-Sham equation,

$$\hat{h}_s \psi_i = \left[-\frac{1}{2}\nabla^2 + v_s(\mathbf{r}) \right] \psi_i = \epsilon_i \psi_i. \tag{45}$$

The ground-state density can be calculated via the occupied ψ_i

$$n(\mathbf{r}) = \sum_{i=1}^{N} |\psi_i|^2 \tag{46}$$

and the Kohn-Sham noninteracting kinetic energy is

$$T_s[n] = \left\langle \Psi_s \left| \sum_{i=1}^{N} -\frac{1}{2}\nabla_i^2 \right| \Psi_s \right\rangle = \sum_{i=1}^{N} \left\langle \psi_i \left| -\frac{1}{2}\nabla^2 \right| \psi_i \right\rangle. \tag{47}$$

Then, the total energy of the real system with interacting electrons is given by

$$\begin{aligned} E &= E_s - U[n] + E_{\mathrm{XC}} - \int d^3\mathbf{r} n(\mathbf{r}) v_{\mathrm{XC}}([n];\mathbf{r}) \\ &= T_s + U[n] + E_{\mathrm{XC}} + \int d^3\mathbf{r} n\, v(\mathbf{r}). \end{aligned} \tag{48}$$

2.7 EXCHANGE ENERGY AND CORRELATION ENERGY

$E_{\mathrm{xc}}[n]$ is the sum of distinct exchange and correlation terms,

$$E_{\mathrm{XC}}[n] = E_{\mathrm{X}}[n] + E_{\mathrm{c}}[n], \tag{49}$$

where $E_{\mathrm{X}}[n]$ is defined as [18]

$$E_{\mathrm{X}}[n] = \left\langle \Phi_n^{\min} \left| \hat{V}_{\mathrm{ee}} \right| \Phi_n^{\min} \right\rangle - U[n]. \tag{50}$$

When $\Phi_n^{\min}$ is a single Slater determinant, Eq. (50) is the usual Fock integral applied to the Kohn-Sham orbitals, that is, it differs from the HF exchange energy only to the extent that the Kohn-Sham orbitals differ from the HF orbitals for a given system or density (in the same way that $T_s[n]$ differs from the HF kinetic energy). Note that

$$\left\langle \Phi_n^{\min} \middle| \hat{T} + \hat{V}_{ee} \middle| \Phi_n^{\min} \right\rangle = T_s[n] + U[n] + E_X[n], \tag{51}$$

and that, in the one-electron ($\hat{V}_{ee} = 0$) limit,

$$E_X[n] = -U[n] \quad (N = 1), \tag{52}$$

and

$$v_X([n];\mathbf{r}) = \frac{\delta E_x}{\delta n(\mathbf{r})} = -u([n];\mathbf{r}) \quad (N = 1), \tag{53}$$

that is, the exchange energy and its potential completely cancel the corresponding Hartree electrostatic self-repulsion, $U[n]$ and $u([n];\mathbf{r})$, respectively.

The correlation energy is defined as,

$$\begin{aligned} E_C[n] &= F[n] - \{T_s[n] + U[n] + E_X[n]\} \\ &= \left\langle \Psi_n^{\min} \middle| \hat{T} + \hat{V}_{ee} \middle| \Psi_n^{\min} \right\rangle - \left\langle \Phi_n^{\min} \middle| \hat{T} + \hat{V}_{ee} \middle| \Phi_n^{\min} \right\rangle \\ &= T_C[n] + V_C[n], \end{aligned} \tag{54}$$

where

$$T_C[n] = \left\langle \Psi_n^{\min} \middle| \hat{T} \middle| \Psi_n^{\min} \right\rangle - \left\langle \Phi_n^{\min} \middle| \hat{T} \middle| \Phi_n^{\min} \right\rangle, \tag{55}$$

is the kinetic energy contribution to the correlation energy which arises from the difference between the real kinetic energy T and the Kohn-Sham noninteracting kinetic energy T_s, and,

$$V_C[n] = \left\langle \Psi_n^{\min} \middle| \hat{V}_{ee} \middle| \Psi_n^{\min} \right\rangle - \left\langle \Phi_n^{\min} \middle| \hat{V}_{ee} \middle| \Phi_n^{\min} \right\rangle \tag{56}$$

is the potential energy contribution to the correlation energy. Since $\Psi_n^{\min}$ is the wavefunction that minimizes $\langle \hat{T} + \hat{V}_{ee} \rangle$ and yields density $n(\mathbf{r})$, Eq. (54) shows that

$$E_C[n] \le 0. \tag{57}$$

Since $\Phi_n^{\min}$ is the wavefunction that minimizes $\langle \hat{T} \rangle$ and yields the same density $n(\mathbf{r})$, we can deduce that

$$T_C[n] \ge 0, \tag{58}$$

and thus

$$V_C[n] \le 0 \tag{59}$$

via Eqs. (54), (57).

2.8 COUPLING-CONSTANT INTEGRATION

In Eqs. (50), (54), which give the formal definitions for E_x and E_c, not much intuitive or physical insight into the exchange and correlation energies is provided. However, the coupling-constant integration (or the adiabatic connection) [19,20], which smoothly connects the KS noninteracting system to the interacting physical system, can shed light on the exchange-correlation energy.

Let $\Psi_n^{\min,\lambda}$ be those normalized antisymmetric wavefunctions, which yield density $n(r)$ and minimize the expectation value of $\hat{T} + \lambda\hat{V}_{ee}$ for each λ between 0 and 1. When $\lambda = 1$, $\Psi_n^{\min,\lambda}$ is $\Psi_n^{\min}$, the true interacting correlated ground-state wavefunction for density $n(r)$, and when $\lambda = 0$, $\Psi_n^{\min,\lambda}$ is $\Phi_n^{\min}$, the noninteracting Kohn-Sham wavefunction for the same density $n(r)$. Varying λ at fixed $n(r)$ amounts to varying the external potential $v_\lambda(r)$. For $\lambda = 0$, $v_\lambda(r)$ is the Kohn-Sham effective potential, and for $\lambda = 1$, $v_\lambda(r)$ is $v(r)$. We normally assume a smooth, "adiabatic connection" between the interacting and noninteracting ground states as λ varies from 1 to 0.

Based on the previous analysis, Eqs. (49), (50), (54) can be rewritten as

$$\begin{aligned} E_{xc}[n] &= \langle\Psi_n^{\min,\lambda}|\hat{T} + \lambda\hat{V}_{ee}|\Psi_n^{\min,\lambda}\rangle|_{\lambda=1} \\ &\quad - \langle\Psi_n^{\min,\lambda}|\hat{T} + \lambda\hat{V}_{ee}|\Psi_n^{\min,\lambda}\rangle|_{\lambda=0} - U[n] \\ &= \int_0^1 d\lambda \frac{d}{d\lambda}\langle\Psi_n^{\min,\lambda}|\hat{T} + \lambda\hat{V}_{ee}|\Psi_n^{\min,\lambda}\rangle - U[n]. \end{aligned} \tag{60}$$

via the Hellmann-Feynman theorem (Eq. 13), the coupling-constant integration [19, 20] of E_{xc} can be expressed as:

$$E_{xc}[n] = \int_0^1 d\lambda\langle\Psi_n^{\min,\lambda}|\hat{V}_{ee}|\Psi_n^{\min,\lambda}\rangle - U[n]. \tag{61}$$

Getting more insight for the exchange-correlation energy necessitates the introduction of one-particle (ρ_1) and two-particle (ρ_2) reduced density matrices, which can be used to evaluate the N-electron expectation value of a sum of one-body operators like $\hat{T}$, or a sum of two-body operators like $\hat{V}_{ee}$.

The one-particle (ρ_1) and two-particle (ρ_2) reduced density matrices are defined in terms of the wavefunction Ψ as:

$$\begin{aligned} \rho_1(r'\sigma, r\sigma) \equiv N \sum_{\sigma_2,\ldots,\sigma_N} \int d^3r_2 \ldots \int d^3r_N \\ \times \Psi^*(r'\sigma, r_2\sigma_2, \ldots, r_N\sigma_N)\Psi(r\sigma, r_2\sigma_2, \ldots, r_N\sigma_N) \end{aligned} \tag{62}$$

and

$$\begin{aligned} \rho_2(r', r) \equiv N(N-1) \sum_{\sigma_1,\ldots,\sigma_N} \int d^3r_3 \ldots \int d^3r_N \\ \times |\Psi^*(r'\sigma_1, r\sigma_2, \ldots, r_N\sigma_N)|^2. \end{aligned} \tag{63}$$

Then the observables of spin density, kinetic energy, and potential energy can be expressed in terms of these one- or two-particle reduced density matrices

$$n_\sigma(r) = \rho_1(r\sigma, r\sigma), \tag{64}$$

$$\langle \hat{T} \rangle = -\frac{1}{2}\sum_\sigma \int d^3r \frac{\partial}{\partial r}\cdot\frac{\partial}{\partial r'}\rho_1(r'\sigma, r\sigma)|_{r'=r}, \tag{65}$$

and

$$\langle \hat{V}_{\text{ee}} \rangle = \frac{1}{2}\int d^3r \int d^3r' \frac{\rho_2(r', r)}{|r'-r|}. \tag{66}$$

Eqs. (65), (66) suggest that the energy of a many-electron system can be determined by the one- and two-particle reduced density matrices. Unfortunately, no convenient principle (analogous to the wavefunction variational principle or the Hohenberg-Kohn-Sham density variational principle used to calculate wavefunctions or densities) has been developed that would allow direct calculation of these density matrices without first requiring calculation of the wavefunction.

From the probability feature implied by the wavefunction, $\rho_2(r', r)d^3r'd^3r$ can be interpreted as the joint probability of finding an electron in d^3r' at r', and an electron in d^3r at r [5]. In the independent events approximation,

$$\rho_2(r', r)d^3r'd^3r = [n(r')d^3r'][n(r)d^3r], \tag{67}$$

which implies the Hartree approximation and reduces the expectation value of $\langle \hat{V}_{\text{ee}} \rangle$ to the Hartree energy $U[n]$. By standard probability theory, the joint probability $\rho_2(r', r)d^3r'd^3r$ is the product of the probability of finding an electron in d^3r at $r(n(r)d^3r)$ and the conditional probability of finding an electron in d^3r at $r, n(r)d^3r$, and $n_2(r, r')d^3r'$, the conditional probability of finding an electron in d^3r' at r', given that there is an electron at r:

$$\rho_2(r', r) = n(r)n_2(r, r'). \tag{68}$$

The quantity $n_2(r, r')$ can be interpreted as the average density of electrons at r', given that there is an electron at r. Clearly then

$$\int d^3r' n_2(r, r') = N - 1. \tag{69}$$

Furthermore, the $n_2(r, r')$ can be divided into two parts

$$n_2(r, r') = n(r') + n_{\text{xc}}^{\lambda}(r, r'), \tag{70}$$

to define $n_{\text{xc}}^{\lambda}(r, r')$, the density at r' of the exchange-correlation hole [20] about an electron at r. Then from Eq. (69) we have

$$\int d^3r' n_{\text{xc}}^{\lambda}(r, r') = -1, \tag{71}$$

which says that, if an electron is definitely at r, it is missing from the rest of the system.

By substituting Eqs. (70), (68) into Eq. (66) and using Eq. (61), we obtain

$$E_{\mathrm{xc}}[n] = \frac{1}{2}\int d^3r \int d^3r' \frac{n(r)\bar{n}_{\mathrm{xc}}(r,r')}{|r-r'|}, \tag{72}$$

where

$$\bar{n}_{\mathrm{xc}}(r,r') = \int_0^1 n_{\mathrm{xc}}^{\lambda}(r,r')d\lambda \tag{73}$$

is the coupling-constant averaged exchange-correlation hole density, which describes how the presence of an electron at r depletes the total density of other electrons at r'. Eq. (72) shows that the exchange-correlation energy is just the electrostatic interaction between each electron and the coupling-constant averaged exchange-correlation hole, which surrounds it. There are three effects responsible for creating the hole: (1) self-interaction correction, a classical effect which guarantees that an electron cannot interact with itself; (2) the Pauli exclusion principle, which tends to keep any two electrons of the same spin apart in space; and (3) the Coulomb repulsion, which tends to keep any two electrons apart in space. Effects (1) and (2) are responsible for the exchange energy, which is present even at $\lambda = 0$, while effect (3) is responsible for the correlation energy, which arises only for $\lambda \neq 0$.

Among all the values of λ, the point $\lambda = 0$ is interesting and important. If $\Psi_n^{\min,\lambda=0}$ is a single Slater determinant, as is typical, then the one- and two-electron density matrices at $\lambda = 0$ can be constructed explicitly from the occupied Kohn-Sham spin-orbitals $\psi_{\alpha\sigma}(r)$:

$$\rho_1^{\lambda=0}(r'\sigma, r\sigma) = \sum_{\alpha=1}^{N} \psi_{\alpha\sigma}^*(r')\psi_{\alpha\sigma}(r), \tag{74}$$

$$\rho_2^{\lambda=0}(r',r) = n(r)n(r') + n(r)n_x(r,r'), \tag{75}$$

where

$$n_{\mathrm{x}}(r,r') = n_{\mathrm{xc}}^{\lambda=0}(r,r') = -\sum_{\sigma} \frac{|\rho_1^{\lambda=0}(r'\sigma, r\sigma)|^2}{n(r)} \tag{76}$$

is the exact exchange-hole density. Eq. (76) shows that

$$n_{\mathrm{x}}(r,r') \leq 0 \tag{77}$$

implying the exact exchange energy

$$E_{\mathrm{x}}[n] = \frac{1}{2}\int d^3r \int d^3r' \frac{n(r)n_{\mathrm{x}}(r,r')}{|r-r'|} \tag{78}$$

is also negative, and can be written as the sum of up-spin and down-spin contributions:

$$E_X = E_X^{\uparrow} + E_X^{\downarrow} < 0. \tag{79}$$

There are several other features about the exchange hole that need to be stressed. The first one is the sum rule:

$$\int d^3r' n_X(r, r') = -1, \tag{80}$$

and the second one is the "on-top" exchange-hole density [21]:

$$n_X(r, r) = -\frac{n_{\uparrow}^2(r) + n_{\downarrow}^2(r)}{n(r)}. \tag{81}$$

Eq. (81) shows that the "on-top" exchange-hole density is determined by just the local spin densities at position r.

After we define the exchange hole, the correlation hole density can be defined by

$$\bar{n}_{XC}(r, r') = n_X(r, r') + \bar{n}_C(r, r'), \tag{82}$$

which satisfies the sum rule

$$\int d^3r' \bar{n}_C(r, r') = 0. \tag{83}$$

This means the correlation hole density $\bar{n}_C(r, r')$ must have positive as well as negative contributions. Moreover, unlike the exchange-hole density $n_X(r, r')$, the exchange-correlation hole density $\bar{n}_{XC}(r, r')$ can be positive.

2.9 UNIFORM ELECTRON GAS AND SLOWLY VARYING DENSITIES

The uniform electron gas [22], a relatively simple system, is a paradigm for solid-state physics, and also for DFT, as the hydrogen atom is for all of atomic physics. In this system, the electron density $n(r)$ is uniform over space, and thus the electron number is infinite. The negative charge of the electrons is neutralized by a rigid uniform positive background. The noninteracting kinetic energy and exchange energy per electron for a spin-unpolarized electron gas of uniform density n are evaluated here. The corresponding energies for the spin-polarized case can then be found using spin scaling relations [23]. Exact analytic expression for $e_c^{unif}(n)$, the correlation energy per electron of the uniform gas, are known only in extreme limits.

The Kohn-Sham potential $v_s(r)$ is taken to be zero since it must be uniform or constant by symmetry. Boundary conditions are imposed within a cube of volume $V \to \infty$, that is, the orbitals are required to be equal on opposite faces of the cube. (Presumably any choice of boundary conditions would give the same answer as $V \to \infty$.) The Kohn-Sham orbitals are then plane waves $\exp(i\mathbf{k} \cdot \mathbf{r})/\sqrt{V}$, with momenta or wavevectors $\mathbf{k}$ and energies $k^2/2$. The number of orbitals of both spins in a volume d^3k of wavevector space is $2[V/(2\pi)^3]d^3k$, by an elementary geometrical argument [24].

Let $N = nV$ be the number of electrons in volume V. These electrons occupy the N lowest Kohn-Sham spin-orbitals, that is, those with $k < k_F$:

$$N = 2\sum_k \theta(k_F - k) = 2\frac{V}{(2\pi)^3}\int_0^{k_F} dk4\pi k^2 = V\frac{k_F^3}{3\pi^2}, \tag{84}$$

where k_F is called the Fermi wavevector. The Fermi wavelength $2\pi/k_F$ is the shortest de Broglie wavelength for the noninteracting electrons. Clearly

$$n = \frac{k_F^3}{3\pi^2} = \frac{3}{4\pi r_s^3}, \tag{85}$$

where we have introduced the Seitz radius r_s, the radius of a sphere which on average contains one electron. The kinetic energy of an orbital is $k^2/2$, and the average kinetic energy per electron is

$$t_s(n) = \frac{2}{N}\sum_k \theta(k_F - k)\frac{k^2}{2} = \frac{3}{5}\frac{k_F^2}{2} = \frac{3}{10}(3\pi^2 n)^{2/3} = \frac{3}{10}\frac{(9\pi/4)^{2/3}}{r_s^2}. \tag{86}$$

To evaluate the exchange energy, we need the Kohn-Sham one-electron density matrix for electrons of spin σ, as defined in Eq. (74):

$$\begin{aligned}\rho_1^{\lambda=0}(\mathbf{r}+\mathbf{u}\sigma, \mathbf{r}\sigma) &= \sum_k \theta(k_F - k)\frac{\exp(-i\mathbf{k}\cdot(\mathbf{r}+\mathbf{u}))}{\sqrt{V}}\frac{\exp(i\mathbf{k}\cdot\mathbf{r})}{\sqrt{V}} \\ &= \frac{k_F^3}{2\pi^2}\frac{\sin(k_F u) - k_F u\cos(k_F u)}{(k_F u)^3}.\end{aligned} \tag{87}$$

The exchange-hole density at distance u from an electron is, by Eq. (76),

$$n_x(u) = -2\frac{|\rho_1^{\lambda=0}(\mathbf{r}+\mathbf{u}\sigma, \mathbf{r}\sigma)|^2}{n(r)}, \tag{88}$$

which ranges from $-n/2$ at $u = 0$ (where all other electrons of the same spin are excluded by the Pauli principle) to 0 (like $1/u^4$) as $u \to \infty$. The exchange energy per electron is

$$e_x^{\text{unif}}(n) = \int_0^\infty du 2\pi u n_x(u) = -\frac{3}{4\pi}k_F = -\frac{3}{4\pi}(3\pi^2 n)^{1/3} = -\frac{3}{4\pi}\frac{(9\pi/4)^{1/3}}{r_s}. \tag{89}$$

Since the self-interaction correction vanishes for the diffuse orbitals of the uniform gas, all of this exchange energy is due to the Pauli exclusion principle.

To characterize the inhomogeneity of the real system, which is missing in the ideal system of the uniform electron gas, we define the reduced density gradient

$$s = \frac{|\nabla n|}{2k_F n} = \frac{|\nabla n|}{2(3\pi^2)^{1/3} n^{4/3}}, \tag{90}$$

with its square

$$p = s^2, \tag{91}$$

and the reduced Laplacian

$$q = \frac{|\nabla^2 n|}{4(3\pi^2)^{2/3} n^{5/3}}. \tag{92}$$

The reduced density gradient s measures how fast and how much the density varies on the scale of the local Fermi wavelength $2\pi/k_F$. The slowly varying density requires not only that the density varies slowly in space, i.e., s is small, but also that q and q/s are small.

For a slowly varying spin-unpolarized density, the exact exchange energy has the fourth-order gradient expansion [25]

$$E_{\mathrm{x}}[n] = \int d^3 r n e_{\mathrm{x}}^{\mathrm{unif}} \left[1 + \frac{10}{81} p + \frac{146}{2025} q - \frac{73}{146} pq + Dp^2\right], \tag{93}$$

where $D \approx 0$. The correlation energy has the second-order gradient expansion,

$$E_{\mathrm{c}}[n_\uparrow, n_\downarrow] = \int d^3 r \left[n e_{\mathrm{c}}^{\mathrm{unif}} + C_{\mathrm{c}}(n_\uparrow, n_\downarrow) \frac{|\nabla n|^2}{n^{4/3}} \right], \tag{94}$$

where $C_{\mathrm{c}}(n_\uparrow, n_\downarrow)$ is a known gradient coefficient [26–28] and

$$C_{\mathrm{c}}(n_\uparrow, n_\downarrow) = \frac{0.066725}{16\phi^2} \left(\frac{\pi}{3}\right)^{1/3} \tag{95}$$

in the high-density or weak-interaction limit [29]. Here

$$\phi(\zeta) = \frac{1}{2}[(1+\zeta)^{2/3} + (1-\zeta)^{2/3}], \tag{96}$$

and ζ is the relative spin polarization

$$\zeta = \frac{n_\uparrow - n_\downarrow}{n_\uparrow + n_\downarrow}. \tag{97}$$

3 APPROXIMATIONS OF EXCHANGE AND CORRELATION ENERGY DENSITY FUNCTIONAL

3.1 APPROACHES OF DEVELOPING APPROXIMATIONS

As was discussed in the previous section, DFT is exact in principle for the ground-state energy and electron density, but in practice the exchange-correlation energy has to be approximated as a functional of electron density. There are two major approaches to developing the approximations. The empirical approach determines parameters of a proposed approximation by fitting to some carefully chosen and calibrated experimental datasets. The nonempirical approach, however, avoids the use of experimental datasets and instead relies on the mathematical properties of the exact functional, maximizing the transferability of the proposed approximation for different properties and systems.

The systematic, nonempirical way to improve approximations to the exact density functional for the exchange-correlation energy can be described by the following connected steps: (i) Prove the existence of the exact functional and derive exact formal expressions for it, as described in the previous section. (ii) Discover mathematical properties of the exact functional (exact constraints), which include limits, scaling relations, equalities, and bounds. (iii) Develop approximate but computationally tractable forms for the approximations at various levels of flexibility, including the local spin density approximation (LSDA) [30–33], generalized gradient approximations (GGAs) [34–42], meta-GGAs [43–55], hyper-GGAs [30,56–74], and random-phase approximation (RPA) [75–77]. These approximations will be discussed in following paragraphs. (iv) Impose the "exact constraints" from step (ii) on each form, as appropriate. (v) If a form still retains some flexibility, fit it to energies/densities of appropriate norms, systems for which the form can be expected to be highly accurate. The period 1965–79 focused on step (i), the 1980s on step (ii), 1965–98 on step (iii), and the 1970s to the present on step (iv), whereas the first appropriate norm for step (v) was the uniform electron gas (1965) [3]. A recent example of steps (i–v) is the strongly constrained and appropriately normed (SCAN) meta-GGA [50], which respects all 17 exact constraints suitable for a meta-GGA and works well for diversely bonded molecules and materials [78–86]. Section 4 will focus on the derivation of SCAN.

The empirical approach, which has become increasingly popular, is to skip or downplay steps (ii), (iv), and (v) and to use experimental data sets for fitting, which are often of greatest interest to the user. Empirical functionals may be enough to interpolate between similar systems and properties, but references [87,88] suggest that only incorporation of the exact constraints can result in reliable prediction over the enormous space of possible molecules, materials, and properties. Machine-learned density functionals [89,90] are even more strongly interpolative than familiar empirical functionals, but they will remain even less predictive until they are taught to respect the exact constraints.

3.2 JACOB'S LADDER OF DENSITY FUNCTIONAL THEORY

According to the ingredients used in approximations to the exchange-correlation energy,

$$E_{\mathrm{xc}}[n] = \int d^3 r n(r) e_{\mathrm{xc}}([n], r), \tag{98}$$

Table 1 Jacob's Ladder of DFT

	Heaven (Chemical Accuracy)	
$n, \nabla n, \nabla^2 n, \tau, \psi_{\text{occ}}, \psi_{\text{unocc}}$	Rung 5	Example: RPA
$n, \nabla n, \nabla^2 n, \tau, \psi_{\text{occ}}$	Rung 4	Example: hybrid functionals
$n, \nabla n, \nabla^2 n, \tau$	Rung 3	Meta-GGA's
$n, \nabla n$	Rung 2	GGA's
n	Rung 1	LSDA
	Earth (Hartree Theory)	

Note: Spin has been suppressed for simplicity of notation.

where $e_{\text{xc}}([n], r)$ is an energy per electron at point r, a Jacob's Ladder stretching from the Hartree world up to the heaven of chemical accuracy was proposed to categorize the approximations [91] as shown in Table 1.

The right-hand column of Table 1 includes the names for each rung, LSDA, GGA, and meta-GGA, etc., while the left-hand column gives the ingredients needed for the approximations of the corresponding rung (e.g., n and ∇n for GGA's). The approximations in the lowest three rungs are semilocal since their ingredients, n, ∇n, $\nabla^2 n$, and τ only requiring the information at local point r or infinitesimally around r. $\tau = \tau_\uparrow + \tau_\downarrow$ is the Kohn-Sham kinetic energy density, where $\tau_\sigma = \frac{1}{2}\sum_{i=1}^{\text{occ}} |\nabla \psi_{i\sigma}(r)|^2$ with $\sigma = \uparrow, \downarrow$. The highest two rungs are nonlocal due to the explicit appearance of the occupied or unoccupied orbitals in the functionals. Climbing from a lower rung to a higher rung usually results in the increase of accuracy. However, from semilocal rungs to nonlocal ones, the computational cost increases dramatically.

The lowest rung in the Jacob's ladder is represented by the simplest approximation for the exchange-correlation energy, the so-called LSDA [3] which uses only local spin densities $n_\uparrow$ and $n_\downarrow$ to calculate the exchange-correlation energy:

$$E_{\text{xc}}^{\text{LSDA}}[n_\uparrow, n_\downarrow] = \int d^3 r n(r) e_{\text{xc}}^{\text{unif}}(n_\uparrow, n_\downarrow), \tag{99}$$

where $e_{\text{xc}}^{\text{unif}}(n_\uparrow, n_\downarrow)$ is the exchange-correlation energy per electron for the uniform electron gas. The exchange part $e_{\text{x}}^{\text{unif}}(n_\uparrow, n_\downarrow)$ in $e_{\text{xc}}^{\text{unif}}(n_\uparrow, n_\downarrow)$ has an analytic expression of Eq. (89). For the correlation part $e_{\text{c}}^{\text{unif}}(n_\uparrow, n_\downarrow)$; however, there exist three widely used parametrizations (PZ [30], VWN [31], and PW92 [32]) that fit to the same high accuracy Quantum Monte Carlo (QMC) data points [92]. Sun et al. [33] and Bhattarai et al. [93] presented a correlation energy of the uniform electron gas from an interpolation between high- and low-density limits with minimum fitting to the QMC data. The LSDA predicts reasonable but too-short lattice constants for solids, good surface energies for simple metals (but with substantial error cancelation between exchange and correlation), and molecular atomization energies that are unacceptably high.

The gradient expansion approximation (GEA) [3,29] might appear to be the natural next step beyond LSDA, which adds the second-order gradient expansion terms of Eqs. (93), (94) to LSDA for the exchange and correlation energies, respectively, and therefore offers systematic corrections to LSDA. However, due to the fact that the exchange-correlation hole of GEA, $\bar{n}_{xc}^{GEA}(r, r+u)$, is far less realistic at large u than the corresponding LSDA model [34,94,95], the added second-order gradient expansion terms deteriorate from E_{xc}^{LSDA}. By removing these spurious large-u behaviors, Perdew and Wang were able to construct the Perdew-Wang (PW) 91 [35,36,96,97] GGA for E_{xc}. Compared with the LSDA, the GGA [94] includes additional semilocal information, the gradients of the spin densities, and can be written as:

$$E_{xc}^{GGA}[n_\uparrow, n_\downarrow] = \int d^3r n(r) e_x^{unif}(n) F_{xc}(n_\uparrow, n_\downarrow, \nabla n_\uparrow, \nabla n_\downarrow), \tag{100}$$

where F_{xc} is the enhancement factor. With the advent of GGAs [34,34–42], DFT has become popular not only in solid-state physics, but also in quantum chemistry. However, the accuracy of the commonly used GGAs is still limited. For example, the standard nonempirical Perdew-Burke-Ernzerhof (PBE) GGA [37] predicts reasonable but too-long lattice constants, surface energies that are better than LSDA for exchange alone and correlation alone but worse for their sum, and improved atomization energies with errors of about 10 kcal/mol on average. Since the commonly used GGAs overcorrect the lattice constant over LSDA in general, there have been efforts to design "GGAs for solids" (e.g., AM05 [98] and PBEsol [99]), which typically predict good lattice constants and surface energies, but rather poor atomization energies. Such effort confirm a dilemma [100], that is, a GGA can be rather accurate for structures or energies, but not both. This dilemma reflects a formal limitation: A GGA cannot satisfy all the known exact constraints appropriate to a semilocal functional, to be discussed in detail in Section 4.

Then, Meta-GGA is a natural way to improve accuracy further by making use of additional semilocal information (e.g., the Laplacian of the density $\nabla^2 n_\sigma$ or the kinetic energy densities τ_σ):

$$E_{xc}^{MGGA}[n_\uparrow, n_\downarrow] = \int d^3r n(r) e_x^{unif}(n) F_{xc}(n_\uparrow, n_\downarrow, \nabla n_\uparrow, \nabla n_\downarrow, \nabla^2 n_\uparrow, \nabla^2 n_\downarrow, \tau_\uparrow, \tau_\downarrow). \tag{101}$$

Usually, MGGA functionals include either the Laplacian of the density [101–103] or the kinetic energy density [43–55]. Some of them, however, include both ∇^2 and τ [42,104,105]. Because the Laplacian of the density becomes singular at the nuclei due to the cusp condition, the kinetic energy density is the preferred choice for meta-GGAs. If not mentioned explicitly, meta-GGAs discussed in this chapter are based only on the kinetic energy density.

There are two formal reasons for the inclusion of τ. First, it arises naturally in the Taylor expansion of the exact spherically averaged exchange hole near the reference point [106]. Second, the use of τ provides a simple and straightforward way to make a correlation functional exactly one-electron self-interaction free [107]. The inclusion of the kinetic energy density also enables meta-GGAs to have the flexibility to satisfy more exact constraints and thus circumvent the "structure or energy" dilemma experienced by GGAs. By using a dimensionless variable $\alpha = (\tau - \tau^W)/\tau^{unif}$, where $\tau^{unif} = (3/10)(3\pi^2)^{2/3} n^{5/3}$ is the kinetic energy density of the uniform electron gas and $\tau^{W} = |\nabla n|^2/8n$ is the von Weizsäcker kinetic energy density that is exact for single-orbital systems, meta-GGAs can recognize the slowly varying densities ($\alpha \approx 1$, characterizing metallic bonds), the single-orbital systems ($\alpha = 0$, characterizing covalent single bonds), and noncovalent bonds with $\alpha \gg 1$ between two closed shells [50,78,108]. For noncovalent bonds, for example the van der Waals bond of the Ar dimer around equilibrium, τ^{W} is zero at the bond center by symmetry, τ^{unif} scales as $n^{5/3}$ which is small, and τ is large due to the occupation of antibonding molecular orbitals and α is thus $\gg 1$ for noncovalent bonds. α is directly related to the electron localization function with $1/(1+\alpha^2)$, and therefore identifies different chemical bonds [109,110]. Due to the semilocal feature of the kinetic energy density in computation, meta-GGAs are comparable to GGAs in terms of efficiency. τ, however, is determined by KS orbitals, which are nonlocal functionals of the electron density, and meta-GGAs, therefore, are intrinsically nonlocal functionals. Based on this observation, Nazarov and Vignale showed that the exchange-correlation kernel derived from meta-GGAs within the adiabatic TDDFT can provide a nonlocality—which is missing in GGAs and needed for accurate descriptions of the excitonic effect in crystals [111].

The recently developed nonempirical SCAN meta-GGA was constructed to take advantage of the previous discoveries. SCAN is the first meta-GGA that is fully constrained, obeying all 17 known exact constraints that a semilocal functional can. It is also exact or nearly exact for a set of "appropriate norms," including rare-gas atoms and nonbonded interactions. SCAN recognizes different chemical bonds via α and treats them appropriately with different GGAs. SCAN thus predicts accurate geometries and energies of diversely bonded molecules and materials (including covalent, metallic, ionic, hydrogen, and van der Waals bonds), significantly and systematically improving at comparable efficiency over its predecessors, the GGAs.

For the highest two rungs in the Jacob ladder of DFT, the Kohn-Sham orbitals are explicitly used in functionals. The hyper-GGA at the fourth rung adds another ingredient to GGA: the exact exchange energy density, a fully nonlocal functional of the occupied Kohn-Sham orbitals:

$$E_{\text{xc}}^{\text{HGGA}}[n_\uparrow, n_\downarrow] = \int d^3r n(r) e_{\text{xc}}^{\text{unif}}(n) F_{\text{xc}}(n_\uparrow, n_\downarrow, \nabla n_\uparrow, \nabla n_\downarrow, \nabla^2 n_\uparrow, \nabla^2 n_\downarrow, \tau_\uparrow, \tau_\downarrow, e_x^{\text{exact}}), \tag{102}$$

where

$$e_x^{\text{exact}}(r) = \frac{1}{2}\sum_\sigma \int d^3r' \left(\frac{n_\sigma}{n}\right) \frac{n_x(r'\sigma, r\sigma)}{|r' - r|} \tag{103}$$

can be calculated from the occupied Kohn-Sham orbitals $\psi_{i\sigma}$ (see Eq. 76). The simplest construction of HGGA is Becke's exact exchange mixing approach [56,57] based on the "adiabatic connection" (also called "coupling-constant integration") [19,20], which significantly improves accuracy to the order of 2 kcal/mole or smaller for atomization energies. The kinds of HGGAs constructed by globally mixing a small fraction of exact exchange in density functional

Table 2 Error Statistics of Different Properties for Various Density Functionals

	LSDA	PBE	SCAN	M06L	B3LYP	HSE/PBE0
Enthalpies of formation (kcal/mol) of the 223 G3 molecules						
ME	−121.4	−21.7	−4.6	−1.6	3.3	−4.9
MAE	121.4	22.2	5.7	5.2	4.8	6.8
Lattice constants (Å) of 20 solids						
ME	−0.081	0.051	0.007	0.015	–	0.031
MAE	0.081	0.059	0.016	0.069	–	0.033
Jellium surface exchange energies (%) for r_s =2, 3, 4, 6						
MRE	45.8	−20.9	−24.3	−75.9	–	−15.7
MARE	45.8	20.9	24.5	75.8	–	15.7
Jellium surface exchange-correlation energies (%) for r_s =2, 3, 4, 6						
MRE	−5.5	−8.13	−0.45	17.6	–	−5.6
MARE	5.5	8.13	1.25	17.6	–	5.6

Notes: ME, MAE, MRE, and MARE denote mean error, mean absolute error, mean relative error, and mean absolute relative error, respectively. The error statistics of B3LYP for solids and surfaces are not given as B3LYP is seldom used for solid studies.
Data are from Table II of Sun J, Ruzsinszky A, Perdew JP. Strongly constrained and appropriately normed semilocal density functional. Phys Rev Lett 2015;115(3):036402. https://doi.org/10.1103/PhysRevLett.115.036402. The G3 of B3LYP and PBE0 is from Sun J, Haunschild R, Xiao B, Bulik IW, Scuseria GE, Perdew JP. Semilocal and hybrid meta-generalized gradient approximations based on the understanding of the kinetic-energy-density dependence. J Chem Phys 2013;138(4):044113. https://doi.org/10.1063/1.4789414. The lattice constants of 30 solids for HSE were from Schimka L, Harl J, Kresse G. Improved hybrid functional for solids: the HSEsol functional. J Chem Phys 2011;134(2):024116. https://doi.org/10.1063/1.3524336.

approximations for the exchange and correlation functional are called global hybrid functionals. Several hyper-exchange-correlation functionals have been suggested [56,57,61,107,112–118].

At the fifth rung, exact exchange can be combined with exact partial correlation, making use not only of the occupied Kohn-Sham orbitals but also the unoccupied. The RPA using Kohn-Sham orbitals [27,119–123] in this rung shows some promising features for solids and surfaces [124] while it generally underestimates the atomization energy for molecules [125]. Other examples are the interaction strength interpolation (ISI) [126–128] and double hybrids [72,73].

4 THE STRONGLY CONSTRAINED AND APPROPRIATELY NORMED META-GENERALIZED GRADIENT APPROXIMATION

Recent years have seen dramatic developments of meta-GGAs [43–55]. Despite the advantages of meta-GGAs, the GGAs and GGA-based nonlocal functionals are still dominant in condensed matter physics and quantum chemistry. There are several possible reasons for this: (1) The meta-GGA appeared much later than the GGA and was then overshadowed by the hybrid-GGA, which has resulted in few meta-GGAs being available in popular computer codes. (2) The kinetic energy density brings complexities with the advantages and there was no clear way to understand and analyze the dependence of meta-GGAs on the kinetic energy density until recently [51]. (3) The semilocal feature of meta-GGAs in computation masks their intrinsic nonlocality. The semilocality is advantageous in computation, but not in concept, because nonlocality is needed to describe long-range interactions, for example, strongly correlated electron systems such as the dissociated H_2 molecule, and the van der Waals interaction between two well-separated objects. The last point drives most of developers and users from GGAs directly to GGA-based nonlocal functionals. It is the second point however, that causes meta-GGAs to be overlooked in favor of more mature GGAs. With their comparable computational efficiency and advanced accuracy, and with the better understanding on the kinetic energy density dependence [50,51,78,108], it is expected that meta-GGAs and meta-GGA-based functionals will replace GGAs and GGA-based functionals in the near future.

For GGAs and meta-GGAs, the enhancement factor is a good tool for understanding the functionals. The exchange part of a GGA or meta-GGA for spin-unpolarized electron densities can be written as,

$$E_{\mathrm{X}}^{\mathrm{sl}} = \int d^3 r n(r) e_{\mathrm{X}}^{\mathrm{unif}}(n) F_{\mathrm{X}}(s, \alpha), \tag{104}$$

where "sl" stands for semilocal. The spin scaling relationship can be applied to the previous equation to get the expression for spin-polarized electron densities [23].

The enhancement factor, $F_{\mathrm{X}}(s)$, for a GGA is only dependent on s. As shown in Fig. 1, the GGA enhancement factors must recover the second-order gradient expansion, that is, $F_{\mathrm{X}} = 1 + \mu s$, for small s. The first-principles coefficient derived by Antoniewicz and Kleinman [129] for densities that vary slowly over space is $\mu_{GE} = 10/81 = 0.12346$. However, $\mu_{PBE} = 0.21951$ was used in PBE, in order to recover the good LSDA linear response of the UEG [37]. It was shown recently that the coefficient μ_{AE} must be about twice μ_{GE} ($\mu_{AE} = 2.109\mu_{GE}$) if a GGA is required to satisfy the large-Z asymptotic expansion of the exchange energy of neutral atoms with nuclear number Z [130]. The B88 GGA [41], which is popular in chemistry, has $\mu_{B88} = 2.208\mu_{GE}$, very close to μ_{AE}. However, it is argued that μ_{GE} is more relevant and must be restored for solids, as in the PBEsol GGA [99]. Fig. 1 also shows that the enhancement factors of GGAs monotonically increase with s to values much larger than 1 (e.g., 1.804 for PBE) for $s \gg 1$ in order to better describe atoms and molecules which are far from having slowly varying densities [37,51]. Such monotonic dependence favors less compact systems than LSDA does, by lowering the energy of the set of isolated atoms relative to that of the corresponding molecules, lowering a surface energy, or enlarging lattice constants. With their choices of the gradient coefficient, PBE and B88 have much stronger dependences on s than PBEsol does. In general PBE performs better than PBEsol for calculating energetics. However, PBE usually overcorrects the lattice constants of solids in comparison to LSDA, while PBEsol yields lattice constants between those

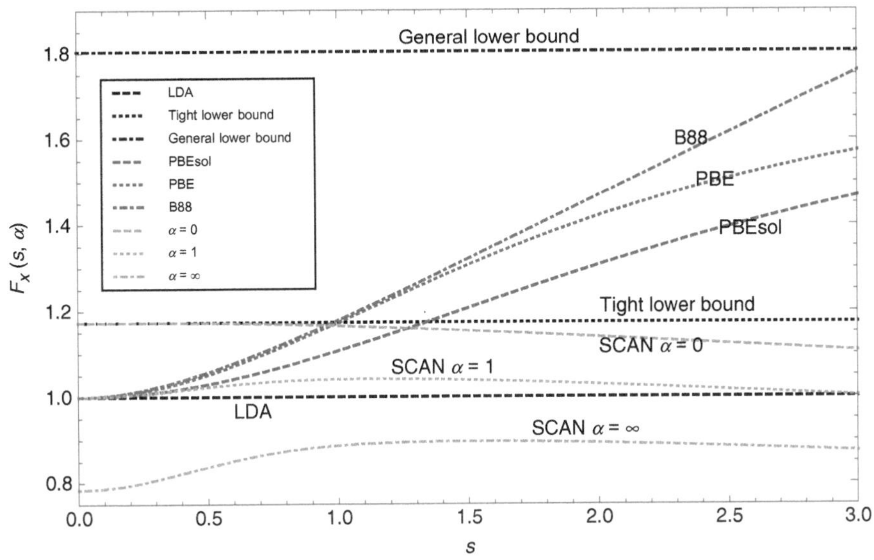

FIG. 1

The enhancement factor of LDA, PBEsol [99], PBE [37], B88 [41], and SCAN [50] for the exchange energy with its tight lower bound and the general lower bound.

of LSDA and PBE and is the best choice for solid studies. This is the fundamental limit to GGAs that leads to the " structure or energy" dilemma, that is, GGAs have a limited form that is not flexible enough to simultaneously satisfy different exact constraints, for example, the second-order gradient expansion (μ_{GE} for PBEsol) and the good LSDA linear response of the UEG (μ_{PBE} for PBE).

For two-electron systems, where $\alpha = 0$, it has been proved [131] that an optimal lower bound on the exchange energy will be satisfied for all possible densities if and only if

$$F_X(s, \alpha = 0) \leq 1.174, \tag{105}$$

and it is conjectured that

$$F_X(s, \alpha) \leq F_X(s, \alpha = 0) \leq 1.174. \tag{106}$$

Eq. (106) is much tighter than the older bound 1.804 [132] that is respected by PBE and PBEsol, as demonstrated by their enhancement factors for $s \to \infty$. At the GGA level, the tight bound of Eq. (106) is not compatible with any second-order gradient expansion ($F_X = 1 + \mu s + \cdots$), since the enhancement factor would be too small in comparison with those of PBE and B88 for a large range of s as shown in Fig. 1. A GGA respecting the tight bound would not be useful for atoms and molecules, even less so than PBEsol [99].

However, at the meta-GGA level, the restriction on the s-dependence can be compensated by the freedom brought by the α-dependence. With this help, the SCAN meta-GGA was able to be designed to satisfy Eq. (106) and 16 other exact constraints, including those satisfied by LSDA and PBE. The 17 exact constraints are suitable for semilocal density functionals, but cannot be simultaneously satisfied at the GGA level as discussed in the previous paragraph. Fig. 1 shows that SCAN is radically different from those previous semilocal density functionals. SCAN's enhancement is monotonically decreasing, rather than increasing as in previous GGAs like PBE, with s, whereas it decreases monotonically with α. It has been shown [51] that the monotonically decreasing α-dependence of an enhancement factor has similar effects on the lattice constant prediction of solids as the monotonically increasing s-dependence.

The 17 exact constraints include [50]: for exchange, (1) negativity, (2) spin scaling, (3) uniform density scaling, (4) fourth-order gradient expansion, (5) nonuniform density scaling, and (6) tight bound for two-electron densities; for correlation, (7) nonpositivity, (8) second-order gradient expansion, (9) uniform density scaling to the high-density limit, (10) uniform density scaling to the low-density limit, (11) zero correlation energy for any one-electron spin-polarized density, and (12) nonuniform density scaling; for both together, (13) size extensivity, (14) general Lieb-Oxford bound, (15) weak dependence upon relative spin polarization in the low-density limit, (16) static linear response of the uniform electron gas, and (17) Lieb-Oxford bound for two-electron densities. PBE satisfies 11 of them, 1–3, 7–10, and 13–16, while LSDA satisfies 9, 1–3, 6–7, and 13–16. It is hard to argue that any one exact constraint is more important than any other as they are concerted to make SCAN highly transferable between different chemical environments and thus gain the predictive power.

SCAN, however, is not the unique form that satisfies the 17 exact constraints. To fix the remaining flexibility, SCAN uses the appropriate norms as guides from one exact constraint across to another. Eq. (72) shows that the exchange-correlation energy is an electrostatic interaction between the electron density at each position $\mathbf{r}$ and the density at position $\mathbf{r}'$ of the exchange-correlation hole around the electron at $\mathbf{r}$. Only when the exact hole is well localized around its electron, can the semilocal approximation of Eq. (101) expected to be accurate. In an atom, the exact hole and its separate exchange and correlation contributions must remain within the density of the atom. So a good semilocal approximation should be accurate for the exchange energy alone and for correlation energy alone in an atom, which is therefore an appropriate norm [50,133] to which a semilocal functional can be fitted. (Spherical atoms with closed magnetic subshells are especially appropriate because their noninteracting wavefunctions are unquestionably single Slater determinants.) The same is true of a uniform density or one that varies slowly over space.

In a multicenter system like a molecule, the exact exchange hole can delocalize over the available centers. But the exact exchange-correlation hole tends to be deeper and more localized around its electron than the exact exchange hole in every system [20,134]. In other words, there is an error cancelation between semilocal exchange and semilocal correlation [20,50,133]. This is why semilocal functionals often work well for exchange and correlation in molecules near equilibrium bond lengths. But this error cancelation is imperfect at best, so molecules are not appropriate norms for a semilocal functional. We regard semilocal functionals fitted to molecules as empirical. Indeed, there are well-known cases where the exact hole is completely delocalized, and no error cancelation can occur, such as the one-electron molecular ion H_2^+ stretched to infinite bond length, for which no semilocal functional can be expected to be accurate. The transition states for chemical reactions are also typically stretched radicals with delocalized exact exchange-correlation holes. Only fully nonlocal functionals can be accurate for such systems.

From the same viewpoint, the appropriate constraints [78] for a semilocal functional are those exact constraints that can be satisfied by a model exchange-correlation hole that is well localized around its electron [19,20]. Inappropriate constraints include asymptotic exchange and correlation potentials that are experienced by an electron that wanders far outside the electron density distribution, leaving its exact hole behind. The first nonempirical density functional for the exchange-correlation energy was the LDA. It was constructed to satisfy an appropriate norm, the electron gas of uniform density, before it was realized that it also satisfied appropriate constraints. Subsequent nonempirical functionals have added more appropriate norms and appropriate constraints. SCAN satisfies the following appropriate norms: (a) uniform and slowly varying densities, (b) the jellium surface energy, (c) the H atom, (d) the He atom and the limit of large atomic number for the rare-gas atoms, plus compressed Ar_2, and (e) the $Z \rightarrow \infty$ limit of the two-electron ion.

Without being fitted to any bonded system, SCAN accurately predicts diverse kinds of bonding [78], including even the effects of intermediate-range van der Waals interaction. SCAN performs much better than PBE for defects in semiconductors [78], surface properties of metals [79], seven phases of ice [78], liquid water [80], liquid and supercooled silicon [81], subtle structural distortions in ferroelectrics [82, 83], the formation energies and structural predictions for solids including transition-metal oxides [85], and the critical pressures for structural phase transitions in semiconductors [135]. For liquid water, SCAN is correct where PBE is qualitatively wrong (for the mass density, diffusion coefficient, atom-atom correlation, etc.). SCAN is the only density functional that correctly predicts the size of the band gap (or absence thereof), size and alignment of the spin moment on Cu, and transition from undoped insulator to doped metal for cuprate high-temperature superconductor materials [84], without free parameters. At a typically much lower computational cost, and without empiricism, the semilocal SCAN functional does some things better than (and others not as well as) the hybrid functionals [78].

For structural predictions, taking ferroelectric oxides as an example [83], SCAN usually predicts the local structure distortion of MO_6 octahedron (M stands for metal elements) characterized by the ratio of the longer bond length of metal-oxygen (stronger bonds) and the shorter one (weaker bonds) much closer to the experimental results than LSDA and PBE, schematically plotted in Fig. 2.

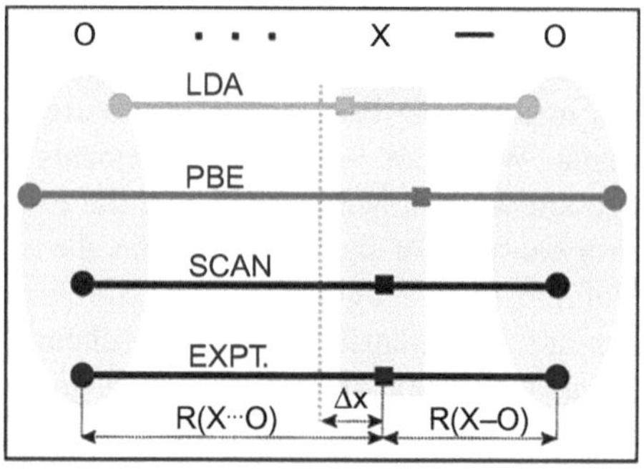

FIG. 2

Schematic diagram for description of the local asymmetric structural distortions using the LDA, PBE, and SCAN functionals.

5 OUTLOOK

Though being highly accurate for various types of bonding in systems around equilibrium geometry, SCAN is limited as a semilocal functional. Semilocal density functionals use a single integral to model the exchange-correlation energy, formally a double integral as Eq. (72) shows, and thus must fail for systems where the exchange-correlation hole is delocalized from its reference electron.

The long-range van der Waals (vdW) interaction is one such nonlocal effect. The vdW interaction originates from dynamic electron correlations, causing a net attraction between fragments [136]. It has minor effects in most bulk systems, becomes noticeable in some bulk solids like soft alkali metals [137], and is significant in sparse matter including molecular complexes, molecular crystals, layered materials, and surface-adsorbate systems. Fully accounting for the vdW interaction is achievable by high-level methods such as QMC [138], coupled-cluster singles and doubles with perturbative triples [CCSD(T)] [139], and adiabatic-connection fluctuation dissipation theorem within the RPA [75]. These methods are, however, only feasible for limited size systems because of high computational cost.

Aiming for a better efficiency, various approaches to account for the vdW interaction have been proposed within the framework of DFT [140–151]. Unlike other approaches, the vdW density functionals (vdW-DFs [144–149], VV10 [150], and rVV10 [151]) are truly first principles as they are only based on the electron density and its first derivative. Comparing with the Rutgers-Chalmers vdW-DFs [144–149], the VV10 and rVV10 nonlocal vdW correlation functionals are more flexible in form. Taking advantage of the flexibility of rVV10 and SCAN's capability of capturing different bonds, including even intermediate-range vdW interactions, SCAN + rVV10 turns out to be a successful versatile vdW density functional, with a high accuracy for different systems, including molecular complexes, bulk solids, benzene adsorbed on metal surfaces, the challenging layered materials, and the surface energies and work functions of transition metals [79,152].

Two other important nonlocal effects suffering semilocal density functionals (including SCAN) are the strong many-electron interaction (SMEI) originating from a degeneracy or near-degeneracy and the self-interaction error (SIE) due to the imperfect cancelation of the spurious classical Coulomb interaction between an electron and itself. These deficiencies prevent the DFT-based simulations from accurately describing key properties and systems important for applications, including band gaps, band alignments, stretched bonds, interactions with transition-metal catalysts, charge transfer, reaction barriers, and their spin dependencies. The H_2^+ and H_2 dissociation curves (Fig. 3) are the canonical examples for SIE and SMEI, respectively. Around the equilibrium separations, PBE and SCAN both give accurate total energies, with SCAN showing slightly higher accuracy. At the dissociation limit, however, both PBE and SCAN give significant errors for both H_2^+ and H_2.

For the one electron H_2^+ molecule the exact exchange-correlation hole reduces to the exact exchange hole, and the HF theory is exact across the whole dissociation curve. At the dissociation limit, the exact exchange hole of H_2^+ becomes delocalized across the two distant protons, whereas the semilocal functionals tend to localize the exchange-correlation hole around the reference electron, resulting in energies that are too negative. It is this localization of the exchange-correlation hole that is the source of the SIE in semilocal functionals. The H_2 molecule is not so straightforward. At the dissociation limit the system should remain in the singlet state, with the exact exchange hole spread across the two distant protons, similar to the exact exchange hole of the dissociated H_2^+. The HF description is fundamentally incorrect for the dissociated H_2 case, however, as the SMEI between the two electrons arising from the spin degeneracy tends to localize one electron around one proton when the

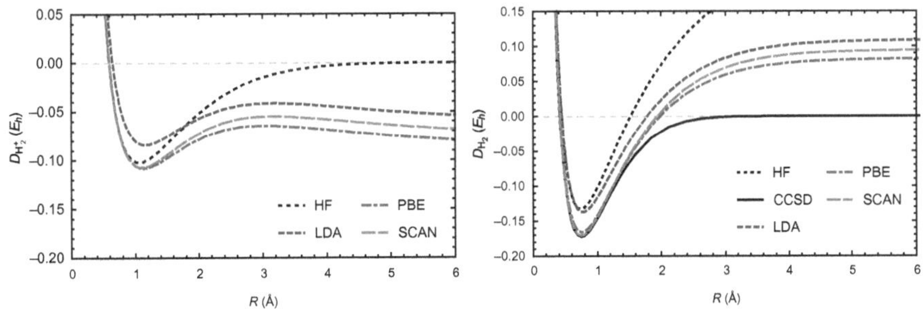

FIG. 3

The binding curves of the H_2^+ and the singlet H_2 molecules calculated using the LSDA/LDA, PBE, and SCAN density functionals plotted against the HF and coupled-cluster single and double (CCSD) (equivalent to full configuration interaction [FCI] for H_2) benchmark curves.

other electron is around the other proton. This strong interaction generates a deep correlation hole around the reference electron at one proton which, by the sum rule must integrate to 0 over all space. A positive hump therefore forms at the other proton, ensuring the sum rule is obeyed. The exchange-correlation holes of semilocal functionals are not deep enough to capture the full exact exchange-correlation hole generated by the SMEI and yield dissociation energies that are less negative than the true values. Spin symmetry breaking can recover the exact dissociation limit at the cost of introducing spin contamination through contributions from the higher lying triplet state that account for some of the SMEI correlation energy. The previous discussion can also be analyzed in terms of fractional charge and fractional spin [153], focusing on one of the component protons as an open system.

The previous examples highlight the errors caused by SIE and SMEI. A number of efforts have been made to treat them [30,60–63,66,74,121]. One of the first to systematically remove the SIE from the underlying semilocal functional was the Perdew and Zunger self-interaction correction (PZ-SIC) [30]. The PZ-SIC removes SIE from a semilocal functional on an orbital by orbital basis, and is thus 1-electron exact. However, the early versions of PZ-SIC were not unitarily invariant and degradations of the good performance are often seen in the underlying functionals for the prediction of thermochemical properties when PZ-SIC corrections are added [154]. The nodal surfaces of the orbital densities seen by semilocal density functionals, which are usually absent in the total electron density, are suspected to be responsible for the degradations [154]. SCAN recognizes single orbitals and treats them much better than previous semilocal density functionals [155]. The recent introduction of Fermi-Loewdin orbitals (FLOs) to SIC results in the FLO-SIC scheme that is unitarily invariant [63]. Thus, the FLO-SIC SCAN is a promising direction for treating SIE while retaining the good performance of SCAN for systems around equilibrium. FLO-SIC SCAN still depends on the spin symmetry breaking to treat SMEI. More recently, a localized orbital scaling correction (LOSC) was developed by Yang et al. [68,74] that shows promising results for treating SIE while preserving the performance of the underlying functional for systems around their equilibrium geometries. Tests of both FLO-SIC and LOSC were focused on molecules, however, and their performance on solids needs to be assessed further, especially for the uniform electron gas limit to check their universality for both molecules and solids.

A more empirical, and more efficient, way to deal with SIE and SMEI in systems with transition-metal elements is to add a parameter, U, that penalizes the delocalization of electrons in subshells of an atom. While effectively removing SIE and treating SMEI, this approximation is fundamentally undesirable as the parameters cannot easily be defined as a function of the system in question [156,157]. Other approaches have been developed to treat SIE and SMEI beyond those already mentioned, including the particle-hole RPA [75,76] and particle-particle RPA [77], for example, though these schemes are limited by their dependence on the unoccupied KS orbitals and an associated dramatic increase in computational cost.

The global hybrid density functional originally proposed by Becke [56,57] is another scheme which aims to treat nonlocal effects by mixing a semilocal functional with the nonlocal exact exchange. In its simplest (one parameter) version [58], the exchange-correlation energy is constructed as,

$$E_{\mathrm{xc}}^{\mathrm{GH}} = aE_{\mathrm{x}}^{\mathrm{exact}} + (1-a)E_{\mathrm{x}}^{\mathrm{sl}} + E_{\mathrm{c}}^{\mathrm{sl}}, \tag{107}$$

where GH denotes "global hybrid." The exact exchange mixing parameter a is determined empirically in the range (0, 1). Although global hybrid functionals have been widely adapted by the DFT user community, there are a number of problems associated with them. The first and foremost problem is that the mixing parameter is a position-independent constant and fixed empirically, which should instead be system dependent [158,159]. For example, SIE and SMEI favor different values of the global hybrid mixing, and so cannot be corrected simultaneously in a global hybrid scheme. The global mixing factor is required to be large to correctly treat SIE, as 100% of the exact exchange is required to exactly cancel the Coulomb self-interaction. This cancelation is clearly seen in HF theory, which includes 100% exact exchange and no correlation and is exact for the one electron H_2^+ molecule. A large fraction of exact exchange energy is disfavored in a global hybrid functional; however, because the semilocal correlation component requires a corresponding semilocal exchange component for accurate thermochemical properties, relying on an error cancelation between exchange and correlation terms. In contrast to SIE the mixing factor must be small to correctly treat SMEI as the more exact exchange is mixed the more the correlation is lost, an effect that can be seen in the HF description of H_2. On top of this, further developments using a global mixing fraction of exact exchange are limited, as the insufficient flexibility of the mixing parameter prevents the satisfaction of additional exact constraints.

The concept of local hybrid functional naturally extends the global hybrid functional by making the mixing factor position dependent [59,69]; proposed about two decades ago it can be expressed as,

$$e_{\mathrm{xc}}^{\mathrm{LH}} = a(\mathbf{r})e_{\mathrm{x}}^{\mathrm{exact}} + (1-a(\mathbf{r}))e_{\mathrm{x}}^{\mathrm{sl}} + e_{\mathrm{c}}^{\mathrm{sl}}, \tag{108}$$

where $a(\mathbf{r})$ is now a position-dependent function and LH stands for "local hybrid." Conceptually, local hybrid functionals solve the first serious problem associated with global hybrid functionals, the position and system independence of the global mixing parameter. Due to the increased complexity of the local mixing factor, however, relatively few local hybrid functionals [59,61,69,70] have been proposed, among which the Perdew-Staroverov-Tao-Scuseria (PSTS) local hybrid [61] stands as one of the most sophisticated, designed for both molecules and solids using the local mixing function to satisfy more exact constraints. The PSTS functional relies on spin symmetry breaking to properly recover the H_2 dissociation limit. In PSTS, two different local mixing factors were designed to recognize different abnormal regions where semilocal functionals are inaccurate, and combined to form the total local mixing factor. The first mixing function was designed for regions where exchange dominates correlation, important for the one electron, nonuniform high density, and rapidly varying limits. The second mixing function is targeted at open subsystems for which electron number fluctuates and the exact exchange-correlation hole integrates to a value greater than -1. The inclusion of e_x^{exact} within the mixing function ensures the satisfaction of exact constraints (including the 1-electron exactness and the asymptotic dependence of the exchange-correlation potential) and guarantees that all the exact constraints satisfied by the semilocal functionals are conserved. The second mixing function is less general and was constructed around the two prototypical model open systems, H_2^+ and H_2. The performance of the PSTS local hybrid for molecules was not satisfactory [160]; however, its accuracy for hydrogen-bonded systems was found to be much worse than existing popular semilocal and global hybrid functionals. This is possibly due to the Tao-Perdew-Staroverov-Scuseria (TPSS) meta-GGA [46] used in the construction of PSTS, which was constructed before the ability of α was fully recognized for meta-GGAs [50,51,78,108]. It is therefore reasonable to expect that using the SCAN meta-GGA as the basc lcad to a local hybrid that can correctly treat SIE and SMEI simultaneously, largely solving the difficult properties and systems affected by SIE and SMEI while at least matching the popular conventional density functionals for other properties.

REFERENCES

[1] Born M, Oppenheimer R. Zur Quantentheorie der Molekeln. Ann Phys 1927;389(20): 457–84. https://doi.org/10.1002/andp.19273892002.

[2] Hohenberg P, Kohn W. Inhomogeneous electron gas. Phys Rev 1964;136(3B):864–71. https://doi.org/10.1103/PhysRevB.7.1912.

[3] Kohn W, Sham LJ. Self-consistent equations including exchange and correlation effects. Phys Rev 1965;140(4A):1133–9. https://doi.org/10.1103/PhysRev.140.A1133.

[4] Parr RG, Weitao Y. Density-functional theory of atoms and molecules. New York, Oxford: Oxford University Press; 1989.

[5] Perdew JP, Kurth S. Density functionals: theory and application. Berlin: Springer; 1998.

[6] Burke K. Perspective on density functional theory. J Chem Phys 2012;136(15):150901. https://doi.org/10.1063/1.4704546.

[7] Becke AD. Perspective: fifty years of density-functional theory in chemical physics. J Chem Phys 2014;140(18):18A301. https://doi.org/10.1063/1.4869598.

[8] Yu HS, Li SL, Truhlar DG. Perspective: Kohn-Sham density functional theory descending a staircase. J Chem Phys 2016;145(13):130901. https://doi.org/10.1063/1.4963168.

[9] Mardirossian N, Head-Gordon M. Thirty years of density functional theory in computational chemistry: an overview and extensive assessment of 200 density functionals. Mol Phys 2017;8976:1362–3028. https://doi.org/10.1080/00268976.2017.1333644.

[10] Runge E, Gross EKU. Density-functional theory for time-dependent systems. Phys Rev Lett 1984;52(12):997–1000. https://doi.org/10.1103/PhysRevLett.52.997.

[11] Onida G, Reining L, Rubio A. Electronic excitations: density-functional versus many–body Green's-function approaches. Rev Mod Phys 2002;74(2):601–59. https://doi.org/10.1103/RevModPhys.74.601.

[12] Maitra NT. Perspective: fundamental aspects of time-dependent density functional theory. J Chem Phys 2016;144(22):220901. https://doi.org/10.1063/1.4953039.
[13] Atkins P, Friedman R. Molecular quantum mechanics. 4th ed. Oxford: Oxford University Press; 2005. ISBN 0-19-927498-3.
[14] Levy M. Universal variational functionals of electron densities, first-order density matrices, and natural spin-orbitals and solution of the V-representability problem. Proc Natl Acad Sci 1979;76(12):6062–5. https://doi.org/10.1073/pnas.76.12.6062.
[15] Levine IN. Quantum chemistry. Englewood Cliffs, NJ: Prentice-Hall; 1991.
[16] Arfken GB. Mathematical methods for physicist. New York, NY: Academic Press; 1980. ISBN 0123846544.
[17] Gelfand IM, Fomin SV. Calculus of variations. Prentice-Hall: Engelwood Cliffs, NJ; 1963. 240 pp. ISBN 0486414485.
[18] Levy M. Recent developments and applications of modern density functional theory. Amsterdam: Elsevier; 1996.
[19] Langreth DC, Perdew JP. The exchange-correlation energy of a metallic surface. Solid State Commun 1975;17(11):1425–9. https://doi.org/10.1016/0038-1098(75)90618-3.
[20] Gunnarsson O, Lundqvist BI. Exchange and correlation in atoms, molecules and solids by this spin-density-functional formalism. Phys Rev B 1976;13(10):4274–98. https://doi.org/10.1103/PhysRevB.13.4274.
[21] Ziegler T, Rauk A, Baerends EJ. On the calculation of multiplet energies by the Hartree-Fock-Slater method. Theor Chim Acta 1977;43(3):261–71. https://doi.org/10.1007/BF00551551.
[22] Fulde P. Electron correlations in molecules and solids. Berlin: Springer; 1993.
[23] Oliver GL, Perdew JP. Spin-density gradient expansion for the kinetic energy. Phys Rev A 1979;20(2):397–403. https://doi.org/10.1103/PhysRevA.20.397.
[24] Ashcroft NW, Mermin N. Solid state physics. New York, NY: Holt, Rinehart and Winston; 1976.
[25] Svendsen P, von Barth U. Gradient expansion of the exchange energy from second-order density response theory. Phys Rev B 1996;54(24):17402–13. https://doi.org/10.1103/PhysRevB.54.17402.
[26] Geldart DJW, Rasolt M. Exchange and correlation energy of an inhomogeneous electron gas at metallic densities. Phys Rev B 1976;13(4):1477–88. https://doi.org/10.1103/PhysRevB.13.1477.
[27] Langreth DC, Perdew JP. Theory of nonuniform electronic systems. I. Analysis of the gradient approximation and a generalization that works. Phys Rev B 1980;21(12):5469–93. https://doi.org/10.1103/PhysRevB.21.5469.
[28] Wang Y, Perdew JP. Spin scaling of the electron-gas correlation energy in the high--density limit. Phys Rev B 1991;43(11):8911–6. https://doi.org/10.1103/PhysRevB.43.8911.
[29] Ma SK, Brueckner KA. Correlation energy of an electron gas with a slowly varying high density. Phys Rev 1968;165(1):18–31. https://doi.org/10.1103/PhysRev.165.18.
[30] Perdew JP, Zunger A. Self-interaction correction to density-functional approximations for many-electron systems. Phys Rev B 1981;23(10):5048–79. https://doi.org/10.1103/PhysRevB.23.5048.
[31] Vosko SH, Wilk L, Nusair M. Accurate spin-dependent electron liquid correlation energies for local spin density calculations: a critical analysis. Can J Phys 1980;58(8):1200–11. https://doi.org/10.1139/p80-159.
[32] Perdew JP, Wang Y. Accurate and simple analytic representation of the electron-gas correlation-energy. Phys Rev B 1992;45(23):13244–9. https://doi.org/10.1103/PhysRevB.45.13244.

[33] Sun J, Perdew JP, Seidl M. Correlation energy of the uniform electron gas from an interpolation between high- and low-density limits. Phys Rev B 2010;81(8):085123. https://doi.org/10.1103/PhysRevB.81.085123.

[34] Perdew JP, Yue W. Accurate and simple density functional for the electronic exchange energy: generalized gradient approximation. Phys Rev B 1986;33(12):8800–2. https://doi.org/10.1103/PhysRevB.33.8800.

[35] Perdew JP, Burke K, Wang Y. Generalized gradient approximation for the exchange–correlation hole of a many-electron system. Phys Rev B 1996;54(23):16533–9. https://doi.org/10.1103/PhysRevB.54.16533.

[36] Perdew JP, Chevary J, Vosko S, Jackson K, Pederson MR, Singh D, et al. Atoms, molecules, solids, and surfaces: applications of the generalized gradient approximation for exchange and correlation. Phys Rev B 1992;46(11):6671–87. https://doi.org/10.1103/PhysRevB.48.4978.2.

[37] Perdew JP, Burke K, Ernzerhof M. Generalized gradient approximation made simple Phys Rev Lett 1996;77(18):3865–8.

[38] Langreth DC, Mehl MJ. Beyond the local-density approximation in calculations of ground-state electronic properties. Phys Rev B 1983;28(4):1809–34. https://doi.org/10.1103/PhysRevB.28.1809.

[39] Lee C, Yang W, Parr RG. Development of the Colle-Salvetti correlation-energy formula into a functional of the electron density. Phys Rev B 1988;37(2):785–9. https://doi.org/10.1103/PhysRevB.37.785.

[40] Hammer B, Hansen LB, Nørskov JK. Improved adsorption energetics within density-functional theory using revised Perdew-Burke-Ernzerhof functionals. Phys Rev B 1999;59(11):7413–21. https://doi.org/10.1103/PhysRevB.59.7413.

[41] Becke AD. Density-functional exchange-energy approximation with correct asymptotic behavior. Phys Rev A 1988;38(6):3098–100. https://doi.org/10.1103/PhysRevA.38.3098.

[42] Becke AD. A multicenter numerical integration scheme for polyatomic molecules. J Chem Phys 1988;88(4):2547–53. https://doi.org/10.1063/1.454033.

[43] Van Voorhis T, Scuseria GE. A novel form for the exchange-correlation energy functional. J Chem Phys 1998;109(2):400–10. https://doi.org/10.1063/1.476577.

[44] Ernzerhof M, Scuseria GE. Kinetic energy density dependent approximations to the exchange energy. J Chem Phys 1999;111(3):911–5. https://doi.org/10.1063/1.479374.

[45] Perdew JP, Kurth S, Zupan A, Blaha P. Accurate density functional with correct formal properties: a step beyond the generalized gradient approximation. Phys Rev Lett 1999;82(12):2544–7. https://doi.org/10.1103/PhysRevLett.82.5179.

[46] Tao J, Perdew JP, Staroverov VN, Scuseria GE. Climbing the density functional ladder: non-empirical meta-generalized gradient approximation designed for molecules and solids. Phys Rev Lett 2003;91(14):146401. https://doi.org/10.1103/PhysRevLett.91.146401.

[47] Zhao Y, Truhlar DG. A new local density functional for main-group thermochemistry, transition metal bonding, thermochemical kinetics, and noncovalent interactions. J Chem Phys 2006;125(19):194101. https://doi.org/10.1063/1.2370993.

[48] Perdew JP, Ruzsinszky A, Csonka GI, Constantin LA, Sun J. Workhorse semilocal density functional for condensed matter physics and quantum chemistry. Phys Rev Lett 2009;103(2):026403. https://doi.org/10.1103/PhysRevLett.103.026403.

[49] Tao J, Mo Y. Accurate semilocal density functional for condensed-matter physics and quantum chemistry. Phys Rev Lett 2016;117(7):1–6. https://doi.org/10.1103/PhysRevLett.117.073001.

[50] Sun J, Ruzsinszky A, Perdew JP. Strongly constrained and appropriately normed semilocal density functional. Phys Rev Lett 2015;115(3):036402. https://doi.org/10.1103/PhysRevLett.115.036402.
[51] Sun J, Xiao B, Ruzsinszky A. Communication: Effect of the orbital-overlap dependence in the meta generalized gradient approximation. J Chem Phys 2012;137(5):051101. https://doi.org/10.1063/1.4742312.
[52] Sun J, Haunschild R, Xiao B, Bulik IW, Scuseria GE, Perdew JP. Semilocal and hybrid meta-generalized gradient approximations based on the understanding of the kinetic-energy-density dependence. J Chem Phys 2013;138(4):044113. https://doi.org/10.1063/1.4789414.
[53] Mardirossian N, Head-Gordon M. Mapping the genome of meta-generalized gradient approximation density functionals: the search for B97M-V. J Chem Phys 2015;142(7):074111. https://doi.org/10.1063/1.4907719.
[54] Vela A, Pacheco-Kato JC, Gázquez JL, Campo JM, Trickey SB. Improved constraint satisfaction in a simple generalized gradient approximation exchange functional improved constraint satisfaction in a simple generalized gradient. J Chem Phys 2012;136(14):144115. https://doi.org/10.1063/1.3701132.
[55] del Campo JM, Gázquez JL, Trickey SB, Vela A. A new meta-GGA exchange functional based on an improved constraint-based GGA. Chem Phys Lett 2012;543:179–83. https://doi.org/10.1016/j.cplett.2012.06.025.
[56] Becke AD. Density-functional thermochemistry. III. The role of exact exchange. J Chem Phys 1993;98(7):5648–52. https://doi.org/10.1063/1.464913.
[57] Becke AD. A new mixing of Hartree-Fock and local density-functional theories. J Chem Phys 1993;98(2):1372. https://doi.org/10.1063/1.464304.
[58] Perdew JP, Ernzerhof M, Burke K. Rationale for mixing exact exchange with density functional approximations. J Chem Phys 1996;105(22):9982–5. https://doi.org/10.1063/1.472933.
[59] Cruz FG, Lam KC, Burke K. Exchange correlation energy density from Virial theorem. J Phys Chem A 1998;102(25):4911–7. https://doi.org/10.1021/jp980950v.
[60] Becke AD. Real-space post-Hartree-Fock correlation models. J Chem Phys 2005;122(6):064101. https://doi.org/10.1063/1.1844493.
[61] Perdew JP, Staroverov V, Tao J, Scuseria GE. Density functional with full exact exchange, balanced nonlocality of correlation, and constraint satisfaction. Phys Rev A 2008;78(5):052513. https://doi.org/10.1103/PhysRevA.78.052513.
[62] Becke AD. Density functionals for static, dynamical, and strong correlation. J Chem Phys 2013;138(7):074109. https://doi.org/10.1063/1.4790598.
[63] Pederson MR, Ruzsinszky A, Perdew JP. Communication: Self-interaction correction with unitary invariance in density functional theory. J Chem Phys 2014;140(12):121103. https://doi.org/10.1063/1.4869581.
[64] Toulouse J, Gori-Giorgi P, Savin A. A short-range correlation energy density functional with multi-determinantal reference. Theor Chem Acc 2005;114(4–5):305–8. https://doi.org/10.1007/s00214-005-0688-2.
[65] Sharkas K, Toulouse J, Savin A. Double-hybrid density-functional theory made rigorous. J Chem Phys 2011;134(6):064113. https://doi.org/10.1063/1.3544215.
[66] Kong J, Proynov E. Density functional model for nondynamic and strong correlation. J Chem Theory Comput 2015;12(1):133–43. https://doi.org/10.1021/acs.jctc.5b00801.
[67] Johnson ER, Contreras-García J. Communication: A density functional with accurate fractional-charge and fractional-spin behaviour for s-electrons. J Chem Phys 2011;135(8):081103. https://doi.org/10.1063/1.3630117.

[68] Li C, Cohen AJ, Mori-Sánchez P, Yang W. Local scaling correction for reducing delocalization error in density functional approximations. Phys Rev Lett 2015;114(5):053001. https://doi.org/10.1103/PhysRevLett.114.053001.

[69] Jaramillo J, Scuseria GE, Ernzerhof M. Local hybrid functionals. J Chem Phys 2003;118(3):1068–73. https://doi.org/10.1063/1.1528936.

[70] Bahmann H, Rodenberg A, Arbuznikov AV, Kaupp M. A thermochemically competitive local hybrid functional without gradient corrections. J Chem Phys 2007;126(1). https://doi.org/10.1063/1.2429058.

[71] Zhang IY, Rinke P, Perdew JP, Scheffler M. Towards efficient orbital-dependent density functionals for weak and strong correlation. Phys Rev Lett 2016;117(13):133002. https://doi.org/10.1103/PhysRevLett.117.133002.

[72] Zhang Y, Xu X, Goddard III WA. Doubly hybrid density functional for accurate descriptions of nonbond interactions, thermochemistry, and thermochemical kinetics. Proc Natl Acad Sci 2009;106(13):4963–8. https://doi.org/10.1073/pnas.0901093106.

[73] Grimme S. Semiempirical hybrid density functional with perturbative second-order correlation. J Chem Phys 2006;124(3):034108. https://doi.org/10.1063/1.2148954.

[74] Li C, Zheng X, Su NQ, Yang W. Localized orbital scaling correction for systematic elimination of delocalization error in density functional approximations. Natl Sci Rev 2017;5:203–15. https://doi.org/10.1093/nsr/nwx111.

[75] Eshuis H, Bates JE, Furche F. Electron correlation methods based on the random phase approximation. Theor Chem Acc 2012;131(1):1084. https://doi.org/10.1007/s00214-011-1084-8.

[76] Harl J, Schimka L, Kresse G. Assessing the quality of the random phase approximation for lattice constants and atomization energies of solids. Phys Rev B 2010;81(11):115126. https://doi.org/10.1103/PhysRevB.81.115126.

[77] Van Aggelen H, Yang Y, Yang W. Exchange-correlation energy from pairing matrix fluctuation and the particle-particle random phase approximation. J Chem Phys 2014;140(18):030501(R). https://doi.org/10.1063/1.4865816.

[78] Sun J, Remsing RC, Zhang Y, Sun Z, Ruzsinszky A, Peng H, et al. Accurate first-principles structures and energies of diversely bonded systems from an efficient density functional. Nat Chem 2016;8(9):831–6. https://doi.org/10.1038/nchem.2535.

[79] Patra A, Bates JE, Sun J, Perdew JP. Properties of real metallic surfaces: effects of density functional semilocality and van der Waals nonlocality. Proc Natl Acad Sci 2017:1–9. https://doi.org/10.1073/pnas.1713320114.

[80] Chen M, Ko HY, Remsing RC, Calegari MF, Santra B, Sun Z, et al. Ab initio theory and modeling of water. Proc Natl Acad Sci 2017;114(41):10846–51. https://doi.org/10.1073/pnas.1712499114.

[81] Remsing RC, Klein ML, Sun J. Dependence of the structure and dynamics of liquid silicon on the choice of density functional approximation. Phys Rev B 2017;96(2):024203. https://doi.org/10.1103/PhysRevB.96.024203.

[82] Paul A, Sun J, Perdew JP, Waghmare UV. Accuracy of first-principles interatomic interactions and predictions of ferroelectric phase transitions in perovskite oxides: energy functional and effective Hamiltonian. Phys Rev B 2017;95(5):054111. https://doi.org/10.1103/PhysRevB.95.054111.

[83] Zhang Y, Sun J, Perdew JP, Wu X. Comparative first-principles studies of prototypical ferroelectric materials by LDA, GGA, and SCAN meta-GGA. Phys Rev B 2017;96(3):035143. https://doi.org/10.1103/PhysRevB.96.035143.

[84] Furness JW, Zhang Y, Lane C, Buda IG, Barbiellini B, Markiewicz RS, et al. An accurate first-principles treatment of doping-dependent electronic structure of high-temperature cuprate superconductors. Commun Phys 2018;1(11):1–6. https://doi.org/10.1038/s42005-018-0009-4.
[85] Zhang Y, Kitchaev DA, Yang J, Chen T, Dacek ST, Sarmiento-Pérez RA, et al. Efficient first-principles prediction of solid stability: towards chemical accuracy. NPJ Comput Mater 2018;4(1):9. https://doi.org/10.1038/s41524-018-0065-z.
[86] Goerigk L, Hansen A, Bauer CA, Ehrlich S, Najibi A, Grimme S. A look at the density functional theory zoo with the advanced GMTKN55 database for general main group thermochemistry, kinetics and noncovalent interactions. Phys Chem Chem Phys 2017;19(48):32184–215. https://doi.org/10.1039/C7CP04913G.
[87] Medvedev MG, Bushmarinov IS, Sun J, Perdew JP, Lyssenko KA. Density functional theory is straying from the path toward the exact functional. Science 2017;355(6320):49–52. https://doi.org/10.1126/science.aah5975.
[88] Medvedev MG, Bushmarinov IS, Sun J, Perdew JP, Lyssenko KA. Response to comment on "Density functional theory is straying from the path toward the exact functional". Science 2017;356(6337):496. https://doi.org/10.1126/science.aam9550.
[89] Snyder JC, Rupp M, Hansen K, Müller KR, Burke K. Finding density functionals with machine learning. Phys Rev Lett 2012;108(25):253002. https://doi.org/10.1103/PhysRevLett.108.253002.
[90] Brockherde F, Vogt L, Li L, Tuckerman ME, Burke K, Müller KR. Bypassing the Kohn-Sham equations with machine learning. Nat Commun 2017;8(1):862. https://doi.org/10.1038/s41467-017-00839-3.
[91] Perdew JP, Schmidt K. Density functional theory and its applications to materials. American Institute of Physics; 2001.
[92] Ceperley DM, Alder BJ. Ground state of the electron gas by a stochastic method. Phys Rev Lett 1980;45(7):566–9. https://doi.org/10.1103/PhysRevLett.45.566.
[93] Bhattarai P, Patra A, Shahi C, Perdew JP. How accurate are the parametrized correlations energies of the uniform electron gas. Phys Rev B 2018;97(19):195128. https://doi.org/10.1103/PhysRevB.97.195128.
[94] Perdew JP. Accurate density functional for the energy: real-space cutoff of the gradient expansion for the exchange hole. Phys Rev Lett 1985;55(16):1665–8. https://doi.org/10.1103/PhysRevLett.55.1665.
[95] Perdew JP, Yue W. Erratum: Accurate and simple density functional for the electronic exchange energy generalized gradient approximation. Phys Rev B 1989;40(5):3399. https://doi.org/10.1103/PhysRevB.40.3399.
[96] Perdew JP. Electronic structure of solids. Berlin: Akadmie Verlag; 1991.
[97] Perdew JP, Burke K, Wang Y. Erratum: Generalized gradient approximation for the exchange-correlation hole of a many-electron system [Phys. Rev. B 54, 16 533 (1996)]. Phys Rev B 1998;57(23):14999. https://doi.org/10.1103/PhysRevB.57.14999.
[98] Armiento R, Mattsson AE. Functional designed to include surface effects in self-consistent density functional theory. Phys Rev B 2005;72(8):085108. https://doi.org/10.1103/PhysRevB.72.085108.
[99] Perdew JP, Ruzsinszky A, Csonka GI, Vydrov OA, Scuseria GE, Constantin LA, et al. Generalized gradient approximation for solids and their surfaces. Phys Rev Lett 2008;100(13):136406. https://doi.org/10.1103/PhysRevLett.100.136406.
[100] Zhang Y, Yang W. Comment on "Generalized gradient approximation made simple". Phys Rev Lett 1998;80(4):890. https://doi.org/10.1103/PhysRevLett.80.890.

[101] Engel E, Vosko SH. Fourth-order gradient corrections to the exchange-only energy functional: importance of 2n contributions. Phys Rev B 1994;50(15):10498–505. https://doi.org/10.1103/PhysRevB.50.10498.

[102] Filatov M, Thiel W. Exchange-correlation density functional beyond the gradient approximation. Phys Rev A 1998;57(1):189–99. https://doi.org/10.1103/PhysRevA.57.189.

[103] Proynov EI, Ruiz E, Vela A, Salahub DR. Determining and extending the domain of exchange and correlation functionals. Proc Int Symp Atomic Mol Condens Matter Theory Comput Methods 1995;56(29):61–78. https://doi.org/10.1002/qua.560560808.

[104] Becke AD, Roussel MR. Exchange holes in inhomogeneous systems: a coordinate-space model. Phys Rev A 1989;39(8):3761–7. https://doi.org/10.1126/science.37.944.183.

[105] Becke AD. Exploring the limits of gradient corrections in density functional theory. J Comput Chem 1999;20(63):010063. https://doi.org/10.1002/(SICI)1096-987X(19990115)20:1<63::AID-JCC8>3.0.CO;2-A.

[106] Becke AD. Hartree-Fock exchange energy of an inhomogeneous electron gas. Int J Quant Chem 1983;23(6):1915. https://doi.org/10.1002/qua.560230605.

[107] Becke AD. A new inhomogeneity parameter in density-functional theory. J Chem Phys 1998;109(6):2092–8. https://doi.org/10.1063/1.476722.

[108] Sun J, Xiao B, Fang Y, Haunschild R, Hao P, Ruzsinszky A, et al. Density functionals that recognize covalent, metallic, and weak bonds. Phys Rev Lett 2013;111(10):106401. https://doi.org/10.1103/PhysRevLett.111.106401.

[109] Becke AD, Edgecombe KE. A simple measure of electron localization in atomic and molecular systems. J Chem Phys 1990;92(9):5397–403. https://doi.org/10.1063/1.458517.

[110] Silvi B, Savin A. Classification of chemical-bonds based on topological analysis of electron localization functions. Nature 1994;371(6499):683–6. https://doi.org/10.1038/371683a0.

[111] Nazarov VU, Vignale G. Optics of semiconductors from meta-generalized-gradient-approximation-based time-dependent density-functional theory. Phys Rev Lett 2011;107(21):1–5. https://doi.org/10.1103/PhysRevLett.107.216402.

[112] Baker J, Muir M, Andzelm J. A study of some organic reactions using density functional theory. J Chem Phys 1995;102(5):2063–79. https://doi.org/10.1063/1.468728.

[113] Becke AD. Density-functional thermochemistry. IV. A new dynamical correlation functional and implications for exact-exchange mixing. J Chem Phys 1996;104(3):1040–6. https://doi.org/10.1063/1.470829.

[114] Becke AD. Density-functional thermochemistry. V. Systematic optimization of exchange-correlation functionals. J Chem Phys 1997;107(20):8554–60. https://doi.org/10.1063/1.475007.

[115] Ernzerhof M, Perdew JP, Burke K. Coupling-constant dependence of atomization energies. Int J Quant Chem 1997;64(3):285–95. https://doi.org/10.1002/(SICI)1097-461X(1997)64:3<285::AID-QUA2>3.0.CO;2-S.

[116] Schmider HL, Becke AD. Optimized density functionals from the extended G2 test set. J Chem Phys 1998;108(23):9624–31. https://doi.org/10.1063/1.476438.

[117] Schmider HL, Becke AD. Density functionals from the extended G2 test set: second-order gradient corrections. J Chem Phys 1998;109(19):8188–99. https://doi.org/10.1063/1.477481.

[118] Heyd J, Scuseria GE, Ernzerhof M. Hybrid functionals based on a screened Coulomb potential. J Chem Phys 2003;118(18):8207–15. https://doi.org/10.1063/1.1564060.

[119] Pitarke JM, Eguiluz AG. Jellium surface energy beyond the local-density approximation: self-consistent-field calculations. Phys Rev B 2001;63(4):045116. https://doi.org/10.1103/PhysRevB.63.045116.

[120] Furche F. Molecular tests of the random phase approximation to the exchange-correlation energy functional. Phys Rev B 2001;64(19):195120. https://doi.org/10.1103/PhysRevB.64.195120.

[121] Furche F. Developing the random phase approximation into a practical post-Kohn-Sham correlation model. J Chem Phys 2008;129(11):114105. https://doi.org/10.1063/1.2977789.

[122] Fuchs M, Gonze X. Accurate density functionals: approaches using the adiabatic-connection fluctuation-dissipation theorem. Phys Rev B 2002;65(23):235109. https://doi.org/10.1103/PhysRevB.65.235109.

[123] Kresse G, Harl J. Accurate bulk properties from approximate many-body techniques. Phys Rev Lett 2009;103(5):056401. https://doi.org/10.1103/PhysRevLett.103.056401.

[124] Schimka L, Harl J, Stroppa A, Grüneis A, Marsman M, Mittendorfer F, et al. Accurate surface and adsorption energies from many-body perturbation theory. Nat Mater 2010;9(9):741–4. https://doi.org/10.1038/nmat2806.

[125] Ruzsinszky A, Csonka GI, Perdew JP. The RPA atomization energy puzzle. J Chem Theory Comput 2010;6(1):127–34. https://doi.org/10.1021/ct900518k.

[126] Seidl M, Perdew JP, Kurth S. Simulation of all-order density-functional perturbation theory, using the second order and the strong-correlation limit. Phys Rev Lett 2000;84(22):5070–3. https://doi.org/10.1103/PhysRevLett.84.5070.

[127] Sun J. Extension to negative values of the coupling constant of adiabatic connection for interaction-strength interpolation. J Chem Theory Comput 2009;5(4):708–11. https://doi.org/10.1021/ct800515w.

[128] Vuckovic S, Irons TJP, Savin A, Teale AM, Gori-Giorgi P. Exchange-correlation functionals via local interpolation along the adiabatic connection. J Chem Theory Comput 2016;12(6):2598–610. https://doi.org/10.1021/acs.jctc.6b00177.

[129] Antoniewicz PR, Kleinman L. Kohn-Sham exchange potential exact to first order in $\rho(K)/\rho 0$. Phys Rev B 1985;31(10):6779–81. https://doi.org/10.1103/PhysRevB.31.6779.

[130] Elliott P, Burke K. Non-empirical "derivation" of B88 exchange functional. Can J Chem 2009;87:1485–91. https://doi.org/10.1139/V09-095.

[131] Perdew JP, Ruzsinszky A, Sun J, Burke K. Gedanken densities and exact constraints in density functional theory. J Chem Phys 2014;140(18):18A533. https://doi.org/10.1063/1.4870763.

[132] Lieb EH, Oxford S. Improved lower bound on the indirect coulomb energy. Int J Quant Chem 1981;19(3):427–39. https://doi.org/10.1002/qua.560190306.

[133] Handy NC, Cohen AJ. Left-right correlation energy. Mol Phys 2001;99(5):403–12. https://doi.org/10.1080/00268970010018431.

[134] Burke K, Perdew JP, Ernzerhof M. Why semilocal functionals work: accuracy of the on-top pair density and importance of system averaging. J Chem Phys 1998;109(10):3760–71. https://doi.org/10.1063/1.476976.

[135] Shahi C, Sun J, Perdew JP. Accurate critical pressures for structural phase transitions of group IV, III–V, and II–VI compounds from the SCAN density functional. Phys Rev B 2018;97(9):094111. https://doi.org/10.1103/PhysRevB.97.094111.

[136] Berland K, Cooper VR, Lee K, Schröder E, Thonhauser T, Hyldgaard P, et al. Van der Waals forces in density functional theory: a review of the vdW-DF method. Rep Prog Phys 2015;78(6). https://doi.org/10.1088/0034-4885/78/6/066501.

[137] Tao J, Perdew JP, Ruzsinszky A. Long-range van der Waals attraction and alkali-metal lattice constants. Phys Rev B 2010;81(23):233102. https://doi.org/10.1103/PhysRevB.81.233102.

[138] Foulkes WMC, Mitas L, Needs RJ, Rajagopal G. Quantum Monte Carlo simulations of solids. Rev Mod Phys 2001;73(1):33–83. https://doi.org/10.1103/RevModPhys.73.33.

[139] Ragahavachari K, Trucks GW, Pople JA, Head-Gordon M. A fifth-order perturbation comparison of electron correlation theories. Chem Phys Lett 1989;157(6):479–83. https://doi.org/10.1016/S0009-2614(89)87395-6.

[140] Grimme S. Semiempirical GGA-type density functional constructed with a long-range dispersion correction. J Comput Chem 2006;28(1):1788–99. https://doi.org/10.1002/jcc.

[141] Grimme S, Antony J, Ehrlich S, Krieg H. A consistent and accurate ab initio parametrization of density functional dispersion correction (DFT-D) for the 94 elements H-Pu. J Chem Phys 2010;132(15). https://doi.org/10.1063/1.3382344.

[142] Tkatchenko A, Scheffler M. Accurate molecular van der Waals interactions from ground-state electron density and free-atom reference data. Phys Rev Lett 2009;102(7):6–9. https://doi.org/10.1103/PhysRevLett.102.073005.

[143] Tkatchenko A, Distasio RA, Car R, Scheffler M. Accurate and efficient method for many-body van der Waals interactions. Phys Rev Lett 2012;108(23):236402. https://doi.org/10.1103/PhysRevLett.108.236402.

[144] Dion M, Rydberg H, Schröder E, Langreth DC, Lundqvist BI. Van der Waals density functional for general geometries. Phys Rev Lett 2004;92(24):22–5. https://doi.org/10.1103/PhysRevLett.92.246401.

[145] Lee K, Murray ÉD, Kong L, Lundqvist BI, Langreth DC. Higher-accuracy van der Waals density functional. Phys Rev B 2010;82(8):3–6. https://doi.org/10.1103/PhysRevB.82.081101.

[146] Cooper VR. Van der Waals density functional: an appropriate exchange functional. Phys Rev B 2010;81(16):1–4. https://doi.org/10.1103/PhysRevB.81.161104.

[147] Klimeš J, Bowler DR, Michaelides A. Chemical accuracy for the van der Waals density functional. J Phys Condens Matter 2010;22(2):022201. https://doi.org/10.1088/0953-8984/22/2/022201.

[148] Klimeš J, Bowler DR, Michaelides A. Van der Waals density functionals applied to solids. Phys Rev B 2011;83(19):195131. https://doi.org/10.1103/PhysRevB.83.195131.

[149] Hamada I. Van der Waals density functional made accurate. Phys Rev B 2014;89(12):121103. https://doi.org/10.1103/PhysRevB.89.121103.

[150] Vydrov OA, Van Voorhis T. Nonlocal van der Waals density functional: the simpler the better. J Chem Phys 2010;133(24):244103. https://doi.org/10.1063/1.3521275.

[151] Sabatini R, Gorni T, De Gironcoli S. Nonlocal van der Waals density functional made simple and efficient. Phys Rev B 2013;87(4):4–7. https://doi.org/10.1103/PhysRevB.87.041108.

[152] Peng H, Yang ZH, Sun J, Perdew JP. Versatile van der Waals density functional based on a meta-generalized gradient approximation. Phys Rev X 2016;6(4):041005. https://doi.org/10.1103/PhysRevX.6.041005.

[153] Cohen AJ, Mori-Sánchez P, Yang W. Challenges for density functional theory. Chem Rev 2012;112(1):289–320. https://doi.org/10.1021/cr200107z.

[154] Perdew JP, Ruzsinszky A, Sun J, Pederson MR. Chapter One—Paradox of self-interaction correction: how can anything so right be so wrong? Adv Atomic Mol Opt Phys 2015;64:1–14. https://doi.org/10.1016/bs.aamop.2015.06.004.
[155] Sun J, Perdew JP, Yang Z, Peng H. Communication: Near-locality of exchange and correlation density functionals for 1- and 2-electron systems. J Chem Phys 2016; 144(19):191101. https://doi.org/10.1063/1.4950845.
[156] Liechtenstein AI, Anisimov VI, Zaanen J. Density-functional theory and strong interactions: orbital ordering in Mott-Hubbard insulators. Phys Rev B 1995;52(8):5467–71. https://doi.org/10.1103/PhysRevB.52.R5467.
[157] Dudarev SL, Botton GA, Savrasov SY, Humphreys CJ, Sutton AP. Electron-energy-loss spectra and the structural stability of nickel oxide: an LSDA+U study. Phys Rev B 1998;57(3):1505–9. https://doi.org/10.1103/PhysRevB.57.1505.
[158] Atalla V, Zhang IY, Hofmann OT, Ren X, Rinke P, Scheffler M. Enforcing the linear behavior of the total energy with hybrid functionals: implications for charge transfer, interaction energies, and the random-phase approximation. Phys Rev B 2016;94(3): 035140. https://doi.org/10.1103/PhysRevB.94.035140.
[159] Skone JH, Govoni M, Galli G. Self-consistent hybrid functional for condensed systems. Phys Rev B 2014;89(19):195112. https://doi.org/10.1103/PhysRevB.89.195112.
[160] Jimenez-Hoyos CA, Janesko BG, Scuseria GE, Staroverov VN, Perdew JP. Assessment of a density functional with full exact exchange and balanced non-locality of correlation. Mol Phys 2009;107(8–12):1077–88. https://doi.org/10.1080/00268970902740555.

FURTHER READINGS

[161] Schimka L, Harl J, Kresse G. Improved hybrid functional for solids: the HSEsol functional. J Chem Phys 2011;134(2):024116. https://doi.org/10.1063/1.3524336.

CHAPTER

Vibrational energies and partition functions

5

Andrew L. Cooksy

Department of Chemistry and Biochemistry, San Diego State University, San Diego, CA, United States

1 INTRODUCTION

In 1738, Daniel Bernoulli's *Hydrodynamica* proposed that energy in matter could be stored in the form of the kinetic energy of microscopic moving particles. As our understanding of molecular structure and intermolecular forces improved, this kinetic energy of atomic motion became generally divided into contributions from translation, rotation, and vibration. The classical vibration of springs had been described by Robert Hooke in 1676, and the extension of Hooke's law to vibrating molecules may at first have appeared straightforward. However, attempts to develop a theory of blackbody radiation could not be reconciled with the classical description of the atomic vibrations of the body.

In the meantime, the science was advancing that would eventually allow us to probe individual vibrational quantum states. In 1800, William Herschel discovered the infrared region of the spectrum, when diffracted light from his prisms warmed areas to the long wavelength side of visible red light. In 1880, inspired by applications of visible light to the spectroscopic identification of compounds, Abney and Festing developed the infrared spectroscopy of chemical compounds, observing highly excited CH stretching states of various organic liquids, establishing the future of IR spectroscopy as a tool of molecular identification. In 1905 William Coblentz published the first atlas of IR spectra and refined the structure of the spectrometers, but it was only with the need to improve compound identification in World War II that attention became focused on chemical applications of infrared spectroscopy. A signal achievement was the use of IR spectroscopy to determine the correct structure of penicillin, cementing the use of the technique in organic analysis for decades to come [1].

As we came to better understand how to solve the electronic Schrödinger equation, it became clear that the motions of the nuclei that fuel chemical processes

Mathematical Physics in Theoretical Chemistry. https://doi.org/10.1016/B978-0-12-813651-5.00005-X

lie at a lower energy scale. The quantum mechanics of molecular vibrations is central to our understanding of chemical thermodynamics and kinetics.

2 MOLECULAR VIBRATIONS AND THE HARMONIC APPROXIMATION

Chemical research now routinely employs electronic structure packages, programs that solve the electronic Schrödinger equation. Typically, the approach of these programs first applies the Born-Oppenheimer approximation, which treats the nuclei as stationary while solving the Schrödinger equation just for the electrons. The cost of this separation of coordinates is that we have to solve the electronic Schrödinger equation over again for each position of the nuclei. This methodology results in an effective potential energy for the nuclei, actually a combination of electronic kinetic energy and electronic and nuclear potential energy.

2.1 HARMONIC APPROXIMATION

The importance of vibrations in many applications stems simply from the fact that in polyatomic molecules, vibrations store nearly all of the chemically available energy. Energy may also be stored in rotational or librational motion (where librations are the hindered rotations that occur in liquids) and in translations (motion of the molecule's center of mass). However, we assume that the energy transferred into the molecule has an equal chance of finding its way to any of the available coordinates of motion. The total number of those coordinates is $3N_{\text{atom}}$, for motion along the three Cartesian coordinates of each of the N_{atom} atoms in the molecule. The bonding of the atoms leads us to regroup these coordinates into translations, rotations, and vibrations, which describe more accurately the kinds of motions we expect to find. Whether or not any given degree of freedom is available for excitation is determined by how the energy gaps between the quantum states compare to the thermal energy $k_B T$, which is a rough measure of the average energy transferred between the particles by collisions.

There are only three translational coordinates (motion of the center of mass along x, y, or z) and at most three rotational coordinates (θ and ϕ to specify the orientation of any given bond axis, and a third angle of rotation about that bond if the molecule is nonlinear). Therefore, rotations and translations account for at most six of the $3N_{\text{atom}}$ coordinates. All of the remaining coordinates ($3N_{\text{atom}} - 5$ if linear or $3N_{\text{atom}} - 6$ if nonlinear) are vibrations. We will call that number N_{vib}, and we will label the vibrational coordinates q_i. Each vibrational coordinate corresponds to a relative motion of the atoms, such as stretching a bond distance, bending a bond angle, or twisting the structure about a chemical bond. For a molecule as small as acetone, with 10 atoms, there are 3 coordinates for translation, 3 for the rotations, and 24 distinct vibrational coordinates. Already, 80% of the coordinates that describe the atomic motions are vibrations, and that percentage increases as the number of atoms increases.

In a classical mechanics problem, we could calculate the motions of the particles by finding how the available forces affect the velocity vectors. The force at any point in space we would obtain by differentiating the corresponding potential energy function. For the vibrational dynamics of the nuclei in a molecule, we also describe the forces working on the nuclei through an effective potential energy function $V(\vec{q})$, where $\vec{q}$ is a shorthand notation for the complete set of values q_i that we need to specify all the values of the N_{vib} vibrational coordinates. All of the information about the vibrational forces is contained in $V(\vec{q})$, which we call the potential energy surface or PES of the molecule.

Along each vibrational coordinate, we assume that the potential energy seen by the nuclei constrains the nuclei to move back and forth around some minimum energy point, which we call the *equilibrium geometry*. Because the equilibrium geometry is the minimum on the N_{vib}-dimensional PES, it has zero first derivative along each vibrational coordinate q_i, and the second derivatives of $V(\vec{q})$ at that geometry are all positive. The simplest smoothly continuous function that gives such a minimum is a parabola:

$$V(\vec{q}) = \frac{1}{2}\left(k_1\, q_1^2 + k_2\, q_2^2 + \cdots + k_{N_{\text{vib}}}\, q_{N_{\text{vib}}}^2\right). \tag{1}$$

We have written Eq. (1) in the form of a *harmonic oscillator* potential, where the force constants k_i are determined by the steepness of the walls of the parabola along coordinate q_i. As long as the energy of the system is near the minimum potential energy (usually meaning that the vibrational energy is far less than what will break the bond) then it is usually a good approximation to assume that the potential energy is harmonic.

Once we have the potential energy function, we can in principle solve the Schrödinger equation. It is a minor paradox of quantum chemistry that the most interesting part of the Hamiltonian is never the quantum mechanical part (the kinetic energy operator $\hat{T}$), but rather the potential energy function $V(\vec{q})$, which we treat classically. That is because $\hat{T}$ always has the same form: only the masses and number of particles change from one problem to another. The potential energy function defines all of the forces working on and between the particles, and only that part of the Hamiltonian changes substantially from one molecular system from another.

Within the harmonic approximation, the vibrational Schrödinger equation for a diatomic molecule AB has the general form

$$\text{Diatomic:} \quad \left[-\frac{\hbar^2}{2\mu}\frac{\partial^2}{\partial R^2} + \frac{1}{2}k(R-R_e)^2\right]\psi_{\text{vib}}(R) = E_{\text{vib}}\psi_{\text{vib}}(R), \tag{2}$$

where R is the bond distance between nuclei A and B, R_e is the equilibrium bond distance, k is the force constant, and μ is the reduced mass, which for a diatomic molecule is given by

$$\text{Diatomic:} \quad \mu = \frac{m_{\text{A}} m_{\text{B}}}{m_{\text{A}} + m_{\text{B}}}. \tag{3}$$

Table 1 The First Six Hermite Polynomials and Normalization Constants for the Harmonic Oscillator Wavefunctions $\eta_v(y) = A_v H_v(y) e^{-y^2/2}$, Where the Dimensionless Coordinate y Is Equal to $(R - R_e)(k\mu/\hbar^2)^{1/4}$

v	A_v	$H_v(y)$
0	$\left(\frac{k\mu}{\hbar^2}\right)^{1/8}\left(\frac{1}{\sqrt{\pi}}\right)^{1/2}$	1
1	$\left(\frac{k\mu}{\hbar^2}\right)^{1/8}\left(\frac{1}{2\sqrt{\pi}}\right)^{1/2}$	$2y$
2	$\left(\frac{k\mu}{\hbar^2}\right)^{1/8}\left(\frac{1}{8\sqrt{\pi}}\right)^{1/2}$	$4y^2 - 2$
3	$\left(\frac{k\mu}{\hbar^2}\right)^{1/8}\left(\frac{1}{48\sqrt{\pi}}\right)^{1/2}$	$8y^3 - 12y$
4	$\left(\frac{k\mu}{\hbar^2}\right)^{1/8}\left(\frac{1}{384\sqrt{\pi}}\right)^{1/2}$	$16y^4 - 48y^2 + 12$
5	$\left(\frac{k\mu}{\hbar^2}\right)^{1/8}\left(\frac{1}{3840\sqrt{\pi}}\right)^{1/2}$	$32y^5 - 160y^3 + 120y$

The wavefunctions that solve this Schrödinger equation are based on the Hermite polynomials $H_v(y)$ discovered by Laplace in 1910 to solve an unrelated problem in diffusion [2]:

$$\psi_v(y) = A_v H_v(y) e^{-y^2/2}, \quad \text{where } y = \left(\frac{k\mu}{\hbar^2}\right)^{1/4} (R - R_e). \tag{4}$$

In this expression, y is a dimensionless, scaled distance parameter, and A_v is a normalization constant. The first several expressions for A_v and $H_v(y)$ are given in Table 1, and the corresponding wavefunctions are graphed against the potential energy curve in Fig. 1.

The energies that solve our Schrödinger equation are proportional to a *vibrational constant* ω_e:

$$E_v = \omega_e\left(v + \tfrac{1}{2}\right), \quad \text{where } \omega_e = \hbar\sqrt{\frac{k}{\mu}},\ v = 0, 1, 2, 3, \ldots. \tag{5}$$

These energy levels are evenly spaced by the vibrational constant ω_e. In Eq. (5), the value of ω_e is defined in proper energy units, but it is more common to use ω_e in cm^{-1}, to conform to the units almost universally used for vibrational spectra:

$$\omega_e\,(\mathrm{cm}^{-1}) = \frac{1}{2\pi c}\sqrt{\frac{k}{\mu}} = 130.28\sqrt{\frac{k\,(\mathrm{N\,m^{-1}})}{\mu\,(\mathrm{amu})}}. \tag{6}$$

The expressions in Eqs. (4), (5) must be modified for polyatomic molecules, but we continue to apply the harmonic approximation to each vibrational coordinate in the molecule. To deal with polyatomics, however, we first need to decide what those vibrational coordinates look like.

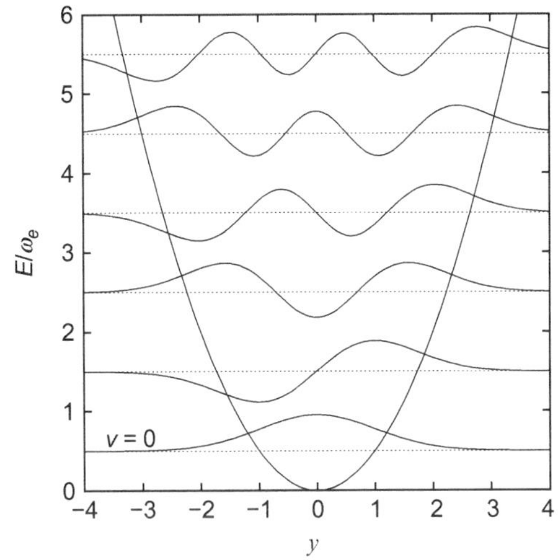

FIG. 1

The six lowest energy harmonic oscillator wavefunctions plotted at the corresponding energy relative to the potential energy curve.

2.2 NORMAL MODES

Given a rough idea what the potential energy function in the system looks like, we can address a key remaining question: What *are* the vibrational coordinates that we should use? In CO_2, for example, should we use separate vibrational coordinates for each C=O bond, or should we treat both bonds at the same time because they are equivalent?

Most often, we use the *normal modes* of the system as the vibrational coordinates. In general, the same term in the Hamiltonian may depend on two different vibrational coordinates. However, if the vibrational coordinates q_i are the normal mode coordinates, then there is only one term for the contribution of each coordinate to the classical kinetic and potential energies of vibration:

$$E_{\text{vib}} = \sum_{i=1}^{N_{\text{vib}}} \left[\frac{\mu_i}{2} \left(\frac{dq_i}{dt} \right)^2 + \frac{1}{2} \kappa_i q_i^2 \right], \tag{7}$$

where the value of a given κ_i depends on the force constant k_i for the vibration. Eq. (5) was written assuming that the coordinates q_i are the normal mode coordinates; otherwise we would have to include terms of the type $k_{ij} q_i q_j$ in the potential energy expression.

In general, during the course of motion along any normal mode coordinate, all the atoms in the molecule move simultaneously, at least a tiny amount. (In molecules with symmetry, some modes may have stationary atoms.) In each normal mode vibration, all the atoms move back and forth at the same frequency, and the molecule always passes through the equilibrium geometry. These motions would accurately describe stable vibrational motions—the resonance motions—in a system of Hooke's law springs connected together.

It is always possible to write the potential energy of coupled harmonic oscillators in terms of the normal mode coordinates, but it can take some linear algebra to find out what they are. As an example, we will identify the normal modes for the stretches of a linear molecule of the form ABC. We place the three atoms A, B, and C on the z-axis and label the positions z_A, z_B, and z_C from left to right. Assigning a force constant k to the stretch of each bond $R_{AB} = z_B - z_A$ and $R_{AB} = z_C - z_B$, the potential energy function may be written

$$\begin{aligned} V(z_A, z_B, z_C) &= \tfrac{1}{2} k_{AB} \left[R_{AB} - R_{eAB}\right]^2 + \tfrac{1}{2} k_{BC} \left[R_{BC} - R_{eBC}\right]^2 \\ &= \tfrac{1}{2} k_{AB} \left[(z_B - z_A) - R_{eAB}\right]^2 + \tfrac{1}{2} k_{BC} \left[(z_C - z_B) - R_{eBC}\right]^2, \end{aligned} \tag{8}$$

where the equilibrium bond distances R_{eAB} and R_{eBC} are constants. Given Eq. (8), we can set up and solve an eigenvalue equation based on the second derivatives of the nuclear motions of the molecule. We accomplish this by diagonalizing the mass-weighted Hessian matrix, a symmetric matrix whose elements are the second derivatives $\partial^2 V/(\partial z_i \partial z_j)$ divided by $\sqrt{m_i m_j}$:

$$\begin{pmatrix} a & b & d \\ b & c & e \\ d & e & f \end{pmatrix}, \tag{9}$$

where

$$a = \frac{1}{m_A} \frac{\partial^2 V}{\partial z_A^2} = \frac{k_{AB}}{m_A}, \qquad b = \frac{1}{\sqrt{m_A m_B}} \frac{\partial^2 V}{\partial z_A \partial z_B} = -\frac{k_{AB}}{\sqrt{m_A m_B}},$$

$$c = \frac{1}{m_B} \frac{\partial^2 V}{\partial z_B^2} = \frac{k_{AB} + k_{BC}}{m_B}, \qquad d = \frac{1}{\sqrt{m_A m_C}} \frac{\partial^2 V}{\partial z_A \partial z_C} = 0,$$

$$e = \frac{1}{\sqrt{m_B m_C}} \frac{\partial^2 V}{\partial z_B \partial z_C} = -\frac{k_{BC}}{\sqrt{m_B m_C}}, \qquad f = \frac{1}{m_C} \frac{\partial^2 V}{\partial z_C^2} = \frac{k_{BC}}{m_C}.$$

Diagonalizing this matrix leads to the eigenvalues

$$0, \quad \frac{\alpha - \beta}{2 m_A m_B m_C}, \quad \frac{\alpha + \beta}{2 m_A m_B m_C}, \tag{10}$$

where

$$\begin{aligned} \alpha &= k_{AB}(m_A m_C + m_B m_C) + k_{BC}(m_A m_B + m_A m_C), \\ \beta &= \Big[k_{AB}\left(m_A^2 m_C^2 - 2 m_A m_B m_C^2 + m_B^2 m_C^2\right) + k_{BC}\left(m_A^2 m_B^2 - 2 m_A^2 m_B m_C + m_A^2 m_C^2\right) \\ &\quad + 2 k_{AB} k_{BC}\left(m_A^2 m_C^2 - m_A^2 m_B m_C - m_A m_B^2 m_C - m_A m_B m_C^2\right)\Big]^{1/2}. \end{aligned}$$

The eigenvalue of zero corresponds to the translational motion of the whole molecule moving to the right or the left. This is a translation, not a vibration. Because the potential energy does not depend on the location of the molecule, the eigenvalue for that motion is zero. The remaining eigenvalues are the squares of the fundamental vibrational frequencies, and the eigenvectors are the normal modes.

If we simplify the expressions in Eq. (10) by making atoms A and C chemically equivalent, so that $k_{AB} = k_{BC} = k$ and $m_A = m_C$, then the nonzero eigenvalues become

$$\frac{k}{m_A}, \quad \frac{(2m_A + m_B)k}{m_A m_B}. \tag{11}$$

We obtain the vibrational constants by taking the square roots:

$$\omega_1 = \sqrt{\frac{k}{m_A}}, \quad \omega_2 = \sqrt{\frac{(2m_A + m_B)k}{m_A m_B}}. \tag{12}$$

The eigenvectors give us the motions of the individual atoms in the normal mode coordinates q_i:

$$\begin{aligned} &q_1\colon \delta z_A = -\delta z_C \quad \delta z_B = 0 \\ &q_2\colon \delta z_A = \delta z_C \quad \delta z_B = -2\sqrt{\frac{m_A}{m_B}}\,\delta z_A. \end{aligned} \tag{13}$$

We call the first solution, q_1, the symmetric stretch. The two terminal atoms A and C move in opposite directions, while the central B atom remains stationary, so that both bonds lengthen and contract simultaneously. The second solution, q_2, gives the antisymmetric stretch, with one bond contracting while the other lengthens.

In a fashion similar to that shown here, electronic structure programs calculate the second derivatives of the effective potential energy with respect to the Cartesian coordinates of all of the atoms, and then diagonalize the Hessian matrix to find the normal modes, excluding the five or six solutions that correspond to rotations and translations. This normal mode analysis, however, is accurate only in the limit of an ideal harmonic oscillator, and we can improve the model of vibrations by considering deviations from harmonicity.

3 VIBRATIONAL ANALYSIS ON ANHARMONIC POTENTIAL SURFACES

The harmonic oscillator approximation serves as a valuable first approximation to vibrational spectra in the lowest vibrational quantum states. However, high-resolution spectroscopy and studies at high vibrational quantum number reveal *anharmonic* effects, shifts from the harmonic values, which we wish to address both to improve the predictive power of the model and to extract data about the potential energy surface of the molecule from the experimental data.

In the simplest example, a diatomic molecule has a vibrational potential energy curve that asymptotically approaches a horizontal line at high R, as the vibrational energy approaches the bond dissociation energy. By contrast, the walls of the harmonic oscillator potential continuously get steeper as we pull away from the equilibrium geometry. Applying the correspondence principle of quantum mechanics, we recognize that as the domain of the vibrating system increases in size, the energy levels will tend to become more closely spaced, approaching the continuum of energy found for a classical system.

This decrease in the energy level spacing at higher vibrational energy can be viewed as the result of an effective decrease in the force constant for the vibration as the potential flattens, approaching dissociation. To aid in the analysis of spectra, the effects of anharmonicity on the vibrational energies of a particular mode are often expressed in terms of a power series in $(v + \frac{1}{2})$:

$$E_v = \omega_e \left(v + \tfrac{1}{2}\right) - \omega_e x_e \left(v + \tfrac{1}{2}\right)^2 + \omega_e y_e \left(v + \tfrac{1}{2}\right)^3 + \cdots . \tag{14}$$

(In the original notation, x_e and y_e would be dimensionless coefficients that expressed the magnitude of the deviation from harmonicity. However, it is more common simply to list the combined values $\omega_e x_e$ and $\omega_e y_e$ in units of cm^{-1}.)

In polyatomic molecules, however, the anharmonicities are more complex, because deviations from the harmonic oscillator model also invalidate the normal mode analysis, meaning that the vibrational coordinates themselves become more poorly defined at higher excitation. In addition to anharmonic corrections within each mode, we need coupling terms which model the extent to which the different modes become mixed as the wavefunctions probe increasingly anharmonic regions of the PES. The introduction of anharmonicity means that the normal modes no longer comprise a completely separable set of coordinates in the Hamiltonian. The vibrational energies are then often expressed in terms of anharmonic coupling coefficients χ_{ij} such that

$$E_{v_1,\ldots,v_k} = \sum_{i=1}^{k} \omega_i \left(v_i + \tfrac{1}{2}\right) + \sum_{j=1}^{i} \chi_{ij} \left(v_i + \tfrac{1}{2}\right)\left(v_j + \tfrac{1}{2}\right) + \cdots . \tag{15}$$

The values of these anharmonicities may be determined by fitting Eq. (15) to vibrational data that include hot bands (transitions from vibrationally excited states), overtones (transitions where $\Delta v_i > 1$), and/or combination bands (transitions where $\Delta v_i \neq 0$ for more than one i).

One common example of anharmonic motion has a distinct simplifying limit. Internal rotations about a single bond, such as rotation about the central C–C bond in butane, are periodic (with the potential energy function repeating every rotation by 2π) and often occur with multiple minima. In general, the distribution of these energy levels is a sensitive function of the number of minima and the relative magnitudes of the harmonic vibrational constant and the energy barrier separating adjacent minima. However, in the limit that there are no barriers, the motion becomes a free rotation along one coordinate. For that motion, the energy levels are given by

$$E_{\text{rotor}} = \frac{\hbar^2}{2I_a} K_a^2, \tag{16}$$

where I_a is the moment of inertia of the rotor measured with respect to the rotation axis, and K_a is the internal rotational quantum number, and may be equal to any integer.

There are a variety of ways to predict the energy levels of vibrational states when the harmonic approximation does not apply. If only small deviations from harmonicity need to be modeled, then the anharmonicities will normally be calculated as corrections to the harmonic oscillator through perturbation theory. If the harmonic oscillator is wholly inappropriate, however, for example, in the case of a potential surface that exhibits several widely separated local minima, then more elaborate approaches may be required. Only a small sample of the available methodologies for addressing anharmonicity will be summarized here.

3.1 EFFECTIVE HAMILTONIANS AND PERTURBATION THEORY

In spectroscopy, an effective Hamiltonian is a reformulation of the Schrödinger equation that allows experimentally measured transition energies to be broken down into contributions from a catalog of molecular constants. Often the effective Hamiltonian is expressed as a set of matrix elements

$$H_{ij} = \langle i \mid \hat{H}_{\text{eff}} \mid j \rangle = \iint \psi_j^* \hat{H}_{\text{eff}} \psi_i d\tau_i d\tau_j \tag{17}$$

that are written as algebraic expressions of the molecular constants and the quantum numbers that determine states i and j. If sufficient spectra are obtained, each molecular constant can be uniquely determined by its dependence on the quantum numbers.

Perturbation theory is the most common tool for approximating the impact of anharmonic terms in the potential energy surface on the vibrational energies. In perturbation theory, a zero-order system (in our case, the harmonic oscillator) is a close approximation to the system we want to study, differing only by the addition of a relatively small perturbation term $\hat{H}'$ to the Hamiltonian. Multiplying the perturbation by an adjustable parameter λ between 0 and 1, we then write the Schrödinger equation as a Taylor series expansion in powers of λ about the zero-order solution ($\lambda = 0$). This separates contributions to the solution into rapidly increasing levels of complexity in integrals of the form

$$\int \psi_i^{(0)*}(\vec{q}) \hat{H}' \psi_j^{(0)}(\vec{q})\, d\vec{q},$$

with $\psi_i^{(0)*}(\vec{q})$ and $\psi_j^{(0)}(\vec{q})$ corresponding to different zero-order (harmonic oscillator) solutions to the wavefunctions.

The perturbation in this case will be higher-order polynomials in the potential energy function [3]:

$$V(\vec{q}) = \frac{1}{2}\sum_{i=1}^{N_{\rm vib}} k_i q_i^2 + \hat{H}' = \frac{1}{2}\sum_{i=1}^{N_{\rm vib}} k_i q_i^2 + \sum_{i,j,k} F_{ijk}\, q_i\, q_j\, q_k + \sum_{i,j,k,l} F_{ijkl}\, q_i\, q_j\, q_k\, q_l + \cdots. \quad (18)$$

Use of the normal modes at the equilibrium geometry ensures that we have set the quadratic off-diagonal terms $F_{i\neq j}$ to zero, leaving only terms in $F_{ii} \equiv \frac{1}{2}k_i$. Although the series in Eq. (18) continues, it is rarely extended beyond the quartic term, if only because computational demands rapidly expand with each term. The harmonic force constants can be calculated analytically for certain electronic structure methods such as Hartree-Fock. However, the higher-order force constants typically have to be evaluated numerically, by taking differences between energies for geometries slightly different from the equilibrium geometry along each vibrational coordinate. Therefore, each anharmonic constant can require the calculation of several additional energies, and for this reason the force constants are not always evaluated among all possible values of the indices i, j, k, l.

Among the numerous examples described in the literature, a relatively recent analysis of the ozone vibrational spectrum [4] employed an effective Hamiltonian that detailed matrix elements for couplings between different quanta of stretching and bending excitations, expressing the magnitude of the matrix elements in terms of the anharmonicities of the PES.

3.2 MODEL POTENTIALS

One option in devising an effective Hamiltonian is to assume a qualitatively correct potential energy curve. In particular, the potential energy for the stretching vibration of a terminal chemical bond should lead to dissociation. One function which combines this requirement with a qualitatively accurate exponential repulsive wall is the *Morse oscillator*, which is defined by the potential energy function

$$V_{\rm Morse}(R) = D_e\left\{\left[1 - e^{-\beta(R-R_e)}\right]^2 - 1\right\}, \quad (19)$$

where D_e and R_e are the equilibrium well-depth and bond length, respectively, and where $\beta = \sqrt{k_e/(2D_e)}$ determines the width of the well. This model potential is contrasted with a harmonic potential in Fig. 2. Unlike the harmonic oscillator, the second derivative of the Morse oscillator is not constant, so the k_e value here is the second derivative of the potential energy curve at its minimum. As written in Eq. (19), the Morse potential approaches zero in the limit that the dissociation fragments are separated by large distance and no longer interacting. At the equilibrium position,

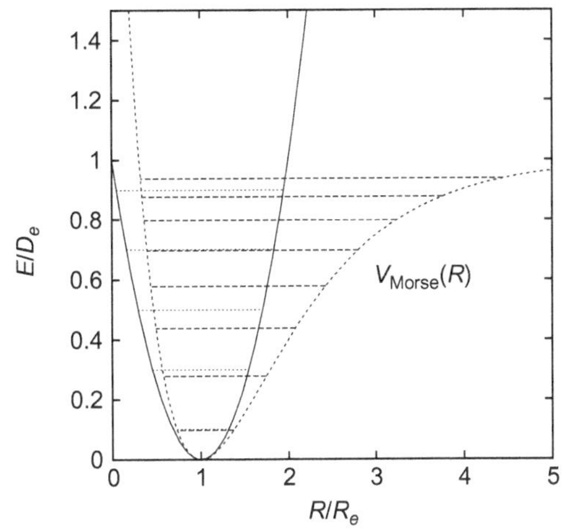

FIG. 2

The Morse potential and sample distribution of the lowest seven vibrational energy levels. A harmonic oscillator potential and energy levels for the same k and ω_e values are plotted for comparison.

$V_{\text{Morse}}(R_e) = -D_e$, and at zero distance $V_{\text{Morse}}(0) = D_e(1 - e^{+\beta R_e})^2$, which is usually in the range of $10D_e$.

The energies obtained for vibrational quantum states on $V_{\text{Morse}}(R)$ then have the form

$$E_v = \omega_e\left(v + \tfrac{1}{2}\right) - \omega_e x_e\left(v + \tfrac{1}{2}\right)^2 \tag{20}$$

$$\omega_e x_e = \frac{\omega_e^2}{4D_e}. \tag{21}$$

So the Morse potential delivers an exact energy expression incorporating the first anharmonic constant and showing how we might infer a value for the dissociation energy of a bond if we have enough vibrational data to successfully measure $\omega_e x_e$. However, it would be helpful to have a more general approach, applicable to a wider range of vibrational motions, including bending and interior stretching motions which do not so clearly lead to dissociation.

The goal of the effective vibrational Hamiltonian is to allow properties of the molecular potential energy surface and other phenomena to be expressed in terms of molecular constants such as the χ_{ij} that can be determined experimentally. The Morse oscillator provides a simple case, where an assumed form for the potential energy function leads to the energy expression Eq. (14) and the interpretation of the fitted constant $\omega_e x_e$ in terms of parameters of the potential energy function given by Eq. (21). An experimental measurement of the anharmonic constant $\omega_e x_e$ would allow one to infer the value of the dissociation energy D_e. Similarly, in a polyatomic system, one interpretation of the anharmonic constants χ is as coupling coefficients between Morse oscillator potentials for different normal modes [5].

3.3 CALCULATION OF POTENTIAL ENERGY SURFACES

If the deviation from harmonicity is relatively small, then we need not explore the PES far from the equilibrium geometry, and the anharmonicity can be treated as a perturbation of the harmonic approximation. However, we often encounter cases where the anharmonicity is large and where other ideal cases (such as the free rotor) also do not apply.

In principle, we can always solve the vibrational Schrödinger equation by numerical integration, provided that we have an accurate PES. Although there are practical concerns with the numerical integration, there is a growing body of work on the characterization of the vibrational PES along several coordinates. In several cases, even full-N_{vib}-dimensional calculations of the PES have been made, including molecules of up to at least seven atoms [6].

As the dimensionality of the surface increases, the calculation of the PES rapidly becomes the most demanding stage of solving the vibrational Schrödinger equation. High-resolution spectroscopy of gas-phase vibrational transition energies provides our most precise and sensitive test of the quality of these solutions, with energies routinely measured to experimental errors of roughly $0.001\ \text{cm}^{-1}$ or a part in 10^7. However, few of these measurements are extended to highly excited vibrational states, which are needed if we are going to probe a substantial domain of the PES. Even so, measurements of overtone transitions and vibrational components of electronic spectra may still give relative energies to a resolution within $1\ \text{cm}^{-1}$, or about $0.01\ \text{kJ mol}^{-1}$. This is nearly two orders of magnitude below the typical expected systematic errors in relative energies of configurational isomers as calculated by fairly demanding coupled cluster calculations.

It is true that the relative energies of the quantum states depend on the entire domain sampled with significant amplitude by the corresponding wavefunctions, and therefore isolated errors in the PES might not be an issue. However, the errors in the PES are not random, but systematic errors of the approximations necessary to solve the electronic Schrödinger equation. Consequently, the errors are more likely to be amplified than averaged away as more of the surface is sampled by the wavefunction. The quality of the resulting solution is normally limited by the quality of the calculated PES, as determined both by the accuracy of the computational methodology and the number of geometries at which the potential energy is calculated. These demands put researchers in the unenviable position of requiring as many individual calculations as possible of the most resource-intensive calculations feasible to generate the PES.

When it is possible to accurately fit an analytical function to the explicitly calculated points of the PES, the resulting expression can greatly simplify the solution of the Schrödinger equation, because some methods rely on the potential energy function being analytically differentiable. This remains perhaps the most appealing option for three or fewer dimensions, although even for three dimensions it may be difficult to find a suitable analytical form without introducing errors from overfitting. (With too many adjustable parameters, it becomes possible to find an expression that passes through all of the calculated points at the expense of generating large and unphysical oscillations in the parts of the function that are not sampled.)

The computational demands of these studies have led groups to search for alternative algorithms that optimize the impact of the individual electronic structure calculations on the quality of the overall PES. As an example, the Interpolating Moving Least Squares (IMLS) approach offers a flexible method for using calculated points on the surface to identify regions of the surface that require additional sampling [7,8]. Briefly, the IMLS method consists of minimizing a function

$$Y_x[p(x)] = \sum_{i=0}^{N} w_i(x)\, [p(x_i) - f_i]^2 , \tag{22}$$

where

$$w_i(x) = \frac{\exp\left[-(x-x_i)^2\right]}{(x-x_i)^k + \epsilon} \qquad p(x) = \sum_{n=0}^{m} a_n\, x^n,$$

and where $p(x)$ is a polynomial used to approximate the calculated values f_i of the PES. Values of k and ϵ have been determined empirically. The weighting function w_i determines how much any given point i contributes to the value of Y_x. The values of the a_n are adjusted to minimize Y_x, at a given value of x. This approach does not require globally fitting the entire surface simultaneously, providing instead a local interpolation among points in the same neighborhood of the PES. The fitted polynomial is still analytical and differentiable, but now can be extended more easily to multiple dimensions.

Moreover, the quality of the fit at different orders n of the polynomial for several positions x_i can be used to determine where on the surface additional points would most benefit the overall fit. The IMLS therefore provides a dynamic tool for fitting a local expression to the calculated potential energies while determining the geometries where the next potential energies should be calculated.

With the PES, we have a complete vibrational Hamiltonian. It then remains to choose among different methods available for solving the Schrödinger equation on that surface.

3.4 VARIATIONAL METHODS

The variational method is arguably the most straightforward strategy for solving the Schrödinger equation in principle, and for that reason is often applied to problems of low dimensionality. However, it tends to be more computationally expensive than other options. A schematic approach to the variational method will involve the following steps:

(i) Write the vibrational wavefunction as a linear combination of basis functions:

$$\psi_k(\vec{q}) = \sum_{i=1}^{N_{\text{basis}}} c_{ik}\, \varphi_i(\vec{q}). \tag{23}$$

(ii) Express the Hamiltonian as a matrix **H** of terms:

$$H_{ij} = \int \varphi_i\, \hat{H}\, \varphi_j\, d\tau. \tag{24}$$

(iii) Diagonalize the matrix by solving the determinant:

$$|\mathbf{H} - E\mathbf{I}| = 0 \tag{25}$$

to obtain a set of N_{basis} solutions, each solution consisting of an energy E_k and a vector of coefficients $\vec{c}_{ik}$ for state k.

The choice of integration method will in general divide these approaches into grid-based and analytical approaches. Grid-based numerical integration divides the space into cells, tiny volumes, and the finer the division the more accurate the numerical solution. The method employs basis sets where each function is defined over a single cell, with some level of continuity required from one cell to the next for the wavefunction (and perhaps its derivatives) as it is calculated. Grid-based methods are computationally extremely expensive, but have the advantage of being generally applicable to any shape of PES. In fact, the PES (which is normally evaluated as a series of points from hundreds or thousands of individual electronic structure calculations) can be left in a point-wise form.

Analytical methods use basis functions of algebraic expressions that span the entire space. Harmonic oscillator wavefunctions (Eq. 4) are one choice for basis functions, although other choices such as Laguerre and Legendre functions (analogous to the radial and angular components, respectively, of the atomic orbital wavefunctions) may be more suitable, depending on the coordinate system and potential surface. Typically, one-dimensional basis functions will be used and multiplied together. If the PES can be expressed in a compatible fashion, say as a linear combination of Gaussians, then the integration itself may be analytical. Obtaining analytical solutions can be much more efficient than grid-based numerical integration.

The closer the basis set is to the correct variational solution, the smaller will be the off-diagonal ($i \neq j$) elements of the Hamiltonian matrix. Although N_{basis} solutions are obtained, only a fraction of these, the lowest energy solutions, are likely to be trustworthy. Higher energy solutions borrow more extensively from the higher energy basis functions, and eventually reach a point where significant character would be drawn from basis functions that are not included in the finite basis set. Beyond that energy, the variational wavefunctions will largely be determined by the constraints on the solution from the truncated basis set size, rather than by the physics of the natural system.

The discrete variable representation method of Light et al. [9] employs continuous functions localized around individual points in the PES to take advantage of analytical solutions to the extent possible (e.g., for the kinetic energy) while retaining the flexibility of grid-based approaches. This approach has been applied to efforts to generate a fully flexible approach to solution of the vibrational Schrödinger equation [10].

3.5 VIBRATIONAL SCF

The vibrational self-consistent field (SCF) approach employs the same fundamental assumption as the Hartree-Fock method applies to electronic states: that the wavefunction is separable (in this case by vibrational coordinate) but is partly determined by an interaction term that integrates over contributions from all the other coordinates. Because each component of the wavefunction depends implicitly on each of the other coordinates, the solution is achieved only through an iterative improvement of the wavefunction over several cycles until the wavefunctions going into the calculation are consistent with the resulting wavefunction; that is, the results are self-consistent.

In this case, the separable wavefunction may be written

$$\Psi(\vec{q}) = \prod_{i=1}^{N_{\text{vib}}} \psi_i(q_i), \tag{26}$$

and the SCF requirement is that the wavefunctions $\psi_i(q_i)$ satisfy the single-mode Schrödinger equation

$$\left[-\frac{1}{2}\frac{\partial^2}{\partial q_i^2} + \bar{V}_i(q_i)\right]\psi_i(q_i) = \epsilon_i\,\psi_i(q_i), \tag{27}$$

where

$$\bar{V}_i(q_i) = \int \left|\prod_{j\neq i}^{N_{\text{vib}}} \psi_j(q_j)\right|^2 V(\vec{q})\,d\vec{q}_{i\neq j}. \tag{28}$$

Note that the integral in $\bar{V}_i(q_i)$ is carried out over all vibrational coordinates except q_i in order to preserve $\bar{V}_i$ as a function of q_i.

The potential energy function in this case may be completely arbitrary; however, for complex systems, it may be limited to a truncated expansion of the type in Eq. (18). Extensively developed by Bowman et al. [11], including extensions to configuration interaction [12] and coupled-cluster theory [13], this is an extremely efficient approach to solving the vibrational Schrödinger equation, applicable to systems of over a hundred vibrational modes [14].

4 VIBRATIONAL TRANSITION INTENSITIES

The transition strength $\mathcal{I}$ for the spectroscopic transition from initial state i to final state f due to interactions with the electric dipole field of radiation is proportional to

$$\mathcal{I} \propto \left| \int \psi_f^* \, \hat{\mu} \, \psi_i \, d\tau \right|^2 , \tag{29}$$

where $\hat{\mu}$ is the transition dipole moment operator, determined by the charge separation in the molecule, perpendicular to the propagation axis of the radiation. For vibrational transitions, the convenient way to look at the transition moment is to expand the expression for the dipole moment around its equilibrium value in a Taylor series:

$$\mu = \mu_e + \left(\frac{\partial \mu}{\partial R}\right)_{R_e} (R - R_e). \tag{30}$$

In the simplest case, we can assume that the dipole moment is given by fixed partial charges on two atoms separated by a bond length R. If the difference q between the two charges (actually this is half the difference) stays constant then μ is proportional to the bond length and the *dipole derivative* $\left(\frac{\partial \mu}{\partial R}\right)_{R_e}$ is a constant. Therefore, the transition strength for a transition between different stretching states becomes

$$\begin{aligned}
\mathcal{I} &\propto \left[\int \psi_f^* \left[\mu_e + \left(\frac{\partial \mu}{\partial R}\right)_{R_e} (R - R_e) \right] \psi_i d\tau \right]^2 \\
&= \left[\left[\mu_e - \left(\frac{\partial \mu}{\partial R}\right)_{R_e} R_e \right] \int \psi_f^* \psi_i d\tau + \int \psi_f^* \left[\left(\frac{\partial \mu}{\partial R}\right)_{R_e} R \right] \psi_i d\tau \right]^2 \\
&= \left[\int \psi_f^* \left[\left(\frac{\partial \mu}{\partial R}\right)_{R_e} R \right] \psi_i d\tau \right]^2 = \left(\frac{\partial \mu}{\partial R}\right)_{R_e}^2 \left[\int \psi_f^* R \psi_i d\tau \right]^2 ,
\end{aligned} \tag{31}$$

where we take advantage of the fact that the eigenstates of the system ψ_f and ψ_i are always orthogonal. The dipole derivative $(\partial \mu / \partial R)_{R_e}$ is a constant because it is evaluated at the equilibrium geometry. The magnitude of the transition intensity therefore depends on the magnitudes of the dipole derivative and the squared integral

$$\left[\int \psi_{v'}(y)^* \, y \, \psi_v(y) \, dR \right]^2 ,$$

where we are using the harmonic oscillator wavefunctions $\psi(y)$ (Eq. 4). (Using dR here instead of dy makes the integral unitless as written.)

For the harmonic oscillator, only transitions where the v quantum number changes by one unit give nonzero values of the integral. For the $v = 0 \rightarrow 1$ transition, we obtain:

$$\left[\int_{-\infty}^{\infty} \psi_0\, y\, \psi_1\, dR\right]^2 = \left[\frac{1}{\sqrt{2\pi}}\int_{-\infty}^{\infty}(2y^2)e^{-y^2}\, dy\right]^2 = \left[\frac{1}{\sqrt{2\pi}}(2)\left(\frac{\sqrt{\pi}}{2}\right)\right]^2 = \frac{1}{2}.$$

The other integrals are exactly zero, and in general there is a selection rule for pure harmonic oscillator wavefunctions requiring the selection rule $\Delta v = \pm 1$. That rule occurs because a recursion relation

$$H_{v+1}(y) = 2yH_v(y) - 2vH_{v-1}(y) \quad \text{or} \quad H_{v+1}(y) + 2vH_{v-1}(y) = 2yH_v(y)$$

allows the integrand to be rewritten as a product of wavefunctions for different values of the vibrational quantum number v:

$$\int e^{-y^2} yH_{v'}H_v\, dR = \int e^{-y^2}\frac{1}{2}\left[H_{v'+1}(y) + 2v'H_{v'-1}(y)\right] H_v\, dR.$$

All of the ψ_v wavefunctions are orthogonal, and therefore this integral is nonzero only if the final state quantum number v' is $v+1$ or $v-1$.

In perturbation theory, we would typically write the effect of anharmonicity on the wavefunctions as mixing in harmonic oscillator basis functions with different values of v. This allows violation of the $\Delta v = \pm 1$ selection rule. If the wavefunction for quantum state v of the anharmonic system includes contributions from harmonic oscillator functions v and $v+1$, for example, then transitions to $v+1$ and $v+2$ will have nonzero strengths, and the strength of the $v \to v+2$ transition will increase with the amount of $v+1$ character in the initial state.

5 VIBRATIONAL PARTITION FUNCTION

The *partition function* extends the results of a quantum mechanical analysis of the energy levels to their impact on the thermodynamics and kinetics of the system. For any degree of freedom in the system (any unique coordinate of motion available to store the energy), the partition function is defined by

$$Z(T) \equiv \sum_{i=0}^{\infty} g(\varepsilon_i)\, e^{-\varepsilon_i/(k_B T)}, \tag{32}$$

where ε_i is the ith excited energy level ($i = 0$ being the ground state), $g(\varepsilon_i)$ is the degeneracy at that energy, k_B is Boltzmann's constant ($1.381 \times 10^{-23}\,\mathrm{J\,K^{-1}}$), and T is the temperature in K. The partition function normalizes the thermal probability distribution $\mathcal{P}(\varepsilon_i)$ for the degree of freedom, so that the probability of finding any randomly selected molecule in a macroscopic sample at energy ε_i is

$$\mathcal{P}(\varepsilon_i) = \frac{g(\varepsilon_i)\, e^{-\varepsilon_i/(k_B T)}}{Z(T)}. \tag{33}$$

When many degrees of freedom are present and the coordinates are completely separable, the overall partition function $Z_{tot}(T)$ is the product of the individual partition functions. In particular, for a gas-phase sample, $Z_{tot}(T)$ is the product of contributions from translational, rotational, vibrational, and electronic energies:

$$Z_{tot}(T) = Z_{trans}(T)\, Z_{rot}(T)\, Z_{vib}(T)\, Z_{elec}(T). \tag{34}$$

In this product, the translational and rotational contributions can normally be approximated by the classical limit, because at 298 K the thermal energy $k_B T$ greatly exceeds the energy necessary to excite translations and typical rotations. In particular, the single-particle translational partition function is given by:

$$\left(\frac{2\pi m k_B T}{h^2}\right)^{3/2} V, \tag{35}$$

where m is the particle mass, h is Planck's constant, and V is the volume of the domain. At the other extreme, excited electronic states normally lie at energies much greater than the available thermal energy. Given a nondegenerate ground state and a lowest excited state at $\varepsilon_1 \gg k_B T$, the electronic partition function simplifies to one:

$$Z_{elec}(T) = \sum_{i=0} g(\varepsilon_i)\, e^{-\varepsilon_i/(k_B T)} \approx (1)\, e^0 + g(\varepsilon_1)\, e^{-\infty} = 1 + 0 = 1. \tag{36}$$

The vibrational partition function, on the other hand, normally needs to be worked out in some detail, reflecting the energy expressions derived from quantum mechanics in the previous sections.

In general, we can calculate the vibrational partition function only if we know all of the vibrational energies of the system ε_{vib}, in which case we can set

$$Z_{vib}(T) = \sum_{\varepsilon_{vib}}^{\infty} e^{-\varepsilon_{vib}/(k_B T)}, \tag{37}$$

the average energy in each vibrational mode can then be expressed in terms of the weighted sum over all of the vibrational states of each vibrational mode j:

$$\langle \varepsilon_{vib} \rangle = \sum_{j=1}^{N_{vib}} \langle \varepsilon_j \rangle, \tag{38}$$

$$\langle \varepsilon_j \rangle = \sum_{v=0}^{\infty} \varepsilon_j\, \mathcal{P}(v) = \sum_{v=0}^{\infty} \varepsilon_j \frac{e^{-\varepsilon_j/(k_B T)}}{Z_{vib,j}(T)}. \tag{39}$$

As long as $\omega_e \gtrapprox k_B T$, only the first few vibrational quantum states will have any significant population, and it is likely that the harmonic approximation will be valid. In that case, to simplify notation, we first adapt the energy expression to set the ground vibrational state $v = 0$ to zero energy:

$$\varepsilon_{\text{vib}} = \omega_e\, v. \tag{40}$$

This removes the zero-point energy from the calculations for convenience; since this simply adjusts the arbitrary position of our energy scale, it has no impact on any observable quantity. The sum that we need to solve for the partition function of any nondegenerate harmonic vibrational mode j,

$$Z_{\text{vib},j}(T) = \sum_{v=0}^{\infty} e^{-\omega_j v/(k_\text{B}T)} = 1 + e^{-\omega_j/(k_\text{B}T)} + e^{-2\omega_j/(k_\text{B}T)} + \cdots \tag{41}$$

is exactly equal to the Taylor series expansion for the function $\left(1 - e^{-\omega_j/(k_\text{B}T)}\right)^{-1}$, so we may write

$$Z_{\text{vib},j}(T) = \frac{1}{1 - e^{-\omega_j/(k_\text{B}T)}}. \tag{42}$$

Similarly, a Taylor series expansion within the harmonic approximation allows us to write the average vibrational energy $\langle\varepsilon_j\rangle$ per molecule in each mode j:

$$\begin{aligned}\langle\varepsilon_j\rangle &= \sum_{v=0}^{\infty} \varepsilon_j\, \mathcal{P}(v) = \sum_{v=0}^{\infty} \omega_j\, v\, e^{-\omega_j/(k_\text{B}T)} \left(1 - e^{-\omega_j/(k_\text{B}T)}\right) \\ &= \frac{\omega_j}{e^{\omega_j/(k_\text{B}T)} - 1}.\end{aligned} \tag{43}$$

Deviations from these expressions are significant for high degrees of excitation, when $k_\text{B}T > \omega_j$, because higher vibrational levels explore more anharmonic regions of the surface, and also for vibrational modes which are highly anharmonic even at low excitation. In this latter category fall internal rotations, such as methyl group rotations in organic molecules. The barriers to torsion range from near 0 (the *free rotor*) to over 100 kJ mol^{-1} for rotation about a double bond, at which point the motion is so strongly hindered it is likely to appear harmonic at low temperatures. In general, these internal rotations present motion across two or more distinct minima, and the high deviations from harmonicity arise from the presence of multiple minima, separated by barriers comparable to or less than the available thermal energy. Let's take a simpler case, where the potential energy curve may be written

$$V(\phi) = \frac{V_0}{2}\left[1 - \cos(n\phi)\right], \tag{44}$$

where ϕ is the torsion angle, n is the number of equivalent minima on the surface, and V_0 is the barrier height relative to the minimum energy. Truhlar [15] has shown that a suitable approximation for the vibrational partition function in such a mode is given by

$$Z_{tors} = \frac{\tanh[\pi V_0/(k_B T)]^{1/2}}{2\sinh(u/2)}, \tag{45}$$

where

$$u = \sqrt{\frac{\hbar^2 n^2 V_0}{2I(k_B T)^3}}, \tag{46}$$

and where I is the moment of inertia of the rotor.

The most direct application of the partition function is in the calculation of vibrational state populations, which, for example, are necessary to predict the absorbance of vibrational hot band transitions. For relatively low-mass, gas-phase molecules at 298 K (which is a thermal energy $k_B T$ of 207 cm^{-1}), vibrational constants ω_e are typically greater than 500 cm^{-1}. The number of molecules in the ground vibrational state is found from

$$\mathcal{P}(v=0) = \frac{e^{-v\omega_e/(k_B T)}}{Z_{vib}(T)} = \frac{e^0}{\left[1 - e^{-\omega_j/(k_B T)}\right]^{-1}} = 1 - e^{-\omega_j/(k_B T)}. \tag{47}$$

Every gain of 500 cm^{-1} in the vibrational energy corresponds to a loss in population of roughly one order of magnitude, so for $^{35}Cl_2$ ($\omega_e = 560$ cm^{-1}) at 298 K, less than 10% of the molecules are in excited vibrational states; whereas for O_2 gas ($\omega_e = 1580$ cm^{-1}) that figure drops to less than 0.1%.

6 APPLICATIONS TO THERMODYNAMICS

The partition function is also a convenient starting point to assess the contribution of vibrations to thermodynamic potentials such as the energy, enthalpy, and free energy.

6.1 EQUIPARTITION PRINCIPLE

A classical analysis of molecular motions in the gas phase finds that the three major contributors to the thermodynamic energy can be cast as quadratic in the corresponding coordinates. For a linear diatomic molecule, for example:

$$\begin{aligned} &\text{Translation:} \quad \varepsilon_{trans} = \tfrac{1}{2} m v^2 = \tfrac{1}{2} m \left(v_x^2 + v_y^2 + v_z^2\right), \\ &\text{Rotation:} \quad \varepsilon_{rot} = \tfrac{1}{2} I \left(v_\theta^2 + \frac{1}{\sin^2\theta} v_\phi^2\right), \\ &\text{Vibration:} \quad \varepsilon_{vib} = \tfrac{1}{2} \mu v_{rel}^2 + \tfrac{1}{2} k (R - R_e)^2. \end{aligned} \tag{48}$$

In these expressions, I is again the moment of inertia and the v values are velocities along various coordinates; in particular v_{rel} is the relative speed of the two nuclei in the diatomic molecule. Assuming that each of these energy contributions can be treated as a continuous parameter, the fact that there is a term of the form cq_i^2 for each degree of freedom i means that the average energy $\langle \varepsilon_i \rangle$ found in each coordinate is given by

$$\begin{aligned}\langle \varepsilon_i \rangle &= \frac{\int_{-\infty}^{\infty} \varepsilon_i \, e^{-\varepsilon_i/(k_B T)} \, dq_i}{\int_{-\infty}^{\infty} e^{-\varepsilon_i/(k_B T)} \, dq_i} \\ &= \frac{\int_{-\infty}^{\infty} cq_i^2 \, e^{-cq_i^2/(k_B T)} \, dq_i}{\int_{-\infty}^{\infty} e^{-cq_i^2/(k_B T)} \, dq_i} \\ &= \frac{\frac{c}{2}\left(\frac{\pi}{c^3/(k_B T)^3}\right)^{1/2}}{\left(\frac{\pi}{c/(k_B T)}\right)^{1/2}} = \frac{k_B T}{2}.\end{aligned} \tag{49}$$

As long as the energy has a quadratic dependence on the coordinate, the contribution to the total energy is $k_B/2$, regardless of the physical nature of the motion. The total energy may then be summarized as

$$E = NN_{\text{ep}}\frac{k_B T}{2} = nN_{\text{ep}}\frac{RT}{2}, \tag{50}$$

where N_{ep} is the sum of the number of degrees of freedom, counting vibrations twice because each vibrational coordinate has both a potential energy and a kinetic energy contribution. This result is the *equipartition principle*, which formed a basis for the classical understanding of energy distributions in translations, rotations, and vibrations through the 19th century.

Unlike translation and rotation, vibration includes both kinetic and potential energy terms, each contributing $k_B T/2$ for each vibrational mode. According to equipartition, therefore, the energy content in vibrational motion per molecule is given by

$$\varepsilon_{\text{vib}} = N_{\text{vib}} k_B T. \tag{51}$$

6.2 HEAT CAPACITIES AND VIBRATIONS

On the basis of this estimate, we can calculate the heat capacity $C_V = \left(\frac{\partial E}{\partial T}\right)_V$ for an ideal crystal of N atoms. Assuming that each atom can only vibrate back and forth along each of the three Cartesian axes, we can ascribe three vibrational coordinates to each atom and arrive at a total energy content of

$$\text{Ideal crystal}: \quad E = 3Nk_B T. \tag{52}$$

The heat capacity would therefore be

$$\text{Ideal crystal}: \quad C_V = \left(\frac{\partial E}{\partial T}\right)_V = 3Nk_B = 3nR, \tag{53}$$

where we have taken advantage of the relationship between the gas constant and Boltzmann's constant, $R = \mathcal{N}_A k_B$, with $\mathcal{N}_A$ equal to Avogadro's number. Dividing by the number of moles n to obtain the molar heat capacity for a monatomic crystal, we obtain the law of Dulong and Petit:

$$C_{V\mathrm{m}} = 3R. \tag{54}$$

This equation successfully describes crystal heat capacities with fair accuracy, but only in the high temperature limit where the thermal energy is sufficient to easily excite vibrational transitions. At lower temperatures, the heat capacity rapidly drops toward zero as the thermal energy becomes small compared to the energy gaps between the vibrational quantum states. Einstein first recognized that quantizing the vibrational energy levels in the solid would have this effect, and applied the energy expression for the *quantum* harmonic oscillator in Eq. (43). Multiplying that energy expression by $3\mathcal{N}_A$ and taking the derivative with respect to temperature gives Einstein's formula for the heat capacity of a solid:

$$C_{V\mathrm{m}} = 3\mathcal{N}_A \left(\frac{\partial \langle \varepsilon_{\mathrm{vib}} \rangle}{\partial T}\right)_V = 3\mathcal{N}_A \left[\frac{\partial}{\partial T} \frac{\omega_{\mathrm{E}}}{e^{\omega_{\mathrm{E}}/(k_B T)} - 1}\right]_V = \frac{3\mathcal{N}_A \omega_{\mathrm{E}}^2 e^{\omega_{\mathrm{E}}/(k_B T)}}{k_B T^2 (e^{\omega_{\mathrm{E}}/(k_B T)} - 1)^2}. \tag{55}$$

The vibrational constant ω_{E} used in this expression is intended to be a single characteristic frequency for the crystal, and is usually empirically obtained for any given substance. Eq. (55) follows the qualitatively correct trend for solids, giving $3R$ in the limit of high temperature and dropping to zero as the temperature approaches zero.

For gases, the heat capacities can usually be estimated by combining the equipartition contributions from translation and rotation with a vibrational contribution similar to Eq. (55) for each vibrational mode:

$$C_{P\mathrm{m}} = \frac{3}{2} + \frac{N_{\mathrm{rot}}}{2} + \sum_{j=1}^{N_{\mathrm{vib}}} \frac{\omega_j^2 e^{\omega_j/(k_B T)}}{k_B T^2 (e^{\omega_j/(k_B T)} - 1)^2}. \tag{56}$$

Sample calculations in Table 2 show that this estimate is fairly successful, given all of the vibrational constants for the molecule.

6.3 FREE ENERGIES

In classical benchtop chemistry, chemical reactions occur in a temperature-regulated container such as a flask in ice or in boiling water, with the reaction mixture exposed to some atmosphere such as air or dry nitrogen. Under these conditions, the thermodynamic potential that becomes the natural focus of chemical thermodynamics is the Gibbs free energy G:

$$G \equiv E - TS + PV. \tag{57}$$

Table 2 Vibrational Constants, Calculated, and Measured Heat Capacities C_{Pm} at 298 K for Selected Gases

Species	ω_j (cm^{-1})	C_{Pm} (Calc.) ($J\,K^{-1}\,mol^{-1}$)	C_{Pm} (Exp.) ($J\,K^{-1}\,mol^{-1}$)
CO	2169.8	29.12	29.15
I_2	214.5	36.58	36.87
OCS	2062, 859 520, 520	41.17	41.49
CH_3OH	3681, 3000, 2960 2844, 1477, 1477, 1455, 1345, 1165, 1060, 1033, 295	45.10	44.06

The convenience of G lies in the form of its derivative, which simplifies under conditions of constant temperature and pressure to yield a dependence only on the changes in the chemical nature of the system. For a system with k chemical components,

$$dG = \sum_{i=1}^{k} n_i \, d\mu_i. \tag{58}$$

It is from Eq. (58) that we obtain the relationship between free energy and chemical equilibrium constant K_{eq}:

$$\ln K_{eq} = -\frac{\Delta_{rxn} G^{\ominus}}{RT}. \tag{59}$$

In order to calculate the free energy for a given species by means of electronic structure packages, we need to relate the terms in Eq. (57) to the computed properties of the molecule. For a gas-phase system, that means including contributions from zero-point energy, thermal energy $\langle\varepsilon\rangle$ (the average energy absorbed by the molecules to reach the given temperature), and the $-TS + PV$ terms. The thermal energy per molecule we can estimate by assuming that at 0 K the molecule would have no translation energy, no rotational energy, and only the zero-point energy ε_{ZPE} in vibration. Therefore, the average energy in translations $(3k_B T/2)$ and in rotations $(N_{rot} k_B T/2)$ we can combine with the average vibrational energies from Eq. (43) for each mode.

In order to write this in terms of the molar quantities appropriate to thermodynamics, we will multiply the molecular quantities by Avogadro's number $\mathcal{N}_A$. The terms we need in order to calculate the molar free energy G_m are therefore:

$$\begin{aligned} G_m &= \mathcal{N}_A \left(\varepsilon_0 + \varepsilon_{ZPE} + \langle\varepsilon\rangle\right) - TS_m + PV_m \\ &= \mathcal{N}_A \left(\varepsilon_0 + \varepsilon_{ZPE} + \frac{(3 + N_{rot})k_B}{2} + \sum_{j=1}^{N_{vib}} \langle\varepsilon_j\rangle\right) - TS_m + RT. \end{aligned} \tag{60}$$

The ideal gas law allowed us to replace PV_m by RT. In this expression, ε_0 is the electronic energy at the equilibrium geometry, and becomes significant if we wish to compare the energy of this geometry to the energy of some other geometry.

The molar entropy S_m is the only remaining term that we need to quantify, and we accomplish that by expressing the entropy in terms of the partition function. We start by assuming that the probability of the molecule being in any accessible quantum state is the same for each state, and therefore equal to $1/\Omega$ where Ω is the number of available states:

$$\mathcal{P}(i) = \frac{1}{\Omega}. \tag{61}$$

We can also use the canonical distribution, however. If we remove the factor of the degeneracy $g(\varepsilon_i)$ from Eq. (33), we can write the probability of being in any single quantum state (as opposed to the probability of being at a given energy level) as follows:

$$\mathcal{P}(i) = \frac{e^{-\varepsilon/(k_B T)}}{Z(T)}. \tag{62}$$

Setting Eqs. (61), (62) equal, we then take the logarithm of both sides to isolate Ω:

$$\Omega = e^{\varepsilon/(k_B T)} Z(T). \tag{63}$$

Boltzmann's law allows us to convert Ω to the molar entropy:

$$S_m = R \ln \Omega = R\left(\frac{\langle\varepsilon\rangle}{k_B T}\right) + R \ln Z(T) = \mathcal{N}_A \frac{\langle\varepsilon\rangle}{T} + R \ln Z_{\text{tot}}(T). \tag{64}$$

Substituting this expression into the equation for G_m reveals considerable simplification,

$$\begin{aligned} G_m &= \mathcal{N}_A\left(\varepsilon_0 + \varepsilon_{\text{ZPE}} + \frac{(3+N_{\text{rot}})k_B}{2} + \sum_{j=1}^{N_{\text{vib}}}\langle\varepsilon_j\rangle\right) - \left[\mathcal{N}_A \frac{\langle\varepsilon\rangle}{T} + R \ln Z_{\text{tot}}(T)\right] T + RT \\ &= E_{m0} - RT \ln Z_{\text{tot}}(T). \end{aligned} \tag{65}$$

This result allows us to express the equilibrium constant in Eq. (59) in terms of the partition functions. Using "prd" to indicate molar properties of the products and "rct" for the reactants, we obtain:

$$\begin{aligned} \ln K_{\text{eq}} &= -\frac{G^{\circ}_{\text{prd}} - G^{\circ}_{\text{rct}}}{RT} \\ &= -\frac{1}{RT}\left\{\left[E_{\text{prd}} - RT \ln Z_{\text{prd}}(T)\right] - \left[E_{\text{rct}} - RT \ln Z_{\text{rct}}(T)\right]\right\} \\ &= -\frac{\Delta_{\text{rxn}} E^{\circ}}{RT} + \ln \frac{Z_{\text{prd}}(T)}{Z_{\text{rct}}(T)}. \end{aligned} \tag{66}$$

7 APPLICATIONS TO KINETICS

In transition state theory, the reaction dynamics of any elementary step are assumed to be controlled by the reactants, an activated complex, and the products. For a bimolecular reaction of molecules A and B, the first step in the reaction is the formation of the activated complex $AB^{\ddagger}$, governed by an equilibrium constant $K^{\ddagger}$ such that

$$K^{\ddagger} = \frac{X^{\ddagger}}{X_A X_B}, \tag{67}$$

where the X values are the mole fractions of each component. Rearranging Eq. (67), we can express the fraction of the activated complex in terms of the reactants:

$$X^{\ddagger} = K^{\ddagger} X_A X_B. \tag{68}$$

The reaction velocity v is the rate of product formation, and we now assume this to be proportional to the concentration of the activated complex:

$$v = f X^{\ddagger} = f K^{\ddagger} X_A X_B = k_{TST}[A][B], \tag{69}$$

where in the last step we have converted the mole fractions to molarities. The transmission factor f gives the rate at which the activated complex is successfully converted to products. The rate constant therefore depends on two values: f and the equilibrium constant $K^{\ddagger}$.

The transmission factor f is approximated to varying levels by different means. Of interest to us here is the evaluation of the equilibrium constant $K^{\ddagger}$, which relies primarily on the vibrational partition functions of the reactants and the activated complex. Applying Eq. (66) to this system results in the expression

$$K^{\ddagger} \approx \frac{Z^{\ddagger}_{tot}}{Z_A Z_B}, \tag{70}$$

where $K^{\ddagger}$ includes contributions from all of the vibrational coordinates of the complex. However, the reaction coordinate is one of those coordinates. When A and B are independent molecules, they approach each other by motions that (in the gas phase) we would characterize as translations. However, on the global potential energy surface which combines the two molecules, the relative motion is along a vibrational coordinate, the reaction coordinate, because it parameterizes changes in the distances between atoms in the system.

However, the reaction coordinate is unlike the other vibrational coordinates in that the activated complex is not bound by this motion. The complex does not lie at a minimum along this coordinate, but instead will spontaneously dissociate back into reactants or move on to form products. To separate its contribution to the overall partition function, we label this term function $Z^{\ddagger}_{rxn}$ and set

$$Z^{\ddagger}_{tot} = Z^{\ddagger} Z^{\ddagger}_{rxn}. \tag{71}$$

Because the motion along the reaction coordinate is not bound, we approximate that contribution to the partition function using an expression for translational partition functions, calculated for a particle in a box:

$$Z^{\ddagger}_{\mathrm{rxn}} \approx \left(\frac{\sqrt{2\pi \mu k_{\mathrm{B}} T}}{h}\right) R_{\mathrm{rxn}}, \tag{72}$$

where R_{rxn} is some characteristic distance along the PES that separates the reactant and product regions. This gives a rough form of the rate constant k_{TST} from transition state theory, showing the dependence on the partition functions:

$$k_{\mathrm{TST}} \approx \left(\frac{k_{\mathrm{B}} T}{h}\right) \frac{Z^{\ddagger}}{Z_{\mathrm{A}} Z_{\mathrm{B}}}. \tag{73}$$

Contributions to the partition function from the reaction coordinate are largely canceled by factors involved in the transmission factor f. The value of $Z^{\ddagger}$ is commonly calculated by using the vibrational properties calculated at the transition state, which is ideally a well-defined maximum on the minimum energy path that connects the reactants and products.

As Eq. (73) indicates, accurate calculation of rate constants from first principles generally rely on accurate vibrational partition functions. Effectively, this contribution to the rate constant is measuring the ratio of the quantum state density of the transition state to the quantum state density of the reactants. The more states that exist in the activated complex at the appropriate energy, the likelier it is that the reactants will channel their internal energy and collision energy into forward motion along the reaction coordinate to products. On the other hand, the greater the density of quantum states in the individual reactants, the likelier it is that the energy will be dissipated among the internal degrees of freedom.

8 CONCLUSION

For many years, as the use of electronic structure programs proliferated and became gradually standardized, vibrational analysis for the prediction of spectra and thermodynamic properties was limited to the harmonic approximation. This was no great burden, as the efficacy of the harmonic approximation proved to be extraordinary. However, computational capabilities and standards of performance have advanced to the level that anharmonic analysis of vibrational modes is routine and increasingly expected. The hope is that improved methods for calculating these properties, especially of intrinsically anharmonic systems such as hindered rotors and surfaces with multiple minima, will lead to significant progress in the predictive power of kinetic models for complex processes like combustion and atmospheric chemistry.

REFERENCES

[1] Jones RN. Some retrospective thoughts on vibrational spectroscopy. Can J Spectrosc 1981;26:1–9.

[2] Jacobsen M. Lapace and the origin of the Ornstein-Uhlenbeck process. Bernoulli 1996;2:271–86.

[3] Barone V. Anharmonic vibrational properties by a fully automated second-order perturbative approach. J Chem Phys 2005;122:014108.

[4] Zuniga JZ, Picón JAG, Bastida A, Requena A. On the use of optimal internal vibrational coordinates for symmetrical bent triatomic molecules. J Chem Phys 2005;122:224319.

[5] Benjamin I, Levine RD, Kinsey JL. High-lying levels of ozone via an algebraic approach. J Chem Phys 1983;87:727–9.

[6] McCoy AB, Carter S, Landeweer MY, Bowman JM. Full-dimensional vibrational calculations for $H_5O_2^+$ using an ab initio potential energy surface. J Chem Phys 2005;122:061101.

[7] Maisuradze GG, Thompson DL. Interpolating moving least-squares methods for fitting potential energy surfaces: illustrative approaches and applications. J Phys Chem A 2003;107:7118–24.

[8] Dawes R, Thompson DL, Wagner AF, Minkoff M. Interpolating moving least-squares methods for fitting potential energy surfaces: a strategy for efficient automatic data point placement in high dimensions. J Chem Phys 2008;128:084107.

[9] Light JC, Carrington T. Discrete-variable representations and their utilization. Advances in Chemical Physics. New York, NY: John Wiley & Sons, Inc.; 2007. p. 263–310.

[10] Mátyus E, Czakó G, Császár AG. Toward black-box-type full- and reduced-dimensional variational (ro)vibrational computations. J Chem Phys 2009;130:134112.

[11] Bowman JM. Self-consistent field energies and wavefunctions for coupled oscillators. J Chem Phys 1978;68:608–10.

[12] Bowman JM, Christoffel K, Tobin F. Application of SCF-SI theory to vibrational motion in polyatomic molecules. J Phys Chem 1979;83:905–12.

[13] Christiansen O. Vibrational coupled cluster theory. J Chem Phys 2004;120:2149–59.

[14] Cheng X, Steele RP. Efficient anharmonic vibrational spectroscopy for large molecules using local-mode coordinates. J Chem Phys 2014;141:104105.

[15] Truhlar DG. A simple approximation for the vibrational partition function of a hindered internal rotation. J Comput Chem 1991;12:266–70.

CHAPTER

Introduction to fixed-node quantum monte carlo

6

Stuart M. Rothstein[*,†], **Egor Ospadov**[*,‡], **Caila Bruzzese**[†]

Department of Physics, Brock University, St. Catharines, Ontario, Canada[*] *Department of Chemistry, Brock University, St. Catharines, Ontario, Canada*[†] *Department of Chemistry, The University of Western Ontario, London, Ontario, Canada*[‡]

1 INTRODUCTION

While quantum mechanics is a remarkably successful theory, it is not without its shortcomings. There are relatively few systems that can be solved exactly using this theory. The problem arises in the complexity of the mathematics required to solve the Schrödinger equation applied to atoms, molecules, soft matter, and solid-state systems. As the size of the system (e.g., number of electrons) grows, the equations associated with the system become progressively more difficult to solve analytically. Initial attempts to solve larger systems involve using approximations, such as the variational method or perturbation theory, to reduce the complexity of the associated equations. While these methods provide us with a modest increase in the size of the system that could be treated using quantum mechanics, they do so at the expense of the accuracy of the results. Ultimately, these methods led to a deeper understanding of atomic and molecular structure, but the goal of being able to solve any possible system analytically eludes us to this day.

After coming to terms with our current inability to solve the Schrödinger equation exactly, the next natural step is to develop approximation methods capable of giving us a solution with desired accuracy. Quantum Monte Carlo (QMC) [1,2], which is an extension of the classical Monte Carlo scheme applied to solve the Schrödinger equation [3,4], is one such widely used and successful approach; see reviews [5–8] Technological advances in the field of high-performance computing have greatly aided this approach by providing the necessary computing power to implement this method [9–14].

QMC methods are used to study the quantum systems that can be described with or even without an input wave function, by obtaining a stochastic solution to the associated Schrödinger equation. Inspection of Anderson's tree of theoretical chemistry shows that QMC comes in several varieties (Fig. 1). To discuss all of

Mathematical Physics in Theoretical Chemistry. https://doi.org/10.1016/B978-0-12-813651-5.00006-1

them would be beyond the scope of this chapter. Instead, we focus upon fixed-node quantum Monte Carlo (FNQMC) [2,15,16], and on a more recently developed pure-sampling quantum Monte Carlo (PSQMC) [17,18], in the context of the ground state of atoms and molecules.

Both algorithms generate a *mixed distribution* of electrons, $\Psi\Phi_0$. The ground-state energy is calculated by stochastically sampling from this distribution. At the same time, a *pure distribution* of electrons, Φ_0^2, is generated by PSQMC. Stochastic sampling from the latter distribution calculates various experimentally important properties such as the dipole moment and other electronic properties with minimal bias (Section 4).

Φ_0 is the unknown, putatively exact, nonrelativistic ground-state wave function, generally within the Born-Oppenheimer and fixed-node approximations. Ψ is an input *importance-sampling or trial* function, which controls statistical fluctuations during mathematical moves that guide electrons away from unimportant regions of configuration space and toward important ones (Section 2.1). There are methods that can sample directly from Φ_0, bypassing the requirement for Ψ, but their range of application is currently limited; see recent review [19].

The concept of electron-exchange nodes is critical to understanding the fixed-node QMC methods described in this chapter. The exact solution to the Schrödinger equation, Φ_0, has an exchange node (where it vanishes) whenever two electrons with the *same* spin coalesce. Except for node-less systems, a minimal bias in calculated properties arises from the mismatch between the nodal hypersurface of the importance-sampling function, Ψ, and that of the exact ground-state wave function. This is the so-called "fixed-node bias." Electron-exchange nodes and the accompanying fixed-node bias are further discussed in Section 2.4.

Exchange nodes are not to be confused with, say, the nodes in atomic orbitals. This is illustrated by the example of Ne atom, taken from Ceperley [20]. The atom has 10 electrons, fully occupying the 1s, 2s, and 2p orbitals. The 1s orbital is node-less, the 2s orbital has a spherical node, and the 2p orbitals have planar nodes through the origin. The exchange nodes belong to a 29-dimensional manifold within the 30-dimensional space of electron coordinates.

Fixed-node diffusion Monte Carlo (FNDMC) (Section 2) is the most widely used quantum Monte Carlo (QMC) method [2,15,16]. As mentioned earlier, it affords sampling of physical properties from the mixed distribution of electrons. Only the ground-state energy of the system can be determined directly using FNDMC, albeit with the fixed-node bias. The fixed-node energy affords an *upper bound* to the exact ground-state energy, with an error commonly believed to be small. A benchmark on first row atoms suggests that this method is capable of recovering 99% of the electron correlation energy, and calculations on the G1 set have mean absolute error in total energy 2 mH and 0.8 kcal/mol in atomization energies [21]. Any property represented by an operator that commutes with the position operator is biased more severely (Section 3.1). This is a direct result of FNDMC's inability to generate and sample from the pure distribution of electrons.

PSQMC [17,18] (Section 3) is a straightforward and computationally efficient method used to sample electronic properties from both mixed and pure distributions of electrons. This algorithm can simultaneously calculate both the ground-state energy (from the mixed distribution, $\Psi\Phi_0$) and ground-state properties associated with operators that commute with the position operator (from the pure distribution, Φ_0^2), without the above-mentioned importance-sampling bias. PSQMC rests on independent Metropolis sampling (Section 3.3), which improves upon the reliability of the calculated properties, suggesting that this method is computationally superior to reptation QMC [22,23], its nearest competitor.

There are other methods beside QMC that are designed to give an approximate solution to the Schrödinger equation for atoms and molecules, shown in Fig. 1. Some of these widely used methods, discussed in Chapters 4 and 5, include density functional theory (DFT) [24–27], coupled-cluster (CC) [28–32], and configuration interaction (CI) [33–37]. QMC methods have intrinsic disadvantages when compared to these methods: statistical error in the calculations [38], fixed-node biases, and importance-sampling bias (in FNQMC), discussed previously, in addition to other more subtle biases [39], which we shall ignore due to their technical nature. At the same time, QMC methods also present a big advantage over these methods. QMC is variationally bound, requires insignificant amount of disk IO and RAM resources, is well-suited to modern parallel computer technology, and has favorable scaling with respect to system size, although it does require a large investment of CPU time.

For the sake of comparison, FNDMC and PSQMC, the algorithms described in this chapter, scale as $O(n^3)$. DFT also scales as $O(n^3)$ [26] while CCSD(T) scales as $O(n^3N^4)$ [32] (neither variationally bound). CISDT, which is variationally bound, scales as $O(n^4N^6)$ [32]. Here, n is the number of occupied orbitals and N is the number of unoccupied orbitals. Setting the issue of accuracy aside, the scaling establishes limitations to the practical applications of these methods. QMC methods have the potential to be applied to very large systems, such as molecular clusters, polymers, and proteins, due to their favorable scaling being on par with DFT. On the other hand, CC and CI methods are usually applied to small molecules, but they become computationally challenging for large systems, demanding exhaustive CPU and IO resources.

2 DIFFUSION MONTE CARLO

The exact ground-state solution to the Schrödinger equation, Φ_0, is generally not known analytically, but under certain conditions by utilizing random numbers, diffusion Monte Carlo (DMC) can sample configuration space from Φ_0; recent review [8]. This is shown in Fig. 2, where 20 DMC determinations of a free particle's position in a one-dimensional box of length L are shown. Of course, in this case, Φ_0 is known, but this information is not used in DMC. Observe that the particle tends to be located near the center of the box and is more rarely seen near its ends, where an infinite potential energy is imposed upon it. The point made by this figure

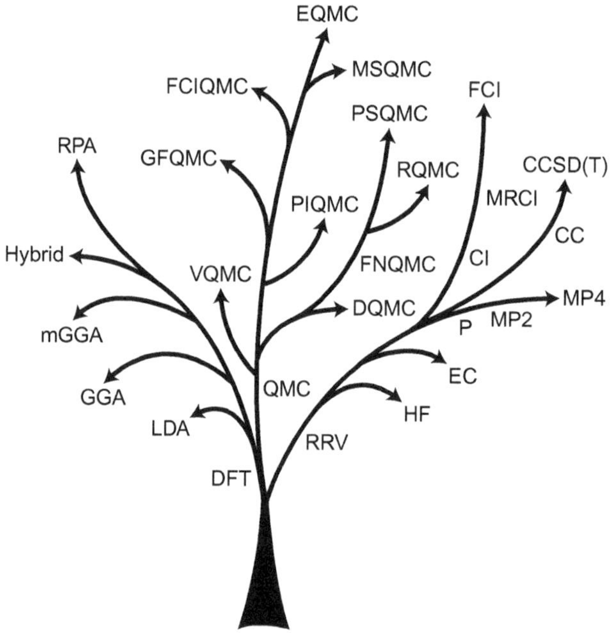

FIG. 1

An updated version of Anderson's tree of theoretical chemistry. The tree has three main branches: density functional theory (DFT), quantum Monte Carlo (QMC), and Rayleigh Ritz variational methods (RRV). Entries on the RRV branch are Hartree-Fock (HF) and Explicitly Correlated (EC) functions. The RRV branch then trifurcates into (A) perturbation theory (P) (which spawns second-order and fourth-order Møller-Plesset theory (MP2 and MP4)); (B) coupled-cluster methods (CC) (spawning coupled cluster with single and double excitations and perturbative triple excitations (CCSD(T))); and (C) configuration interaction (CI) (spawning multireference CI [MRCI] and full configuration interaction [FCI]). Density functional approximations include local density approximation (LDA); generalized gradient approximation (GGA); metageneralized gradient approximation (mGGA); and hybrid GGAs and mGGAs (hybrid). The QMC main branch has two subbranches: (A) variational quantum Monte Carlo (VQMC) and (B) still another subbranch that includes diffusion Monte Carlo (DQMC) and fixed-node quantum Monte Carlo (FNQMC). The latter spawns reptation quantum Monte Carlo (RQMC) and pure-sampling quantum Monte Carlo (PSQMC). The main branch continues by including path-integral quantum Monte Carlo (PIQMC), Green's function quantum Monte Carlo (GFQMC), full configuration interaction quantum Monte Carlo (FCIQMC), model state quantum Monte Carlo (MSQMC), and exact quantum Monte Carlo (EQMC). Methods on the higher branches are more accurate. This chapter deals with diffusion quantum Monte Carlo, fixed-node diffusion Monte Carlo, and pure-sampling quantum Monte Carlo.

(Anderson's tree modified with permission from Anderson JB. Quantum Monte Carlo: atoms, molecules, clusters, liquids, and solids, vol. 13, chap. 3. New York, NY: Wiley-VCH, John Wiley and Sons, Inc.; 1999. p. 133–82.)

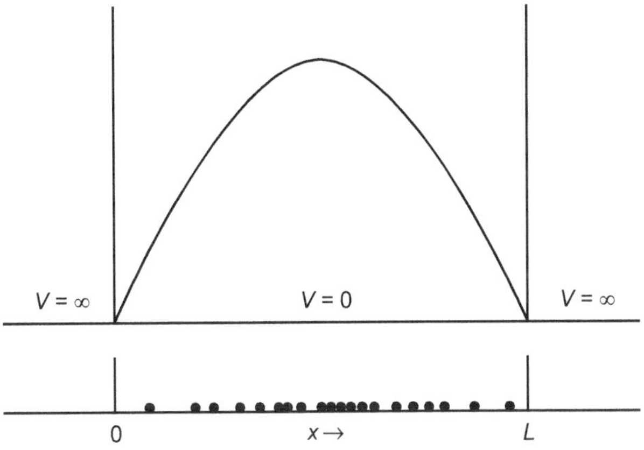

FIG. 2

Diffusion Monte Carlo treatment of a particle in a one-dimensional box of length L. The particle is constrained within the box by infinite potentials applied at its endpoints.

(Reproduced from Anderson JB. Fixed-node quantum Monte Carlo. Int Rev Phys Chem 1995;14(1):85–112. https://doi.org/10.1080/01442359509353305, with permission of Taylor & Francis Ltd.)

is that despite not having an analytic expression for Φ_0, with DMC, one can draw random samples of electron positions from it. For example, the expected value of the particle's position is simply its average value: $\langle x \rangle \approx \sum x_i/N \approx L/2$, where in this case, $N = 20$. The notation stresses that there is a statistical error associated with this determination. The error vanishes in the limit of infinite N.

Now some theory: In DMC diffusion corresponds to the change in distance of the electron's position with relation to imaginary time. The time-dependent Schrödinger equation is solved numerically as imaginary time (t) increases, yielding a simulation that samples configuration space from the lowest energy-state wave function of any given symmetry.

In particular, the time-dependent Schrödinger equation is viewed as the following diffusion equation, expressed in atomic units ($\hbar = m_e = e = 1$)

$$-\frac{\partial \Phi(\mathbf{R}, t)}{\partial t} = \left[-\frac{1}{2}\nabla^2 + V(\mathbf{R}) - E_0\right]\Phi(\mathbf{R}, t) \tag{1}$$

where $\Phi(\mathbf{R}, t)$ represents the exact wave function of the system, $V(\mathbf{R})$ is the system's potential energy, E_0 is the system's exact ground-state energy, and $\mathbf{R}$ is the set of $3n$ coordinates of n electrons. This equation is solved numerically (simulated) by a kinetic branching term, $\left[V(\mathbf{R}) - E_0\right]\Phi(\mathbf{R}, t)$, and a diffusion term, $-\frac{1}{2}\nabla^2\Phi(\mathbf{R}, t)$. These are the essential factors in DMC: the diffusion of electrons in imaginary time, and the branching of electron positions, relating to their multiplication or elimination. The ground-state solution to the time-dependent Schrödinger equation is achieved after a large number of these conceptual moves of the electrons reach a steady state.

Statistical error and failure to account for Fermi statistics are the main problems with this approach, restricting its use to Bose and node-less systems. Two techniques, importance sampling and treatment of Fermi statistics, have aided in reducing these errors in calculations of atomic and molecular structures.

2.1 IMPORTANCE SAMPLING

Attempts to solve Eq. (1) by stochastic processes involving branching would cause the branching term to diverge because it is proportional to the potential energy, which can approach to $+\infty$ or $-\infty$. Diverging branching rates would further lead to a slow convergence of the energy because the statistical weight assigned to each "configuration," $\mathbf{R}$, would vary over an immense range.

Importance sampling generates the *imaginary time-dependent* mixed probability distribution

$$f(\mathbf{R},t) = \Psi(\mathbf{R})\Phi(\mathbf{R},t) \tag{2}$$

instead of $\Phi(\mathbf{R},t)$. $\Psi(\mathbf{R})$ is a known trial wave function [16,40].

Eq. (1) can then be recast in terms of $f(\mathbf{R},t)$ to yield the *modified Schrödinger Equation*, derived in Section 7.1.

$$-\frac{\partial f(\mathbf{R},t)}{\partial t} = -\frac{1}{2}\nabla^2 f + (E_L(\mathbf{R}) - E_o)f + \nabla\cdot(f\mathbf{F}(\mathbf{R})) \tag{3}$$

where E_L, defined by Eq. (4), is the local energy of the trial wave function, $[E_L(\mathbf{R}) - E_o]$ is the excess local energy, and $\mathbf{F}(\mathbf{R})$ is the drift velocity.

$$E_L = \frac{\hat{H}\Psi(R)}{\Psi(R)} \tag{4}$$

$$\mathbf{F}(\mathbf{R}) = \frac{1}{2}\nabla \ln|\Psi(\mathbf{R})|^2 = \nabla\Psi(\mathbf{R})/\Psi(\mathbf{R}) \tag{5}$$

This drift is dependent upon $\Psi(\mathbf{R})$, which allows it to control sampling so that in regions with a high probability (where $\Psi(\mathbf{R})$ is large), the drift is small, but in low probability regions the drift is large, directing particles away. $\frac{\partial f(\mathbf{R},t)}{\partial t}$, unlike $\frac{\partial \Phi(\mathbf{R},t)}{\partial t}$, is simulated by configurations undergoing diffusion, drift, and branching [40].

In Eq. (3) the $E_L(\mathbf{R}) - E_0$ term replaces the $V(\mathbf{R})$ term from Eq. (1). As long as a good trial function is chosen, $E_L(\mathbf{R}) - E_0$ will not have the singularities that $V(\mathbf{R})$ does because E_L will be as consistent as possible in an attempt to minimize branching.

The time-dependent exact wave function can be expanded using the power series, which gives rise to the following equation

$$\begin{aligned}\Phi(\mathbf{R},t) &= \sum_{i=0}^{\infty}\Phi_i(\mathbf{R})e^{-(E_n-E_0)t}\\ &= \Phi_0(\mathbf{R}) + \Phi_1(\mathbf{R})e^{-(E_1-E_0)t} + \Phi_2(\mathbf{R})e^{-(E_2-E_0)t} + \cdots\end{aligned} \tag{6}$$

where Φ_0 corresponds to the ground-state wave function, Φ_1 corresponds to the first excited state wave function, etc. In the asymptotic limit, $\Phi(\mathbf{R}, t \to \infty) \equiv \Phi_0(\mathbf{R})$, because as $t \to \infty$, every term in Eq. (6) approaches to zero except for the ground-state wave function.

2.2 SHORT-TIME GREEN'S FUNCTION

We obtain the asymptotic limit of the mixed-distribution function, $f(\mathbf{R}, t \to \infty)$, and hence $\Psi(\mathbf{R})\Phi_0(\mathbf{R})$, by iterating toward the limit of large t via small *time-steps* (imaginary-time intervals), τ, until the local energy "equilibrates," showing no further systematic changes.

Formally, when the simulation time for one iteration increases by τ, the mixed-distribution function changes as dictated by the short-time Green's function.

$$f(\mathbf{R}', t+\tau) = \int d\mathbf{R} f(\mathbf{R}, t) G(\mathbf{R} \to \mathbf{R}', \tau) \tag{7}$$

where $\mathbf{R}'$ is the new set of coordinates for the electrons. The short-time Green's function, $G(\mathbf{R} \to \mathbf{R}', \tau)$, describes the probability that the set of coordinates $\mathbf{R}$ will be moved to $\mathbf{R}'$ [16]. Since one can assume that $E_L(\mathbf{R})$ and $\mathbf{F}(\mathbf{R})$, Eqs. (4), (5) are constant and independent of $\mathbf{R}$ for small time-steps, this transition probability is approximated by the equation

$$G(\mathbf{R} \to \mathbf{R}', \tau) = Ne^{-(\mathbf{R}'-\mathbf{R}-\tau\mathbf{F}(\mathbf{R}))^2/2\tau} e^{-(E_L(\mathbf{R})+E_L(\mathbf{R}'))\tau/2} \tag{8}$$

where N is a normalization constant. The algorithm is independent of the value of N. (Not having to normalize either the Green's function or the importance-sampling function, Ψ, are two of the great advantages of the QMC methods described in this chapter.)

The first factor in the short-time Green's function, Eq. (8), is simulated by the following drift and diffusion moves (which are of no physical significance):

$$\mathbf{R}' = \mathbf{R} + \tau\mathbf{F}(\mathbf{R}) + \sqrt{\tau}\chi \tag{9}$$

where $\tau\mathbf{F}(\mathbf{R})$ represents the drift and $\sqrt{\tau}\chi$ diffusion of the electrons [16]. Here, diffusion is modeled by a standard Gaussian probability distribution, with a standard deviation of $\sqrt{\tau}$. Any other probability distribution can be employed, provided it has zero mean and unit first moment, in agreement with the standard Gaussian distribution [41]. The second factor is normally simulated by kinetic (or physical) branching, where the electron coordinates (or configuration) $\mathbf{R}$ is retained with a branching probability proportional to $\exp[-(E_L - E_0)\tau]$, as described in the following section.

2.3 KINETIC BRANCHING

The FNDMC algorithm normally works with an ensemble of many "walkers," or electron configurations, each of which is a replicate of the system under study, with its n electrons distributed among the nuclei (normally fixed) in configuration space. Walkers are denoted here and for the remainder of this section by the notation $\mathbf{R}$. The multiplicity term corresponds to branching, simulating the second exponential term in Eq. (8), and determines the number of times $\mathbf{R}$ will appear in the new ensemble of configurations after one τ increase in simulation time. The integer value obtained for the branching probability (B_m) determines the number of copies of the mth walker that will appear in the new set. B_m is calculated from the following equation [42]

$$B_m = e^{-(E_L(\mathbf{R})+E_L(\mathbf{R}'))\tau/2} \tag{10}$$

where $E_L(\mathbf{R})$ and $E_L(\mathbf{R}')$ correspond to the local energies of the mth walker at its initial and final positions, respectively.

In order to ensure that the number of walkers in one iteration does not either explode (approach ∞) or implode (approach 0), due to positive or negative singularities in the local energy, the following equation is used,

$$N_m = \mathrm{INT}[(B_m/\langle B\rangle) \cdot (M/M') + \zeta] \tag{11}$$

where N_m is the number of copies of the mth walker, $\langle B\rangle$ is the average branching probability of the entire set of walkers ($\langle B\rangle = \sum_{m=1}^{M'} \frac{B_m}{M'}$), M is the target number of configurations desired in the new set, and M' is the number of walkers currently in the new set. ζ represents a uniform random number between 0 and 1 that is used to account for the correction to the number of configurations in each set [42]. The INT symbol in this equation truncates the output value to an integer number. It is important to note from Eq. (11) that the number of copies of each walker, in terms of energy, only depends upon the local energies of the new set and the previous set of walkers.

The distribution of electrons is generated repeatedly until the diffusion and multiplicity components produce average local energy values that have obtained a steady state (reached an equilibrium). This branching is vital in order to decrease probability density in unfavorable regions (large potential energy) and increase it in regions of low potential energy [16].

The initial walker coordinates (or configurations) may be random as long as the system has enough time to diffuse, drift, and multiply to a steady state, Eqs. (9), (11), respectively, so that the same distribution is achieved from any set of initial coordinates [40]. The ensemble-averaged energy values are equilibrated to ignore values of E_L that were calculated before the system was in its steady state.

In turn, the ensemble-averaged local energy is averaged over the number of iterations performed at the given time-step.

$$E(\tau) = \sum_{i=1}^{I} \frac{\hat{E_{Li}}(\tau)}{I} \tag{12}$$

where I is the number of iterations performed. This is a time-step biased Monte Carlo estimator of the ground-state energy; $E(\tau) = \frac{\int d\mathbf{R} E_L(\mathbf{R})\Psi\Phi_o}{\int d\mathbf{R}\Psi\Phi_o}$, as shown in Section 7.2.

The previous notion stresses that these computations have a time-step bias; however, extrapolating data from various values of τ allows one to estimate the fixed-node energy at $\tau = 0\,E_h^{-1}$. This bias comes from the use of the short-time Green's function, which does not give the mixed-distribution any time-step except for $\tau = 0\,E_h^{-1}$. The produced data can most often be extrapolated to the fixed-node energy by fitting with a polynomial function [16,41]. An alternative approach that virtually eliminates the time-step bias in FNDMC has been published [39].

2.4 FIXED-NODE APPROXIMATION

Exchange nodes are present in Ψ due to the antisymmetry requirement that is necessary for wave functions of fermions, such as electrons. Antisymmetry causes electrons with like spin to repel one another, consistent with the Pauli exclusion principle. Bosons, on the other hand, require a symmetric wave function, so that there is a higher probability density of finding particles close to one another. Any system with one electron or two electrons with opposite spin is not affected by the Pauli exclusion principle.

As mentioned previously in Section 1, electron-exchange nodes, where they exist, occupy a $3n - 1$-dimensional manifold in the $3n$-dimensional space occupied by n electrons. There are positive ($\Psi > 0$) and negative ($\Psi < 0$) nodal regions. Simulations of Eq. (3) work within the fixed-node approximation, where no adjustment is made to these regions during the course of the calculations. Some algorithms "kill" walkers that cross a nodal boundary [16], but the error introduced by nodal crossings vanishes in the $\tau = 0\,E_h^{-1}$ limit, where the fixed-node energy and other properties are evaluated. The PSQMC algorithm, described in the following section, ignores nodal crossing.

Since $\Psi = 0$ at its nodes, the efficiency of FNDMC and PSQMC calculations is solely determined by the variance in E_L at regions away from the nodal boundaries. The error in the fixed-node approximation, dependent upon the nodal regions of Ψ, is the only remaining variable within the fixed-node approximation that cannot be corrected for, or controlled [43].

3 A PURE-SAMPLING QUANTUM MONTE CARLO METHOD

3.1 WHY DO PURE-SAMPLING?

The above-described FNDMC method provides a way to sample configuration space from the mixed probability density, the product of an input trial solution to the Schrödinger equation for the ground state, Ψ, and its unknown exact solution, Φ_0. Calculations that sample the local energy and properties represented by operators that commute with the Hamiltonian are exact, save for a small fixed-node bias, which was described in the previous section. However, properties represented by operators that commute with the position operator, such as the dipole moment and other electrical properties are also of interest. When calculated by sampling from the mixed distribution, these properties, A, are much more strongly biased by the input trial function.

$$\langle A_{\text{mixed}} \rangle = \frac{\langle \Psi | \hat{A} | \Phi_0 \rangle}{\langle \Psi | \Phi_0 \rangle} \tag{13}$$

instead of the desired

$$\langle A_{\text{pure}} \rangle = \frac{\langle \Phi_0 | \hat{A} | \Phi_0 \rangle}{\langle \Phi_0 | \Phi_0 \rangle} \tag{14}$$

This section describes an algorithm, and data analysis associated with it, that generates expectation values over the pure distribution, Φ_0^2, that are unbiased, exccpt for the exchange node error. Thereby, one accurately calculates physical properties such as the dipole and other electrical moments, electrical response properties of molecules, and particle distribution functions for clusters.

One of us surveyed the results of calculations that employ other pure-sampling methods through what has been published in 2012 [19]. His review also touched on truly exact sampling methods.

3.2 A PURE-SAMPLING ALGORITHM

PSQMC is shown in Fig. 3.

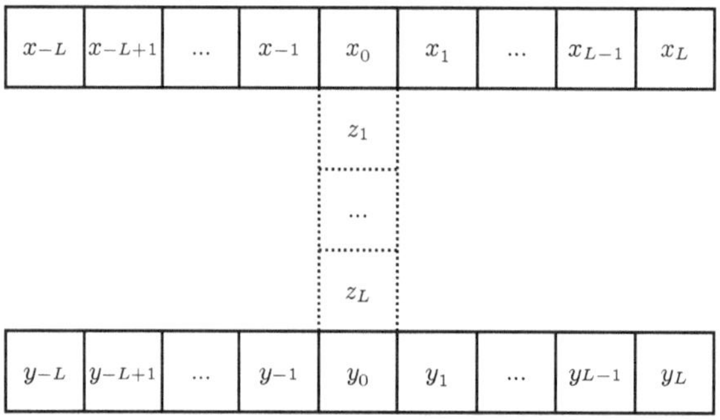

FIG. 3

Electron configurations in pure-sampling Quantum Monte Carlo (PSQMC) algorithm. Each configuration, represented by a *lower-case Roman letter*, is a $3n$-dimensional vector comprising a stochastic sample of the coordinates of the system's n electrons.

(Reproduced from Ospadov E, Rothstein SM. J Chem Phys 2015;142:024114, with the permission of AIP Publishing.)

In contrast to FNDMC, with its imaginary-time evolution comprised of a sequence of drift-diffusion-branching moves, here there are L drift and diffusion moves, Eq. (9), separating from their end points the middle electron configurations of the current (**X**) and proposed paths (**Y**). As well, there are L such moves to construct auxiliary path (**Z**) separating the middle configuration of these two paths. If the Metropolis decision, Eq. (15), accepts the proposed path, then its middle configuration, y_0, is a random sample drawn from the pure distribution, Φ_0^2, and the proposed path becomes the current one in the next iteration. On the other hand, if the Metropolis decision rejects the proposed path, then the middle of the current path, x_0, is counted again as a random draw from Φ_0^2. The mixed distribution, $\Psi\Phi_0$, is sampled at the end configurations on the accepted path.

The pure distribution being sampled at the middle of the accepted path, and the mixed distribution at its ends, is explained as follows: All configurations along the path are variationally (Ψ^2) distributed, but the end ones carry an additional statistical weight equal to Φ_0/Ψ. The mixed distribution is the net result: $\Psi^2 \times \Phi_0/\Psi = \Psi\Phi_0$. Similarly, there is a statistical weight equal to Φ_0/Ψ introduced at the middle configuration due to its imaginary-time evolution to both ends. The pure distribution is the net result: $\Psi^2 \times \Phi_0/\Psi \times \Phi_0/\Psi = \Phi_0^2$.

3.3 INDEPENDENT METROPOLIS

The expression for the Metropolis decision is

$$A(X \to Y) = \min\left\{1, e^{(S(\mathbf{X})-S(\mathbf{Y}))\tau}\right\} \tag{15}$$

where $S(\mathbf{X})$ is the sum of local energies, Eq. (4), along Path **X**, and double counting the local energy evaluated at the middle of the Path, x_0. $S(\mathbf{Y})$ is defined analogous to $S(\mathbf{X})$. The proposed path of electron configurations is accepted if the value of the exponential expression to the right of the comma is greater than ζ, which is determined by a random draw from the uniform distribution on interval $[0, 1]$ each time a Metropolis decision is performed. At this point, let us discuss the significance of the intermediate path that links the current and proposed paths. Despite its electron configurations not entering into the Metropolis decision, this auxiliary path is designed to make the current and proposed paths statistically independent. With statistical independence, the Metropolis decision is rigorous, unlike in reptation QMC [22,23], its nearest competitor, where it is an approximation. Furthermore, serial correlation of properties sampled from the pure distribution (at y_0) is reduced whenever the trial path is accepted.

We illustrate independent Metropolis by considering the problem of generating a gamma distribution from exponentially distributed observations. First, we start with the following expression for the Metropolis decision in its most general form:

$$A(X \to Y) = \min\left\{1, \frac{\hat{\Pi}(Y)W(Y \to X)}{\hat{\Pi}(X)W(X \to Y)}\right\} \tag{16}$$

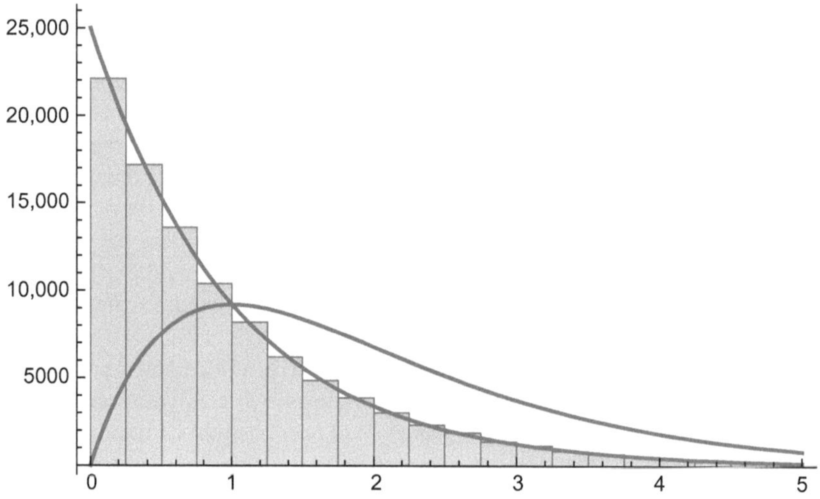

FIG. 4

Monte Carlo generation of gamma distribution from exponentially distributed variables. Zeroth iteration.

Here, $\hat{\Pi}(X)$ is the probability density of random variable X, and $W(X \rightarrow Y)$ is the transition probability for X to Y. The remaining symbols are defined analogously to these. (In the context of PSQMC, X is the current path and Y is the proposed path.)

Now, with X and Y independent, $\hat{\Pi}(X) = xe^{-x}$; $\hat{\Pi}(Y) = ye^{-y}$; $W(X) = e^{-x}$; and $W(Y) = e^{-y}$. Upon substituting these expressions into Eq. (16):

$$A(X \rightarrow Y) = \min\left\{1, \frac{y}{x}\right\} \tag{17}$$

Figs. 4–11 display the results observed from 100,000 independent samples. The algorithm converges after a handful of iterations and is stable thereafter (Vrbik, J. Private Communication).

This example shows that independent Metropolis sampling is rapidly convergent and stable.

3.4 REMOVING BIASES

In Section 5, an application to the ethene molecule, we demonstrate that by doing a sequence of small moves in imaginary time (with increments, time-step, τ) and evoking the Metropolis decision, Eq. (15), we can sample from the mixed and pure distributions of electrons to calculate, respectively, its ground-state energy and other electronic properties. In addition to the fundamental fixed-node bias, this proposed solution has two inherent biases: it is only valid in the limit of zero time-step and infinite simulation time. To remove these biases, we perform Monte Carlo runs for a

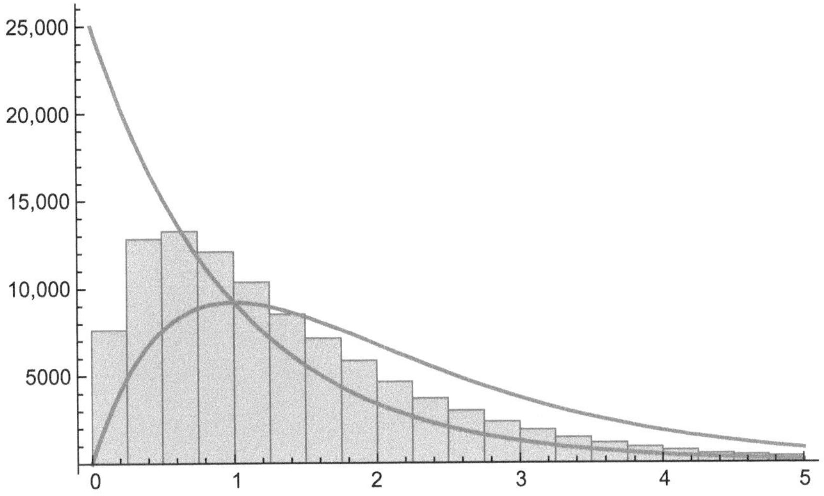

FIG. 5

First iteration.

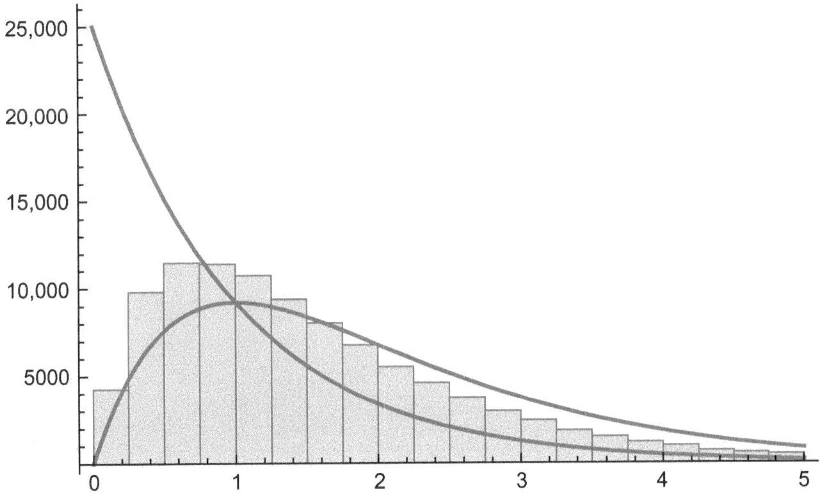

FIG. 6

Second iteration.

range of different time-steps, each having τ-dependent current and trial path-lengths, Eq. (34). The path-lengths become infinitely long in the limit of zero time-step. After sampling our properties at each individual time-step, we remove the time-step bias by extrapolating their average values to the limit of zero time-step.

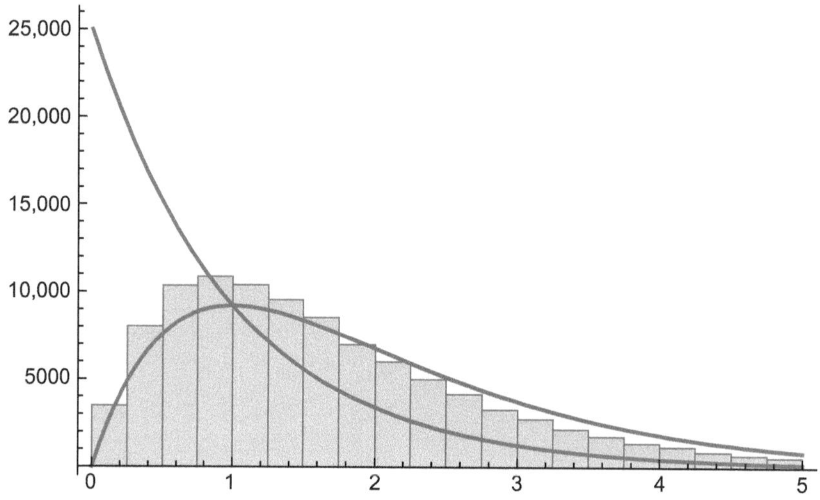

FIG. 7

Third iteration.

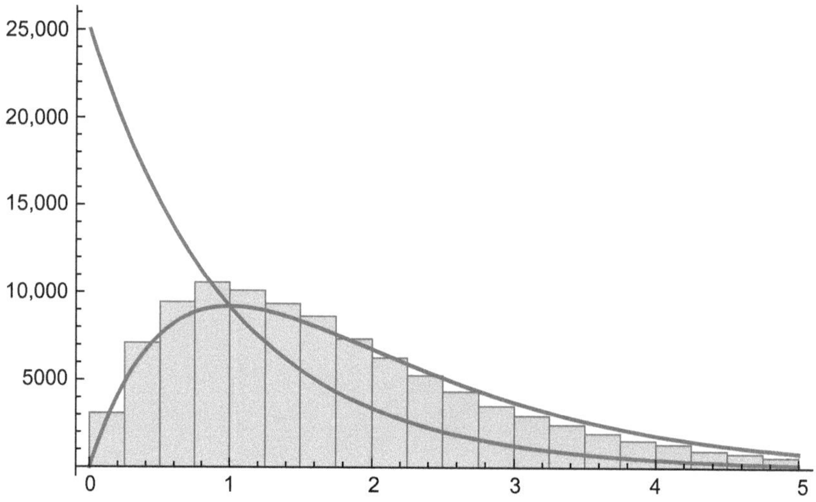

FIG. 8

Fourth iteration.

4 MOLECULAR PROPERTIES

4.1 FIXED-NODE ENERGY

The pure-sampling algorithm provides us with a straightforward method to compute the fixed-node energy of a system in the ground state. This is done by sampling the

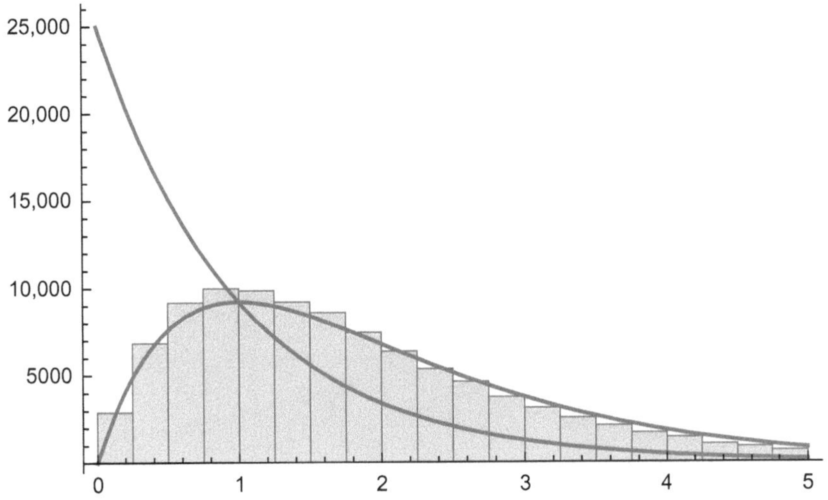

FIG. 9

Fifth iteration.

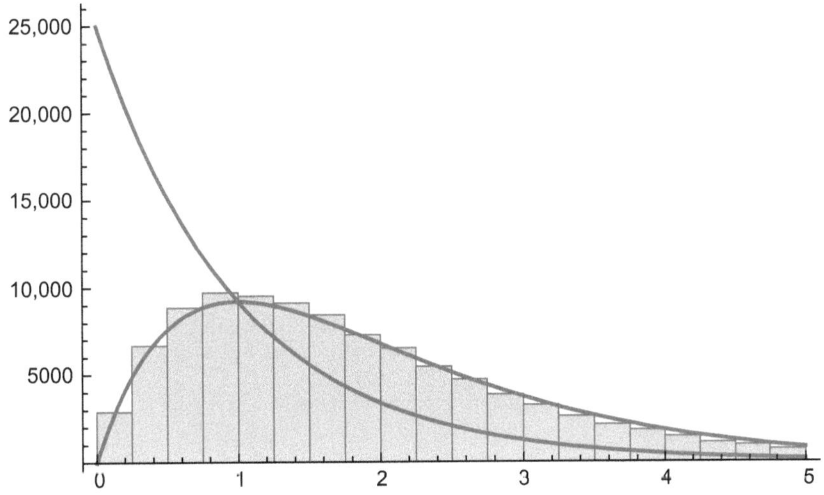

FIG. 10

Sixth iteration.

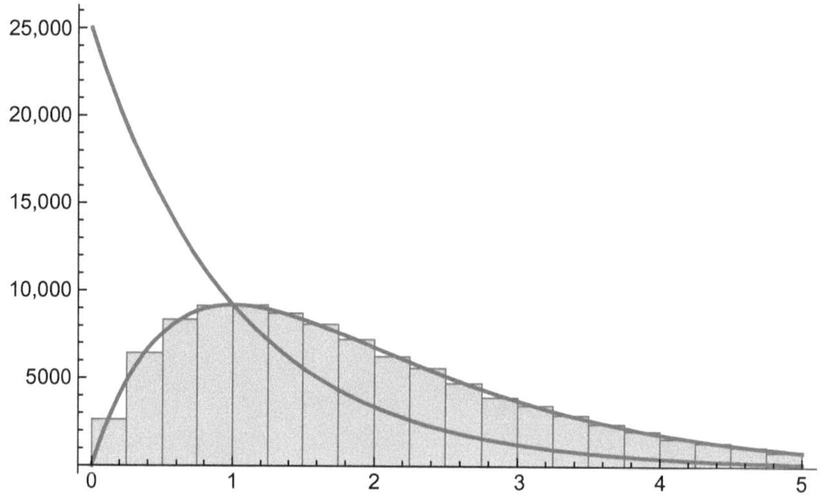

FIG. 11

Sixteenth iteration.

local energy from a mixed distribution of electrons and averaging it over the number of iterations of the algorithm.

PSQMC has two independent locations along the selected path where we can perform sampling from the mixed distribution. We utilize both of them to provide a more accurate sample of the fixed-node energy by taking their average. Our final expression for the time-step-biased fixed-node energy is:

$$E_{FN}(\tau) = \frac{1}{I}\sum_{i=1}^{I} \hat{E}_{Li} \quad (18)$$

where on the *i*th iteration, if the proposed path is accepted by the Metropolis decision

$$\hat{E}_{Li} = \frac{E_L(y_{-L}) + E_L(y_L)}{2} \quad (19)$$

otherwise

$$\hat{E}_{Li} = \frac{E_L(x_{-L}) + E_L(x_L)}{2} \quad (20)$$

where I is the number of iterations of the algorithm. When extrapolated to zero time-step, this technique provides us with a very good estimate of the fixed-node energy, as evident by the results for ethene (Section 5) and other small molecules in the literature [17].

4.2 OTHER ELECTRONIC PROPERTIES

Calculating properties of experimental importance, such as the dipole moment, is extremely straightforward with PSQMC. This is due to our ability to directly generate and sample from the pure distribution of electrons at the middle of the selected path. We can sample any property, which commutes with the position operator, from the pure distribution using the following formula:

$$A(\tau) = \frac{1}{I}\sum_{i=1}^{I} \hat{A}_i(x_0 \text{ or } y_0) \tag{21}$$

where I is the number of iterations of the algorithm, and $\hat{A}_i$ is the property of interest sampled from the pure distribution at a configuration located at the middle of the accepted path on the ith iteration.

4.2.1 Electric moments

Consider a system of wholly or partially charged particles that is overall electrically neutral, but contains within itself a group of fully or partially positively charged particles at one end and a group of wholly or partially negatively charged particles at the opposite end. Such a system is an *electric dipole*, with an *electric dipole moment*, a quantity that measures the degree of separation of positive and negative charges inside the system. The dipole moment is said to be *permanent* if it arises from a static configuration of partially or wholly charged particles in the system, or *induced* if it stems from an applied external electric field.

An asymmetrical molecule that has a permanent dipole moment due to the arrangement of its atoms is called a *polar* molecule. Polar molecules can interact with each other via attraction between their oppositely charged regions. This effect is called the *dipole-dipole interaction*. Water molecule is a good example of this. The oxygen atom in the water molecule is more negative than the hydrogen atoms. Thus, individual water molecules can form a so-called hydrogen bond with each other through the dipole-dipole interaction between the oxygen atom on one molecule and a hydrogen atom on another molecule. This effect gives rise to many unique properties of water.

Mathematically, the dipole moment arises from an expansion of the electric potential for a system of charged particles. This expansion produces other important terms besides the dipole moment. Using Buckingham's notation [44], we can write down equations that define the dipole ($\boldsymbol{\mu}$), quadrupole ($\boldsymbol{\Theta}$), and octupole ($\boldsymbol{\Omega}$) moments in the following form:

$$\mu_\alpha = \sum_i q_i r_{i\alpha} \tag{22}$$

$$\Theta_{\alpha\beta} = \frac{1}{2}\sum_i q_i \left(3 r_{i\alpha} r_{i\beta} - r_i^2 \delta_{\alpha\beta}\right) \tag{23}$$

$$\Omega_{\alpha\beta\gamma} = \frac{1}{2}\sum_i q_i \left(5 r_{i\alpha} r_{i\beta} r_{i\gamma} - r_i^2 \left(r_{i\alpha}\delta_{\beta\gamma} + r_{i\beta}\delta_{\gamma\alpha} + r_{i\gamma}\delta_{\alpha\beta}\right)\right) \tag{24}$$

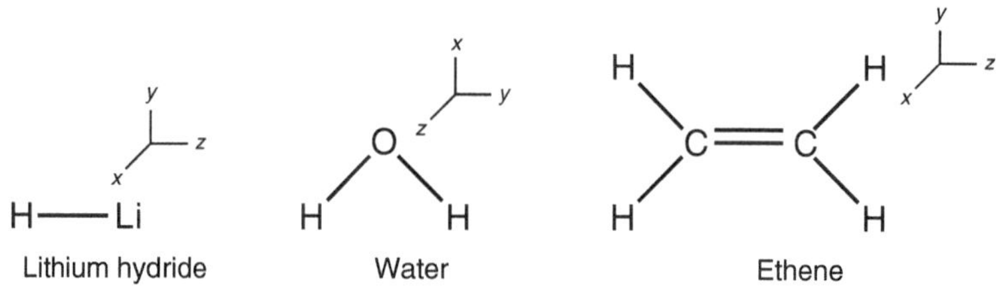

FIG. 12

Illustration of geometries for lithium hydride, water, and ethene.

Here, the Greek letter subscripts α, β, and γ represent the Cartesian coordinates x, y, or z. The δ symbol vanishes unless its indices are the same. The summations occur over all charged particles in the molecule, both electrons and nuclei: q_i represents the charge of the ith particle and r_i denotes its position with respect to the center of mass of the molecule. Furthermore, $\Theta_{\alpha\beta}$ and $\Omega_{\alpha\beta\gamma}$ are symmetric under the interchange of their indices.

Due to the symmetry of lithium hydride, the only nonzero component of the dipole moment is μ_z (Fig. 12). The following tensor properties [45] for $\boldsymbol{\Theta}$ and $\boldsymbol{\Omega}$ apply to lithium hydride:

$$\Theta_{xx} = \Theta_{yy}, \Theta_{zz} = -\left(\Theta_{xx} + \Theta_{yy}\right) = -2\Theta_{xx} \tag{25}$$

$$\Omega_{xxz} = \Omega_{yyz}, \Omega_{zzz} = -\left(\Omega_{xxz} + \Omega_{yyz}\right) = -2\Omega_{xxz} \tag{26}$$

There is a similar set of rules that apply to the water molecule (Fig. 12). Due to its symmetry, μ_z is the only nonzero component of the dipole moment. The following tensor properties for $\boldsymbol{\Theta}$ and $\boldsymbol{\Omega}$ apply to the water molecule:

$$\Theta_{zz} = -\left(\Theta_{xx} + \Theta_{yy}\right) \tag{27}$$

$$\Omega_{zzz} = -\left(\Omega_{xxz} + \Omega_{yyz}\right) \tag{28}$$

Due to the symmetry of the ethene molecule (Fig. 12), its dipole and octupole moments are both zero. The following tensor property for $\boldsymbol{\Theta}$ applies to the ethene molecule:

$$\Theta_{zz} = -\left(\Theta_{xx} + \Theta_{yy}\right) \tag{29}$$

The aforementioned symmetry requirements are useful in various important ways. In the PSQMC approach, the components of all electric moments are calculated independently, so that one can confirm the validity of the results by checking to see if they obey the symmetry rules. Furthermore, one can improve the quality of the results by taking a variance-weighted average of symmetry-equivalent components, where applicable.

4.2.2 Diamagnetic shielding and susceptibility

Other properties of interest, diamagnetic shielding and diamagnetic susceptibility, are expressed by Eqs. (30), (31), respectively.

$$1/r_A = \sum_{i}^{n} \frac{1}{|r_i - R_A|} \tag{30}$$

$$r^2 = \sum_{i}^{n} |r_i - R_{CM}|^2 \tag{31}$$

Here, n is the number of electrons in the molecule. R_A is the location of the nucleus A for which the diamagnetic shielding is being calculated. R_{CM} represents the location of the center of mass.

4.2.3 Electric fields and electric field gradients

In a molecule at equilibrium geometry, the total electric field felt by each nucleus is zero. One can examine the quality of the results and the accuracy of chosen geometry by calculating the electric fields and observing any significant deviations from zero, in addition to electric field gradients that measure the rate at which the electric field changes at a location of a specific nucleus. One can use electric field gradients to calculate the quadrupole coupling constant, a useful experimental quantity, by combining it with a nuclear quadrupole moment [46].

The expressions for the intramolecular electric fields and electric field gradients are given by Eqs. (32), (33), respectively.

$$q\alpha/r_A^3 = -\sum_{\substack{i=1 \\ i\neq A}}^{N} \frac{Z_i\,(R_{i\alpha} - R_{A\alpha})}{|R_i - R_A|^3} + \sum_{\underset{\sim}{i=1}}^{n} \frac{(r_{i\alpha} - R_{A\alpha})}{|r_i - R_A|^3} \tag{32}$$

$$q_A(\alpha\beta) = -\sum_{\substack{i=1 \\ i\neq A}}^{N} \frac{Z_i\left[3\,(R_{i\alpha} - R_{A\alpha})\,(R_{i\beta} - R_{A\beta}) - \delta_{\alpha\beta}\,|R_i - R_A|^2\right]}{|R_i - R_A|^5} + \sum_{\underset{\sim}{i=1}}^{n} \frac{\left[3\,(r_{i\alpha} - R_{A\alpha})\,(r_{i\beta} - R_{A\beta}) - \delta_{\alpha\beta}\,|r_i - R_A|^2\right]}{|r_i - R_A|^5} \tag{33}$$

Again, the Greek letter subscripts α and β represent the Cartesian coordinates x, y, or z. Variable A represents the nucleus for which the quantities are being calculated. N and n are the number of nuclei and electrons in the molecule, respectively. The $\sim$ symbol on the bottom of the sum indicates that the $1/r$ type singularity has been replaced with an appropriate truncation [47]. This and other truncations described in the literature are designed to vanish in the $\tau \to 0$ limit, so that it does not introduce any bias into the extrapolated values.

5 APPLICATION TO ETHENE

To illustrate the performance of PSQMC, we apply the method to ethene molecule to calculate its ground-state energy and electronic properties within the fixed-node approximation (Section 2.4). Applications to lithium hydride and water are in the literature [17].

Importance sampling (Section 2.1) was performed at the experimental geometry [48] by a single-determinant SCF wave function with a QZ4P STO basis set, generated by using Amsterdam Density Functional [49–51] software. The calculations were carried out at various time-steps and path-lengths in accordance with Eq. (34).

$$L(\tau) = L_0 \times \left(\frac{\tau_0}{\tau}\right) \tag{34}$$

where L_0 is the initial (smallest) path-length and τ_0 is the initial (largest) time-step; refer to Fig. 3 for details. The path-length (L) increases as the time-step (τ) decreases from the τ_0 value, approaching infinity as τ approaches zero. The zero time-step value is where the energy and other properties are calculated. It is at this time-step the short-time Green's function, Eq. (8), is free from bias.

Figs. 13 and 14 show the plots of the ground-state energy and Θ_{yy} of ethene molecule as a function of time-step and path-length parameter (L_0), respectively.

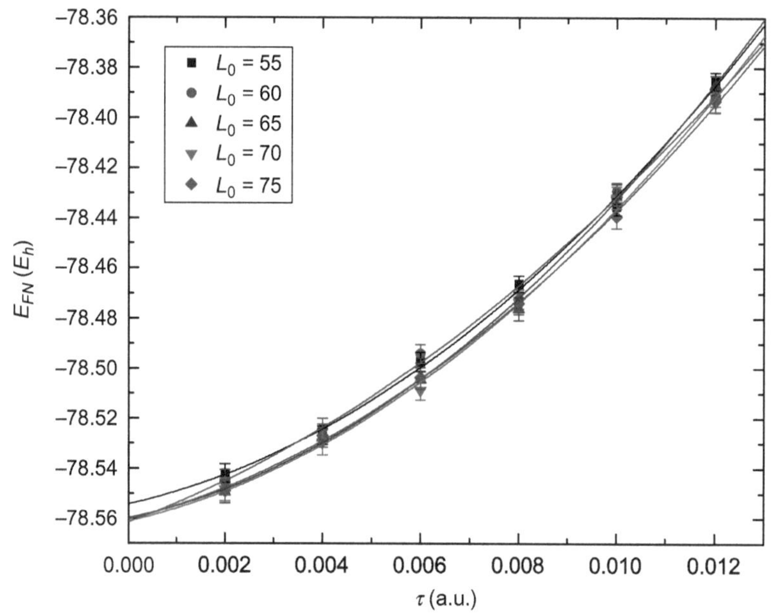

FIG. 13

Ground-state energy of ethene as a function of the initial path-length parameter (L_0). Path-lengths increase as time-steps decrease in accordance with Eq. (34), where $\tau_0 = 0.012$ a.u. Convergence occurs for $L_0 = 60\ldots75$ [5].

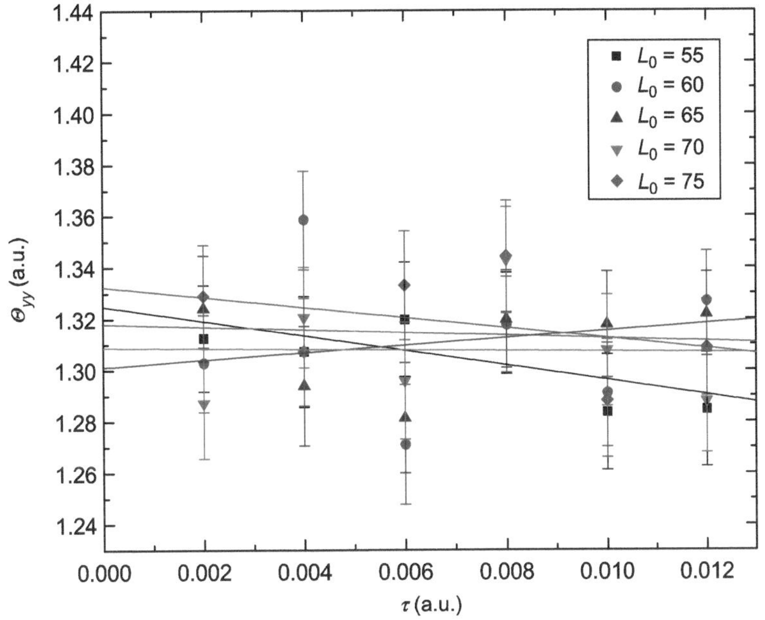

FIG. 14

Quadrupole moment of ethene. Quadrupole moment of ethene as a function of the initial path-length parameter (L_0). Path-lengths increase as time-steps decrease in accordance with Eq. (34), where $\tau_0 = 0.012$ a.u. Convergence occurs for $L_0 = 60\ldots75$ [5].

As L_0 is systematically increased, a regression model is chosen such that its $\tau = 0$ intercept (determined using variance-weighted regression) is as statistically consistent with that of the previous L_0. As the path-length increases and passes a critical value, $L_0 = 60$, the ground-state energy and Θ_{yy} converge to a consistent value. Any further increase of the path-length beyond a critical value yields statistically equivalent results, albeit with an increased computational cost associated with an increased number of electron configurations being generated along the paths. However, the advantage of this approach is variance-weighted averaging of converged results provides us with an accurate estimate of the energy and other properties of interest that is free from time-step and regression-model bias (see Table 1).

Table 2 contains the electronic properties results for ethene. The ground-state energy is in good agreement with Nemec et al. [52], who used an identical trial wave function. The PSQMC energy is above the exact ground-state energy, as expected from its obedience of the variational upper bound, but we still manage to recover more electron correlation energy than the CCSD(T) literature value, which is not variationally bound. It is difficult to say how accurate the quadrupole moment is, since the literature presents a diverse set of results and the theoretical

Table 1 Ground-State Energy and *yy*-Component of the Quadrupole Moment of Ethene in the Limit of Zero Time-Step (τ) and as a Function of Path-Length Parameter (L_0)

L_0	E_0	Θ_{yy}
55	−78.5542(67)[a]	1.325(12)[b]
60	−78.5615(87)	1.318(30)
65	−78.5598(57)	1.301(17)
70	−78.5612(66)	1.309(23)
75	−78.5604(38)	1.332(19)
Average[c]	−78.5605(27)	1.314(10)

[a] *Not included in the weighted average due to a lack of convergence with the rest of the data.*
[b] *Not included in the weighted average because the ground-state energy at this path-length did not converge.*
[c] *Variance-weighted average of the property for all included path-lengths.*
Note: All entries are in atomic units (a.u.).

methods employed different geometries in their calculations. Nonetheless, the PSQMC results are consistent with symmetry restrictions, Eq. (29), and do not show any irregularities. The electric fields agree with their theoretical values, except for the x and y components on the hydrogen nuclei, which may reflect our use of an experimental geometry, instead of one that minimizes the Hartree-Fock energy of the importance-sampling function. To the best of our knowledge, diamagnetic shielding, diamagnetic susceptibility, and electric field gradients of ethene have not been calculated previously.

6 CONCLUSION

We introduced the reader to QMC by describing two algorithms that rest on the Fixed-Node (FN) approximation: FNDMC and PSQMC. The FN approximation, which stems from the mismatch of the nodal hypersurface of the importance-sampling function with that of the exact ground-state wave function, introduces a positive bias to the calculated ground-state energy, as well as biasing all other calculated physical properties.

FNQMC is by far the most commonly used algorithm—its major virtue is its ability to calculate the ground-state energy without bias, save for the FN bias, generally regarded as small, and some more subtle ones, which we have not discussed. On the other hand, there is a substantial bias when calculating ground-state properties represented by operators that commute with the position operator, such as the electrical moments, electric field gradients, etc.

Table 2 Ground-State Energy and Electronic Properties of Ethene

Property	PSQMC	Literature
E_0	−78.561(3)	−78.567(0.4)[a], −78.5526[b], −78.5888[c]
Θ_{xx}	−2.81(1)	−2.444[d], −2.5988[e], −2.8419[f] −2.35(14)[g], −2.45[h]
Θ_{yy}	1.31(1)	1.196[d], 1.2332[e], 1.2399[f] 1.10(7)[g], 0.97[h]
Θ_{zz}	1.500(8)	1.247[d], 1.3656[e], 1.6020[f] 1.24(14)[g], 1.49[h]
$\langle 1/r_C \rangle_e$	18.484(7)	–
$\langle 1/r_H \rangle_e$	6.2049(4)	–
$\langle r^2 \rangle_e$	83.23(3)	–
$\langle qx/r_C^3 \rangle$	−0.002(1)	0.0
$\langle qy/r_C^3 \rangle$	−0.002(1)	0.0
$\langle qz/r_C^3 \rangle$	−0.002(1)	0.0
$\langle qx/r_H^3 \rangle$	−0.0021(3)	0.0
$\langle qy/r_H^3 \rangle$	−0.0039(9)	0.0
$\langle qz/r_H^3 \rangle$	−0.0003(7)	0.0
$q_C(xx)$	0.297(6)	–
$q_C(yy)$	−0.094(7)	–
$q_C(zz)$	−0.206(7)	–
$q_H(xx)$	−0.164(2)	–
$q_H(yy)$	0.305(1)	–
$q_H(zz)$	−0.142(1)	–

[a] *Fixed-node diffusion quantum Monte Carlo [52].*
[b] *CCSD(T) [53].*
[c] *Exact nonrelativistic electronic energy [54].*
[d] *CC3 [55].*
[e] *MP2 [56].*
[f] *SCF [56].*
[g] *Experiment [57].*
[h] *Experiment [58].*
Notes: The z-axis is along the double bond. All entries are in atomic units (a.u.). The tensor q_H was diagonalized.

By contrast, PSQMC, coupled with the data analysis techniques described earlier in detail, is completely free of bias, both for energy and electrical properties, except, again, for the FN bias. We presented results for the ethene molecule to illustrate the power of this approach.

Other methods that we foresee enjoying more attention and development in the future are path-integral ground state [59–61], full configuration interaction quantum Monte Carlo (FCIQMC) [62–65], and Model State Quantum Monte Carlo (MSQMC) [66,67] all of which are capable of pure-sampling. The latter two methods are especially worthwhile because they do not rest upon the FN approximation and are capable of calculating excited states, albeit at the cost of basis set completion issues and substantially increased IO demands on the calculations.

Note added in proof. By using a more recent version of ADF software, we published improved values for the fixed-node energy and quadrupole moments for ethene. See Ospadov E, Rothstein SM, Baer R, Quantum Monte Carlo assessment of density functionals for Π-electron molecules: ethylene and bifuran. Mol Phys 2018. doi:10.1080/00268976.2018.1517905.

7 APPENDIX

7.1 DERIVATION OF THE MODIFIED SCHRÖDINGER EQUATION

To derive the modified Schrödinger equation presented in this work, let us start with the well-known, time-dependent Schrödinger equation:

$$\hat{H}\Phi(\mathbf{r},t) = i\hbar\frac{\partial}{\partial t}\Phi(\mathbf{r},t) \tag{35}$$

$$\left(-\frac{\hbar^2}{2m_e}\nabla^2 + V(\mathbf{r})\right)\Phi(\mathbf{r},t) = i\hbar\frac{\partial}{\partial t}\Phi(\mathbf{r},t) \tag{36}$$

Next, rewrite this equation in imaginary time (it) and atomic units. At the same time, introduce a constant shift in the zero-point energy, denoted by E_T. This will prove useful once we look for the ground-state solution to the modified Schrödinger equation. Accordingly, we obtain the following expression:

$$\left(-\frac{1}{2}\nabla^2 + V(\mathbf{r}) - E_T\right)\Phi(\mathbf{r},t) = -\frac{\partial}{\partial t}\Phi(\mathbf{r},t) \tag{37}$$

At this point, multiply both sides of this equation by the known guiding wave function, $\Psi(\mathbf{r})$. Let us also suppress the dependence on $\mathbf{r}$ and t for the sake of notational simplicity. By doing so, we arrive at:

$$\Psi\left(-\frac{1}{2}\nabla^2 + V - E_T\right)\Phi = -\Psi\frac{\partial\Phi}{\partial t} \tag{38}$$

$$\Psi\left(-\frac{1}{2}\nabla^2 + V - E_T\right)\Phi = -\frac{\partial}{\partial t}(\Phi\Psi) + \Phi\frac{\partial\Psi}{\partial t} \tag{39}$$

Since $\Psi(\mathbf{r})$ is time-independent, then $\Phi\frac{\partial\Psi}{\partial t} = 0$. Using this result, we can write the aforementioned equation as:

$$\Psi\left(-\frac{1}{2}\nabla^2 + V - E_T\right)\Phi = -\frac{\partial}{\partial t}(\Phi\Psi) \tag{40}$$

Using a vector identity, $\nabla^2(\Phi\Psi) = \Psi\nabla^2\Phi + \Phi\nabla^2\Psi + 2\nabla\Phi\cdot\nabla\Psi$, we express the previous equation as:

$$-\frac{1}{2}\left(\nabla^2(\Phi\Psi) - \Phi\nabla^2\Psi - 2\nabla\Phi\cdot\nabla\Psi\right) + V\Phi\Psi - E_T\Phi\Psi = -\frac{\partial}{\partial t}(\Phi\Psi) \tag{41}$$

$$-\frac{1}{2}\nabla^2(\Phi\Psi) + \frac{1}{2}\Phi\nabla^2\Psi + \nabla\Phi\cdot\nabla\Psi + V\Phi\Psi - E_T\Phi\Psi = -\frac{\partial}{\partial t}(\Phi\Psi) \tag{42}$$

Applying yet another vector identity, $\nabla\cdot(\Phi\nabla\Psi) = \nabla\Phi\cdot\nabla\Psi + \Phi\nabla^2\Psi$, we write it as:

$$-\frac{1}{2}\nabla^2(\Phi\Psi) - \frac{1}{2}\Phi\nabla^2\Psi + \nabla\cdot(\Phi\nabla\Psi) + V\Phi\Psi - E_T\Phi\Psi = -\frac{\partial}{\partial t}(\Phi\Psi) \tag{43}$$

Finally, by rearranging some terms in this equation we obtain:

$$-\frac{1}{2}\nabla^2(\Phi\Psi) + \nabla\cdot\left(\Phi\Psi\frac{\nabla\Psi}{\Psi}\right) + \left(-\frac{1}{2}\frac{\nabla^2\Psi}{\Psi} + V - E_T\right)\Phi\Psi = -\frac{\partial}{\partial t}(\Phi\Psi) \tag{44}$$

By substituting the following three equations into the previous equation,

$$f(\mathbf{r},t) = \Phi(\mathbf{r},t)\Psi(\mathbf{r}) \tag{45}$$

$$\mathbf{F}(\mathbf{r}) = \frac{\nabla\Psi(\mathbf{r})}{\Psi(\mathbf{r})} \tag{46}$$

$$E_L(\mathbf{r}) = -\frac{1}{2}\frac{\nabla^2\Psi(\mathbf{r})}{\Psi(\mathbf{r})} + V(\mathbf{r}) = \frac{\hat{H}\Psi(\mathbf{r})}{\Psi(\mathbf{r})} \tag{47}$$

we obtain the Modified Schrödinger Eq. (3):

$$-\frac{1}{2}\nabla^2 f(\mathbf{r},t) + \nabla\cdot(f(\mathbf{r},t)\mathbf{F}(\mathbf{r})) + (E_L(\mathbf{r}) - E_T)f(\mathbf{r},t) = -\frac{\partial f(\mathbf{r},t)}{\partial t} \tag{48}$$

It follows from Eq. (6) that the mixed-distribution is sampled in the limit of long simulation time:

$$f(t\to\infty,\mathbf{r}) = \Phi_0(\mathbf{r})\Psi(\mathbf{r}) \tag{49}$$

Furthermore, as the simulation reaches a steady state in that limit: $-\frac{\partial f(\mathbf{r},t)}{\partial t} = 0$.

7.2 DERIVATION OF THE ENERGY ESTIMATOR FOR DIFFUSION MONTE CARLO

Here, we show why sampling from the mixed distribution would yield the fixed-node ground-state energy. We start by writing the time-independent Schrödinger equation:

$$\hat{H}|\Phi_0\rangle = E_0|\Phi_0\rangle \tag{50}$$

Next, we multiply both sides by $\langle\Psi|$ and rearrange for E_0:

$$\langle\Psi|\hat{H}|\Phi_0\rangle = \langle\Psi|E_0|\Phi_0\rangle \tag{51}$$

$$\langle \Psi | \hat{H} | \Phi_0 \rangle = E_0 \langle \Psi | \Phi_0 \rangle \tag{52}$$

$$E_0 = \frac{\langle \Psi | \hat{H} | \Phi_0 \rangle}{\langle \Psi | \Phi_0 \rangle} \tag{53}$$

We can rewrite this in integral representation, while suppressing the complex conjugates since our wave functions are real.

$$E_0 = \frac{\int \Psi(\mathbf{r}) \hat{H} \Phi_0(\mathbf{r}) \mathrm{d}\mathbf{r}}{\int \Psi(\mathbf{r}) \Phi_0(\mathbf{r}) \mathrm{d}\mathbf{r}} = \frac{\int \Psi(\mathbf{r}) \left(\frac{\hat{H}\Psi(\mathbf{r})}{\Psi(\mathbf{r})} \right) \Phi_0(\mathbf{r}) \mathrm{d}\mathbf{r}}{\int \Psi(\mathbf{r}) \Phi_0(\mathbf{r}) \mathrm{d}\mathbf{r}} = \frac{\int \Psi(\mathbf{r}) E_L(\mathbf{r}) \Phi_0(\mathbf{r}) \mathrm{d}\mathbf{r}}{\int \Psi(\mathbf{r}) \Phi_0(\mathbf{r}) \mathrm{d}\mathbf{r}} \tag{54}$$

$$= \frac{\int \Psi(\mathbf{r}) \Phi_0(\mathbf{r}) E_L(\mathbf{r}) \mathrm{d}\mathbf{r}}{\int \Psi(\mathbf{r}) \Phi_0(\mathbf{r}) \mathrm{d}\mathbf{r}} = \frac{\int f(\mathbf{r}) E_L(\mathbf{r}) \mathrm{d}\mathbf{r}}{\int f(\mathbf{r}) \mathrm{d}\mathbf{r}} \tag{55}$$

where we used the fact that $E_L(\mathbf{r}) = \frac{\hat{H}\Psi(\mathbf{r})}{\Psi(\mathbf{r})}$ and $f(\mathbf{r}) = \Psi(\mathbf{r})\Phi_0(\mathbf{r})$. Finally, we can use the law of large numbers with the result from earlier to arrive at:

$$E_0 = \frac{1}{I} \sum_{i=1}^{I} E_L(\mathbf{r}_i) \tag{56}$$

where I is the number of samples (iterations) of E_L obtained from a mixed distribution of electrons $\mathbf{r}_i$. This expression is advantageous since it does not require us to calculate the action of $\hat{H}$ on the unknown Φ_0.

This equation applies only to node-less systems. Otherwise it is the formula for the fixed-node energy, which is an upper bound to E_0.

REFERENCES

[1] Kalos MH. Monte Carlo calculations of the ground state of three- and four-body nuclei. Phys Rev 1962;128:1791–5.

[2] Anderson JB. A random-walk simulation of the Schrödinger equation: H3+. J Chem Phys 1975;63:1499–503.

[3] Metropolis N, Ulam S. The Monte Carlo method. J Am Stat Assoc 1949;44:335–41.

[4] Yung MH, Aspuru-Guzik A. A quantum-quantum metropolis algorithm. Proc Natl Acad Sci USA 2012;109(3):754–9. https://doi.org/10.1073/pnas.1111758109.

[5] Hammond BL, Lester Jr WA, Reynolds PJ. Monte Carlo methods and ad initio quantum chemistry. Singapore: World Scientific; 1994.

[6] Huang P, Carter EA. Advances in correlated electronic structure methods for solids, surfaces, and nanostructures. Annu Rev Phys Chem 2008;59(1):261–90. https://doi.org/10.1146/annurev.physchem.59.032607.093528.

[7] Luchow A, Petz R. Electron structure quantum Monte Carlo. Chemical modelling applications and theory-specialist periodical report, vol. 7. The Royal Society of Chemistry. ISBN 978-1-84755-075-0; 2010. p. 237–60. https://doi.org/10.1039/9781849730884-00237.

[8] Toulouse J, Assaraf R, Umrigar CJ. Introduction to the variational and diffusion Monte Carlo methods. Advances in quantum chemistry, vol. 73. ISBN 978-0-12-803061-5; 978-0-12-803060-8; 2016. p. 285–314. https://doi.org/10.1016/bs.aiq.2015.07.003.

[9] Anderson AG, Goddard III WA, Schröder P. Quantum Monte Carlo on graphical processing units. Comput Phys Commun 2007;177:298–306.

[10] Uejima Y, Terashima T, Maezono R. Acceleration of a QM/MM-QMC simulation using GPU. J Comput Chem 2011;32:2264–72.

[11] Esler KP, Kim J, Ceperley DM, Shulenburger L. Accelerating quantum Monte Carlo simulations of real materials on GPU clusters. Comput Sci Eng 2012;14:40–51.

[12] Scemama A, Caffarel M, Oseret E, Jalby W. Quantum Monte Carlo for large chemical systems: implementing efficient strategies for petascale platforms and beyond. J Comput Chem 2013;34(11):938–51. https://doi.org/10.1002/jcc.23216.

[13] Lutsyshyn Y. Fast quantum Monte Carlo on a GPU. Comput Phys Commun 2015;187:162–74.

[14] Rothstein SM, Ospadov E. Advances in quantum Monte Carlo—past, present, and reflections on its future. ACS symposium series, vol. 1234; 2016. p. 155–63.

[15] Anderson JB. Quantum chemistry by random-walk – H2P, H + 3 D3H 1A'1, H − 2 3 − $\sum$+U, H − 4 1 − $\sum$+G, Be 1S. J Chem Phys 1976;65:4121–7.

[16] Reynolds PJ, Ceperley DM, Alder BJ, Lester Jr WA. Fixed-node quantum Monte-Carlo for molecules. J Chem Phys 1982;77:5593–603.

[17] Ospadov E, Rothstein SM. A pure-sampling quantum Monte Carlo algorithm. J Chem Phys 2015;142:024114.

[18] Vrbik J, Ospadov E, Rothstein SM. Note: A pure-sampling quantum Monte Carlo algorithm with independent Metropolis. J Chem Phys 2016;145(2):026101. https://doi.org/10.1063/1.4958340.

[19] Rothstein SM. A survey on pure sampling in quantum Monte Carlo methods. Can J Chem 2013;91:505–10.

[20] Ceperley DM. Fermion nodes. J Stat Phys 1991;63:1237–67.

[21] Morales MA, McMinis J, Clark BK, Kim J, Scuseria GE. Multideterminant wave functions in quantum Monte Carlo. J Chem Theory Comput 2012;8(7):2181–8. https://doi.org/10.1021/ct3003404.

[22] Baroni S, Moroni S. Reptation quantum Monte Carlo: a method for unbiased groundstate averages and imaginary-time correlations. Phys Rev Lett 1999;82:4745–8.

[23] Baroni S, Moroni S. Quantum Monte Carlo methods in physics and chemistry. In: Nightingale MP, Umrigar CJ, editors. Proceedings of the NATO Advanced Study Institute, vol. 525. Ithaca, NY; 1999. p. 313–41.

[24] Hohenberg P, Kohn W. Inhomogeneous electron gas. Phys Rev 1964;136:B864–71.

[25] Kohn W, Sham LJ. Self-consistent equations including exchange and correlation effects. Phys Rev 1965;140:A1133–8.

[26] Kohn W, Becke AD, Parr RG. Density functional theory of electronic structure. J Phys Chem 1996;100:12974–80.

[27] Becke AD. Perspective: fifty years of density-functional theory in chemical physics. J Chem Phys 2014;140:18A301.

[28] Coester F, Kümmel H. Short-range correlations in nuclear wave functions. Nucl Phys 1960;17:477–85.

[29] Čížek J. On correlation problem in atomic and molecular systems. Calculation of wave function components in Ursell-type expansion using quantum-field theoretical methods. J Chem Phys 1966;45:4256–66.

[30] Paldus J, Shavitt I, Čížek J. Correlation problems in atomic and molecular systems. 4. Extended coupled-pair many-electron theory and its application to BH3 molecule. Phys Rev A 1972;5:50–67.

[31] Purvis III GD, Bartlett RJ. A full coupled-cluster singles and doubles model: the inclusion of disconnected triples. J Chem Phys 1982;76:1910–8.

[32] Bartlett RJ, Musiał M. Coupled-cluster theory in quantum chemistry. Rev Mod Phys 2007;79:291–352.

[33] Bacher RF. The interaction of configurations sd-p(2). Phys Rev 1933;43:264–9.

[34] Ufford CW. Configuration interaction in complex spectra. Phys Rev 1933;44:732–9.

[35] Shavitt I. The history and evolution of configuration interaction. Mol Phys 1998;94: 3–17.

[36] Sherrill CD, Schaefer III HF. The configuration interaction method: advances in highly correlated approaches. Adv Quant Chem 1999;34:143–269.

[37] Szalay PG, Müller T, Gidofalvi G, Lischka H, Shepard R. Multiconfiguration self-consistent field and multireference configuration interaction methods and applications. Chem Rev 2012;112:108–81.

[38] Rothstein SM, Vrbik J. Statistical error of diffusion Monte-Carlo. J Comput Phys 1988;74(1):127–42. https://doi.org/10.1016/0021-9991(88)90073-3.

[39] Umrigar CJ, Nightingale MP, Runge KJ. A diffusion Monte-Carlo algorithm with very small time-step errors. J Chem Phys 1993;99:2865–90.

[40] Anderson JB. Quantum-chemistry by random-walk: higher accuracy. J Chem Phys 1980;73:3897–9.

[41] DePasquale MF, Rothstein SM, Vrbik J. Reliable diffusion quantum Monte-Carlo. J Chem Phys 1988;89:3629–37.

[42] Cuthbert HL, Rothstein SM. Quantum chemistry without wave functions: diffusion Monte Carlo applied to H and H-2(+). J Chem Educ 1999;76(10):1378–9.

[43] Austin BM, Zubarev DYu, Lester Jr WA. Quantum Monte Carlo and related approaches. Chem Rev 2012;112:263–88.

[44] Buckingham AD. Molecular quadrupole moments. Quart Rev Chem Soc 1959;13: 183–214.

[45] Kielich S, Zawodny R. Tensor elements of molecular electric multipole moments for all point group symmetries. Chem Phys Lett 1971;12:20–4.

[46] Smith JAS. Nuclear quadrupole resonance spectroscopy: general principles. J Chem Educ 1971;48:39–48.

[47] Vrbik J, Legare DA, Rothstein SM. Infinitesimal differential diffusion quantum Monte–Carlo—diatomic molecular properties. J Chem Phys 1990;92:1221–7.

[48] Herzberg G. Electronic spectra and electronic structure of polyatomic molecules. No. III in Molecular spectra and molecular structure, New York: D. Van Nostrand Co, Inc.; 1966.

[49] te Velde G, Bickelhaupt FM, Baerends EJ, Guerra CF, van Gisbergen SJA, Snijders JG, et al. Chemistry with ADF. J Comput Chem 2001;22:931–67.

[50] Guerra CF, Snijders JG, te Velde G, Baerends EJ. Towards an order-N DFT method. Theor Chem Acc 1998;99:391–403.

[51] SCM. ADF2010. Theoretical Chemistry. Amsterdam, The Netherlands: Vrije Universiteit. Available from: http://www.scm.com.

[52] Nemec N, Towler MD, Needs RJ. Benchmark all-electron ab initio quantum Monte Carlo calculations for small molecules. J Chem Phys 2010;132:034111.

[53] Machado FBC, Davidson ER. The ground state of ethylene. J Mol Struct Theochem 1997;400:169–76.
[54] O'Neill DP, Gill PMW. Benchmark correlation energies for small molecules. Mol Phys 2005;103:763–6.
[55] Hald K, Halkier A, Jørgensen P, Coriani S. Orbital nonrelaxed coupled cluster singles and doubles with perturbative triples corrections calculations of first-order one-electron properties. J Chem Phys 2002;117:9983–90.
[56] Russell AJ, Spackman MA. An ab initio study of vibrational corrections to the electrical properties of ethylene. Mol Phys 2000;98:855–65.
[57] Majer W, Lutzmann P, Hüttner W. The molecular electric quadrupole tensor of ethene from the rotational Zeeman effect of CH2=CD2. Mol Phys 1994;83(3):567–78. https://doi.org/10.1080/00268979400101431,
[58] Bose TK, Boudjarane K, Huot J, Starnaud JM. Refractivity virial-coefficients of C_2H_4 and C_2H_4-AR mixtures. J Chem Phys 1988;89:7435–40.
[59] Sarsa A, Schmidt KE, Magro WR. A path integral ground state method. J Chem Phys 2000;113:1366–71.
[60] Cuervo JE, Roy PN, Boninsegni M. Path integral ground state with a fourth-order propagator: application to condensed helium. J Chem Phys 2005;122:114504.
[61] Cuervo JE, Roy PN. Path integral ground state study of finite-size systems: application to small (parahydrogen)(N) ($N = 2 - 20$) clusters. J Chem Phys 2006;125:124314.
[62] Booth GH, Thom AJW, Alavi A. Fermion Monte Carlo without fixed nodes: a game of life, death, and annihilation in Slater determinant space. J Chem Phys 2009;131:054106.
[63] Cleland D, Booth GH, Alavi A. Communications: Survival of the fittest: accelerating convergence in full configuration-interaction quantum Monte Carlo. J Chem Phys 2010;132:041103.
[64] Cleland D, Booth GH, Alavi A. A study of electron affinities using the initiator approach to full configuration interaction quantum Monte Carlo. J Chem Phys 2011;134:024112.
[65] Blunt NS, Smart SD, Booth GH, Alavi A. An excited-state approach within full configuration interaction quantum Monte Carlo. J Chem Phys 2015;143:134117.
[66] Ten-no S. Stochastic determination of effective Hamiltonian for the full configuration interaction solution of quasidegenerate electronic states. J Chem Phys 2013;138:164126.
[67] Ohtsuka Y, Ten-no S. A study of potential energy curves from the model space quantum Monte Carlo method. J Chem Phys 2015;143:214167.

FURTHER READINGS

[68] Anderson JB. Quantum Monte Carlo: atoms, molecules, clusters, liquids, and solids, vol. 13, chap. 3. New York, NY: Wiley-VCH, John Wiley and Sons, Inc.; 1999. p. 133–82.
[69] Anderson JB. Fixed-node quantum Monte Carlo. Int Rev Phys Chem 1995, 14(1):85–112. https://doi.org/10.1080/01442359509353305.

CHAPTER

Personal computers in computational chemistry

7

Errol Lewars

Department of Chemistry, Trent University, Peterborough, ON, Canada

PREFACE

The approach of this account will be to outline the story of the appearance of the personal computer as a serious tool for computational chemistry, then to illustrate this concretely with several examples of its application to problems of well-defined theoretical significance. We will occasionally use the abbreviation PC without any proprietary allusion.

Although the adjective in the term literally implies any computer that is not communally owned, we will soon see that this is much too loose a definition for our purpose here. The other part of the term designates, of course, an electronic digital information-processing device, that is, electronic rather than mechanical, digital rather than analog, and information processing rather than merely calculating. It was not thus in the earliest days of even quite large quantum mechanical calculations; wavefunctions for atoms (rather than molecules) were determined in the 1920s, if not quite by hand, with the aid of mechanical calculators. The laborious nature of these calculations is documented in some detail in the biography of a pioneer whose name in almost synonymous with a fundamental concept in modern quantum chemical computations, the self-consistent method of obtaining wavefunctions [1], which was devised by Douglas Hartree [2] (and elaborated by Vladimir Fock, who took into account electron spin and the antisymmetry of electronic wavefunctions [3]). These early calculations on atoms were for the most part ground out essentially manually by Hartree and his father [2].

Our very brief survey of computer hardware seeks the approximate threshold at which personal computers could perform research-level quantum chemical calculations on molecules. Research-level because for years prior "minicomputers" as they were called had been used pedagogically to illustrate quantum calculations on very small molecules like H_2 using very limited software. Quantum chemical because molecular mechanics, which requires orders of magnitude less computer power, had for some years been practical on PCs. The emphasis is on ab initio rather

Mathematical Physics in Theoretical Chemistry. https://doi.org/10.1016/B978-0-12-813651-5.00007-3

than semiempirical quantum chemical calculations because the former are nowadays the tools for most decisive (insofar as theory can be decisive) theoretical chemical calculations and, generally speaking, they are by far the most computationally demanding theoretical tools for the study of molecules and reactions. The very popular method of density functional theory (DFT) is, we will see, considerably less demanding computationally than ab initio. On molecules because besides the obvious feature of greater size, problems of geometry and molecular spectroscopy exist here. These tools (molecular mechanics, ab initio, semiempirical, and DFT) are explained in introductory books on computational chemistry. The other main tools of computational chemistry are molecular dynamics and Monte Carlo methods. These simulate the behavior of large collections of molecules and can provide in effect a movie of molecules in motion (some molecular dynamics methods are able to simulate bond breaking and making, and so go beyond mere molecular motion). The situation with many-body problems like molecular dynamics and Monte Carlo could soon undergo a dramatic change due to the application of graphical processing units [4], and these methods will not be discussed here.

COMPUTING, COMPUTERS, AND THE PERSONAL COMPUTER

A sketch of the history of scientific computing might begin by pointing out that the word "computer" originally (c.1900–40) meant a person doing calculations; for scientific computing in particular this was a member of an army of people, usually unacknowledged women, performing the arithmetic needed to transform astronomical theories and observations into concrete predictions [5]. These people would have perhaps used slide rules or, starting about 1900, mechanical calculators [6]. The transition to "autonomous" machines, that is machines able to execute a prescribed program, marks the beginning of computers proper. The first such machines were digital electromechanical rather than electronic, using relays in lieu of the later vacuum tubes and then transistors [7]. An anomaly in chemical computing was the projected use of an analog computer for calculating wavefunctions. This particular instrument was a device in which the number of turns of a wheel corresponds to the area arising from integration of a differential equation. The impetus for its construction came from the theoretical chemist Lennard-Jones on his hearing of such a machine that had been built by Hartree (but evidently not used by the latter for chemical work). The Lennard-Jones instrument was not used as envisaged because World War II had begun the month before it became available and other projects took priority over wavefunctions [8]. By the end of the war digital computers had eclipsed analog machines [7].

Among the first prominent chemists to apply electronic computers to quantum chemistry was John Slater [9]. He "was connected with the field [quantum calculations on atoms, molecules and solids] through the whole 50 years from 1923 to 1973," that is, from before the computer era until well into the years when computers were indispensable for such work [9, p. 5]. It is difficult to determine the year when digital electronic computers were first used in computational chemistry, but it must have been in the 1950s, and these would have been so-called mainframes, filling a room [10]. Some indication of specific machines and approximate dates comes from M.J.S. Dewar, the pioneer of semiempirical quantum methods, in his biography [11]. Dewar implied that his first attempts at such calculations were on an IBM 7090 [11, p. 107], a computer introduced in 1959, costing around 3 million dollars ($25 million in 2017 dollars, using the official US consumer price index); it could be rented for $6400 a month ($54,000 in 2017 dollars) [12]. Nearly 20 years later, the Dewar group acquired a VAX 11/780, "the first really powerful minicomputer" [11, p. 115], introduced in 1977, described as "legendary" by another early worker in the field [13]. This was smaller than the IBM 7090, but by no means a desktop and, more important, cheaper, costing $120,000–$160,000 ($960,000–$1.28 million in 2017 dollars) [14]. So from c.1960 to 1980 we see research-level computers getting smaller and cheaper, becoming perhaps affordable for well-heeled research groups rather than only institutions, but still beyond personal purchase (by most of us).

Between "minicomputers" (a term scarcely heard nowadays) and truly personal research-level computers lay, in cost and chronology, desktop machines called workstations (the word seems to have been first used to denote a kind of computer in the 1980s, although its use simply for an area devoted to work goes back decades earlier). Especially powerful machines of this kind, "superworkstations," appeared "as early as 1988," "threatening to change the computing environment" [13]. An example was the DEC CCQC 3100, bought in 1990 as part of a discount deal for around $14,000 each [13]. Another milestone in desktop computational chemistry was the IBM RS/6000 family, first offered in 1990 [15], which in quantity purchase were obtained by the Schaefer group for $46,000 each in 1991 [13]. It is the author's impression from the literature that IBM RS/6000 machines were particularly popular among computational chemists in the 1990s. We may note parenthetically that it was a cluster of RS/6000 machines that first defeated a reigning chess champion, in 1997 [16].

The relevance of the development of personal computers to computational chemistry is two-pronged: consider molecular mechanics and quantum chemistry. Molecular mechanics, which one should not forget is an important tool in the discipline, is enormously less demanding of computer power than quantum mechanical (ab initio, and even semiempirical and DFT) methods, and worked tolerably well on the PCs available in the 1980s. One popular molecular mechanics program, PCModel, was described in Simulation Conference Proceedings in 1989 [17].

Quantum chemical, mainly ab initio, calculations began to slip onto PCs in the 1980s as teaching aids, rather than serious research tools [18]. This changed in 1994 with the porting of the powerful program suite Gaussian 94 [19] into a version for Windows (rather than for UNIX-type mainframe and workstation machines), that is, a version for PCs, G94W [20]. Of course this had to be preceded by the availability of sufficiently powerful PCs to make the execution of the code in this environment practical. Such machines were the IBM-architecture-based 486, and the somewhat faster Intel Pentium series [21]. A table in reference [20] indicates that G94 ran about 30 times slower on a 486 PC than on an IBM RS/6000 workstation, and it is the author's impression that on the PC the program was primarily useful for single-point energy calculations at the Hartree-Fock level, rather than correlated, post-HF work. Nevertheless, G94W did make it possible, even practical, to do on a PC ab initio calculations that constituted part of a serious research project. Another powerful program suite that made the transition from UNIX workstations to PCs (running under either LINUX or Windows) at about the same time as Gaussian 94 was Spartan [22]. GAMESS (General Atomic and Molecular Electronic Structure System), another widely used ab initio package [23], became available for PCs around this time ("We acquired a 266 MHz Pentium-II PC running RedHat Linux in August 1997, and found it performed flawlessly" [24]; GAMES is also available for Windows).

By the end of the 1990s it was recognized that the personal computer had come into its own for computational chemistry: Nicklaus et al. carried out, c.1998, an extensive series of comparisons of PCs ("commodity-type computers," because they are for a mass market, not for a few scientific specialists; this makes them cheap) with workstations and supercomputers, using DFT calculations with G94 [25]. They concluded that "The most powerful of today's [c.1998] commodity-type processors surpassed current supercomputers in speed" and that "the user is now able to purchase... a machine costing only a few thousand dollars that rivals in performance the last generation of supercomputers, whose costs were maybe 3 orders of magnitude higher." Similarly, Schaefer concluded (c.2001) that "it makes no economic sense today to purchase workstations to carry out routine DFT computations." These conclusions seem to be formulated in terms of DFT calculations, rather than the more demanding and, some would say, more rigorous wavefunction methods; purists reserve "ab initio" for the latter methods. Nevertheless the author's experience is that, as will be illustrated, demanding ab initio methods is widely applicable on personal computers, although we will see that, as implied by Schaefer [13], electron-correlated-level calculations on medium-to-large molecules can begin to run into trouble.

EXAMPLES OF CALCULATIONS ON PERSONAL COMPUTERS

1 AN INSTRUCTIVE EXAMPLE OF THE GROWTH OF COMPUTER SPEED

1,3,5-triamino-2,4,6-trinitrobenzene may be chemically trivial but is of some interest to computational chemistry, by virtue of the collapsing times over the years needed for a single-point calculation (energy only, no geometry optimization, no frequencies) and for a geometry optimization, at the Hartree-Fock (i.e., SCF) level (no electron

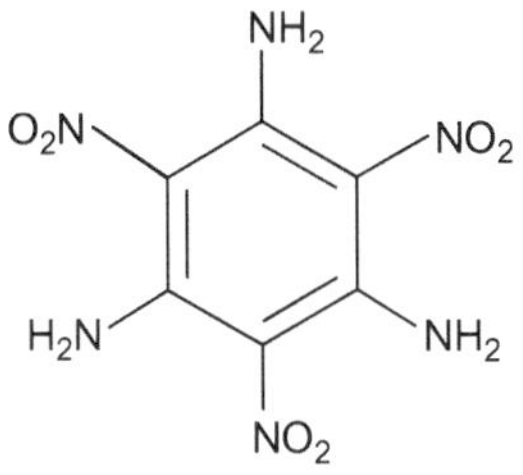

1,3,5-triamino-2,4,6-trinitrobenzene.

correlation beyond the mutual aversion of electrons of the same spin [1(b)]) with the 6-31G** basis set [26] (300 basis functions).

Times for this calculation are:

(a) 1967, on a CDC 1604
single point, 200 years (this is an estimated time) [27].
(b) 1992, on a 486 DX PC
single point, 20 h [27].

Calculations by the author, on this molecule with D3 symmetry; each of these PCs cost around $2000 at the time:

(c) 2004-model PC, with Gaussian 03 [19]
single point, 97 s
geometry optimization, 48 min
(d) 2008-model PC, with Gaussian 03 [19]
single point, 21 s
geometry optimization, 6.3 min
(e) 2012-model PC, with Gaussian 09 [19]
single point, 32 s
geometry optimization, 4.2 min

We see that there is a dramatic (to put it mildly) fall in the times needed for the calculations. The 21 s/32 s "anomaly" for the 2008/2012 single-point jobs is trivial.

1.1 A NOTE ON FACTORS AFFECTING CALCULATION TIMES

In the following paragraphs examples are given of calculations on a PC probing structures and properties of various molecules of theoretical interest; where possible the times required are compared with those for much more expensive machines, mainframes, or workstations. Times depend of course on the electron correlation level and the size of the basis set, and these will always be specified in each case.

But they also vary with how closely the input structure resembles the final optimized geometry (and in one case, [2.2.2]propellane, a nondefault force constant matrix or Hessian was needed for the geometry optimization). The input structures here were routinely molecular mechanics structures from the Merck Molecular Force Field, and occasionally, especially for transition structures, from the semiempirical AM1 method, both utilized in Spartan [22]. The times for obtaining the input structures were normally of the order of a few seconds.

2 PLANAR TETRACOORDINATE CARBON IN AN OTHERWISE NORMAL SATURATED HYDROCARBON

2.1 DIMETHANO[2.2]OCTAPLANE, 1

Our next example is not chemically trivial: dimethano[2.2]octaplane **1** (Fig. 1) is of considerable theoretical interest and at the time it was first investigated (c.1998 [28]) the calculations verged on being mammoth. This and related molecules, and planar tetracoordinate carbon in general, have been reviewed [29].

This molecule was considered to be a good showcase system for the power of modern (c. the mid-2010s; modern for a couple of years only, probably!) personal computers, because (1) the calculations address an intriguing theoretical question, the possibility of planar tetracoordinate carbon in a neutral hydrocarbon; (2) the molecule(s) are of substantial, or at least nontrivial, size; and (3) the pioneering work [28] was for all practical purposes well beyond the abilities of personal computers of the time (c. the mid-1990s).

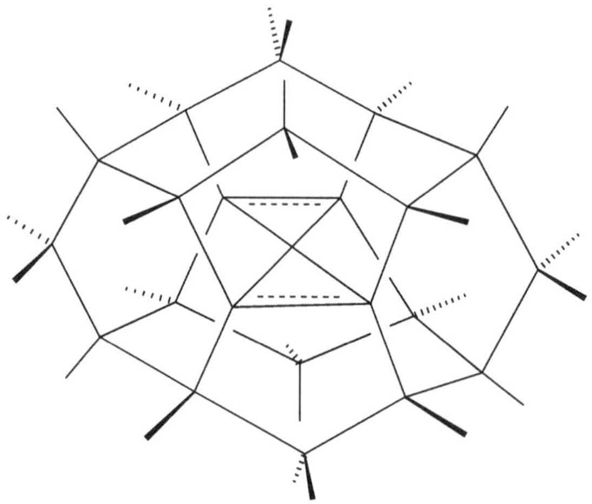

FIG. 1

Dimethanospiro[2.2]octaplane, **1**. This structure was calculated at the MP/6-31G(d) level.

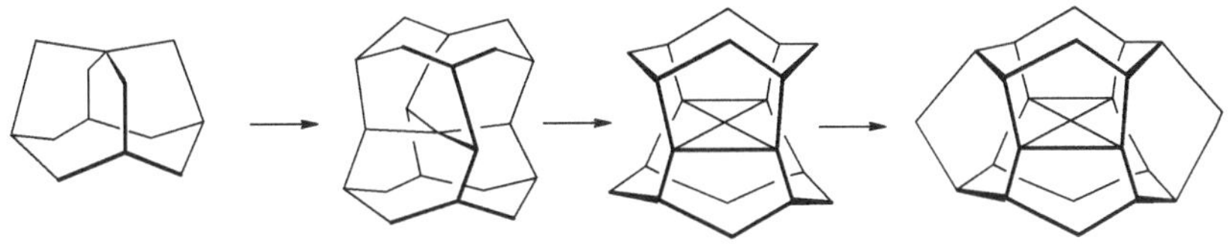

FIG. 2

Subjects of "computer experiments" on the road from bowlane to dimethanospiro[2.2]octaplane.

The theoretical interest of planar carbon has been reviewed [28,29]. Of interest to us in connection with dimethano[2.2]octaplane is tetracoordinate carbon that does not use any tricks like an exotic liaison with lithium or boron [29] to flatten the carbon, but is honestly coplanar in a neutral saturated hydrocarbon: an unabashedly anti-van't Hoff organic molecule [30]. This is a demanding request because planar methane was reported to be about 530 kJ mol^{-1} [29(a)] to 520 kJ mol^{-1} [29(b)] higher than the tetrahedral (and was known of course not to be a local minimum). A carbon atom in this configuration is required to sequester an electron pair in a 2p orbital and reserve for tetracoordination the 2s and two coplanar 2p orbitals. The molecule in Fig. 1 was designed by Radom and Rasmussen by a series of very systematic computer "experiments" extending over 6 years [28(a)]; these finally arrived at dimethano[2.2]octaplane. The most salient features of this quest are shown in Fig. 2. The strategy was to start with bowlane, which has a flattened but nonplanar-carbon atom. Recognizing that symmetry does not permit this to have a planar "central" carbon because the top and bottom of the molecule are different (i.e., because bowlane could not have a symmetry plane through even a perfectly flat carbon and its four neighbors) these workers experimented with constructing equivalent rings above and below the candidate planar-carbon atom, giving first octaplane. Imposing on octaplane two lateral constraints led through spiro[2.2]octaplane to dimethano[2.2]octaplane, **1**, which potentially has bonds to a perfectly flat central carbon atom and was calculated to have no imaginary frequencies. This means that (at the level of theory used!) this is a local minimum (i.e., a real molecule) rather than a mere transition state or some even more ephemeral species [31].

The result of this quest was the first candidate which, with reasonable confidence, could be expected to possess a planar tetracoordinate carbon atom in a neutral saturated hydrocarbon. Note the caveat above: **1** is a local minimum on the potential energy surface at the level of theory used. For reasonable confidence that we have really found a planar-carbon molecule the level should be reasonably high. This means using reasonable electron correlation and a reasonably big basis set. The most widely used post-HF electron correlation method (but by no means the most accurate one) is MP2 [32]. The popular 6-31G* (6-31G(d)) basis set is not particularly big;

triply splitting the carbon valence shell and adding diffuse and extra d functions takes us from 15 or 14 (depending on whether one uses six or five d functions) to the 6-311+G(2d) basis for carbon. This has a core 1s function, three sets of 1 × s and 3 × p functions, two sets of 5 × d functions, and four diffuse (+) functions, for $1 + (3 \times 4) + (2 \times 5) + 4 = 1 + 12 + 10 + 4 = 27$ basis functions. This basis set on small carbon-containing model test species was found to lower the tetrahedral-planar energy difference and to be the "smallest basis set which adequately approaches the basis set limit for this energy difference" [28(b)]. Since **1** is $C_{23}H_{24}$, giving all the carbon atoms 27 basis functions and contenting each H with the modest 6-31G* basis (one s function and one p function) would load the molecule with $23(27) + 24(2) = 669$ functions, and even today this might be challenging for an MP2 optimization cum frequencies calculation. So instead a compromise was accepted, using what we may call a multiple basis set: the 6-311+G(2d) basis on the critical carbon, a 6-311+G(d) basis (5 d functions short of that for 6-311+G(2d), 22 functions) on each of the four attached carbon atoms, and 6-31G* (5 d functions, for 14 functions) on all the other 18 carbon atoms, with 6-31G* on each H; this gives a total of $1(27) + 4(22) + 18(14) + 24(2) = 415$ basis functions. This is not that large a number compared to simply using the popular 6-31G* basis (with the standard 6 d functions on C, $23(15) + 24(2) = 393$ *functions*), but these 415 basis functions have been judiciously apportioned, with the five critical core atoms getting $1(27) + 4(22)$ rather than $1(15) + 4(15)$, that is, 115 rather than only 75 functions. So both the level of electron correlation and the size of the basis set may be adequate. It is interesting to note that at the MP2/6-311+G(2d) level tetrahedral methane lies 592 kJ mol^{-1} below the planar configuration (trivial calculations on the PC). We now compare the response of the computational resources used in c. the mid-1990s with the response of personal computers available in subsequent years to the challenge of optimizing structure 1 and calculating its frequencies, at the level explained earlier.

After Radom and Rasmussen had cornered **1** as a suspect at lower levels of computation, they subjected it to optimization/frequency calculations as described earlier [28]. The one thing that **1** has going for the amelioration of this computational task is its symmetry, D_{2h}. It has what we might be called horizontal (top = bottom), longitudinal (left = right), and lateral (front = back) symmetry planes. This makes many C and H atoms equivalent (the reader could amuse him/herself identifying these) and so greatly reduces the number of integrals that need be computed and thus reduces the computational time, compared to that required for a molecule of lower symmetry. Nevertheless, much of this study was evidently performed with supercomputers of the time ("We thank the Australian National University Supercomputer Facility and the Maui High Performance Computer Center for generous allocations of time on the Fujitsu VPP 300 and IBM SP supercomputers, respectively" [28(b)]. We concentrate on the frequency calculations for dimethano[2.2]octaplane, **1**, because these were among the most challenging for the computer resources, and they also reveal a quite interesting potential problem with such calculations.

The times and the platform reported for c.1995 frequency calculations on **1** with the 415 functions basis set are: "It took 256 hours on 128 nodes of a Cray

T3E Supercomputer, equivalent to 32,768 hours on a single processor of this computer . . . about 3.7 years" [28(a)]. We learn (Wikipedia) that "The Cray T3E was Cray Research's second-generation massively parallel supercomputer architecture, launched in late November 1995" and that the cost was "about \$1 million for a 16-processor version" [33]. Internet pictures indicate that it would not have fit on a desk. Let us compare the times for this frequency calculation on the Cray with that on a vintage 2012 PC. Both calculations used numerical, rather than analytical, frequencies. Frequencies are obtained from vibrational force constants (a measure of the stiffness of a bond viewed as a spring) by "mass-weighting" these constants. For the simplest case, a diatomic molecule AB, the connection of force constant and atomic masses is given by the simple classical equation:

$$\tilde{\nu} = \frac{1}{2\pi c}\left(\frac{k}{\mu}\right)^{1/2}$$

where $\tilde{\nu}$ is the vibrational "frequency," the wavenumber, in cm^{-1}; c is the velocity of light; k is the force constant for the vibration; μ is the reduced mass of the molecule $= (m_A m_B)/(m_A + m_B)$, where m_A and m_B are the masses of A and B.

Force constants are the second derivatives of energy with respect to geometric displacements, $\partial^2 E/\partial^2 q$, where differential q is the displacement of a vibrating atom with respect to the rest of the system (e.g., the other atom), for a diatomic molecule. For a polyatomic molecule the frequencies are calculated with the aid of a second-derivative matrix (a Hessian) and a matrix involving the masses of all the atoms; the details of the calculation of the values of the frequencies and their "direction vectors" are somewhat complex [34]. The challenge for frequency calculations at correlated levels with big basis sets comes from the second derivatives. These can in principle be found analytically or numerically. An analytic frequency calculation uses an algorithm based on the fact that since the energy of a molecule is a function of the geometric coordinates of the atomic nuclei, it should be possible to differentiate this function to get the force constants. That the energy is a function of nuclear coordinates (of geometry) is obvious, but the mathematical form of the second-derivative function and its implementation as an algorithm are not necessarily obvious, and in fact analytical second derivatives are not available for some higher-level methods. The alternative to analytical calculation of force constants, and thus frequencies, is numerical calculation. Here second derivatives are approximated as ratios of finite increments:

$$\partial^2 E/\partial^2 q \approx \Delta(\partial E/\Delta\partial q)/\Delta q$$

The second derivative is approximately the change in the first derivative divided by the corresponding change in the displacement. First derivatives $\partial E/\Delta\partial q$, called gradients, are easily calculated and are available for almost all methods. The average gradient is taken from the displacement in one direction and in the opposite direction (double displacement). Numerical frequencies are usually as accurate as analytical, but the calculations are slower. However, they must be used when analytical frequencies are not available (because no one has implemented the algorithm), or because the computer cannot implement the algorithm for a molecule of the requested size (e.g., because the data generated in the frequency run overflows the addressing capability of the disk files; this problem does not arise with first derivatives). This second limitation is less of a problem now, since the 64-bit version of Gaussian 16 overcomes this deficiency of the 32-bit version [35].

We can now compare and discuss calculation of numerical frequencies for **1** from the Cray T3E supercomputer above, and from a PC:

Cray T3E supercomputer, c.1995, approximately 250 h (10.4 days) [28(a)]. Using GAMESS [24], L. Radom, private communication, July 12, 2017.
PC, 2012-model, 183 h (7.6 days). Using Gaussian 09 [19], author's work.

The time ratio is 250/183 = 1.4, with the Cray slower. The cost ratio is greater than 1,000,000/2000 = 500 (computer cost dollar ratio). This is a pretty good illustration of how far computers have come in PC-power from c.1995 to 2012, 17 years. Presumably the times would have been significantly shorter for both platforms had the machines and programs been able to calculate the frequencies analytically. Incidentally, the optimization time on the PC was 2.5 h, shorter than the frequency time by a factor of 73.

This work alerted the author to a caveat that should be borne in mind when using numerical frequencies (calculations with **1** and related molecules here routinely used Gaussian 09 [36]). To give an idea of the size of the jobs for which analytical frequencies failed: analytical frequencies worked for MP2/6-31G* on D_{2h} $C_{13}H_8$, 211 basis functions, but failed for MP2/6-31G* on D_{2h} $C_{17}H_{16}$, 287 basis functions. In comparing for **1** a PC today with the Cray supercomputer of 1995, the numerical frequencies were initially calculated using the Gaussian 09 default step size of $\Delta q = 0.001$ Å (0.0001 nm, 0.1 pm). Dismayingly, the canonical ostensibly planar-carbon cycloalkane **1** thereupon showed four imaginary frequencies! With a tightly optimized structure, from Gaussian's stricter optimization convergence criteria, the result was worse, five imaginary frequencies. These frequencies, about −200 to $-700\ \text{cm}^{-1}$ (the fifth, $-31\ \text{cm}^{-1}$), corresponded to twisting of the scaffolding around the coplanar C_5 core (Gaussian denotes imaginary frequencies as negative; strictly speaking these are 200*i*, etc.). This made no sense, in view of the results of Radom and Rasmussen [28], and the fact that MP2/6-31G*, and DFT calculations with several different functionals [37], uniformly gave only one imaginary frequency. For example, TPSS with the cc-pVTZ basis [26], 1026 functions, gave $-15\ \text{cm}^{-1}$, and DFT is known to sometimes give spurious imaginary frequencies due to numerical integration grid errors [38]. The solution to this problem was suggested by Gaussian [39]: the molecule sits in a shallow depression on the potential energy surface, so a small geometric displacement leads to only a very small change in energy and thus in the gradient. Therefore, the difference in gradients, $\Delta(\partial E/\Delta\partial q$, obtained by subtracting one analytical gradient from the other, is prone to considerable error, subjecting the force constant to error. A vibrational mode whose path on the multidimensional potential energy hypersurface is so flat that this inaccuracy arises can be called a soft mode [39]. The problem was overcome by using a larger displacement, 0.005 Å rather than 0.001 Å. With this there were no imaginary frequencies, the lowest frequency being 45 cm^{-1} (corresponding to twisting of the planar five-carbon atom core in the direction of a standard CC/CC orthogonal spiropentane structure):

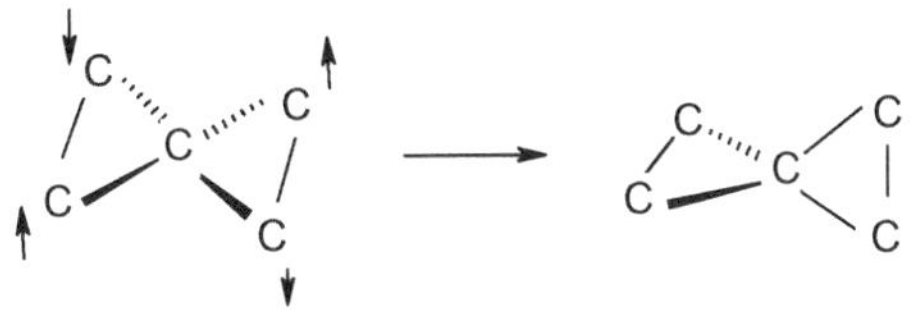

Of course the restraints of the scaffolding may not permit the attainment of a truly orthogonal structure. DFT calculations on **1** and molecules of similar size are much faster than ab initio MP2 ones, even for bigger basis sets than with MP2. On a 2012-model PC the TPSS functional [40] with the large cc-pVTZ basis [41], 1026 functions with **1**, needed only 4.4 h for optimization and 32 h for frequencies, compared to the 415 basis function MP2 calculation which demanded 2.5 and 183 h (the apparent anomaly of a longer optimization time for DFT is more than compensated for by the faster calculation of frequencies). The TPSS structure is very similar to the MP2 one, but has a small imaginary frequency of 15*i*, which as implied above could be spurious. So for PCs DFT certainly might seem more attractive than ab initio for calculations on relatively big molecules. The problem with DFT is: when can you trust the results? This is a much-discussed problem with the method: that the theoretical basis of ab initio calculations is well-understood while that of DFT is, dare one say much shakier, need not be discussed here one more time [42]. In fact, it has been claimed that since 2000 DFT has been headed in the wrong direction, "sacrificing physical rigor for the flexibility of empirical fitting" [43]. DFT has even been called "AM1 for the 21st century," implying that while fast and useful, it is of questionable reliability for nonroutine calculations [44]. The TPSS functional [40] that was used here for **1** is nonempirical, in contrast to some popular hyperparameterized functionals (which, it is readily conceded, *can* give accurate geometries and relative energies), and so may not be tainted by the problems that seem to have arisen for empirical functionals. For an exotic molecule like **1** this enhances the degree of trust one can place in the TPSS results. Judicious use of DFT on the PC is not unreasonable.

2.2 A SMALLER PLANAR-CARBON MOLECULE THAN DIMETHANO[2.2]OCTAPLANE: *3A*

Radom and Rasmussen finally arrived at dimethano[2.2]octaplane, **1**, by top-bottom symmetrizing bowlane, varying the capping rings, and adding bracing carbon chains (Fig. 2) [28]. The author tried a reverse process, removing carbon atoms from **1** to see if a smaller, more tractable (computationally and synthetically) molecule housing a planar carbon might be found. Carbons were deleted from the C_{23} molecule **1** in increments of 2 or a multiple of 2, maintaining the symmetry requirements for a planar central carbon atom and leaving the C_5 core formally undisturbed (Fig. 3). Below the C_{23} milestone molecule one encounters five molecules that a priori (because of symmetry) might contain a planar $C(C_4)$ core. Omitting **2a** and **2b**, which had been examined by Radom and Rasmussen, we consider the diminished structures **3a** and **3b**, **4**, and **5**. Structure **4** could not be "located": attempts to obtain an ab initio-optimized structure corresponding to this structural formula gave structures with different atomic connectivities, for example, a structure with four double bonds which could be viewed as resulting from opening of the two capping cyclobutane rings. The species **4** may not be a stationary point [31] on the potential energy surface Structure **5** can be quickly dismissed: a planar-carbon neopentane as a local minimum on the potential energy surface (a real molecule) should have long ago been recognized computationally, which would invalidate the claim (1999) that "...no structures with a planar tetracoordinate $C(C)_4$ substructure have been found" [45].

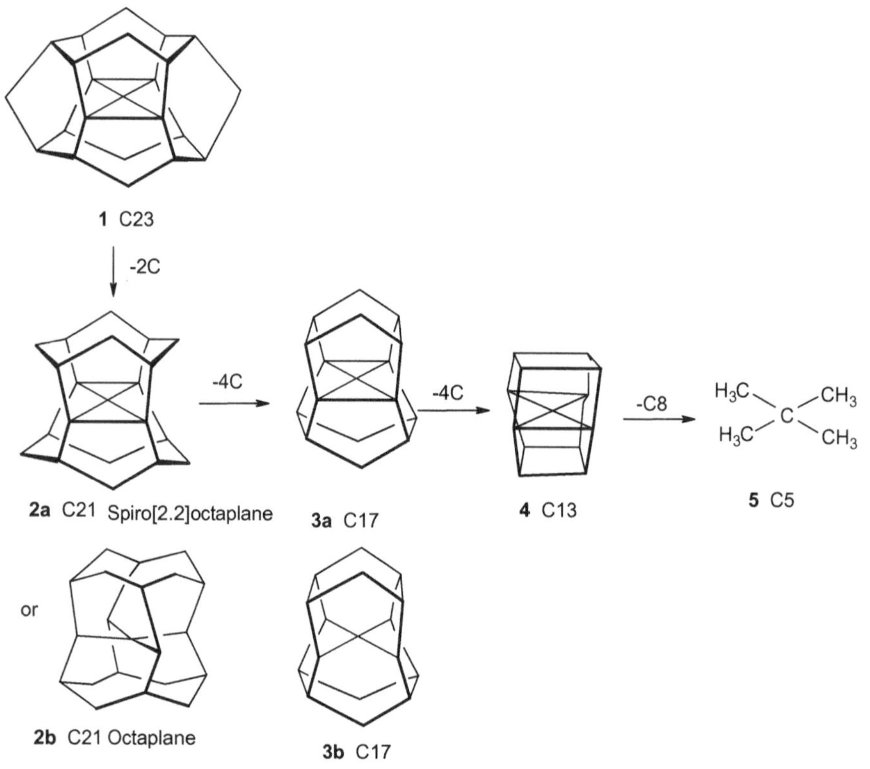

FIG. 3

All these structures have D_{2h} symmetry.

Structure **3a** is intriguing. This was optimized and checked by frequency calculations for imaginary frequencies. The results for various methods on a 2012-model PC were:

MP2/6-31G* 287 basis functions
Optimization 56 min
Frequencies (numerical) 16 h, no imaginary frequencies (default step size of 0.001 Å)
MP2/multiple basis (cf. calculations on **1**, above) 315 basis functions
Optimization 39 min
Frequencies (numerical) 32 h, no imaginary frequencies (this was with step size 0.005 Å; as with **1**, with the default 0.001 Å step size there were imaginary frequencies (two)).

So by the same criteria as were applied to **1**, molecule **3a** is a minimum on the potential energy surface! This is quite interesting because **3a**, with only 17 carbon atoms would be the smallest theoretically pinpointed neutral saturated hydrocarbon with a planar tetracoordinate carbon atom. We can compare this with **1**, $C_{23}H_{24}$ with (potentially) such a carbon, and with **2a** and **2b**, $C_{21}H_{24}$, which evidently are not potential energy surface local minima [28]. We note the resemblance of **3a**, $C_{17}H_{16}$, to a similar structure, **3b**, $C_{17}H_{20}$ (denoted **5** in reference [28(a)]). Structure **3a** is more restrained, because it has two CC bonds lacking in **3b**, by virtue of its two fewer pairs of H atoms, H_{16} (cf. H_{20}). If **3a** is indeed a “real” molecule, then it is quite significant: we may be near or at the smallest cycloalkane which can shelter a planar tetracoordinate carbon atom. And being significantly smaller than **1**, **3a** may be easier to synthesize. Apparently none of these planar-carbon compounds has yet (mid-2017) been made.

We see that a project that in 1995 required a supercomputer could in 2012 be accomplished on a PC, although some frequency calculations might have needed about a week. The calculations in c.2017 would of course be even more feasible for execution on a PC. While this whole particular project will not, of course, be repeated on a PC, the limited recapitulation reported here shows again the considerable utility of personal computers as research tools in computational chemistry.

3 PYRAMIDAL TETRACOORDINATE CARBON IN AN OTHERWISE NORMAL SATURATED HYDROCARBON

By a pyramidal tetracoordinated carbon atom we mean one whose four bonds are directed toward the corners of a rectangle. For proximate comparison, Fig. 4 shows methane in the pyramidal, planar, and tetrahedral configurations. Structures containing pyramidal carbon atoms have been reviewed [46].

Before focusing on a specific candidate for a real molecule with pyramidal carbon let us examine the energy needed to distort methane from the planar to the

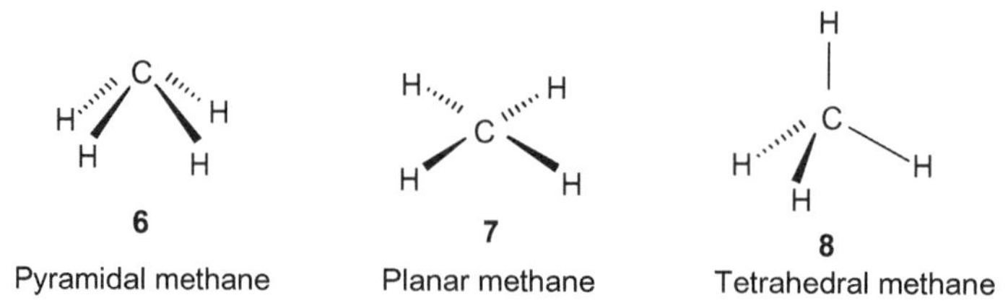

FIG. 4

Methane in the pyramidal, planar, and tetrahedral configurations.

pyramidal configuration, and compare these two with the tetrahedral. This gives a feel for the energy demand of these forms containing distorted carbon atoms. Fig. 5 plots energy against HCH angle (all HCH angles are the same in the tetrahedral, the planar, and these various pyramidal configurations) at the MP2/6-311++G(2d) level. For a hydrocarbon this is a slightly bigger basis set than was used for the central carbon of the canonical planar-carbon molecule **1** (MP2/6-311+G(2d), because it adds one diffuse function to each hydrogen atom. There really should be no significant difference between the results from these two different basis sets, because the hydrogens atoms here are "not doing anything," for example, hydrogen

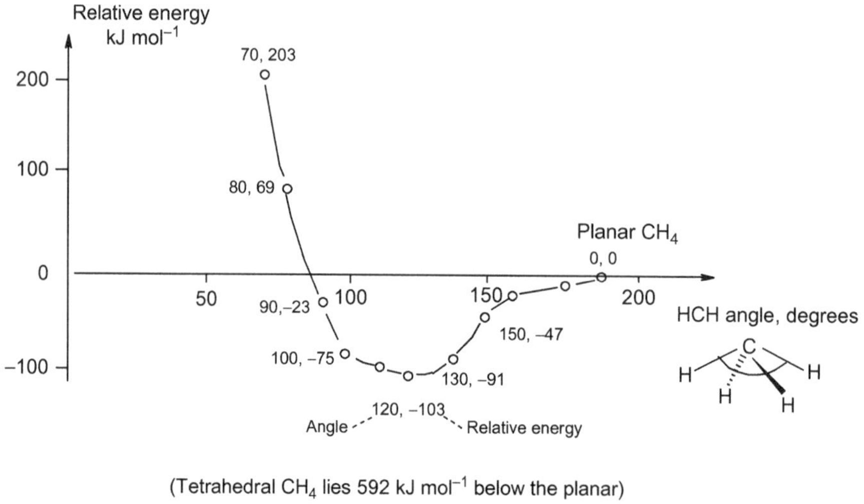

FIG. 5

Pyramidal carbon flattening to a planar structure. Variation of the energy of pyramidal CH_4 relative to the planar configuration, at the MP2/6-311++G(2d) level. The lowest-energy pyramidal structure for CH_4 has an HCH angle (*inset*) of approximately 120 degrees.

bonding, and it must be confessed that saying the bigger basis gives better results is, in this case, almost pedantic.

For CH_4 the computational demands on a 2012-PC are trivial and multiple basis sets are not needed to alleviate the basis load on the other atoms. The figure shows that at this level the lowest-energy pyramidal CH_4 structure lies at the approximately 120 degree angle, approximately 100 kJ mol^{-1} below the planar structure (the flanking 110 degree and 130 degree HCH angles correspond to −99 and −91 kJ mol^{-1}). So the increasing order of stability (decreasing relative energy, planar highest, MP2/6-311++G(2d) level), of the methane configurations, examined is: planar (relative energy 0 kJ mol^{-1}), pyramidal (approximately −100 kJ mol^{-1}), and tetrahedral (−592 kJ mol^{-1}). Pyramidal methane lies approximately 490 kJ mol^{-1} above tetrahedral, and planar methane "only" another approximately 100 kJ mol^{-1} higher. Compared with tetrahedral the two kinds of unnatural tetracoordinate carbon are both roughly equally bad.

In seeking a neutral saturated hydrocarbon with a planar-carbon atom, Radom and Ramussen (above) performed a series of computer "experiments" starting with a somewhat flattened carbon (Fig. 2). By contrast, in looking for a candidate molecule with a pyramidal tetracoordinated carbon atom, a simple possibility immediately presents itself:

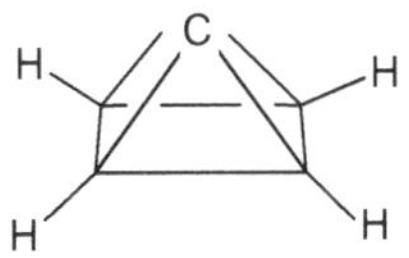

9

Intuitively, this is the simplest alkane/cycloalkane which might realistically be expected to harbor a pyramidal tetracoordinate carbon atom. The "colloquial" name of **9** is pyramidane. This structure has erroneously been called tetrahedrane [47], but that is actually $(CH)_4$, a pyramid with a triangular, rather than a square, a base, and a hydrogen atom on each carbon atom. Semiempirical and small basis set HF-level ab initio calculations on **9** were reported in 1978–82 [48]; HF/6-31G* geometry optimizations and frequencies and MP2/6-31G* energies on the HF geometries in 1988 [49]; and higher-level ab initio investigations in 1998, 2000, and 2001 [50–52]; pyramidane was reviewed in 2008 [46].

At the MP2/6-311++G(2d) level, pyramidane is a relative minimum on the potential energy surface, a real molecule. The electronic nature of the unusual apical carbon was discussed in [50] and again in [46]: it can be approximated as having a lone pair in an s orbital and employing three p orbitals to form a bond to each of the four basal carbon atoms. In accord with this, pyramidane was predicted to be very basic for a hydrocarbon, stronger than pyridine [46,50]. The small size of this molecule makes it fairly easy to do something that has not been done for dimethano[2.2]octaplane (**1**), namely put a number on its likely stability. A potential energy surface to test the stability of **9** can be created by starting with the intuitively reasonable idea that likely decomposition modes are breaking bonds to opposite sides of the base, and to adjacent sides:

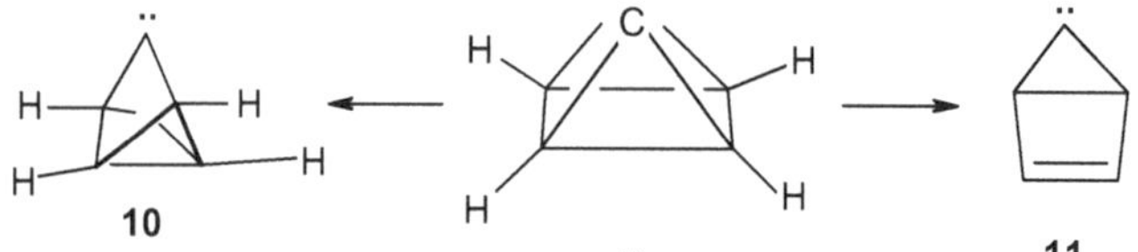

This leads, respectively, to the tricyclic carbene **10** and the bicyclic carbene **11**. To locate the transition states leading to these we use input structures in which the breaking bonds are stretched by, say, about 50%, and forming bonds are implied by decreasing the distance between pairs of atoms. A transition state optimization tends to be more finicky than optimization to a relative minimum, and to get good starting structures it was sometimes necessary to use various tricks, for example, first optimizing to a minimum with a stretched-bond constraint. The calculated pyramidane potential energy surface is shown in Fig. 6. Transition state **ts10(1)** leads from pyramidane to the tricyclic carbene **10** and transition state **ts10(2)** interconverts mirror image structures of **10**; transition state **ts11** leads from pyramidane to the bicyclic carbene **11**. Some license has been exercised in calling **11** bicyclic, as the transannular CC distance is calculated at approximately 1.9 Å; nevertheless, CC single bonds much longer than approximately 1.5 Å are known (e.g., 1.71 Å) [53]; in Fig. 6 this bond is shown dashed. The symmetry designations (C_s, C_{2v}, C_{4v}) are relevant in revealing the number of symmetry planes. The trio of numbers accompanying each species are relative (to pyramidane) energies in kJ mol^{-1}. Here I have put first (left to right) the highest-level of the three, the CCSD(T)/DZP of Kenny et al. [52], then our CCSD(T) single-point calculations, then our MP2/6-311++G(2d) calculations. Discussion of these is warranted.

The CCSD(T)/DZP energies (0, 98 kJ mol^{-1}, etc.) are from geometry optimization at this level [52]: the electron correlation level is CCSD(T) [54] and the basis set is double zeta with polarization, DZP, a correlation-consistent basis [55]. Coupled-cluster singles doubles approximate triples are the highest-level electron correlation calculations routinely performed (as of the mid-2017) on more than very small molecules. Correlation-consistent basis sets are designed specifically for use with post-HF calculations and ideally, at least, are supposed to give with increasing size systematically improved results, so that energies can be extrapolated to an infinite basis set. In the analogous figure in [52] the DZP energies shown in Fig. 6 were augmented in three cases (see below) with a triple zeta two *d* polarization functions (TZ2P) basis optimization, without showing much change in the energy or geometry. For some more information on these basis sets see reference [52]. Our the CCSD(T) single-point calculations refer to single-point energies from the CCSD(T)/6-311++G(2d) level on MP2/6-311++G(2d)-optimized geometries, designated CCSD(T)/6-311++G(2d)//MP2/6-311++G(2d). These single-point

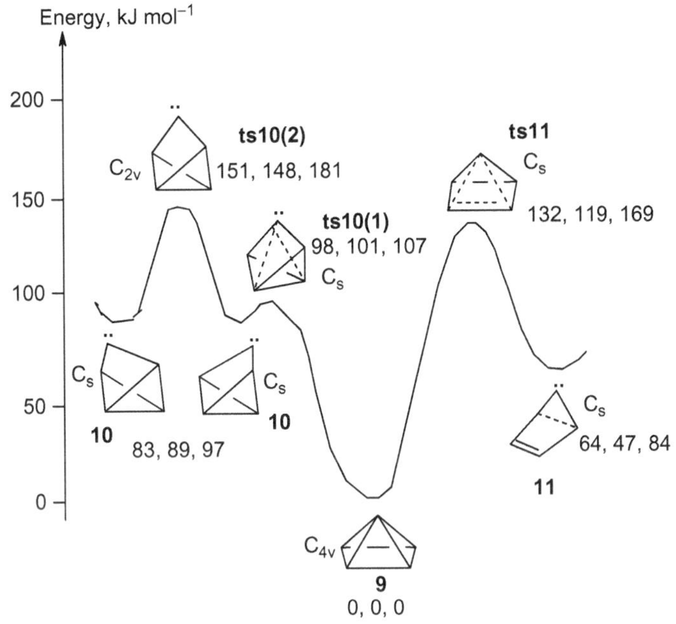

FIG. 6

Pyramidane potential energy surface, kJ mol^{-1}, at the levels (left to right for each trio) CCSD(T)/DZP, CCSD(T)/6-311++G(2d)//MP2/6-311++G(2d), and MP2/6-311++G(2d). The three species **9**, **ts 10(1)**, and **ts11** are critical to the stability of pyramidane; they were also calculated at the CCSD(T)/TZ2P (i.e., triple zeta with two polarization functions) level, and showed practically no change compared with CCSD(T)/DZP (see text).

calculations are much less computationally demanding than CCSD(T) *optimizations* like Kenny et al. did. Energies from such single-point calculations are usually similar to those from the more rigorous regime of optimization at the higher correlation level (although here two species, **ts11** and **11**, show distinct deviations, 132 vs. 119 kJ mol^{-1} and 64 vs. 47 kJ mol^{-1}); the main advantage of higher correlation level optimization-plus-frequencies is reassurance that we can trust the result for the curvature of the potential energy surface, that is, for confidence that a specie is a minimum or merely a transition state, *gratis* the number of imaginary frequencies. On this score our MP2/6-311++G(2d) PES is in full agreement with the CCSD(T)/DZP one of Kenny et al. [52]. Finally, the MP2/6-311++G(2d) energies are simply from those shown at the end of an optimization at that level; for some reason these are uniformly higher than the others, but yield qualitatively the same PES (in Fig. 6 the positions on the curve of the minima and the transition states are drawn to match the vertical energy scale for the CCSD(T)/DZP energies). These results leave little doubt that pyramidane is a true relative minimum on the PES and that it would be (as of the mid-2017 it had not been made) at least reasonably stable to unimolecular decomposition: the lowest barrier to such decomposition, to the tricyclic carbene, is calculated to be approximately 100 kJ mol^{-1}, and from experience a unimolecular barrier of about 100 kJ mol^{-1} is the threshold for stability at room temperature [56].

The calculations whose results are summarized in Fig. 6 were performed, for the CCSD(T)/6-311++G(2d)//MP2/6-311++G(2d), and MP2/6-311++G(2d) levels (the author's calculations), on a 2012-vintage PC. To remind ourselves that this exposition is intended primarily to illustrate the power of personal computers, it would be nice to compare some features of our calculations with the CCSD(T) calculations of c.2001 [52]. Unfortunately, details of the computer used and the times for the various calculations seem to be no longer available. We can, however, make at least a modicum of comparison of the performance of a current PC with the machine used in the 2001 work. The three species **9**, **ts 10(1)**, and **ts11** are critical to the stability of pyramidane because they determine the heights of what are believed to be the two lowest-energy barriers to decomposition. Optimization by Kenny et al. [52] at the particularly high CCSD(T)/TZ2P level gave geometries and relative energies almost the same as at their CCSD(T)/DZP level (triply-split vs. doubly-split valence shells). The barriers (kJ mol^{-1}) for CCSD(T)/TZ2P were:

Barrier over **ts 10(1)** 100.8 (TZ2P), 97.5 (DZP)
Barrier over **ts11** 139.7 (TZ2P), 131.8 (DZP)

Each of these calculations, evidently done on a workstation (an IBM RS/6000?) was said to have taken several days. Our single-point CCSD(T) calculations on MP2/6-311++G(2d) geometries (admittedly much less demanding than CCSD(T) optimizations) using the large cc-pVTZ basis gave these results:

Pyramidane (C_{4v}), **9**, 21 min; **ts 10(1)** (C_s), 47 min; **ts 11** (C_s), 42 min
Barrier over **ts 10(1)** 101.1 kJ mol^{-1}
Barrier over **ts 11** 134.2 kJ mol^{-1}

We see that these CCSD(T)/cc-pVTZ//MP2/6-311++G(2d) barriers agree well with the best ones from the Kenny et al. [52] CCSD(T)/TZ2P optimizations. The geometries we used, MP2/6-311++G(2d), took for these C_5H_4 species on a 2012 PC about 20 min each including frequencies, and the single-point CCSD(T)/cc-pVTZ energies on these geometries about 30 min each. So one could characterize each of the six stationary points on the PES of Fig. 6 for about an hour of PC CPU time, at the CCSD(T)/cc-pVTZ//MP2/6-311++G(2d) level. These energy differences (cf. pyramidane) provide relative energies similar to those from CCSD(T)/TZ2P optimizations, which required in c.2001 several days for the energy of each point.

Of course the beauty of optimization at the CCSD(T) level, in contrast to single-point CCSD(T) energies on a lower-correlation geometry (e.g., MP2) is that the coupled-cluster calculations provide an energy strictly for *that* correlated geometry and, if frequencies were calculated (which can be particularly time consuming), the calculation has characterized the stationary point as a minimum or whatever. So why have we been so coy about CCSD(T) optimizations with a PC? The problem appears to be twofold: a software availability and a hardware architecture restriction. Efficient, indeed practical (in terms of the time for a calculation) geometry optimizations, and even more so frequencies, almost *demand* analytical gradients (first derivatives); analytical second derivatives are nice too, but gradients are already enough of a challenge for high-level electron correlation. These are a real challenge to implement in a geometry optimization program, and the only currently available program suite with this feature seems to be Advanced Concepts in Electronic Structure (ACES) [57]. Installing ACES and getting it to work is evidently not trivial, and is a task for experts working on workstations, clusters, or mainframes [58]. CCSD(T) *single-point* calculations with large basis sets on up to moderately large molecules are entirely practical on a 2012-vintage PC: a study of annulenes (cyclic conjugated polyenes), all the ones here of C_2 symmetry, gave these results ("aug" augments the cc-pVTZ basis by adding diffuse functions) for single-point times:

$(CH)_4$	Basis aug-cc-pVTZ	276 Basis functions	1.7 h
$(CH)_5$	Basis aug-cc-pVTZ	315 Basis functions	6.1 h
$(CH)_6$	Basis cc-pVTZ	264 Basis functions	3.3 h
$(CH)_6$	Basis aug-cc-pVTZ	414 Basis functions	Failed
$(CH)_7$	Basis cc-pVTZ	308 Basis functions	5.8 h
$(CH)_7$	Basis aug-cc-pVTZ	483 Basis functions	Failed
$(CH)_8$	Basis cc-pVTZ	352 Basis functions	Failed

Up to $(CH)_5$ was accessible with the very big aug-cc-pVTZ basis, and up to $(CH)_7$ with the big cc-pVTZ.

Thus it emerges that on a PC the core of a research project on pyramidane, the reliable mapping of the PES, could conceivably be accomplished in hours. Naively, in about 6 h, but this neglects time spent on constructing input structures that work and on reflecting on the results! All this is if we gracefully accept optimizations and frequencies at the MP2 correlation level, and single-point CCSD(T) energies thereon.

4 HALF-PLANAR CARBON, BUTTERFLY CARBON: POLYPRISMANES OR PRISMANES

We have considered the possibility of the normally tetrahedral tetracoordinate carbon adopting planar and pyramidal configurations. Another unconventional stereochemical prospect is half-planar carbon, illustrated here with methane thus contorted:

The H_1CH_2 angle is 180 degrees or approximately 180 degrees and in the molecules we focus on the H_3CH_4 angle can be variously 60 degrees, 90 degrees, etc., depending on whether the carbon belongs to a ring with three, four, etc. carbons. Half-planar is an appropriate designation because in conceptually converting normal tetrahedral carbon to the planar configuration, structure **12** is halfway along the transformation:

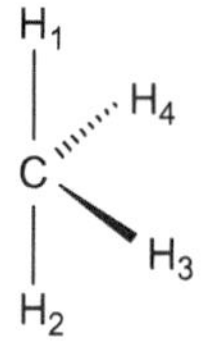

12

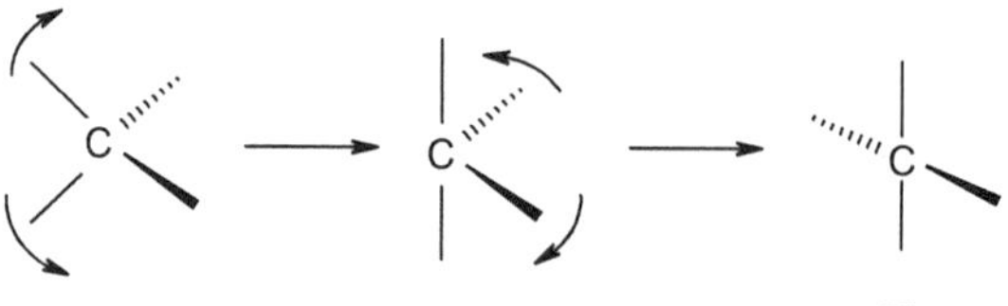

The configuration of carbon in **12** has also been called bisphenoidal, and butterfly. The obscure word bisphenoidal evidently comes from the Greek *sphenoeides*, "wedgelike." It is a highly technical geometrical term (more often disphenoidal, rather than bi-), and appears to have been used since at least c.1900 to describe crystal shape, and later the symmetry of atomic arrangements in X-ray crystallography. The word is, besides being obscure, confusing because a chemist might well interpret it as bis-phenoidal, "double phenol-like," rather than bi-sphenoidal. Its relevance to half-planar carbon is revealed in this Wikipedia entry for the sphenoidal skull bone: "Its shape somewhat resembles that of a butterfly or bat with its wings extended." From this follows the clearer, charmingly whimsical butterfly appellation:

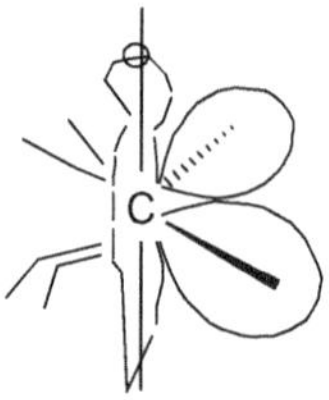

Nevertheless we shall stick to the term half-planar because it is directly apprehended and conforms to the geometrical nomenclature used for planar and pyramidal carbon.

Fig. 7 shows the energy variation as normal methane is distorted to half-planar with one HCH angle going from 109 degrees to 180 degrees and the angle in the orthogonal HCH plane is kept at 109 degrees. Fig. 8 shows the energy variation for half-planar CH_4 as the other (the nonlinear) angle is varied from 60 degrees to 120 degrees. In both figures the energies shown are relative to normal methane and the calculation level is MP2/6-311++G(2d). By comparison, Fig. 5 shows the energy

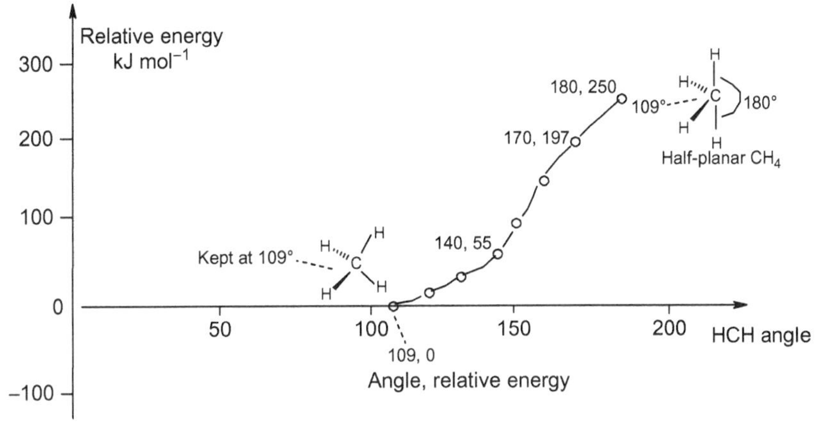

FIG. 7

Tetrahedral to half-planar carbon. Variation of the energy of as tetrahedral CH_4 is distorted to the half-planar configuration with one HCH angle kept at 109 degrees, at the MP2/6-311++G(2d) level.

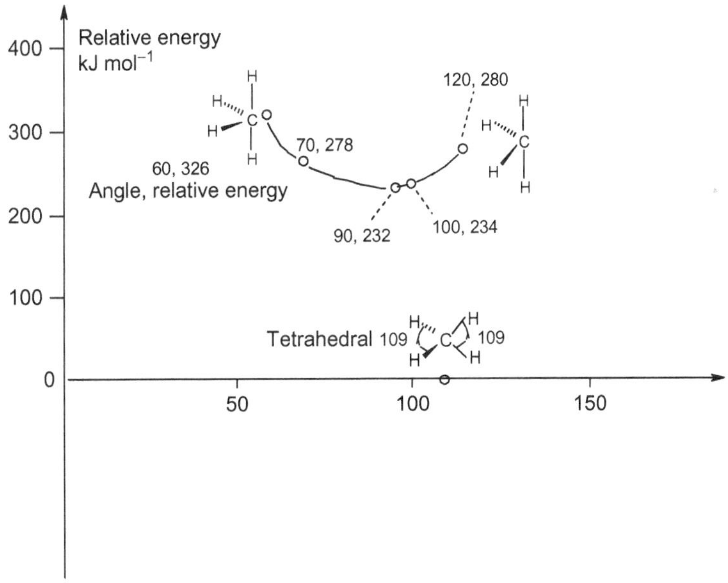

FIG. 8

Energy variation for half-planar CH_4 as the nonlinear angle is distorted from 60 degrees to 120 degrees. Calculated at the MP2/6-311++G(2d) level.

variation as pyramidal CH_4 is flattened to planar. These graphs may give a feel for the energy changes accompanying the distortion of tetracoordinate carbon away from its normal stereochemical proclivity.

To continue illustrating the role of PCs in probing exotic compounds, we now move from planar carbon in species like octaplane (**1**), through pyramidal carbon in pyramidane (**9**) and focus on half-planar carbon in a family of molecules that have been called polyprismanes. These were presented in 2003 by Minyaev et al. [59]. Although half-planar carbon appears in the literature before that work (see references therein), the 2003 paper (Minyaev, Minkin, Gribanova, Starikov, Hoffmann) seems to be the first study of a set of molecules that represent straightforward embodiments of the concept. References to theoretical analysis of half-planar carbon and to molecules that skirt this configuration are given in [59], and the subject is reviewed and discussed further in [60]. Representative polyprismanes are shown in Fig. 9, and here we touch again on nomenclature. In the figure [2,3]prismane, [2,4]prismane, etc., shortened here for convenience to **2,3**, **2,4**, etc., refer to this author's suggested designations for these compounds [60]. This is a simplification, hopefully more transparent, of the naming proposed by Minyaev et al. They designate **2,3** as [3]prismane and **3,3** as bi[3]prismane, regarding **2,3** as a "base" molecule and the ring stacked on top to give **3,3** as affording the first polyprismane in this cyclopropane series. Structures **3,3**, **3,4**, **3,5** with stacks of three rings were called bi[n]prismanes, $n = 3, 4, 5$, and **4,3** with a stack of four rings would be tri[3]prismane, and so on. My nomenclature simply gives the stack height by a first number and the ring size by a second number: structure **3,3** is [3,3]prismane, structure **3,4** is [3,4]prismane, etc. Prismane itself (C_6H_6) is thus [2,3]prismane (**2,3** in Fig. 9) and cubane (C_8H_8) is [2,4]prismane (**2,4** in Fig. 9). Logically cyclopropane is [1,3]prismane and cyclobutane [1,4]prismane; awkward-sounding, but a small price to pay for simplicity and consistency. Minyaev et al. reserve "polyprismane" for a structure with three or more stack layers, since this is required for half-planar carbons, but we will usually simply call all these structures prismanes. We will examine calculations along the homologous series **2,4**, **3,4**, **4,4**, **5,4**, and **6,4**. This will suffice to indicate how well a PC can handle prismanes. Stacked cyclobutane rings were chosen as providing the smallest ring size that might avoid the very special reactivity of cyclopropane and still be time-friendly (of course all prismanes perforce have *walls* made of cyclobutane rings).

Optimized geometries and frequencies were obtained for **3,4**, **4,4**, **5,4**, and **6,4**. The results, given in Table 1, were from Gaussian 09 [19,36] on a year 2012 PC. The correlation methods were MP2 and DFT (TPSS and PBE0) and the basis sets were 6-31G* and 6-311++G(2d). The former basis is the smallest, and thus fastest, for meaningful results with a correlated method like MP2, and the latter is that used for the central carbon atom of dimethano[2.2]octaplane (**1**) and for the calculations on pyramidane (**9**) (++ cf. one+ denotes a diffuse function on H, irrelevant for the central C of **1**). These DFT functionals are discussed in [37]. They were chosen because they are nonempirical: TPSS [61] and PBE0 [62]. The author considers nonempirical functionals more reliable for exotic molecules than

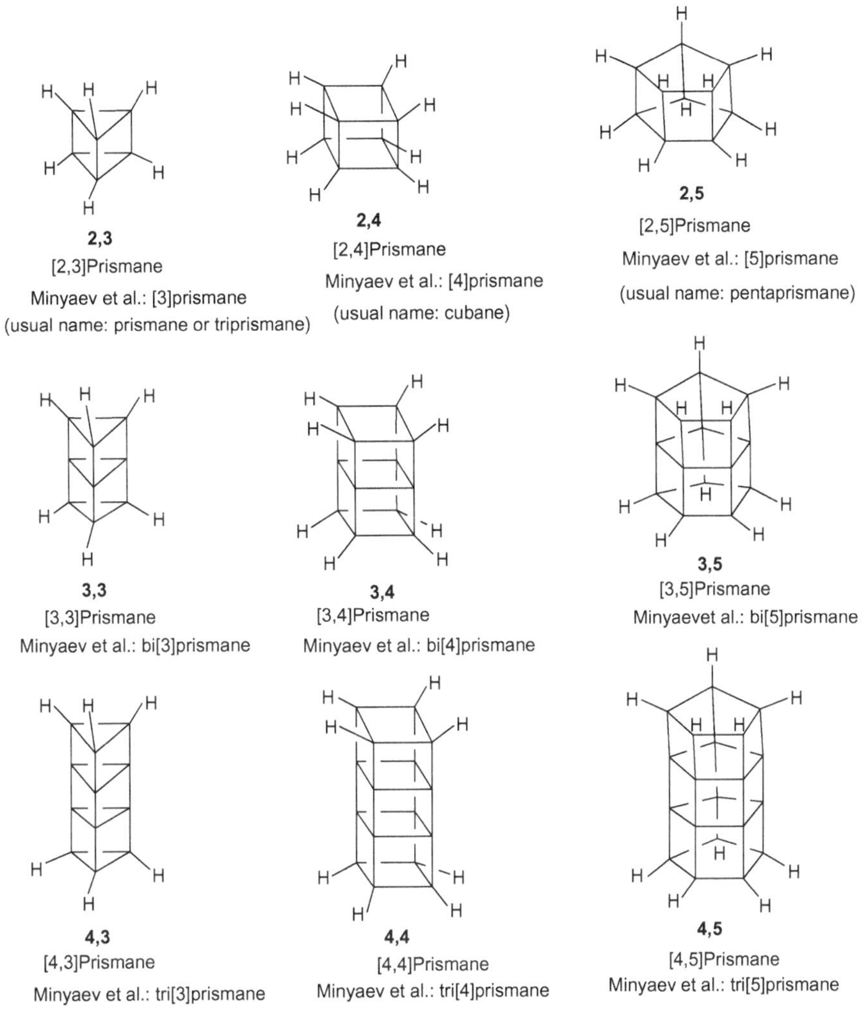

FIG. 9

Examples of polyprismanes (prismanes) (see text).

functionals parameterized, for the most part, against prosaic molecules. It is claimed that since 2000, better DFT energies have *not* corresponded to better wavefunction properties, due to the use of hyperparameterized functionals [63]. Perdew and coworkers have made a case for nonempirical functionals [64].

We will not dwell on the predicted properties of these molecules (the same goes for the planar and pyramidal carbon species discussed earlier), since this has been done in the references to the original work. The focus is, rather, on the applicability

Table 1 Calculations on Prismanes With a Year 2012 PC

Molecule	Method/Basis, Times	Comments
3,4	**(1)** MP2/6-31G*, 196 basis f Opt 3 m, a freq 2.1 h, num freq 1.3 h **(2)** MP2/6-311++G(2d), 356 basis f Opt 42 m, num freq 15 h **(3)** TPSS/6-311++G(2d), 356 basis f Opt 5 m, a freq 38 m **(4)** PBE0/6-311++G(2d), 356 basis f Opt 6 m, a freq 1.1 h	**(1)** No imag freq **(2)** No imag freq **(3)** No imag freq **(4)** No imag freq
4,4	**(1)** MP2/6-31G*, 256 basis funcs Opt 11 m, a freq failed, num freq 3.7 h **(2)** MP2/6-311++G(2d), 464 basis f Opt 2.0 h, a freq 2.4 days **(3)** TPSS/6-311++G(2d), 464 basis Opt 11 m, a freq 1.6 h **(4)** PBE0/6-311++G(2d), 464 basis Opt 14 m, a freq 2.9 h	**(1)** No imag freq **(2)** 13 imag freqs, (a) see text **(3)** No imag freq. Middle CC 1.634 Å, (b) see text **(4)** No imag freq
5,4	**(1)** MP2/6-31G*, 316 basis f Opt 53 m, num freq 11 h **(2)** MP2/6-311++G(2d), 572 basis f Opt incomplete after 3.7 days, (c) see text **(3)** TPSS/6-311++G(2d), 572 basis f Opt 1.1 h, a freq 3.8 h **(4)** PBE0/6-311++G(2d) Opt 1.7 h, a freq 7.4 h	**(1)** No imag freq **(2)** Middle CCC narrows from 157 degrees to 123 degrees, (d) see text **(3)** 4 imag freqs, seems to want to distort to lower symmetry, (e) see text **(4)** No imag freq. Middle CCC narrows from 157 degrees to 108 degrees, (f) see text
6,4	**(1)** MP2/6-31G*, 376 basis f Opt 1.3 h, num freq 23 h **(2)** MP2/6-311++G(2d), 680 basis f Not attempted **(3)** TPSS/6-311++G(2d), 680 basis f **(4)** PBE0/6-311++G(2d), 680 basis f Opt 1.5 h, a freq 13 h	**(1)** 4 imag freqs, seems to want to distort to lower sym. **(2)** Not attempted **(3)** Failed, "rare condition," (g) see text **(4)** No imag freq, no geometric distortion

Notes: All these molecules had D_{4h} symmetry. Times refer to geometry optimization, analytical frequencies (where possible), and numerical frequencies.

*(a, b) With **4,4** the MP2/6-311++G(2d) frequency calculation (necessarily numerical) showed 13 imaginary, the "lowest" 5 being approximately 2500 cm^{-1}. This was with a step size of 0.005 Å (see the discussion of frequencies for the planar-carbon molecule **1** earlier). With the default step size of 0.001 Å there were 25 imaginary frequencies. This bizarre result probably arises from errors in the MP2/6-311++G(2d) force constants due to the D_{4h} structure not sitting securely in a relative minimum on the PES at this level. It may be relevant to this that the TPSS/6-311++G(2d) geometry showed the four middle (along the chain) CC bonds distinctly elongated, 1.634 Å.*

*(c, d) With **5,4** the MP2/6-311++G(2d), optimization was incomplete after 3.7 days, with none of the criteria for completion being satisfied. This optimization would likely have required weeks if it worked at all. In this incomplete optimization the four central CC bonds had bulged out, the angle narrowing from the 157 degrees of the molecular mechanics input geometry to 123 degrees.*

*(e) The TPSS/6-311++G(2d) frequency calculation (analytical frequencies worked for all these DFT calculations) on **5,4** showed four imaginary frequencies. These are probably valid, unlike the 13 or 25 imaginary frequencies for **4,4** earlier, and animating them suggested that at this level the molecule wants to distort to a lower symmetry; however, attempted optimization of D_{2h} or C_{2v} structures failed ("rare condition"), for obscure reasons connected with the algorithm.*

*(f) In the PBE0/6-311++G(2d) geometry of **5,4** the four central CC bonds had bulged out, the angle narrowing from 157 degrees of the molecular mechanics input geometry to 108 degrees, compatible with the significant distortion incipient in the incomplete MP2/6-311++G(2d) optimization.*

*(g) For TPSS/6-311++G(2d) on **6,4** attempted optimization failed ("rare condition"), for obscure reasons connected with the algorithm.*

of PCs to systems of this size at these computational levels. In this spirit we address the results in Table 1. At the MP2 electron correlation level with the 6-31G* basis set optimized geometries and frequencies can be secured for all these molecules, in reasonable times. With the MP2/6-311++G(2d) basis **4,4**, **5,4**, and **6,4** were recalcitrant: **4,4** gave a bizarre, spurious frequency result (notes on Table 1(a, b)); the **5,4** optimization was extremely slow and may even have ground to a halt (notes on Table 1(c, d)); and the **6,4** optimization was not even attempted. The DFT correlation method was much time-friendlier to these molecules than was the MP2—that, after all, is its main attraction versus wavefunction methods. The 6-311++G(2d) basis and analytical frequencies were here used uniformly with DFT, and the longest DFT calculation was with PBE0 for **6,4** frequencies (13 h). PBE0 may be somewhat better behaved here than TPSS, which failed for **6,4** (although this "rare condition" problem could perhaps be coaxed away with certain keywords). Apparently no prismanes with a stack height of three or more, that is with half-planar carbon, are known as of the mid-2017.

General conclusions from this work are that MP2 with a big basis set is useful for moderately large members of this series: **4,4**, $C_{16}H_8$, 464 functions with 6-311++G(2d), yielded an optimized structure and frequencies without incident in 2.5 days, a reasonable time for an ab initio calculation on a molecule of substantial size. As expected, DFT calculations were far faster than MP2: **6,4**, $C_{24}H_8$, 680 basis functions with PBE0/6-311++G(2d), completed in only about 15 h. Experienced computational chemists tend to quickly judge how formidable a task is from the number of "heavy atoms," these being anything beyond helium; almost subconsciously perhaps, they tend to take into account the basis set and the level of the calculation. The 24 carbon atoms of **6,4** signal a significant task for an ab initio calculation, but these molecules have high symmetry, C_{4h}: with 24 heavy atoms **6,4** has only 8 degrees of freedom. In estimating how feasible a calculation would be one should consider the number of basis functions and the symmetry. Some are queasy about DFT: recall the above-mentioned characterization (Section 2.2) as "AM1 for the 21st century" [44], but if one is willing to trust it, the range of systems which can be investigated computationally is considerably extended beyond the reach of wavefunction methods.

5 PROPELLANE CARBON

We have seen examples of the ability of personal computers to perform research-level calculations on exotic compounds of real theoretical interest: dimethano[2.2]octaplane (**1**), pyramidane (**9**), and prismanes (Fig. 9). These molecules embody, respectively, planar, pyramidal, and half-planar tetracoordinate carbon atoms. There is a fourth configuration for tetracoordinate carbon, the propellane one; its conceptual connection to tetrahedral carbon is shown in Fig. 10.

The name propellane comes from the shape of a class of molecules that resemble the propulsive artifact (Fig. 11), and was coined c.1969 by Ginsberg [65].

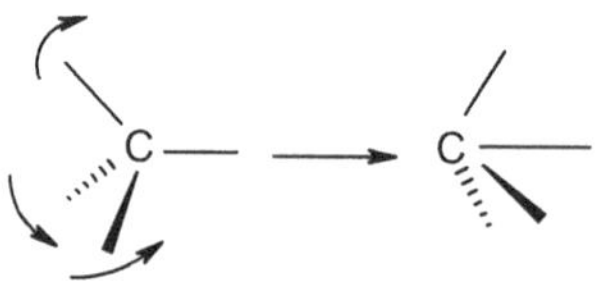

FIG. 10

Conceptual connection of tetrahedral and propellane carbon.

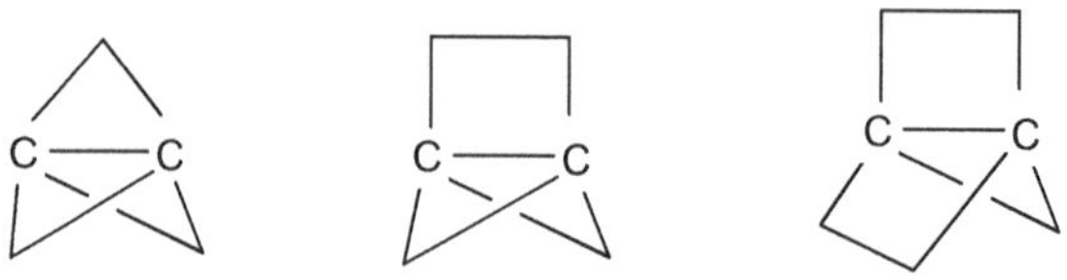

13 **14** **15**

FIG. 11

Examples of propellanes.

The central CC bond drawn in these molecules emanates from two carbon atoms which, formally at least, have their bonds in the propellane configuration of Fig. 10. The cautionary word "formally" is appropriate here because the actual spatial disposition of the bonds is a subtle question, on which there may still be no consensus [66,67].

We have seen energy versus angle graphs showing the calculated response of CH_4 to distortion involving these configurations: pyramidal flattening to planar (Fig. 5); tetrahedral opening to half-planar (Fig. 7); half-planar with various values of the nonlinear angle (Fig. 8). Fig. 12 shows the calculated result of distorting methane to propellane configurations.

Unlike the case of planar, pyramidal, and half-planar tetracoordinate carbon, where actual molecules housing the distorted carbon are so far only theoretical, propellanes are long-known compounds, and surprisingly, [1.1.1]propellane **13**, the flagship exemplar of this configuration, is the stablest of the family [67]. This fact prompts a personal anecdote: at one of the legendary Harvard evening seminars of Robert Burns Woodward, the author (the speaker) mentioned the propellane concept, whereupon Woodward, with a dismissive wave of the hand, declared that "all the bonds point in the same direction." This was 12 years before the reported synthesis of **13** [68].

Our main purpose here is to glean an appreciation of the ability of personal computers to perform research-level computations. Here are some times for geometry optimizations followed by calculation of vibrational frequencies:

[1.1.1]propellane, D_{3h} symmetry	
MP2/6-31G*	Optimization 18 s, frequencies 1.6 min
MP2/6-311++G(2d)	Optimization 2.1 min, frequencies 22 min
[2.2.2]propellane, C_3 symmetry	
MP2/6-31G*	Optimization 33 min, frequencies 26 min
MP2/6-311++G(2d)	Optimization 2.6 h, frequencies 13.2 h (the slower numerical method was necessary here: 264 cf. 144 basis functions with 6-31G*; recall the discussion in Section 2.1)

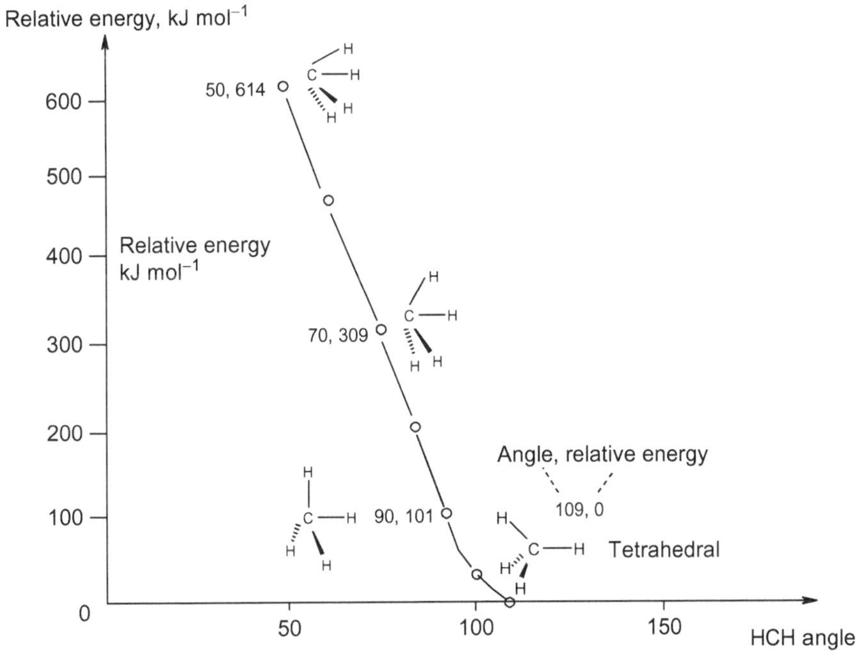

FIG. 12

Change in energy as tetrahedral CH_4 is distorted to propellane configurations. Calculated at the MP2/6-311++G(2d) level.

The geometry of [1.1.1]propellane with the fairly small 6-31G* and the reasonably big 6-311++G(2d) basis set are shown in Fig. 13. The differences are small and would have no effect on conclusions about the electronic structure, but for such an unusual molecule one feels more comfortable with the results from the bigger basis.

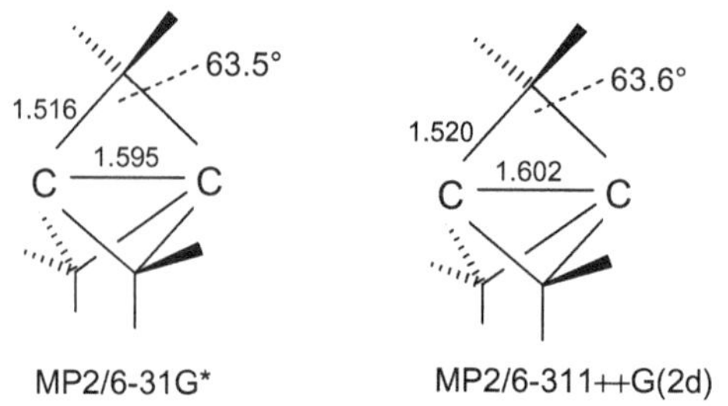

FIG. 13

The geometry of [1.1.1]propellane with different basis sets. Hydrogens are omitted and the apical carbons are emphasized.

One might suspect à la Woodward, earlier, that [1.1.1]propellane docs not really have four bonds pointing in the wrong direction, but is, rather, a diradical:

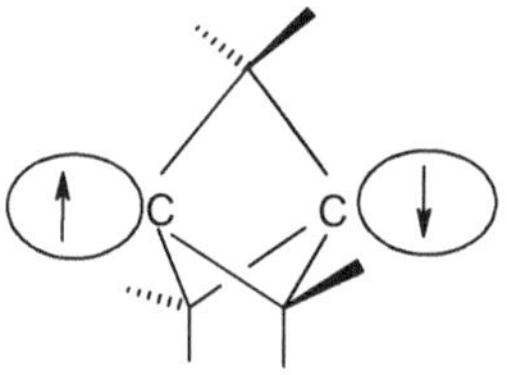

Here a singlet diradical is shown, with the unpaired electrons having opposite spins. Although a *triplet* diradical, with same spin electrons, cannot be casually ruled out as the ground state of the molecule, the stability of [1.1.1]propellane, particularly its resistance to polymerization [68], argues against this. The structure shown does not require geometric distortion of the apical carbon atoms: approximate sp^3 hybridization with each unpaired electron in an sp^3 orbital satisfies them. Calculations on a singlet diradical can be tricky (although the times required for molecules of this particular size are trivial even on a PC). The problem with these species is that "conventional" ab initio methods like the Hartree-Fock and MP2 use a wavefunction which is a single Slater determinant (or a few Slater determinants) composed of those molecular orbitals each of which is occupied by a pair of electrons, but in contrast the diradical has two singly occupied molecular orbitals. So there is a fundamental error in treating such a species with a closed-shell determinant wavefunction. The standard method for singlet diradicals is a form of configuration interaction (CI), a process in which the wavefunction is composed of a sum of determinants, the basic closed-shell one followed by determinants in each of which one or more electrons have been "excited" into a virtual (i.e., previously unoccupied) MO [69]. We use a limited form of CI which gives the wavefunction diradical character by virtue of those determinants which include singly occupied

MOs (this also provides some electron correlations by giving a few electrons more opportunity for avoiding one another). Except for very small molecules full CI would require too many determinants for all possible excitations, so the standard method for calculations on singlet diradicals is a limited form of CI in which we focus on the critical orbitals from which and into which electrons need to be excited and perform complete CI just with these few orbitals. This is called the complete active space SCF method, CASSCF [70]. The critical orbitals (i.e., the active space) depend on the process of interest. Here the process is the relationship between a propellane and its corresponding diradical. For comparison of a propellane with its putative diradical counterpart we expect our minimum active space to be composed of a possible bonding orbital between the apical carbons and the corresponding antibonding orbital (a virtual MO). Electrons are to be excited from the bonding to the antibonding orbital, making such a calculation a CASSCF(2,2) one: two active electrons distributed between two active orbitals.

CASSCF(2,2) calculations on [1.1.1]- and [2.2.2]propellane will be described. This is done first for the [2.2.2] since here the result was more straightforward. The actual commands used in each step vary with the particular program used. Here this was Gaussian 09 [36]. For the diradical:

(1) A reasonable-looking guess geometry was created with molecular mechanics, by optimizing a [2.2.2]propellane structure with the apical carbons constrained to a separation of 2.5 Å, presumably too long for a CC bond but not too long to permit reasonable values of the other CC bonds (approximately 1.58 Å). For this input structure an HF/STO-3G wavefunction was created with localized MOs (#p HF/STO-3G pop=savenbo). This requests a wordy (p for prolix) output, usually unnecessary, but which occasionally makes it easier to see what the problem was if something went wrong. The algorithm localizes and saves the MOs of the HF/STO-3G wavefunction, that is, of the Slater determinant. Localization means that the MOs initially created (the canonical MOs) are converted to linear combinations of these original orbitals that are designed to correspond more or less, when visualized, to orbitals expected from elementary valence ideas, Lewis bonds, and lone pairs. By contrast, the canonical orbitals are usually delocalized over part or all of the molecule and are of little use, when visualized, in deciding which ones are relevant to the process being considered. This particular localization method, nbo, is natural bond orbital localization [71]. Manipulating a wavefunction determinant like this has no effect on the values of observables like electron density which can be calculated from the function. The active space wanted here is composed of the apical CC bonding and antibonding orbitals, and visualization of the output of step 1 showed that these were the HOMO and the LUMO, respectively. In a CASSCF(2,2) calculation, this is as desired: by default the algorithm will use for CASSCF(*nn*) the $n/2$ highest-occupied and the $n/2$ lowest-unoccupied MOs. There are many programs that will visualize orbitals; here GaussView 5.0.9 was used [72].

(2) Orbitals can be switched if visualization of the orbitals from Step 1 showed that any desired orbitals were not in the active space. For example, if the apical bonding CC was below the HOMO, MO 30, say MO 25, then these MOs can be switched, "altered" (#p HF/STO-3G guess=(read, only, alter, save); specify the switch). This step was not needed here.

(3) One creates a CASSCF/STO-3G wavefunction from the information from Step 1 (altered in Step 2 if that was needed), (#p CAS(2,2)/STO-3G guess=read).

(4) The CAS(2,2)/STO-3G wavefunction is used to create a CAS(2,2)/6-31G* wavefunction (6-31G* is the smallest basis set usually considered adequate for a correlated calculation) and a geometry optimization is done with this (#p CAS(2,2)/6-31G* USE=L506 guess=read Opt). The USE=L506 keyword, which is not always essential, requests a diverted route through the algorithm, one more favorable for open-shell species. The orbitals should be checked visually after this step to be sure the active space has not changed significantly on going from the HF to the CASSCF method.

(5) *Frequencies*. This step can be combined with the optimization one by simply including the keyword Freq: it is sound practice to perform a frequency calculation on a structure that has been located, to characterize it as a relative minimum, a transition state, or some (usually unwanted) higher-order saddle point (#p CAS(2,2)/6-31G* USE=L506 guess=read Freq).

Steps 1–3 each took only seconds, and the optimization and frequency times were short, 23 min for optimization and 5 min for frequencies.

The result of this search for the [2.2.2]propellane diradical was structure **16** in Fig. 14.

The procedure used to find **16** was also applied to an input with the structure of [2.2.2]propellane (rather than a structure with a stretched CC distance; CASSCF can be applied to closed-shell molecules). The result was structure **17**. The apical CC distance in **16** is distinctly too long for any reasonable CC covalent bond [73], and the frequencies are all real, as expected for a relative minimum (a "real molecule") on the potential energy surface, so it seems that we have indeed found the desired singlet diradical, whereas **17**, also a relative minimum from the frequencies, appears

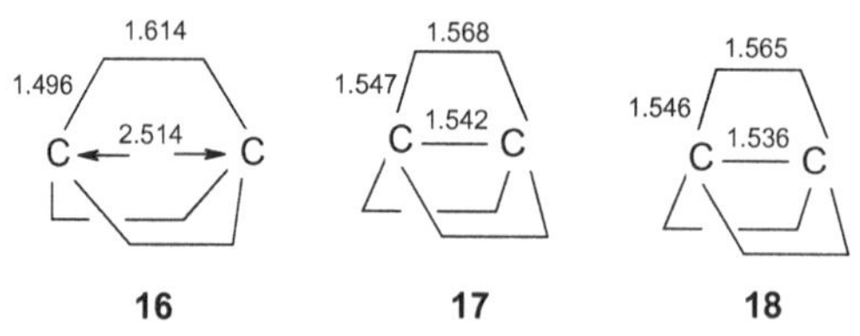

FIG. 14

Result (**16**) of search for [2.2.2]propellane diradical with CASSCF(2,2)/6-31G*, and of CASSCF(2,2)/6-31G* geometry optimization on [2.2.2]propellane (**17**); for comparison **18** is the MP2/6-311++G(2d) structure of the propellane.

to be the CASSCF(2,2)/6-31G* structure of [2.2.2]propellane—the bond lengths are very similar to the MP2/6-311++G(2d) ones (**18**), although at this latter level the CCCC and HCCH dihedral angles were approximately 10 degrees and 17 degrees, respectively (the structure was twisted), rather than essentially zero as with the CASSCF method.

We can compare the energies of **16** and **17**: at the CASSCF(2,2)/6-31G* level the relative energies were:

[2.2.2]propellane, **17** 0 kJ mol^{-1}
[2.2.2]propellane diradical, **16** −21.5 kJ mol^{-1}

Perhaps unexpectedly, since going from **17** to **16** evidently breaks a bond, the diradical from this calculation appears to be at lower energy than the closed-shall molecule. Before commenting on this, another energy calculation is in order. CASSCF calculations are at the SCF (Hartree-Fock) level for electrons outside the active space and so electron correlation for these is rudimentary [1]. The gold standard for correcting this seems to be complete Active Space Perturbational Treatment Second Order (CASPT2N) [74]. This is analogous to improving an SCF energy with a single-point (no geometry change) MP2 calculation. CASPT2N was not available on this PC, so single-point CCSD(T)/6-311++G(2d) calculations were resorted to. That these can perform well on singlet diradical energies was indicated by the fact that CCSD(T)/6-311+G** energies on singlet diradical geometries gave reasonably good results for the reaction profile of the Bergman reaction, which involves a 1,4-benzyne (C_6H_4) diradical [75]. This is despite the fact that CCSD(T) rests on a single-determinant wavefunction. The relative energies of **16** and **17** at the CCSD(T)/6-311++G(2d)//CASSCF(2,2)/6-31G* level were:

[2.2.2]propellane, **17** 0 kJ mol^{-1}
[2.2.2]propellane diradical, **16** −13.7 kJ mol^{-1}

According to this the diradical is indeed of (slightly) lower energy than the propellane. This contrasts with the relationship of a normal cycloalkane to its corresponding diradical, for example, cyclopentane compared to 1,5-pentanediyl; consider these CCST(T)/6-31G*//CASSCF(2,2)/6-31G* results [76]:

cyclopentane 0 kJ mol^{-1}
1,5-pentanediyl 330 kJ mol^{-1}

The diradical here is higher in energy by about the CC bond energy of cyclopentane, which is about 345 kJ mol^{-1} [77]; this is as expected. If the anomaly for [2.2.2]propellane is real, it may be due to relief of strain on breaking the central CC bond approximately canceling this bond energy input. On the question of bonding and strain in propellanes, there are many references [66–68]. The CASSCF energy difference of 21.5 kJ mol^{-1} is not compelling, as CASSCF is not particularly accurate from the energy viewpoint, and energy differences of isomeric species here could easily be in error by as much a 20 kJ mol^{-1} (approximately 5 kcal mol^{-1}) [78]. The CCSD(T) single-point 13.7 kJ mol^{-1} energy difference is likely more trustworthy, as far as *direction* goes (diradical below propellane) with the caveat that despite the good result with the Bergman reaction [75], as alluded to above improving CASSCF energies with CCSD(T) is of shady theoretical propriety [79].

Another ab initio method which has been used on singlet diradicals is the broken symmetry method. In this alternative to CASSCF, the unrestricted Hartree-Fock (UHF) method is applied to open-shell singlets. This is in contrast to its usual use, which is the application to normal free radicals with just one unpaired electron (i.e., doublet species). In the UHF method, alpha (α) and beta (β) electrons are assigned different spatial orbitals instead of being restricted pairwise to the same spatial orbitals [80]. Orbitals holding electrons with α spin and those with β spin can now assume different spatial relationships with respect to the unpaired electron and can thus lower the energy of the system compared with a situation in which α- and β-electrons do not have separate orbitals. The traditional UHF algorithm, for species with an odd number of electrons, was published by Pople and Nesbet in 1954 [81], and has been adapted to open-shell species with an even number of electrons, such as the diradicals examined here. Broken symmetry refers to the fact that α- and β-electrons need not be in energetically symmetrically disposed orbitals, but rather, as in traditional UHF these orbitals are varied independently. One example of broken symmetry in studying reactions mechanisms is the work of Carpenter et al. [82]. Papers that describe work in which broken symmetry is employed do not always use the term, but do stress that the ansatz is UHF.

Here the results of broken symmetry calculations on [2.2.2]propellane and its corresponding diradical are considered. In Gaussian 09 the program should be alerted to the fact that although a singlet (multiplicity 1) is specified, a UHF calculation is wanted. To encourage (if we are lucky, to persuade) the program to carry out the geometry optimization/frequency calculation with separate orbitals for α- and β-electrons, one possible input is:

```
#P umpw1pw91/6-31G* guess=(Mix, Always) Opt Freq
```

We seek a geometry optimization and frequencies on this structure, with umpw1pw91, an unrestricted version of the Perdew-Wang pw1pw91 functional as modified by Adamo and Barone [83], and the basis set 6-31G*, to create an initial guess UHF wavefunction for geometry optimization by *mixing* the HOMO and LUMO, thus breaking the default orbital symmetry (evidently there is no specific literature reference to this variation of the Pople and Nesbet [81] procedure). *Always* requests a new initial guess for each point in the optimization. This particular DFT functional is presented here because it gave good results with the Bergman reaction [75], but it, and indeed the DFT method, is not necessarily the best choice in every case. This calculation took 27 min for the guess and optimization, and 4 min for calculation of the frequencies, essentially the same CPU time as for the CASSCF calculations described earlier. However, CASSCF requires stepwise individual attention; in an introduction to the procedure the reader was even urged "Do not be discouraged..." [84], and the human time (vs. CPU time) needed for a CASSCF result is thus expected to be considerably longer than for a broken symmetry one at

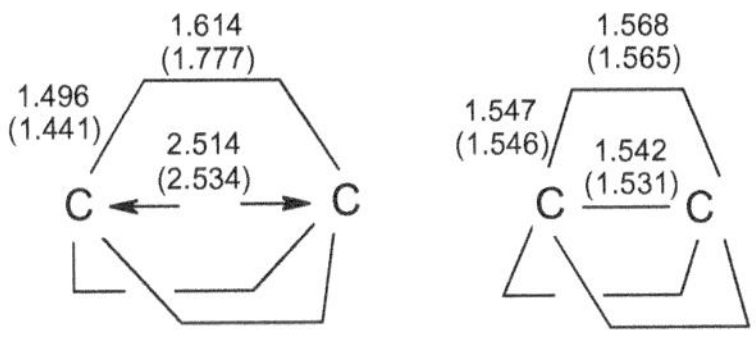

FIG. 15

Comparison of the geometries of [2.2.2]propellane diradical (**16**) and [2.2.2]propellane (**17**) form CASSCF(2,2)/6-31G* and, in parentheses, from broken symmetry umpw1pw91/6-31G*.

the same level. Broken symmetry is almost what Pople called a model chemistry, a black box method that, once an input structure, correlation level and basis have been chosen, will give the same result for all users [85]. The idea was to eliminate judgments along the calculational path and so make results from different workers more uniform and easier to compare. The method is not quite a model chemistry, because it does leave some room for judgment in invoking the unrestricted guess [86]. CASSCF, not broken symmetry, is the standard procedure for obtaining open-shell singlet structures: the latter is "sometimes controversial" because despite frequent good agreement with experiment, the wavefunction is usually strongly contaminated with contributions from higher spin states than the singlet (see Ref. [82, p. 10558]).

We compare geometries and energies of the [2.2.2]propellane species from CASSCF and from broken symmetry in Fig. 15.

The geometries from the two methods were not very different. The relative energies at the broken symmetry level were:

[2.2.2]propellane, **17** 0 kJ mol^{-1}
[2.2.2]propellane diradical, **16** 8.0 kJ mol^{-1}

In contrast to the CASSCF(2,2)/6-31G* energies, where as we saw the diradical lies 21.5 kJ mol^{-1} below the propellane, here broken symmetry places the diradical marginally higher (note that these are kJ, not kcal). Single-point CCSD(T)/6-311++G(2d) energies on the broken symmetry geometries agree with the CASSCF results, however, in putting the diradical significantly lower in energy:

[2.2.2]propellane, **17** 0 kJ mol^{-1}
[2.2.2]propellane diradical, **16** −45.1 kJ mol^{-1}

In the case of [1.1.1]propellane, in contrast to the [2.2.2] system all attempts to locate a diradical counterpart of the propellane failed. With reasonable-looking input structures both CASSCF and broken symmetry led only to the propellane. A possibility is that the propellane itself is a diradical; for a diversity of relevant arguments see Refs. [34–40] on p. 233 of Ref. [66], and five more recent papers [87]. In a nod in this direction, the wavefunctions from Gaussian 09 [36] CASSCF

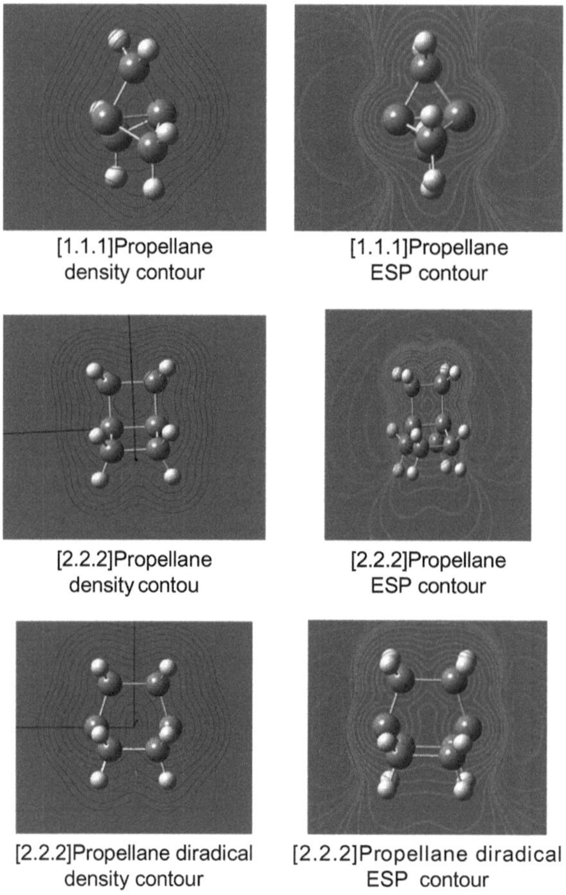

FIG. 16

Electron density and electrostatic potential slices through propellanes and a diradical. Visualized with GaussView from CASSCF calculations.

calculations were used here with GaussView [72] to visualize the electron density and the electrostatic potential (ESP, the net charge from electrons and nuclei) [88] in longitudinal slices through [1.1.1]propellane and [2.2.2]propellane and its diradical (Fig. 16). The results are perhaps not as informative as one might have hoped, but comparison of the three electron density contours (which seem more informative than the ESP) indicates a similarity between the electron distribution in [1.1.1]propellane and the [2.2.2] diradical, rather than between [1.1.1]propellane and [2.2.2]propellane. The bulges at the apical carbon atoms may betray the presence of lone pairs, or rather lone singles. This suggests that the [1.1.1] compound is a diradical or has diradical character. This analysis is a bit pallid compared to those in the references just given, but the aim here is limited largely to illustrating the utility of personal computers in obtaining computational chemical information; we refrain from extensive interpretation of this information.

6 POSTSCRIPT

We have seen a selection of research-level projects that can be readily handled with an inexpensive (around $2000 or less) personal computer. Most of them involved molecules that were of moderate size, and all the calculations were at a reasonably high level. It may be appropriate to close with some thoughts on tasks for which a PC may *not* be suitable. What used to be called mainframes are evidently on the way out, at least as far as scientific computing goes; they are/were used mainly by government and commerce for tasks like census evaluation, consumer statistics, and monitoring transactions. The alternatives to PCs are clusters, the cloud, and supercomputers. The relative merits of these will not be argued here, because they depend on the situation of the particular researcher; is one of these services free? What software is available? One is even inclined to say: try them and see. To find concrete examples of tasks for which one might move beyond ones desktop, we can focus on the just-mentioned attributes of moderate size and reasonably high level. To illuminate this, some tasks from the literature will be considered. For each citation the (sometimes tentative) conclusions that were evoked are given.

(1) Houk KN, Liu F. Acc Chem Res 2017;50:539.

The authors of this paper present their ideas of what the Holy Grails of computational organic chemistry and biochemistry are, and, intriguingly, dare to predict when these will be attained. Main ideals are: "a highly accurate density functional, ideally one that has universal chemical accuracy," accurate polarizable forcefields, methods to efficiently handle massive calculations in molecular dynamics, and treatment of flexible molecules like proteins; the principal Holy Grail is a routine computational method for the prediction and design of multicomponent, often heterogeneous, "functional systems and devices" and the related "materials and device design."

DFT is widely touted as a faster ("lower cost" in the days when there was little alternative to paying for computer time) surrogate for correlated wavefunction methods. Interestingly, there is no guarantee that the above-desired "highly accurate density functional, ... one that has universal chemical accuracy" would be faster than a wavefunction method of similar accuracy. Indeed, there seems to be no guarantee that it would execute in a practical length of time. In any case, we could, ironically, be forced to run such "divine functional" [89] calculations on a faster platform than a PC. Massive molecular dynamics calculations and protein folding are kinds of calculations that would until recently have been considered necessary to sequester to clusters or even supercomputers. Both involve calculation of huge numbers of "points"—iterations or conformations. However, in recent years graphics processing units (GPUs), because of their parallel-processing superiority over standard CPUs, may be bringing such problems back within the reach of personal computers: examples have been reported of the application of GPUs to many-body problems akin to molecular dynamics [90], and to protein folding [91]; in both cases, the results were dramatic. In work on a quantum multibody problem, a PC with a GPU designed for gaming took 15 min to accomplish work normally requiring 2–3 days of supercomputer time. Simulating 1 ms of protein folding might take 3000 years with conventional CPUs, but with the GPU months [91]. Of course with technical improvements, the times may shrink to centuries versus days, but the ratio will likely remain big. An imaginative long-term (Houk and Liu predict its realization c.2075) goal is "materials and device design." One can anticipate that here, as integrated circuits helped to make PCs powerful, improved information-processing devices will have a similar effect, in a synergistic process in which computers (including PCs?) will aid the development of more powerful devices.

(2) Hush MR. Science 2017;355:580.

The introductory paper; the "technical" paper, Carleo G, Troyer M, Science 2017;355:602. An artificial neural network was used to learn the dynamics of a quantum many-body system and find the ground state. This method performed better than current numerical simulation methods. This suggests that artificial intelligence (AI) could take on not just human games (like go [92]) but conundrums posed by nature (i.e., scientific problems). The AI program which played the game of go at grandmaster level required "a huge amount of computing power—four of the specialized chips called tensor processing units, which has been estimated to be US$25 million of hardware." This kind of computing power may represent the last refuge of clusters and supercomputers (recall too that the chess champion Deep Blue ran on a cluster [16]).

(3) (a) Mullin R. Chem Eng News 2016;19, January 25: "Computational Comeback." Computationally assisted drug discovery. (b) Wilson E. Chem Eng News 2011;10, December 5: "A Hit in Silico." A potent reverse-transcriptase inhibitor against HIV-1 activity in human cells.

Designing with computer assistance biologically active molecules for drug use is challenging. Early work appears to have been disappointing, but there are grounds for renewed optimism; after all, all diseases are molecular. Part of the problem is connected with the size and flexibility of the biological macromolecules the drugs may have to interact with. This is akin to the protein conformation problem Houk and Liu, earlier, identified as a Holy Grail. Here too PCs and the currently more powerful platforms may be for some time in a race.

(4) (a) Mullin R. Chem Eng News 2016;26, October 24. "Or, how the pharmaceutical research sector learned to stop worrying and love the cloud." (b) Nature 2015;523:128, 9 July. Editorial. "Opposition to storing vast scientific data sets on cloud-computing platforms is weakening." (c) Fox A. Science 2011;331:406. "...what's in it for me as a scientist?"

Storing vast amounts of data "offshore," in the "cloud," is becoming acceptable. Pharmaceutical companies, understandably once leery of entrusting their data to storage by others, will now (sometimes) entrust their data to cloud facilities. These (Mullin, 2016) "[eliminate] the need for expensive internal data systems." The financial industry has for years used external storage of customer's information. By contrast "the pharmaceutical industry was largely unwilling to risk housing its research data anywhere other than behind its own walls." Breaches of financial walls in recent years might make one uneasy about the security of some pharmaceutical data.

Not just data storage, but actual computing may be gaining popularity on the cloud (Fox, 2011): "Privately owned cloud computers—large data centers filled with computers that mainly run their company's software—are now becoming available to outside users ... If your problem can be computed in a way that takes advantage of parallel processing, you can now get the answer 1000 times as fast for the same amount of money." It may be that tasks involving the storage and manipulation of very large data sets will, for the foreseeable future anyway, be advantageously carried out in the cloud, or on supercomputers.

Research is underway on what one might call postclassical computation, using photons instead of electrons [93] or utilizing completely nonintuitive quantum mechanical phenomena like superposition and entanglement [94]. If such methods outpace our prosaic current electronics, and can be implemented on PCs, then a new era in scientific computing on personal computers will have arrived. Nevertheless, that today one can seriously ponder whether a PC would be a reasonable alternative to a supercomputer (or some similar platform) for a project involving serious quantum mechanical or molecular dynamics calculations, shows how powerful personal computers have become.

REFERENCES

[1] The SCF method is outlined in most introductory books on computational chemistry, for example.
(a) Cramer CJ. Essentials of computational chemistry. 2nd ed. Chichester, England; 2004. Section 4.5.2.
(b) Lewars E. Computational chemistry. 3rd ed. Switzerland: Springer; 2016. Section 5.2.2.

[2] Froese Fischer C. Douglas Rayner Hartree: his life in science and computing. Singapore: World Scientific; 2003.

[3] (a) The Hartree-Fock SCF method is explained in some detail in: Lewars E. Computational chemistry. 3rd ed. Switzerland: Springer; 2016. Sections 5.2.3.1–5.2.3.5.
(b) The modern implementation of the Hartree-Fock method for molecular calculations, using atomic basis functions, is shown in some detail in: Lewars E. Computational chemistry. 3rd ed. Switzerland: Springer; 2016. Section 5.2.3.6.

[4] Pomerantsev VN, Kukulin VI, Rubtsova OA, Sakhiev SK. Comput Phys Commun 2016;204:121.

[5] Grier DA. When computers were human. Princeton University Press; 2007.

[6] An instrument with wheels and cogs turned by hand or, later, by an electric motor. Rudimentary models appeared in the 1600s and 1700s and more practical models were used c.1900–70 (till approximately 30 years after electronic computers existed). https://en.wikipedia.org/wiki/Mechanical_calculator. In his engaging reminiscences of the heady early days of quantum mechanics, Hylleraas cites, evidently in connection with his pioneering (c.1928) electron correlation work on the helium atom, the use of an "automatic electric desk computer." This was of course a motor-driven mechanical calculator: Hylleraas E. Rev Mod Phys 1963;35(421):427.
[7] See, for example, (a) *Turing's Cathedral:* Dyson G. The origins of the digital universe. Pantheon Books; 2012.
(b) Hally M. Electronic brains: stories from the dawn October 25, 2017 of the computer age. Granta; 2005. This is a look at computers c.1930–60.
[8] Froese Fischer C. Douglas Rayner Hartree: his life in science and computing. Singapore: Singapore World Scientific; 2003. p. 98, 99.
[9] (a) Slater JC. Solid-state and molecular theory: a scientific biography. Wiley; 1975.
(b) Slater JC. Solid-state and molecular theory: a scientific biography. Wiley; 1975. p. 5.
[10] For information on the kinds of problems and the nature of the computers first applied to chemistry, particularly quantum chemistry, see:
(a) The development of computational chemistry in the United States, Bolce JD, Hermann RB. Reviews in computational chemistry, vol. 5. VCH; 1994 [Chapter 1].
(b) The development of computational chemistry in the United Kingdom. Smith SJ, Sutcliffe BT. Reviews in computational chemistry , vol. 10. VCH; 1997 [Chapter 5].
[11] (a) Dewar MJS. A semiempirical life. Washington, DC: American g54 Chemical Society; 1992.
(b) Dewar MJS. A semiempirical life. Washington, DC: American g54 Chemical Society; 1992. p. 107.
(c) Dewar MJS. A semiempirical life. Washington, DC: American g54 Chemical Society; 1992. p. 115.
[12] Available from: https://en.wikipedia.org/wiki/IBM_7090.
[13] Schaefer III HF. Theochem 2001;573:129.
[14] Available from: http://www.computerhistory.org/revolution/mainframe_computers/7/182/736.
[15] Available from: https://www_03.ibm.com/ibm/history/documents/pdf/rs6000.pdf.
[16] For example, Newborn M. Kasparov versus Deep Blue: computer chess comes of age. Springer; 1997.
[17] Simulation Conference Proceedings, December 4–6, 1989. Available from: http://ieeexplore.ieee.org/document/718801/?reload=true.
[18] (a) Reed LH, Murphy AR. J Chem Educ 1987;64:789. A program that performs a Hartree-Fock-Roothaan calculation for a two-electron diatomic molecule on Apple II Plus and Apple IIe computers.
(b) Colwell SM, Handy NC. J Chem Educ 1988;65:21. A standard quantum chemistry package adapted to run on a microcomputer. It is shown that, although microcomputers are slow, they can be useful to quantum chemists especially for teaching.
(c) Rogers DW, McLafferty FJ. Am Lab 1991;20. Ab initio molecular orbital calculations using the PC.
[19] The first in the series was Gaussian 70 and the latest (2016) is Gaussian 16. Gaussian Inc., Wallingford, CT. Available from: Info@gaussian.com.
[20] Hinchliffe A. Electron J Theor Chem 1996;1.

[21] Available from: https://www.youtube.com/watch?v=_7lcgBGjkiE.
[22] Wavefunction Inc., http://www.wavefun.com, 18401. Von Karman, Suite 370, Irvine CA 92715, USA. As of mid-2017, the latest version of Spartan was Spartan '16, available in several versions.
[23] GAMESS is maintained by members of the Gordon research group at Iowa State University. Available from: http://www.msg.ameslab.gov/gamess/.
[24] Available from: http://www.msg.ameslab.gov/gamess/GAMESS_Manual/iron.pdf.
[25] Nicklaus MC, Williams RW, Bienfait B, Billings ES, Hodošček M. J Chem Inf Comput Sci 1998;38:893.
[26] Lewars E. Computational chemistry. 3rd ed. Switzerland: Springer; 2016 [Section 5.3].
[27] Gaussian News. 1993;4:1. Available from: http://gaussian.com/.
[28] (a) Radom L, Rasmussen DR. Pure Appl Chem 1998;70:1977.
(b) Rasmussen DR, Radom L. Angew Chem Int Ed Engl 1999;38:2875.
[29] Lewars EG. Modeling marvels. Computational anticipation of novel molecules. Springer; 2008 [Chapter 1].
[30] van't Hoff JH. Bull Soc Chim Fr II 1875;23:295.
[31] Lewars E. Computational chemistry. 3rd ed. Switzerland: Springer; 2016 [Chapter 2].
[32] Lewars E. Computational chemistry. 3rd ed. Switzerland: Springer; 2016 [Section 5.4.2].
[33] Commentary by Norris Parker Smith, editor_at_large. December 1, 1995. Available from: https://www.hpcwire.com/1995/12/01/crays_t3e_prices_cut_sci_added_air_cooling_and_more/.
[34] (a) Ochtersky JW. Vibrational analysis in Gaussian. Gaussian White Paper. Available from: http://gaussian.com/vib/.
(b) A bare outline of the method is given in: Lewars E. Computational chemistry; 2016 [Section 2.5].
[35] Gaussian Inc., Personal communication.
[36] Gaussian 09, Revision C.01, Frisch MJ, Trucks GW, Schlegel HB, Scuseria GE, Robb MA, Cheeseman JR, et al. Wallingford, CT; 2016.
[37] Lewars E. Computational chemistry. 3rd ed. Switzerland: Springer; 2016 [Chapter 7].
[38] Cremer C. Comp Chem List 2007. June 14.
[39] From Clemente FR. Technical Support, Gaussian, Inc; 2017. June 15. "...I believe there are a few soft modes where the changes in energy and gradient are very small, so if the step size for numerical differentiation is too small, the numerical error in those derivatives would be too large. More than likely the numerical frequency calculation that was done at the time of the original publication used a longer step size than 0.001 Angstrom. Although a longer step size means that the displacement is not so 'infinitesimal,' so there is a reduction in accuracy with increasing step size, too small of a step size will produce very small energy and gradient changes, and so the relative numerical errors in the derivatives can be larger... you may use something like 'Freq=(Numer,Step=50),' which means using a step size of 0.005 Angstrom."
[40] Lewars E. Computational chemistry. 3rd ed. Switzerland: Springer; 2016. p. 514. Footnote 3.
[41] Lewars E. Computational chemistry. 3rd ed. Switzerland: Springer; 2016 [Section 5.3.3.6].
[42] See, for example, an engagingly whimsical review which while sympathetic to DFT does point out some of these problems: Burke K. J Chem Phys 2012;136:150901. Among the *bons mots*: "...DFT is some algorithm meaning unreliable, while *ab initio* is Latin for *too expensive*."

[43] (a) Hammes-Schiffer S. Science 2017;355:28.
(b) Medvedev MG, Bushmarinov IS, Sun J, Perdew JP, Lyssenko KA. Science 2007;355:49.
[44] W.T. Borden in interview with S.M. Bachrach: Bachrach SM. Computational organic chemistry. Hoboken, NJ: Wiley; 2007. p. 193.
[45] Ding BW, Keese R, Stoeckli-Evans H. Angew Chem Int Ed Engl 1999;38:375.
[46] Lewars EG. Modeling marvels. Computational anticipation of novel molecules. Springer; 2008 [Chapter 2].
[47] Feng Y, Liu L, Wang J-T, Zhao SW, Guo Q-X. J Org Chem 2004;69:3129, 3133.
[48] Minkin VI, Minyaev RM, Zakharov II, Avdeev VI. Zh Org Khim 1978;14:3.
Minkin VI, Minyaev RM. Zh Org Khim 1979;15:225.
Minkin VI, Minyaev RM, Orlova GA. J Mol Struct (Theochem) 1984;110:241.
Minyaev RM, Minkin VI, Zefirov NS, Zhdanov YA. Zh Org Khim 1979;15:2009.
Minyaev RM, Minkin VI, Zefirov NS, Natanzon VI, Kurbatov SV. Zh Org Khim 1982;18:3.
[49] Balaji V, Michl J. Pure Appl Chem 1988;60:189.
[50] Lewars E. J Mol Struct (Theochem) 1998;423:173.
[51] Lewars E. J Mol Struct (Theochem) 2000;507:165.
[52] Kenny JP, Krueger KM, Rienstra-Kiracofe JC, Schaefer HF. J Phys Chem A 2001;105:7745.
[53] Fokin AA, Chernish LV, Gunchenko PA, Tikhonchuk EY, Hausmann H, Serafin M, et al. J Am Chem Soc 2012;134:13641.
[54] For example: Lewars E. Computational chemistry. 3rd ed. Switzerland; 2016 [Section 5.4.3].
[55] For example: Lewars E. Computational chemistry. 3rd ed. Switzerland; 2016 [Section 5.3.3.6].
[56] Some barriers—room temperature half lives for unimolecular decompositions: (a) The pentazole anion, *cyclo*-N_5^-, 106 kJ mol^{-1}—2.2 days: Bazanov B, Geiger U, Carmieli R, Grinstein D, Weiner S, Haas Y. Angew Chem Int Ed Engl 2016;55:13233.
(b) A stable salt of the pentazole anion, *cyclo*-N_5^-: Christe KO. Science 2017;355:351.
(c) Decomposition of pentazole and its conjugate base, 75 kJ mol^{-1}—10 minutes and 106 kJ mol^{-1}—2.2 days, respectively: Benin V, Kaszynski P, Radziszki JG. J Org Chem 2002;67:1354.
(d) Decomposition of $(CF_3CO)OOO(COCF_3)$, 86.5 kJ mol^{-1}—1 minute. Ahsen SV, Garciá P, Willner H, Paci MB, Argüello G. Chem Eur J 2003;9:5135.
[57] The latest available version (August 2017) is evidently ACES III (version 4 is under development). Available from: http://www.qtp.ufl.edu/ACES/pubs.shtml [Accessed August 2017].
[58] Hardware requirements for running ACES III. Available from: http://www.qtp.ufl.edu/ACES/download.shtml. A hint at possible problems: "Please contact ACES QC Consulting services if you like to purchase the service of having ACES II or ACES III installed and tested on your computer system by one of our experts" [Accessed August 2017].
[59] Minyaev RM, Minkin VI, Gribanova TN, Starikov AG, Hoffmann R. J Org Chem 2003;68:8588.
[60] Lewars EG. Modeling marvels. Computational anticipation of novel molecules. Springer; 2008 [Chapter 12].

[61] TPSS was said to be "close to being the best nonempirical functional so far": its leading developer, Perdew JP. 2009. November 7, a position still held 2015 February 25.
[62] PBE0 has evidently being mistaken for being empirical, as pointed out by: Mawhinney RC, Goddard JD. Theochem 2003;629:263. Their results for PBE0 on oxirene matched well those for CCSD(T).
[63] Introduction: Hammes-Schiffer S. Science 2017;355(28):2017.
"Technical" paper: Medvedev MG, Bushmarinov IS, Sun J, Perdew JP, Lyssenko KA. Science 2017;355:49.
[64] Perdew JP, Ruzsinszky A, Tao J, Starverov VN, Scuseria GE, Csonka GI. J Chem Phys 2005;123:062201.
[65] (a) Ginsberg D. Acc Chem Res 1969;2:121.
(b) In this paper Ginsberg assures readers that his propellane nomenclature is "slowly but surely" being accepted: Ginsberg D. Acc Chem Res 1972;5:249.
[66] Lewars EG. Modeling marvels. Computational anticipation of novel molecules. Springer; 2008. p. 197–201, 233.
[67] A review: Dilmaç AM, Spuling E, de Meijere A, Bräse S. Angew Chem Int Ed Engl 2017;56:5684.
[68] Wiberg KB, Walker FH. J Am Chem Soc 1982;104:5239.
[69] Lewars E. Computational chemistry. 3rd ed. Switzerland: Springer; 2016 [Section 5.4.3].
[70] Lewars E. Computational chemistry. 3rd ed. Switzerland: Springer; 2016 [Section 8.2.2].
[71] See, for example, Jensen F. Introduction to computational chemistry. England: Wiley; 2007.
[72] GaussView: Gaussian Inc. Bldg. 6, Pittsburgh, PA 15106, USA: Carnegie Office Park.
[73] The longest known CC bonds are approximately 1.7 Å: (a) Allinger NL, Lii JH, Schaefer III HF. J Chem Theory Comput 2016;12:2774.
(b) Fokin AA, Chernish LV, Gunchenko PA, Tikhonchuk EY, Hausmann H, Serafin M, et al. J Am Chem Soc 2012;134:13641.
[74] (a) Karlstrom G, Lindh R, Malmqvist PÅ, Roos Ryde BO, Veryazov V, Widmark PO, et al. Comput Mater Sci 2003;28:222.
(b) Anderssson K, Malmqvist PÅ, Roos BO. J Chem Phys 1992;96:1218.
[75] Lewars E. Can J Chem 2014;92:378.
[76] Lewars E. Computational chemistry. 3rd ed. Switzerland: Springer; 2016 [Section 8.2.3.2].
[77] The activation enthalpy for opening cyclopentane, which should be close to the bond enthalpy, has been estimate at 344.8 kJ mol^{-1}: Sirjean B, Glaude PA, Ruiz-Lopez MF, Fournet R. J Phys Chem A 2006;110:12693.
[78] "Chemical accuracy," sometimes defined as accuracy to within 10 kJ mol^{-1}, requires special methods to be reliably attained: Lewars E. Computational chemistry. 3rd ed. Switzerland: Springer; 2016 [Section 5.5.2.3.2].
[79] Virtues of CCSD(T) and a warning about its theoretical inadequacy for open shell molecules: Bally T, Borden WT. Reviews in computational chemistry, vol. 14 New York: VCH; 1999. p. 38, 39 [Chapter 1].
[80] See, for example, Hehre WJ, Radom L, Schleyer PVR, Pople JA. Ab initio molecular orbital theory. New York: Wiley-Interscience; 1986 [Sections 2.6.2, 4.2.2, 6.2.8a].
[81] Pople JA, Nesbet RK. J Chem Phys 1954;22:571.
[82] Carpenter BK, Pittner J, Veis L. J Phys Chem A 2009;113:10557.
[83] Adamo C, Barone V. J Chem Phys 1998;108.

[84] Foresman JB, Frisch Æ. Exploring chemistry with electronic structure methods. Pittsburgh, PA; 1996.
[85] Pople JA. Acc Chem Res 1970;3:217.
[86] Suggestions: Radon M. Computational chemistry list (an online service); 2008. July 32.
[87] (a) Lobayan RM, Bochicchio RC. Chem Phys Lett 2013;557:154.
(b) Shaik S, Chen Z, Wu W, Stanger A, Danovich D, Hiberty PC. Chem Phys Chem 2009;10:2658.
(c) Shaik S, Danovich D, Wu W, Hiberty PC. Nat Chem 2009;1:443.
(d) Wu W, Gu J, Song J, Shaik S, Hiberty PC. Angew Chem Int Ed Engl 2009;48:1407.
(e) Irikura. J Org Chem 2008;73:7906.
[88] Lewars E. Computational chemistry. 3rd ed. Switzerland: Springer; 2016 [Section 5.5.4.4].
[89] Mattson AM. Science 2002;298:759.
[90] Pomerantsev VN, Kukulin VI, Rubtsova OA, Sakhiev SK. Comput Phys Commun 2016;204:121. Online report: EurekaAlert (from the AAAS). Available from: https://www.eurekalert.org/pub_releases/2016_06/lmsu_nni062816.php.
[91] Wolf LK. Chem Eng News 2010;27. November 1.
[92] Gibney E. Nat News 2017. October 18. https://doi.org/10.1038/nature.2017.22858.
[93] See, for example, Caulfield HJ, Dolev S. Nat Photonics 2010;4:261.
[94] Sugisaki K, Yamamoto S, Nakazawa S, Toyota K, Sato K, Shiomi D, et al. J Phys Chem A 2016;120:6459 [and references therein.]

CHAPTER

Chemical applications of graph theory

8

Kimberly Jordan Burch

Department of Mathematics, Indiana University of Pennsylvania, Indiana, PA, United States

1 INTRODUCTION

Chemical graph theory applies this branch of mathematics to model molecules in order to study their various physical properties. A *graph* $G = (V,E)$ consists of a set V of vertices (or nodes) and a set E of unordered pairs of distinct elements of V, which are the edges. In chemistry, the atoms of a molecule are represented by the vertices and the chemical bonds are represented by the edges. The resulting graph is often called a *chemical graph*. When studying alkanes which have the chemical formula C_nH_{2n+2}, the hydrogen atoms are removed from the graph resulting in what is known as a *hydrogen-depleted molecular graph* or a *carbon tree*. Since each carbon has four bonds and each hydrogen has one bond, no information about the molecule is lost by removing the hydrogen atoms. The resulting graph is in fact easier to study since the geometric structure of the alkane is more apparent. For example, a rendering of 2,2,4-trimethylpentane is given in Fig. 1. The chemical graph of 2,2,4-trimethylpentane is given in Fig. 2.

The vertex set of the graph in Fig. 2 is $V = \{a,b,c,d,e,f,g,h\}$, and the edge set of this graph is given by $E = \{ab,bc,cd,de,bf,bg,dh\}$. The *degree* of a vertex v is the number of edges attached to vertex v and is denoted deg(v). In a molecular compound, this is the number of bonds an atom has and is defined as the *valency* of an atom. In Fig. 2, deg(a)= deg(f)=deg(g)=deg(h)=deg(e)=1, deg(b)=4, deg(c)=2, and deg(d)=3.

Given a graph G with vertices $u, v \in V(G)$, we define a ***u*–*v*** *path* as a sequence of edges from G of the form $uv_1, v_1v_2, v_2v_3, \ldots, v_nv$ where each vertex $v_i \in V(G)$ for $1 \le i \le n$ is distinct. The *length* of the given u–v path is n, the number of edges in the path. The *shortest path* from vertex u to v is the path for which n, the number of edges in the path, is minimal. A path from vertex g to h in Fig. 2 is given by gb, bc, cd, dh and has length four. This is also the shortest path from g to h since this is the only g–h path. Given that 2,2,4-trimethylpentane is an acyclic hydrocarbon, its chemical graph contains no cycles. A *cycle* in a graph is a path that begins and ends at the

Mathematical Physics in Theoretical Chemistry. https://doi.org/10.1016/B978-0-12-813651-5.00008-5

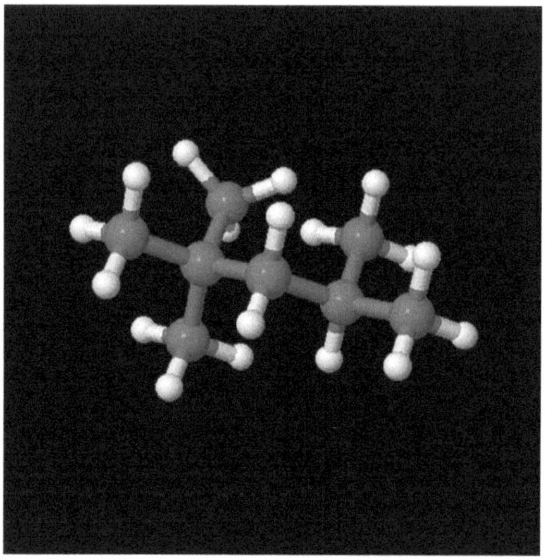

FIG. 1

Rendering of 2,2,4-trimethylpentane molecule.

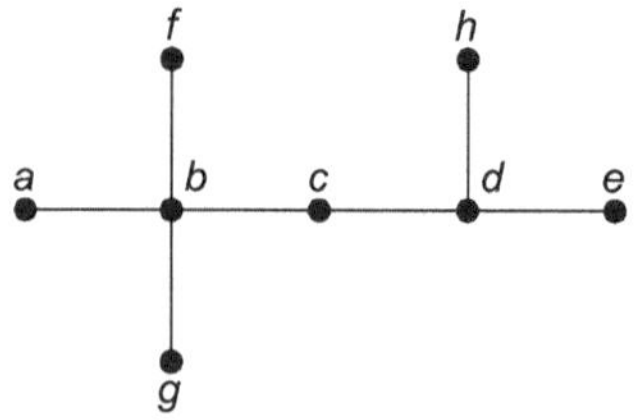

FIG. 2

Chemical graph of 2,2,4-Trimethylpentane.

same vertex. Since the graph of 2,2,4-trimethylpentane contains no cycles, the path between any pair of vertices in the graph will always be the shortest path between the two vertices.

In graph theory, vertices are traditionally labeled with letters so as not to confuse the label of a vertex with the degree of a vertex. However, when the compounds studied are alkanes (also known as acyclic saturated hydrocarbons, paraffins, and the methane series), it is traditional to label the "straight" chain of the hydrocarbon with the numbers 1 to n and all branching carbons with the labels $n+1, n+2$, etc. as needed in the chemical graph. For instance, the chemical graph of 2,2,4-trimethylpentane would be given as seen in Fig. 3. All chemical graphs of alkanes will be labeled in this manner from this point on in this chapter.

A graph is said to be *connected* if there is a path between every pair of vertices in the graph. Otherwise, the graph is said to be *disconnected*. If a graph is connected, the

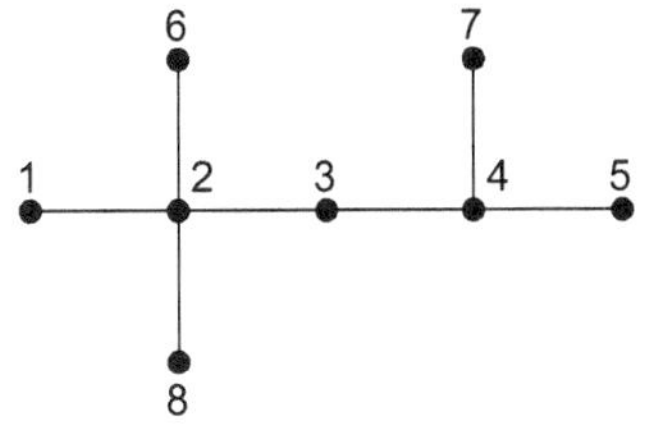

FIG. 3

2,2,4-Trimethylpentane with traditional labeling.

distance $d(u, v)$ from u to v is the length of the shortest path from u to v. The *distance matrix* of a graph G having n vertices is the $n \times n$ matrix whose i–jth entry is $d(i,j)$ and is denoted $D(G)$. For example, the distance matrix for 2,2,4-trimethylpentane is given by

$$D(G) = \begin{bmatrix} 0 & 1 & 2 & 3 & 4 & 2 & 2 & 4 \\ 1 & 0 & 1 & 2 & 3 & 1 & 1 & 3 \\ 2 & 1 & 0 & 1 & 2 & 2 & 2 & 2 \\ 3 & 2 & 1 & 0 & 1 & 3 & 3 & 1 \\ 4 & 3 & 2 & 1 & 0 & 4 & 4 & 2 \\ 2 & 1 & 2 & 3 & 4 & 0 & 2 & 4 \\ 2 & 1 & 2 & 3 & 4 & 2 & 0 & 4 \\ 4 & 3 & 2 & 1 & 2 & 4 & 4 & 0 \end{bmatrix}$$

When the molecular compounds being studied are alkanes as is the case with this example, the entries are the lengths of the shortest paths between any two pairs of carbons. The *adjacency matrix*, $A(G)$, of a labeled graph G having n vertices is the $n \times n$ vertex whose entries are given by

$$A_{ij} = \begin{cases} 1 & \text{if vertices } i \text{ and } j \text{ are adjacent} \\ 0 & \text{otherwise} \end{cases}$$

For example, the adjacency matrix for 2,2,4-trimethylpentane is given by

$$A(G) = \begin{bmatrix} 0 & 1 & 0 & 0 & 0 & 0 & 0 & 0 \\ 1 & 0 & 1 & 0 & 0 & 1 & 0 & 1 \\ 0 & 1 & 0 & 1 & 0 & 0 & 0 & 0 \\ 0 & 0 & 1 & 0 & 1 & 0 & 1 & 0 \\ 0 & 0 & 0 & 1 & 0 & 0 & 0 & 0 \\ 0 & 1 & 0 & 0 & 0 & 0 & 0 & 0 \\ 0 & 0 & 0 & 1 & 0 & 0 & 0 & 0 \\ 0 & 1 & 0 & 0 & 0 & 0 & 0 & 0 \end{bmatrix}$$

Edges are said to be *independent* if they do not share a common vertex. A *matching* of a graph G is a (possibly empty) set of independent edges of G. A *k-matching* is a matching which has exactly k edges. Letting a_k denote the

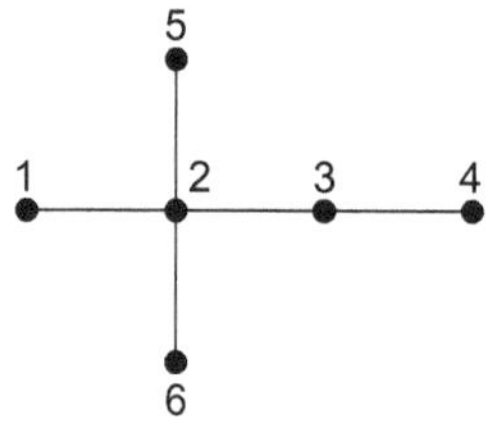

FIG. 4

2,2-Dimethylbutane.

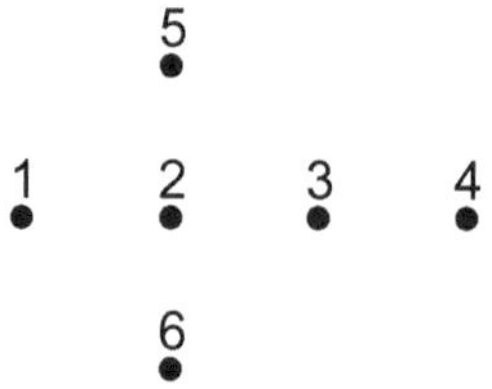

FIG. 5

0-Matching for 2,2-dimethylbutane.

number of k-matchings, the *simple matching polynomial* of a graph with p vertices is given by the formula $\sum_{k=0}^{[p/2]} a_k \omega^{p-k}$. We will demonstrate how to compute the simple matching polynomial of 2,2-dimethylbutane using the chemical graph of this molecule as seen in Fig. 4.

There is only one way to choose zero edges for a matching ($a_0 = 1$) as seen in Fig. 5. There are five ways to form a 1-matching since the graph of 2,2-dimethylbutane has five edges ($a_1 = 5$). These 1-matchings are seen in Fig. 6. We exhibit the three possible 2-matchings of 2,2-dimethylbutane in Fig. 7. These are the only possible 2-matchings of the graph of 2,2-dimethylbutane since any other choice of two edges will not be independent ($a_2 = 3$). There are no k-matchings for $k \geq 3$ since any choice of 3, 4, 5, etc. edges will not be independent. Using the values of a_k as explained previously in the formula for the simple matching formula, the simple matching polynomial for 2,2-dimethylbutane is $\omega^6 + 5\omega^5 + 3\omega^4$.

2 TOPOLOGICAL INDICES

Physical properties of chemical compounds are highly correlated to the geometric structure of the compounds. Using indices which can accurately depict the shape of the molecules allows one to model these physical properties with surprising precision. Topological indices are graph invariant in which they do not change if the graph is drawn or labeled differently. Several topological indices are given with corresponding examples. The notation used for the indices is in agreement with the notation used by Todeschini and Consonni [1].

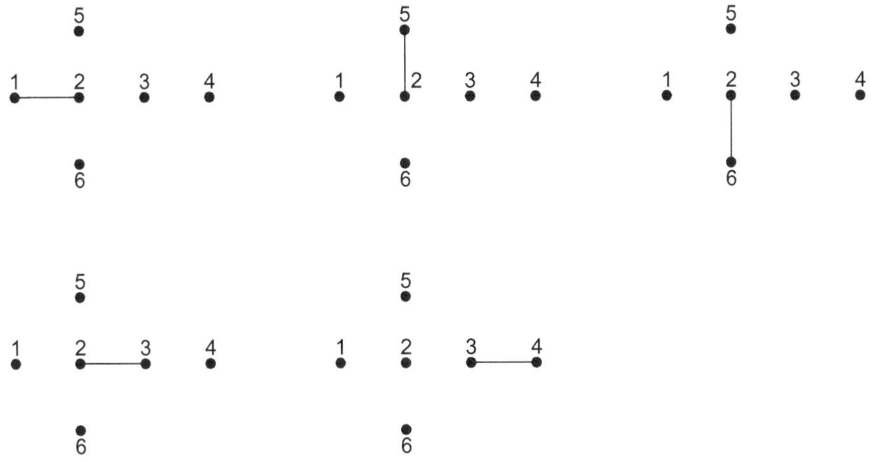

FIG. 6

1-Matchings for 2,2-dimethylbutane.

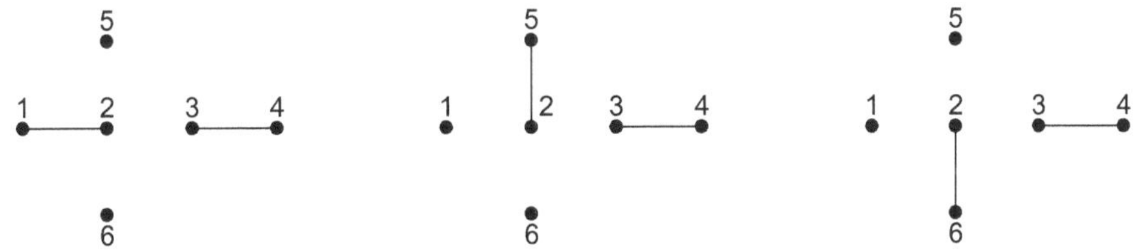

FIG. 7

2-Matchings for 2,2-dimethylbutane.

Relative molecular mass (M_r) is defined as the total mass of a compound and is equal to the sum of the individual atomic masses of each atom in the molecule. Let n be the number of carbons in an alkane. The relative molecular mass of an alkane is given by the formula

$$M_r(n) = 12.01115n + 1.00797(2n + 2)$$

For example, the relative molecular mass of 2,2-dimethylbutane is 86.17848.

Perhaps the best known topological index is the relative molecular mass. For alkanes with similar geometric structure, the relative molecular mass alone may be used to determine a model of the boiling points of these alkanes.

Hosoya index (Z) is the sum of the coefficients of the simple matching polynomial. The Hosoys index was introduced by Haruo Hosoya in 1971 [2] as he studied the correlation between the boiling points of alkanes and the molecular descriptor which he defined. Recall that the chemical graph of 2,2-dimethylbutane, as seen in Fig. 4, has exactly one matching with no edges, five 1-matchings, and three 2-matchings. These matchings were shown in Figs. 5–7, respectively. Hence, the simple matching polynomial of 2,2-dimethylbutane is $\omega^6 + 5\omega^5 + 3\omega^4$ so that the Hosoya index of 2,2-dimethylbutane is $1 + 3 + 5 = 9$.

In [3] the authors present a graph of the boiling points of alkanes having six, seven, and eight carbon atoms versus the natural logarithm of the Hosoya index. This particular graph was the inspiration for the paper [4] which further investigated modeling the boiling points of alkanes. Fig. 8 is the graph of the alkanes having 6 to 10 carbon atoms versus the natural logarithm of the Hosoya index. The best linear fit is given by the equation $f(\ln Z) = 50.8287 \ln Z - 51.9346$. The coefficient of determination for this function, denoted R^2, is $R^2 = 0.933748$. The standard deviation is 7.1°C.

Wiener number (W) is the sum of the distances between all pairs of vertices in a graph. The index was introduced in 1947 by Wiener [5] and is the oldest of the topological indices associated to chemical graph theory. Many subsequent topological indices were defined based on Wiener's work. The Wiener number is given by the formula

$$W = \frac{1}{2} \sum_i \sum_j d_{ij}$$

where d_{ij} is the ijth entry in the distance matrix, $D(G)$ [6].

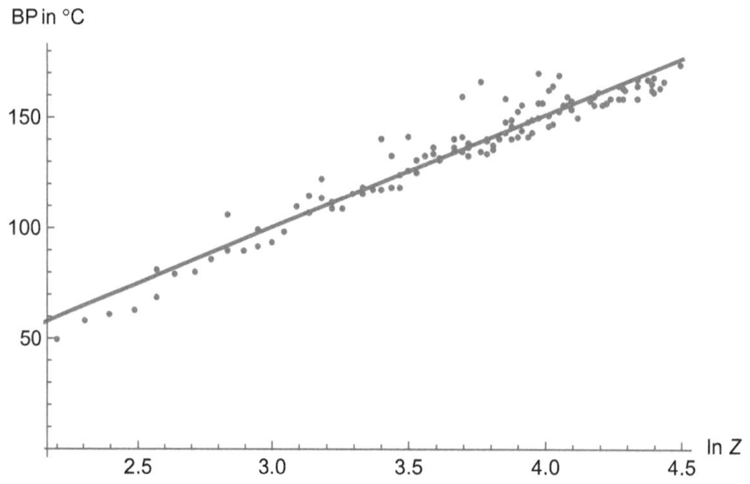

FIG. 8

ln Z versus BP of alkanes having 6–10 carbons.

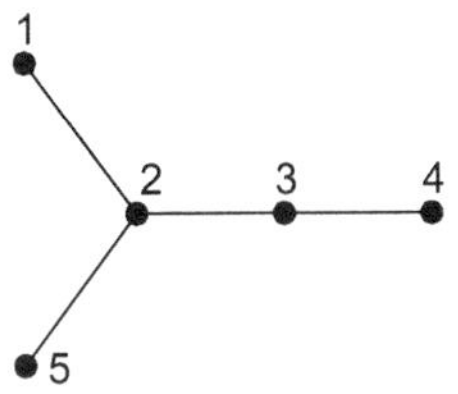

FIG. 9

2-Methylbutane.

We consider the alkane 2-methylbutane whose chemical graph is given in Fig. 9. The distance matrix of 2-methylbutane is given by

$$D(G) = \begin{bmatrix} 0 & 1 & 2 & 3 & 2 \\ 1 & 0 & 1 & 2 & 1 \\ 2 & 1 & 0 & 1 & 2 \\ 3 & 2 & 1 & 0 & 3 \\ 2 & 1 & 2 & 3 & 0 \end{bmatrix}$$

Therefore, the Weiner number for 2,2-dimethylbutane is

$$W = \frac{8(1) + 8(2) + 4(3)}{2} = 18$$

Mean Wiener index ($\overline{W}$) is the average of the distances between all pairs of vertices in a graph. For a graph with n vertices,

$$\overline{W} = \frac{W}{\binom{n}{2}}$$

For 2,2-dimethylbutane, $\overline{W} = \frac{18}{10} = 1.8$.

Wiener Path numbers (${}^1P, {}^2P, \ldots, {}^9P$) are the numbers ${}^iP$ where ${}^iP$ is the number of pairs of vertices in the graph separated by i edges (chemical bonds). Again using the distance matrix, ${}^iP$ can be obtained by counting the number of times i appears in the upper triangular part of the matrix. For example, 2-methylbutane has

$${}^1P = 4,\ {}^2P = 4,\ {}^3P = 2,\ {}^4P = 0,\ {}^5P = 0, \ldots, {}^9P = 0$$

When studying alkanes, ${}^iP$ is the number of shortest paths of length i that occur in the alkane. This means that ${}^1P$ always equals the number of bonds in an alkane.

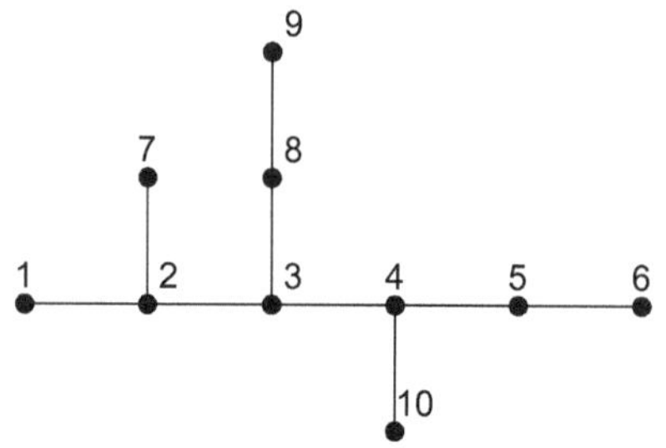

FIG. 10

3-Ethyl-2,4-dimethylheptane.

Methyl (*Mth*) is defined to be the number of degree one vertices, which are adjacent to a vertex of degree three or greater. This is not equivalent to the number of methyls in the IUPAC name.

Consider the chemical graph of 3-ethyl-2,4-dimethylheptane as seen in Fig. 10. We see that 3-ethyl-2,4-dimethylheptane has $Mth = 3$ which differs from the two methyls given in the IUPAC name. The three methyls that are used in the computation of the methyl index are the edges $\{1, 2\}$, $\{2, 7\}$, and $\{4, 10\}$ as seen in Fig. 10.

The methyl index and the ethyl index, which will be defined shortly, were introduced in Burch et al. [4]. Since methyl is an electron donor group, the hydrogen atoms have a slightly positive charge which causes the carbon atoms where the methyl group attaches to be slightly negative. This change in polarity will have an effect on certain physical properties of the alkanes such as the boiling point.

Ethyl (*Eth*) is defined to be the number of degree one vertices that are adjacent to a degree two vertex, which is adjacent to a vertex of degree three or higher. This is not equivalent to the number of ethyls in the IUPAC name.

For example, 3-ethyl-2,4-dimethylheptane has $Eth = 2$. The first ethyl is the path $\{3,8\}$, $\{8,9\}$ and the second is the path $\{4,5\}$, $\{5,6\}$ as seen in Fig. 10.

Randić index (χ) is defined for a graph G with $n - 1$ edges by the formula

$$\chi = \sum_{q \in E(G)} \frac{1}{\text{Geometric mean of } q}$$

where the geometric mean of an edge q joining vertices v_i and v_j is $\sqrt{\deg(v_i) * \deg(v_j)}$ [7]. The Randić index was originally introduced as the *connectivity index* in 1975 by the chemist Milan Randić [8]. Randić noticed that this index was highly correlated with several physical properties of the

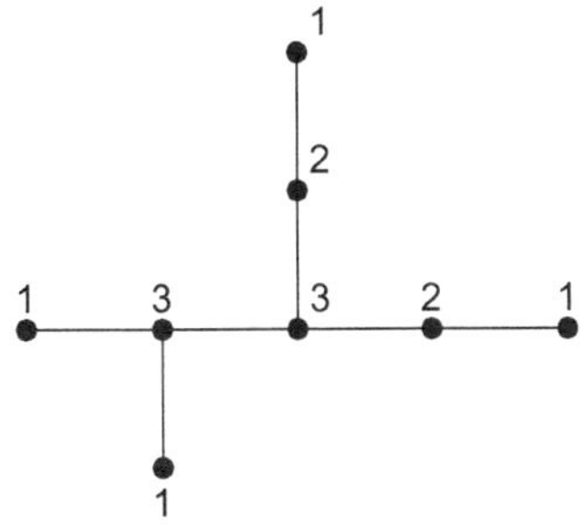

FIG. 11

3-Ethyl-2-methylpentane.

alkanes, including their boiling points. Extensive research was done in the following 25 years based on the Randić index.

We compute the Randić index for the alkane 3-ethyl-2-methylpentane. The chemical graph of 3-ethyl-2-methylpentane is given in Fig. 11 where the vertices are labeled with their corresponding degrees. Therefore, the Randić index for 3-ethyl-2-methylpentane is given by

$$\chi = 2\frac{1}{\sqrt{3}} + \frac{1}{3} + 2\frac{1}{\sqrt{6}} + 2\frac{1}{\sqrt{2}} = 3.7187$$

Balaban index (J) is defined for a graph G with n vertices, m edges, and c connected components by

$$J = \frac{m}{\gamma + 1} \sum_{q \in E(G)} (D_i D_j)^{-1/2}$$

where $\gamma = m - n + c$ is the *cyclomatic number* of the graph G and D_i is the sum of all the entries in the ith row of the distance matrix of G, $D(G)$. The cyclomatic number of a graph is the minimum number of edges that must be removed from the graph in order for the graph to be acyclic. Since the alkanes are all represented by acyclic carbon trees, their corresponding cyclomatic number is zero.

Professor Alexandru Balaban introduced this index in 1982 in order to differentiate between isomers, since indices such as the Randić index failed to do so [9]. The index was originally called the *average distance sum connectivity index* and was based on the Randić index. Where the Randić index uses the degrees of the vertices, the Balaban index uses the distance between the vertices. Hence, this "highly discriminating distance-based topological index" as defined in Balaban's paper of the same name, was critical in the advancement of chemical graph theory.

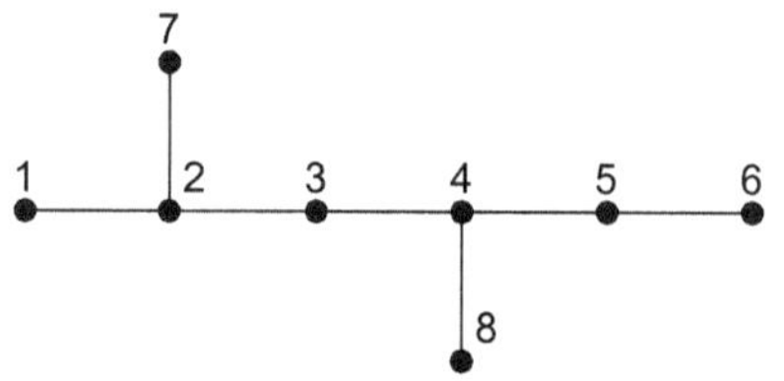

FIG. 12

2,4-Dimethylhexane.

We compute the Balaban index for the alkane 2,4-dimethylhexane whose chemical graph is given in Fig. 12. The distance matrix for the chemical graph of 2,4-dimethylhexane and the corresponding D_i are

$$D(G) = \begin{bmatrix} 0 & 1 & 2 & 3 & 4 & 5 & 2 & 4 \\ 1 & 0 & 1 & 2 & 3 & 4 & 1 & 3 \\ 2 & 1 & 0 & 1 & 2 & 3 & 2 & 2 \\ 3 & 2 & 1 & 0 & 1 & 2 & 3 & 1 \\ 4 & 3 & 2 & 1 & 0 & 1 & 4 & 2 \\ 5 & 4 & 3 & 2 & 1 & 0 & 5 & 3 \\ 2 & 1 & 2 & 3 & 4 & 5 & 0 & 4 \\ 4 & 3 & 2 & 1 & 2 & 3 & 4 & 0 \end{bmatrix} \quad \begin{matrix} D_1 = 21 \\ D_2 = 15 \\ D_3 = 13 \\ D_4 = 13 \\ D_5 = 17 \\ D_6 = 23 \\ D_7 = 21 \\ D_8 = 19 \end{matrix}$$

Using the fact that $\gamma = 7 - 8 + 1 = 0$ and $m = 7$ for the graph of 2,4-dimethylhexane, we compute that the Balaban index for 2,4-dimethylhexane is given by

$$\begin{aligned} J &= 7\left(\frac{2}{\sqrt{15 \cdot 21}} + \frac{2}{\sqrt{15 \cdot 13}} + \frac{1}{\sqrt{13 \cdot 19}} + \frac{1}{\sqrt{13 \cdot 27}} + \frac{1}{\sqrt{17 \cdot 23}}\right) \\ &= 3.0988 \end{aligned}$$

Harary index (H) is the sum of the inverse of the distance between all pairs of vertices in a graph. The index was introduced by Plavšić et al. in 1991 at a conference held in honor of Professor Frank Harary's 70th birthday and was thus named the Harary index [10]. This index attempts to model the fact that atoms closer together in a molecule should have a greater influence on each other than atoms that are separated by a large distance. The formula for the Harary index of a graph G having n vertices is

$$H = \frac{1}{2}\sum_{i=1}^{n}\sum_{j=1}^{n}(D^r)_{ij}$$

where D^r is the *reciprocal distance matrix* of a graph G defined by

$$(D^r)_{ij} = \frac{1}{D_{ij}}; \quad i \neq j$$

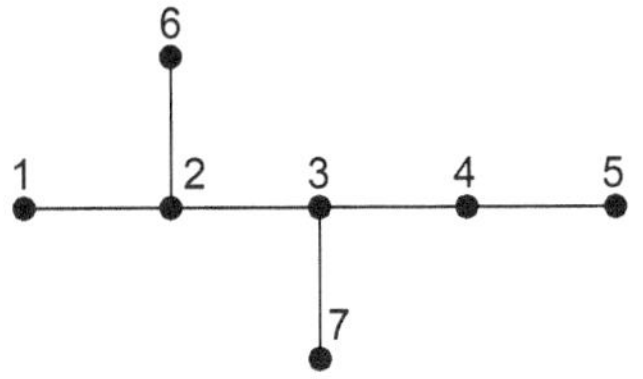

FIG. 13

2,3-Dimethylpentane.

We compute the Harary index for the alkane 2,3-dimethylpentane whose chemical graph is given in Fig. 13.

The distance matrix and the reciprocal distance matrix for the chemical graph of 2,3-dimethylpentane are as follows:

$$D(G) = \begin{bmatrix} 0 & 1 & 2 & 3 & 4 & 2 & 3 \\ 1 & 0 & 1 & 2 & 3 & 1 & 2 \\ 2 & 1 & 0 & 1 & 2 & 2 & 1 \\ 3 & 2 & 1 & 0 & 1 & 3 & 2 \\ 2 & 1 & 2 & 3 & 4 & 0 & 3 \\ 3 & 2 & 1 & 2 & 3 & 3 & 0 \end{bmatrix} \quad D^r(G) = \begin{bmatrix} 0 & 1 & \frac{1}{2} & \frac{1}{3} & \frac{1}{4} & \frac{1}{2} & \frac{1}{3} \\ 1 & 0 & 1 & \frac{1}{2} & \frac{1}{3} & 1 & \frac{1}{2} \\ \frac{1}{2} & 1 & 0 & 1 & \frac{1}{2} & \frac{1}{2} & 1 \\ \frac{1}{3} & \frac{1}{2} & 1 & 0 & 1 & \frac{1}{3} & \frac{1}{2} \\ \frac{1}{2} & 1 & \frac{1}{2} & \frac{1}{3} & \frac{1}{4} & 0 & \frac{1}{3} \\ \frac{1}{3} & \frac{1}{2} & 1 & \frac{1}{2} & \frac{1}{3} & \frac{1}{3} & 0 \end{bmatrix}$$

Therefore, the Harary index for 2,3-dimethylpentane is

$$H = \frac{12(1) + 14\left(\frac{1}{2}\right) + 12\left(\frac{1}{3}\right) + 4\left(\frac{1}{4}\right)}{2} = 12$$

Schultz index or Molecular Topological index (MTI) is defined for a graph G having n vertices by

$$MTI = \sum_{j=1}^{n} e_j$$

where

$$e_j = \sum_{i=1}^{n} v_i(A_{ij} + D_{ij})$$

where v_i is the degree (valence) of vertex i in the graph G.

We compute the Schultz index for the alkane 2,3-dimethylbutane whose chemical graph is given in Fig. 14. The adjacency matrix for the chemical graph of 2,3-dimethylbutane is

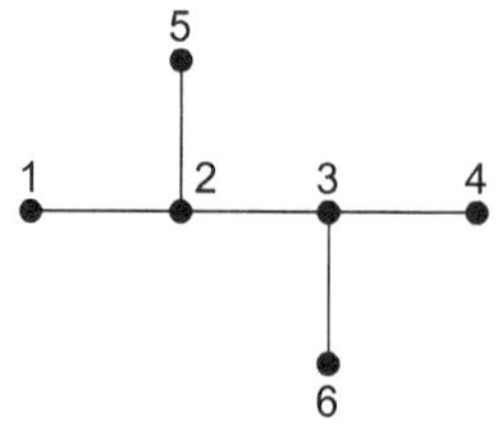

FIG. 14

2,3-Dimethylbutane.

$$A(G) = \begin{bmatrix} 0 & 1 & 0 & 0 & 0 & 0 \\ 1 & 0 & 1 & 0 & 1 & 0 \\ 0 & 1 & 0 & 1 & 0 & 1 \\ 0 & 0 & 1 & 0 & 0 & 0 \\ 0 & 1 & 0 & 0 & 0 & 0 \\ 0 & 0 & 1 & 0 & 0 & 0 \end{bmatrix}$$

The distance matrix for 2,3-dimethylbutane is

$$D(G) = \begin{bmatrix} 0 & 1 & 2 & 3 & 2 & 3 \\ 1 & 0 & 1 & 2 & 1 & 2 \\ 2 & 1 & 0 & 1 & 2 & 1 \\ 3 & 2 & 1 & 0 & 3 & 2 \\ 2 & 1 & 2 & 3 & 0 & 3 \\ 3 & 2 & 1 & 2 & 3 & 0 \end{bmatrix}$$

The valency vector for 2,3-dimethylbutane is $\begin{bmatrix} 1 & 3 & 3 & 1 & 1 & 1 \end{bmatrix}$ To compute each of the e_i from earlier we take

$$v(G)\,[A + D]\,(G) = \begin{bmatrix} 1 & 3 & 3 & 1 & 1 & 1 \end{bmatrix} \begin{bmatrix} 0 & 2 & 2 & 3 & 2 & 3 \\ 2 & 0 & 2 & 2 & 2 & 2 \\ 2 & 2 & 0 & 2 & 2 & 2 \\ 3 & 2 & 2 & 0 & 3 & 2 \\ 2 & 2 & 2 & 3 & 0 & 3 \\ 3 & 2 & 2 & 2 & 3 & 0 \end{bmatrix}$$

$$= \begin{bmatrix} 20 & 14 & 14 & 20 & 20 & 20 \end{bmatrix}$$

Hence, the *MTI* for 2,3-dimethylbutane is $4 \cdot 20 + 2 \cdot 14 = 108$.

Methyl Locant (*loc*) is defined for alkanes having one and only one methyl branch. It is the number of the carbon to which the methyl branch is attached to the straight chain whose carbons are numbered 1 to $n - 1$ consistent with IUPAC nomenclature. The methyl locant number is defined as 0 for the normal alkanes. For example, 4-methylheptane has $loc = 4$. This index was defined to model the melting points of a set of alkanes having a single methyl group which are used in synthetic fuel as seen in Section 3.

3 MODELS

Extensive research had been conducted using the topological indices described here to model physical properties of many different types of chemical molecules. Using combinations of these indices has produced multivariable models of physical properties which may be used to determine said physical properties which are not available by direct measurement.

Experimental boiling point data are available for most alkanes having less than or equal to 10 carbon atoms [11,12]. It is unlikely that comprehensive boiling point data will be available for all alkanes having more than 10 carbon atoms. This has motivated several studies to determine successful models for the boiling points which may be used to predict the boiling points for which no data are available. The boiling point of an alkane depends strongly on the geometric structure of the alkane since boiling points are a measure of the forces of attraction between individual atoms or molecules. Given that alkanes are essentially nonpolar compounds, these forces are London dispersion forces due to instantaneous dipole-induced dipole attractions. Since the forces are weak, they depend on contact between molecules. This means that the forces increase with surface area and therefore the boiling point of an alkane will increase as the relative molecular mass increases. The geometry of the alkane also plays a pivotal role in determining its boiling point. The dependence is complex, but the boiling point of alkanes with the same relative molecular weight will generally decrease as the compactness of the molecules increases. Balaban noted in his 2001 paper that boiling points decreased for alkanes with increased branching for the same relative molecular mass [13].

We examine this phenomenon by viewing the structure of 2,2,4-trimethylpentane and octane. Both acyclic hydrocarbons are composed of eight carbons and therefore have the same relative molecular mass. The geometric structure of these two alkanes differ greatly as seen by their chemical graphs in Fig. 15. We note that 2,2,4-trimethylpentane is clearly a more compact molecule than the straight chain octane and we therefore expect the boiling point of 2,2,4-trimethylpentane to be lower than that of octane. This is indeed the case since the boiling point of 2,2,4-trimethylpentane is 99.3°C whereas the boiling point of octane is 125.6°C.

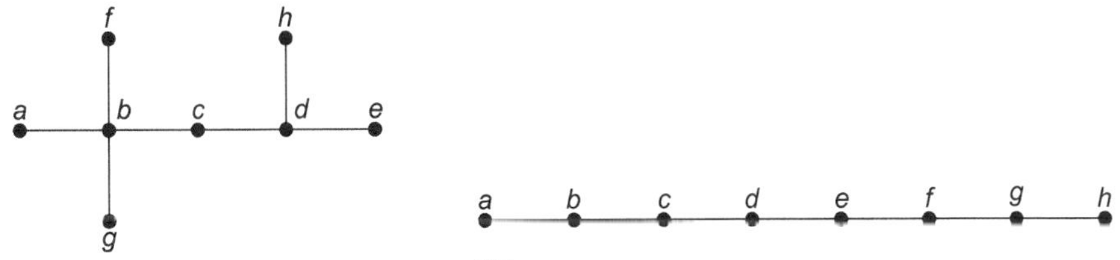

(A) (B)

FIG. 15

Comparing the geometric structure of 2,2,4-trimethylpentane and octane.
(A) 2,2,4-Trimethylpentane; (B) octane.

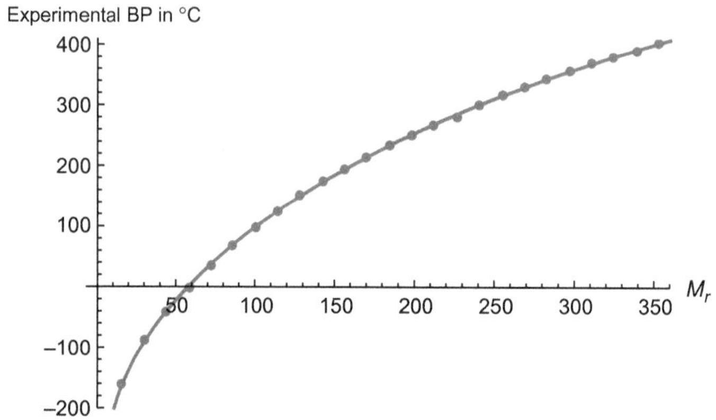

FIG. 16

M_r versus BP of *n*-alkanes for $1 \leq n \leq 25$.

When studying a family of alkanes with similar geometric structure, a single topological index can be used to model the boiling point of the alkanes. The normal alkanes or *n*-alkanes are one such family in which their molecular graphs are the path graphs having n vertices. Although these molecules are simple, their boiling points do not increase in a linear fashion with respect to their length. The boiling points of the first few normal alkanes increase rapidly because the London forces increase with the relative molecular weight which are the forces holding the alkane together in liquid form. As the number of carbons increase, the chain will fold back on itself causing the London forces to be directed toward themselves and therefore diminish the dependence of the boiling point on the relative molecular weight. Therefore, a model using the relative molecular mass fits the boiling points of the *n*-alkanes for $1 \leq n \leq 25$ quite well as seen in Fig. 16. This model was developed in Burch et al. [4] using Mathematica. The equation of the model is $f(M_r) = -240.103 - 330.713M_r^{-0.797084} + 43.9249M_r^{0.793242} - 26.3154M_r^{0.854845}$ where $R^2 = 0.999925$ and the standard deviation is 1.2°C.

Developing models to predict the physical properties of all the alkanes necessitates using a combination of the topological indices described in order to effectively represent the geometric structures of the different alkanes. Multivariable models were developed to model the boiling points of alkanes having 6–12 carbon atoms. These models were then used to predict the boiling points of alkanes having 13–22 carbon atoms. Similar research was conducted using neural nets to predict the boiling points of the alkanes having 6–10 carbon atoms [14]. Using neural nets to predict physical properties does not provide equations for the models. Therefore, models were developed in Burch et al. [4] using the NEOS solver FilterSQP available online from Argonne National Laboratory [15]. FilterSQP is a sequential quadratic programming algorithm that optimizes the nonlinear function by using a “filter” to promote global convergence. FilterSQP uses second derivative information and the constraints in such a way that it is less prone to converge to shallow local minima. By using FilterSQP, we were able to construct equations for our boiling point models. These models can be used to predict the boiling points of alkanes for which no experimental boiling point data exist.

The first boiling point model was developed using all the alkanes having 1–12 carbon atoms for which boiling point data were available and the normal alkanes having 13–22 carbons. This range of normal alkanes was included in the model's dataset because the models were used to predict the boiling points of a set of miscellaneous alkanes having 13–22 carbon atoms. Table 1 gives the boiling point data and corresponding source for the alkanes having 1–12 carbon atoms. Model 1 was developed using the Wiener path numbers from one to eight and the methyl indices. The number of alkanes in the dataset for Model 1 is 195 and $R^2 = 0.995322$. The standard deviation is 4.3°C.

Model 1. This model uses the Wiener ${}^{i}P$ numbers ($1 \leq i \leq 8$) and the Methyl index. The formula is

$$
\begin{aligned}
&f_{\text{Model 1}}({}^{1}P, {}^{2}P, \ldots, {}^{8}P, Mth) \\
&= -167.49997 + 84.28344({}^{1}P)^{0.46072} + 15.94534({}^{2}P)^{0.00348} \\
&\quad + 17.42198({}^{3}P)^{0.53517} + 11.16515({}^{4}P)^{0.00164} + 4.74582({}^{5}P)^{0.00089} \\
&\quad + 5.23270({}^{6}P)^{0.00143} + 6.67018({}^{7}P)^{0.14687} + 5.27829({}^{8}P)^{0.96677} \\
&\quad - 5.53723 Mth^{0.00072}
\end{aligned}
$$

The absolute boiling point deviation, denoted BP dev., is given by the formula BP dev. = |Model BP – Experimental BP|. Table 2 gives the number and percentage of alkanes whose absolute boiling point deviations determined by Model 1 are within the specified ranges.

By adding the Hosoya index as one of the variables in the boiling point model, the standard deviation improved by over a degree. Model 2 uses the same 195 alkanes as its dataset and yields $R^2 = 0.997648$. The standard deviation of Model 2 shows an improvement by 3.0°C.

Model 2. This model uses the Wiener ${}^{i}P$ numbers ($1 \leq i \leq 8$), the Methyl index, and the Hosoya index. The formula is

$$
\begin{aligned}
&f_{\text{Model 2}}({}^{1}P, {}^{2}P, \ldots, {}^{8}P, Mth, Z) \\
&= -157.39801 - 1.71578({}^{1}P)^{2.06741} + 0.38684({}^{2}P)^{1.98738} \\
&\quad - 3.53937({}^{3}P)^{0.73904} - 4.69814({}^{4}P)^{0.04224} - 2.31936({}^{5}P)^{0.85500} \\
&\quad - 0.00004({}^{6}P)^{5.13758} + 1.57097({}^{7}P)^{0.57937} \\
&\quad + 0.23048({}^{8}P)^{2.70854} + 1.20993 Mth^{1.32159} + 108.36449(\ln Z)^{1.00972}
\end{aligned}
$$

Table 3 gives the number and percentage of alkanes whose absolute boiling point deviations determined by Model 2 are within the specified ranges.

Table 1 Boiling Point Data

Alkane	BP (C)	Source	Alkane	BP (C)	Source	Alkane	BP (C)	Source
Methane	−162.15	NIST	*n*-Decane	174.15	NIST	*n*-Dodecane	215.85	NIST
Ethane	−88.55	NIST	2-Methylnonane	166.94	NIST	2,2,4,6,6-Pentamethylheptane	177.85	NIST
Propane	−42.15	NIST	3-Methylnonane (DL)	167.75	NIST	2-Methylundecane	210.05	NIST
Butane	−0.15	NIST	4-Methylnonane (DL)	166.85	NIST	3,3,6,6-Tetramethyloctane	189.75	NIST
2-Methylpropane	−11.15	NIST	5-Methylnonane	165.11	NIST	2,2,4,7-Tetramethyloctane	184.7	Beilstein
n-Pentane	36.05	NIST	3-Ethyloctane	166	Beilstein	2,2,7,7-Tetramethyloctane	186.85	NIST
2-Methylbutane	27.03	Beilstein	4-Ethyloctane v168	163.36	NIST	2,2,4,4,6-Pentamethylheptane	185.05	NIST
2,2-Dimethylpropane	9.45	NIST	2,2-Dimethyloctane	154	Beilstein	3-Ethyl-2,3,5-trimethylheptane	197.8	Beilstein
n-Hexane	68.75	NIST	2,3-Dimethyloctane	164.24	NIST	3-Ethyl decane	208.85	Beilstein
2-Methylpentane	60.85	NIST	2,4-Dimethyloctane	155.9	NIST	4-Ethyldecane	204.5	Beilstein
3-Methylpentane	63.25	NIST	2,5-Dimethyloctane	156.93	NIST	3-Methylundecane	211.2	Beilstein
2,2-Dimethylbutane	49.75	NIST	2,6-Dimethyloctane	158.85	NIST	4-Methylundecane	207.1	Beilstein
2,3-Dimethylbutane	58.05	NIST	2,7-Dimethyloctane	160.05	NIST	5-Methylundecane	204.6	Beilstein
n-Heptane	98.45	NIST	3,3-Dimethyloctane	161.2	NIST	4,5-Diethyloctane	200.5	Beilstein
3-Ethylpentane	93.45	NIST	3,5-Dimethyloctane	158.85	NIST	2,4,5,7-Tetramethyloctane	186.25	NIST
3-Methylhexane	91.82	Beilstein	4,4-Dimethyloctane	157.5	NIST	2,2-Dimethyldecane	199	Beilstein
2-Methylhexane	90.05	NIST	4,5-Dimethyloctane v162.13	162.42	NIST	2,3-Dimethyldecane	203.4	Beilstein
2,3-Dimethylpentane	89.75	NIST	4-Propylheptane	161.85	NIST	2,4-Dimethyldecane	198.5	Beilstein
3,3-Dimethylpentane	86.05	NIST	4-Isopropylheptane	158.8	Beilstein	5,6-Dimethyldecane	197.5	Beilstein

2,4-Dimethylpentane	80.55	NIST	3-Ethyl-3-methylheptane	163.85	NIST	3-Ethyldecane	208.8	Beilstein
2,2-Dimethylpentane	79.15	NIST	5-Ethyl-3-methylheptane	158.3	Beilstein	2,6-Dimethyl-3-isopropylheptane	186.85	NIST
2,2,3-Trimethylbutane	80.95	NIST	4-Ethyl-4-methylheptane	159	Beilstein	2,2,3,5,6-Pentamethylheptane	188.8	NIST
n-Octane	125.55	NIST	2,2,4-Trimethylheptane	148.85	NIST	2,2,6,6-Tetramethyloctane	182.75	NIST
3-Methylheptane	118.85	NIST	2,2,5-Trimethylheptane	146	Beilstein	2,4,4,7-Tetramethyloctane	183.85	NIST
3-Ethylhexane	118.55	NIST	2,2,6-Trimethylheptane	148.33	NIST	2,2,3,4,5,5-Hexamethylhexane	191.85	NIST
3-Ethyl-3-methylpentane	118.35	NIST	2,3,3-Trimethylheptane	160.05	NIST	2-Methyl-5-propyloctane	188.85	NIST
3,4-Dimethylhexane	117.85	NIST	2,3,4-Trimethylheptane	155.85	NIST	2-Methyl-3-isopropyloctane	197.7	Beilstein
4-Methylheptane	117.75	NIST	2,3,6-Trimethylheptane	149.85	NIST	2,2,4-Trimethylnonane	191.6	Beilstein
2-Methylheptane	117.55	NIST	2,4,4-Trimethylheptane	149.55	NIST	2,4,6-Trimethylnonane	186	Beilstein
3-Ethyl-2-methylpentane	115.65	NIST	2,4,6-Trimethylheptane v144.8	146.75	NIST	2,5,8-Trimethylnonane	189.85	NIST
2,3-Dimethylhexane	115.85	NIST	2,5,5-Trimethylheptane	152.8	NIST	3,6-Diethyloctane	202.85	NIST
2,3,3-Trimethylpentane	114.65	NIST	3,3,5-Trimethylheptane	155.67	NIST	3,3-Diethyloctane	204.85	NIST
2,3,4-Trimethylpentane	113.65	NIST	3,3-Diethylhexane	164.48	NIST	4-Ethyl-5-methylnonane	194.5	Beilstein
3,3-Dimethylhexane	111.95	NIST	3,4-Diethylhexane	162.75	NIST	5-Ethyl-5-methylnonane	196	Beilstein

(Continued)

Table 1 Boiling Point Data–Cont'd

Alkane	BP (C)	Source	Alkane	BP (C)	Source	Alkane	BP (C)	Source
2,2,3-Trimethylpentane	109.95	NIST	2,2-Dimethyl-4-ethylhexane	147.05	NIST	4-Propylnonane	202.5	Beilstein
2,4-Dimethylhexane	108.85	NIST	2,4-Dimethyl-4-ethylhexane	158	Beilstein	5-Propylnonane	204.5	Beilstein
2,5-Dimethylhexane	108.95	NIST	2,2,3,3-Tetramethylhexane	158.85	NIST			
2,2-Dimethylhexane	106.85	NIST	2,2,3,4-Tetramethylhexane	156.65	NIST			
2,2,3,3-Tetramethylbutane	106.45	NIST	2,2,3,5-Tetramethylhexane	148.65	NIST			
2,2,4-Trimethylpentane	99.25	NIST	2,2,4,5-Tetramethylhexane	148.04	NIST			
n-Nonane	150.65	NIST	2,2,5,5-Tetramethylhexane	137.17	NIST			
3-Ethylheptane	143.27	NIST	2,3,3,4-Tetramethylhexane	164.59	NIST			
4-Ethylheptane	141.3	NIST	2,3,3,5-Tetramethylhexane	152.85	NIST			
3-Methyloctane	143.85	NIST	2,3,4,4-Tetramethylhexane	162.25	NIST			
4-Methyloctane	141.85	NIST	2,3,4,5-Tetramethylhexane	155.45	NIST			
3,3-Diethylpentane	146.15	NIST	3,3,4,4-Tetramethylhexane	170.32	NIST			
4-Ethyl-3-methylhexane	140.4	NIST	3-Isopropyl-2,4-dimethylpentane	157.04	NIST			
2-Methyloctane	142.95	NIST	2,2,4-Trimethyl-3-ethylpentane	155.3	NIST			

3,4-Dimethylheptane	140.45	NIST	2,3,4-Trimethyl-3-ethylpentane	169.44	NIST			
3-Ethyl-2-methylhexane	138	NIST	2,2,3,3,4-Pentamethylpentane	166.05	NIST			
3,5-Dimethylheptane	135.61	NIST	2,2,3,4,4-Pentamethylpentane	159.29	NIST			
3-Ethyl-3-methylhexane	140.6	NIST	Undecane	194.85	NIST			
2,3-Dimethylheptane	139.85	NIST	2-Methyldecane	189.15	NIST			
4-Ethyl-2-methylhexane	133.8	NIST	3-Methyldecane	190.9	Beilstein			
2,5-Dimethylheptane	134.85	NIST	4-Methyldecane	186.95	NIST			
2,3,4-Trimethylhexane	138.85	NIST	5-Methyldecane	186.8	Beilstein			
3,3-Dimethylheptane	136.85	NIST	4,4-Diethylheptane	183.35	NIST			
2,4-Dimethylheptane	132.85	NIST	4-Ethyl-2,4-dimethylheptane	148.85	NIST			
3-Ethyl-2,3-dimethylpentane	141.6	NIST	2,2,3,5,5-Pentamethylhexane	165.15	NIST			
2,6-Dimethylheptane	134.85	NIST	2,2,4,6-Tetramethylheptane	162.05	NIST			
3,3,4-Trimethylhexane	140.16	NIST	2,2,6,6-Tetramethylheptane	160.3	Beilstein			
3-Ethyl-2,4-dimethylpentane	136.66	NIST	2,2,3,3,5-Pentamethylhexane	173.05	NIST			
4,4-Dimethylheptane	134.93	NIST	2,2,4-Trimethyloctane	170.05	NIST			
2,2-Dimethylheptane	131.85	NIST	2,4,6-Trimethyloctane	168	Beilstein			

(Continued)

Table 1 Boiling Point Data–Cont'd

Alkane	BP (C)	Source	Alkane	BP (C)	Source	Alkane	BP (C)	Source
2,3,5-Trimethylhexane	130.85	NIST	2,2-Dimethylnonane,	186	Beilstein			
2,3,3-Trimethylhexane	137.08	NIST	2,3-Dimethylnonane	186.8	Beilstein			
3-Ethyl-2,2-dimethylpentane	133.75	NIST	2,4-Dimethylnonane	178	Beilstein			
2,2,3-Trimethylhexane	133.12	NIST	3-Isopropyl-2-methyl-heptane	178.1	Beilstein			
2,4,4-Trimethylhexane	130.85	NIST	2,3,4,6-Tetramethylheptane	173.55	NIST			
2,3,3,4-Tetramethylpentane	141.5	NIST	2,2,3,3-Tetramethylheptane	179.65	NIST			
2,2,4-Trimethylhexane	126.45	NIST	4-Ethyl-3,4-dimethylheptane	180.85	NIST			
2,2,5-Trimethylhexane	123.85	NIST	3-Ethyl-3-methyloctane	185.75	NIST			
2,2,3,4-Tetramethylpentane	133	NIST	3,3-Diethylheptane	186.75	NIST			
2,2,3,3-Tetramethylpentane	140.27	NIST	4-Propyloctane	182.6	Beilstein			
2,2,4,4-Tetramethylpentane	122.25	NIST						

Table 2 Alkanes and Their Corresponding Absolute BP Deviations for Model 1

BP Dev. (degrees)	No. of Alkanes	Alkanes (%)
0–1	49	25.1
1–2	35	17.9
2–4	52	26.7
4–6	32	16.4
6–9	22	11.3
9–11	2	1.0
>11	3	1.5

Table 3 Alkanes and Their Corresponding Absolute BP Deviations for Model 2

BP Dev. (degrees)	No. of Alkanes	Alkanes (%)
0–1	65	33.3
1–2	54	27.7
2–4	53	27.2
4–6	18	9.2
6–9	4	2.1
9–10	0	0
>10	1	0.5

In examining Models 1 and 2, we found that the first eight alkanes having one to five carbon atoms greatly constrained the intercept and the coefficients and exponents of some of the indices. This was especially true with the indices ${}^{1}P, {}^{2}P$, and Z. This motivated us to exclude these eight alkanes from our data set and to construct Model 3. The dataset for this model is therefore the set of alkanes having 6–12 carbon atoms for which boiling point data are available and consists of 187 alkanes. The model converged with zero coefficients for the last two Wiener path numbers used in the previous models, ${}^{7}P$, and ${}^{8}P$, so these two indices were eliminated from Model 3. The coefficient of determination for Model 3 is $R^2 = 0.997068$ and the further improved standard deviation is 2.1°C.

Model 3. This model uses the Wiener ${}^{i}P$ numbers ($1 \leq i \leq 6$), the Methyl index, and the Hosoya index. The formula is

$$\begin{aligned} f_{\text{Model 3}}({}^{1}P, {}^{2}P, \ldots, {}^{6}P, Mth, Z) &= 847.41474 + 221.61698({}^{1}P)^{0.49420} \\ &- 1182.20853({}^{2}P)^{0.03689} + 0.00125({}^{3}P)^{3.39724} - 3.02445({}^{4}P)^{0.93751} \\ &- 2.16070({}^{5}P)^{1.01631} - 0.56366({}^{6}P)^{1.38233} - 2.10575Mth^{0.5695} - 9.61075Z^{0.19907} \end{aligned}$$

Table 4 gives the number and percentage of alkanes whose absolute boiling point deviations determined by Model 3 are within the specified ranges. The improvement

Table 4 Alkanes and Their Corresponding Absolute BP Deviations for Model 3

BP Dev. (degrees)	No. of Alkanes	Alkanes (%)
0–1	84	44.9
1–2	48	25.7
2–4	46	24.6
4–6	5	2.7
6–9	3	1.6
>9	1	0.5

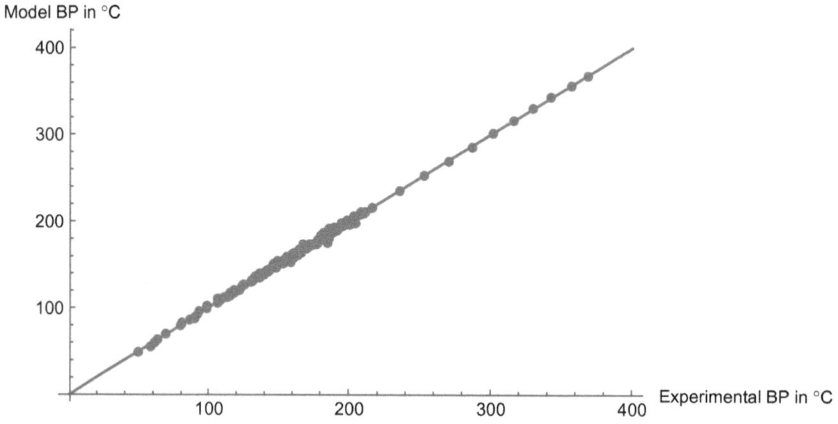

FIG. 17

Experimental BP versus model BP for Model 3.

in the boiling point estimation of Model 3 is demonstrated by the fact that 70% of the alkanes have an error less than 2°C.

Fig. 17 shows a plot of the experimental boiling points versus the model boiling points determined by Model 3. The straight line represents the exact values.

These multivariable models were used to predict the boiling points of a set of 52 alkanes having 13–22 carbon atoms. Table 5 gives the experimental boiling point data for these 52 alkanes, the boiling points predicted by Models 2 and 3, and the corresponding boiling point deviations for each model. Model 3 performs particularly well in predicting the boiling points of these 52 alkanes in that 24 of the 52 alkanes have an absolute boiling point deviation less than 2°C. Fig. 18 shows a plot of the experimental boiling point versus the boiling point predicted by Model 3 for the additional set of 52 alkanes. The straight line again represents the exact values.

Other physiochemical properties can also be modeled using similar topological indices presented in this chapter. Melting point models of the alkanes were developed using the same techniques that were used with the previous boiling point models [16]. Modeling the melting points of alkanes is a more challenging task than that

Table 5 Predictive Ability of Models 2 and 3

		Model 2		Model 3	
Alkane	**Exp. BP**	**Calc. BP**	**BP Dev.**	**Calc. BP**	**BP Dev.**
2,5,9-Trimethyldecane	211.1	210.2	0.9	212.7	1.6
2,5-Dimethylundecane	215.0	214.5	0.5	218.4	3.4
3,3-Diethylnonane	222.1	222.5	0.4	228.9	6.8
2-Methyldodecane	229.5	227.2	2.3	228.2	1.3
2,3-Dimethylundecane	226.3	228.3	2.0	226.7	0.4
3-Methyldodecane	230.2	231.0	0.8	230.0	0.2
2,2-Dimethylundecane	220.0	222.3	2.3	221.4	1.4
4,5-Diethyl-2,7-dimethyloctane	215.9	202.2	13.6	212.9	2.9
2,4,4,5,5,7-Hexamethyloctane	216.9	216.2	0.6	228.1	11.2
4-Propylundecane	231.0	242.8	11.9	239.0	8.0
2-Methyltridecane	247.9	245.5	2.4	246.6	1.3
3-Methyltridecane	248.4	249.3	0.9	248.7	0.3
4-Methyltridecane	246.5	246.7	0.2	246.4	0.1
2,3-Dimethyl-dodecane	245.9	247.4	1.5	246.0	0.1
2,6,10-Trimethyldodecane	252.0	234.7	17.2	251.2	0.7
4-Methyl-6-propylundecane	242.6	220.7	21.9	239.8	2.8
2,3-Dimethyltridecane	263.7	266.0	2.3	264.5	0.8
4-Methyltetradecane	262.7	264.5	1.8	264.2	1.5
3-Methyltetradecane	265.0	267.2	2.2	266.5	1.5
2-Methyltetradecane	265.4	263.3	2.1	264.1	1.3
7,8-Dimethyltetradecane	269.9	273.0	3.2	272.3	2.5
3-Methylpentadecane	282.0	284.5	2.5	283.4	1.4
2,4-Dimethyltetradecane	266.0	278.5	12.5	275.1	9.1
2-Methylpentadecane	281.9	280.6	1.3	280.7	1.2
4-Methylpentadecane	279.7	281.8	2.1	281.1	1.4
2,3-Dimethyltetradecane	280.2	284.1	3.9	282.1	1.9
2,3-Dimethylpentadecane	295.5	301.7	6.2	298.9	3.4
4-Methylhexadecane	294.3	298.7	4.4	297.3	3.0
3-Methylhexadecane	294.0	301.4	7.4	299.5	5.5
2-Methylhexadecane	297.7	297.4	0.3	296.4	1.3
2-Methylheptadecane	312.0	313.8	1.8	311.5	0.5
3-Methlheptadecane	310.0	317.7	7.7	314.9	4.9
4-Methylheptadecane	307.1	315.1	8.0	312.8	5.7
2,3-Dimethyl-hexadecane	310.0	318.8	8.8	315.2	5.2
2,3-Dimethyl-heptadecane	324.9	335.2	10.3	330.8	5.9
4-Methyloctadecane	322.5	330.7	8.2	327.6	5.1
3-Methyloctadecane	323.5	333.4	9.9	329.8	6.3
2-Methyloctadecane	327.0	329.4	2.4	325.9	1.1
3-Ethyloctadecane	340.9	349.6	8.8	347.9	7.1

(Continued)

Table 5 Predictive Ability of Models 2 and 3–Cont'd

		Model 2		Model 3	
Alkane	**Exp. BP**	**Calc. BP**	**BP Dev.**	**Calc. BP**	**BP Dev.**
2,6,11,15-Tetramethylhexadecane	318.9	257.2	61.6	324.6	5.7
2,2-Dimethyloctadecane	330.0	349.1	19.1	336.7	6.7
2,3-Dimethyloctadecane	338.8	350.5	11.7	345.9	7.1
4-Methylnonadecane	336.0	345.2	9.2	341.9	5.9
3-Methylnonadecane	336.0	347.9	11.9	344.0	8.0
2-Methylnonadecane	341.0	343.9	2.9	339.6	1.4
2-Methyleicosane	354.0	356.9	2.9	352.8	1.2
3-Methyleicosane	347.0	360.9	13.9	357.6	10.6
4-Methyleicosane	348.0	358.3	10.3	355.6	7.6
2,3-Dimethylnonadecane	351.8	364.2	12.4	360.4	8.6
2-Methylheneicosane	367.0	367.8	0.8	365.4	1.6
3-Methylheneicosane	358.0	371.8	13.8	370.8	12.8
4-Methylheneicosane	359.0	369.2	10.2	368.8	9.8

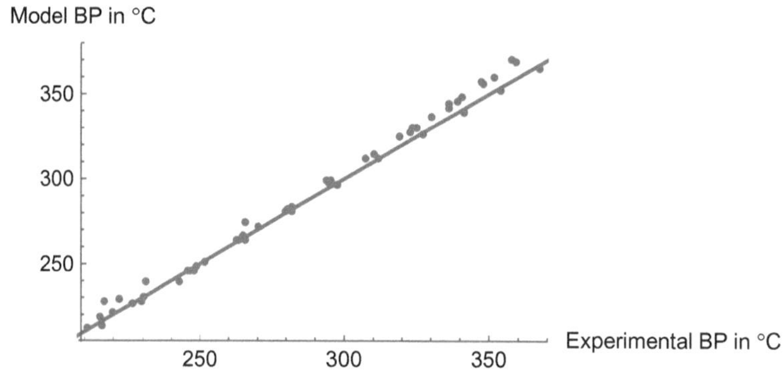

FIG. 18

Experimental BP versus model BP for Model 3 on the 52 additional alkanes.

of boiling points. As previously mentioned, boiling points are mainly determined by the London dispersion forces between the molecule in liquid form. When studying melting points, the molecules are in a solid state which introduces many additional factors in determining the melting point since the alkane is a rigid 3D structure. Attempting to model all the alkanes as was done with the boiling points produces models with very high error rates. Therefore, a select group of alkanes was chosen in order to more effectively model the melting points and use the models to predict the melting points of alkanes for which no experimental data exist.

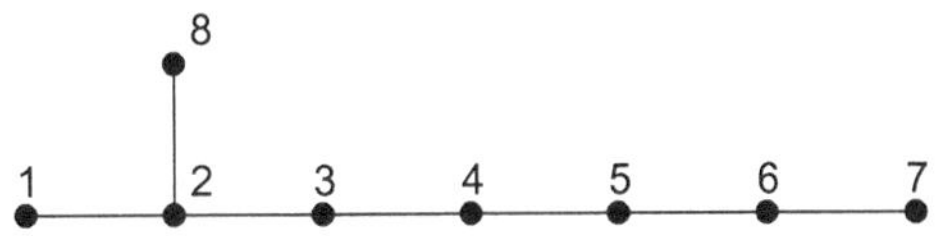

FIG. 19

2-Methylheptane.

Professors Dr. John Tierney and Dr. Irving Wender of the Chemical and Petroleum Engineering Department at the University of Pittsburgh first approached us asking if we could model a select set of alkanes having between 10 and 20 carbon atoms. These alkanes were of particular interest to them since they comprise most of the molecules in jet and diesel fuels. Normal alkanes have a high cetane number which is environmentally desirable but they also have a high melting point. Since jet and diesel fuels are often exposed to cold temperatures, the high melting points can be problematic in which the fuels produced with these alkanes could possible freeze. Methyl-substituted alkanes have a lower, but still acceptable, cetane number and have lower melting points than their straight chain counterparts. The methyl-substituted alkanes specifically studied in our research were alkanes that had one and only one methyl group attached by a single bond to one of the carbons in the straight chain of the alkane. An example of the chemical graph of a methyl-substituted alkane used in this study is given in Fig. 19.

Fuels containing methyl-substituted alkanes have lower pour points than fuels based on normal alkanes. Therefore, the fuels remain in liquid form even when subjected to cold temperatures. This is a distinct advantage for jet fuel since it is exposed to very low temperatures at high altitudes. Vehicular transportation also benefits from such improvements to the pour point of diesel fuels in colder climates.

While this improvement in melting points using methyl-substituted alkanes is known, there are several methyl-substituted alkanes for which no melting point data exist. Therefore, we again used the NEOS solver FilterSQP from Argonne National Laboratory to develop nonlinear models for the melting points of the alkanes having between 10 and 20 carbon atoms and a single methyl group. Experimental melting point data are available for 69 of the 80 alkanes in this selected group [11,12]. The melting points of these single methyl group alkanes are given in Table 6. All data are given in Kelvin (K).

The nonlinear models were constructed using the single methyl group alkanes presented with the exception of small group of data that were withheld to test the predictive ability of the models. This predictive set was chosen using a random number generator for each of the models we developed. We then used the models to predict the melting points of the 11 alkanes for which no experimental melting point data existed.

Model 4 was developed using 62 of the 69 single methyl alkanes with the remaining seven withheld to serve as a predictive dataset. The indices used for this model include the Wiener ${}^{i}P$ numbers ($1 \leq i \leq 4$) and the mean Wiener index.

Table 6 Melting Point Data

Alkane	T_{fus}/K	Source	Data Points	Average	Std. Dev.
Decane	243.7	Beilstein	13[a]	243.4	0.7
2-Methylnonane	198.6	Beilstein	4	198.6	0.2
3-Methylnonane	186.6	NIST	1	NA	NA
4-Methylnonane	174.5	Beilstein	1	NA	NA
5-Methylnonane	186.2	Beilstein	3	186.2	0.6
Undecane	247.4	Beilstein	7	247.4	0.4
2-Methyldecane	224.0	Beilstein	2	224.0	0.5
3-Methyldecane	189.9	Beilstein	2	189.9	5.3
4-Methyldecane	180.3	Beilstein	1	NA	NA
5-Methyldecane	183.2	Beilstein	1	NA	NA
Dodecane	263.6	Beilstein	15[a]	263.4	0.6
2-Methylundecane	227.4	Beilstein	2	227.4	0.3
3-Methylundecane	216.9	Beilstein	2	216.8	2.3
4-Methylundecane	205.3	Beilstein	2	205.3	1.6
5-Methylundecane	204.3	Beilstein	2	204.3	1.6
6-Methylundecane	209.3	Beilstein	1	NA	NA
Tridecane	267.6	Beilstein	9	267.6	0.3
2-Methyldodecane	246.4	Beilstein	2	246.4	1.1
3-Methyldodecane	213.0	Beilstein	2	213.0	1.0
4-Methyldodecane	222.0	Beilstein	2	222.0	2.3
5-Methyldodecane	203.9	Beilstein	2	203.9	0.4
6-Methyldodecane	215.0	Beilstein	1	NA	NA
Tetradecane	279.0	Beilstein	13	278.9	0.2
2-Methyltridecane	247.0	Beilstein	2	247.0	0.4
3-Methyltridecane	232.4	Beilstein	2	232.4	4.6
4-Methyltridecane	228.5	Beilstein	2	228.5	1.6
5-Methyltridecane	222.1	Beilstein	2	222.1	3.6
6-Methyltridecane	222.6	Beilstein	1	NA	NA
7-Methyltridecane	234.5	Beilstein	2	234.5	2.1
Pentadecane	282.9	Beilstein	11[a]	282.9	0.2
2-Methyltetradecane	264.6	Beilstein	2[a]	264.6	0.4
3-Methyltetradecane	237.2	Beilstein	1	NA	NA
4-Methyltetradecane	241.5	Beilstein	2	241.5	1.7
5-Methyltetradecane	226.5	Beilstein	2	226.5	3.7
6-Methyltetradecane	237.8	Beilstein	1	NA	NA
7-Methyltetradecane	225.2	Beilstein	1	NA	NA
Hexadecane	291.7	Beilstein	30[b]	291.4	0.9
2-Methylpentadecane	263.3	Beilstein	3	263.3	0.9
3-Methylpentadecane	251.8	Beilstein	4	251.8	2.7
4-Methylpentadecane	250.9	Beilstein	3	250.9	1.8
5-Methylpentadecane	242.1	Beilstein	3	242.1	2.3

(Continued)

Table 6 Melting Point Data–Cont'd

Alkane	T_{fus}/K	Source	Data Points	Average	Std. Dev.
6-Methylpentadecane	241.8	Beilstein	2	241.8	1.1
7-Methylpentadecane	243.1	Beilstein	2	243.1	1.0
8-Methylpentadecane	245.8	Beilstein	2[a]	245.7	2.7
Heptadecane	295.2	Beilstein	18[a]	294.9	0.9
2-Methylhexadecane	278.1	Beilstein	1	NA	NA
3-Methylhexadecane	258.5	Beilstein	1	NA	NA
4-Methylhexadecane	258.4	Beilstein	1	NA	NA
5-Methylhexadecane	241.3	Beilstein	1	NA	NA
Octadecane	301.7	Beilstein	32[a]	301.3	0.8
2-Methylheptadecane	278.5	Beilstein	2	278.5	0.5
3-Methylheptadecane	267.0	Beilstein	1	NA	NA
4-Methylheptadecane	264.9	Beilstein	1	NA	NA
5-Methylheptadecane	253.3	Beilstein	1	NA	NA
9-Methylheptadecane	265.0	NIST	1	NA	NA
Nonadecane	305.1	Beilstein	23	304.9	0.7
2-Methyloctadecane	284.4	Beilstein	2	285.2	1.2
3-Methyloctadecane	274.5	Beilstein	2	274.5	1.2
4-Methyloctadecane	270.4	Beilstein	2	270.4	2.5
5-Methyloctadecane	257.4	Beilstein	2	258.5	1.6
6-Methyloctadecane	269.2	Beilstein	1	NA	NA
7-Methyloctadecane	256.9	Beilstein	1	NA	NA
8-Methyloctadecane	263.2	Beilstein	1	NA	NA
9-Methyloctadecane	256.7	Beilstein	1	NA	NA
Eicosane	311.0	Beilstein	40	309.8	1
2-Methylnonadecane	291.9	Beilstein	3	291.9	1.1
3-Methylnonadecane	280.8	Beilstein	1	NA	NA
4-Methylnonadecane	271.9	Beilstein	1	NA	NA
5-Methylnonadecane	266.2	Beilstein	1	NA	NA

[a] *One additional bad data point omitted in average and standard deviation calculations.*
[b] *Two additional bad data points omitted in average and standard deviation calculations.*

The coefficient of determination for Model 4 is $R^2 = 0.97659$ and the standard deviation is 5.1 K.

Model 4.

$$f_{\text{Model 4}}({}^1P, {}^2P, {}^3P, {}^4P, \overline{W}) = 1625 + 2971({}^1P)^{0.1428} - 4460({}^2P)^{0.05842} - 236.7({}^3P)^{0.3554} - 0.01762({}^4P)^{2.727} + 16.11\overline{W}^{1.047}$$

Fig. 20 shows a plot of the experimental melting points versus the model melting points determined by Model 4. The straight line represents an exact prediction.

Table 7 gives the number and percentage of alkanes whose absolute melting point deviations determined by Model 4 are within the specified ranges. Table 8 illustrates the predictive ability of Model 4 for the given set of seven withheld alkanes.

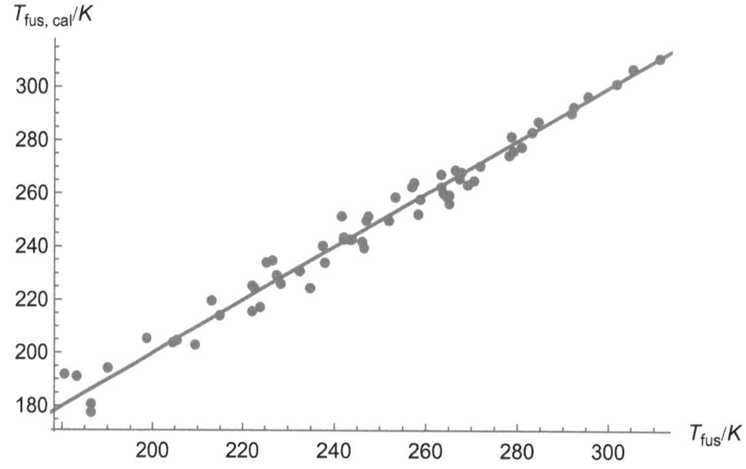

FIG. 20

Experimental MP versus model MP for Model 4.

Table 7 Alkanes and Their Corresponding Absolute MP Deviations for Model 4

$\Delta T/K$	No. of Alkanes	Alkanes (%)
0–1	11	17.7
1–2	9	14.5
2–4	18	29.0
4–6	6	9.7
6–9	14	22.6
>9	4	6.5

Table 8 Predictive Ability of Model 4

Alkane	T_{fus}/K	$T_{fus,cal}/K$	$\Delta T/K$
4-Methylnonane	174.5	178.0	3.5
3-Methylundecane	216.9	207.6	9.3
5-Methyldodecane	203.9	214.9	11.1
4-Methyltetradecane	241.5	236.1	5.4
4-Methylpentadecane	250.9	244.7	6.2
3-Methyloctadecane	274.5	272.1	2.4
9-Methyloctadecane	256.7	262.1	5.4

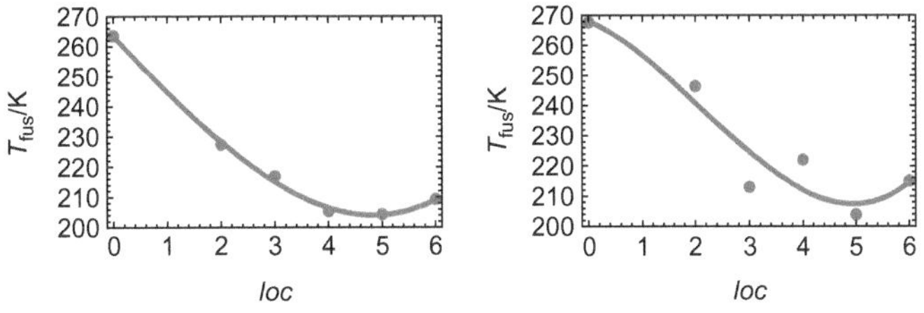

FIG. 21

(A) MP of the single branch dodecanes. (B) MP of the single branch tridecanes.

Two of the alkanes have a high melting point deviation which made us examine the data more closely. Upon analyzing graphs of the melting points of the single methyl alkanes versus their methyl locant indices, we found that the melting point of alkanes having an odd number of carbons oscillated along the cubic regression as the methyl locant changed. In contrast, the alkanes having an even number of carbons closely followed the cubic regression. Fig. 21 shows this phenomenon with the graphs of the melting points of the single branch dodecanes and single branch tridecanes versus their respective methyl locant indices. This difference in the behavior of the melting points based on the parity of the number of carbons is a well-known fact dating back to the 19th century [17].

Based on these findings, the melting point data were split based on the parity of the number of carbon atoms. New predictive sets using a random number generator were chosen for each model. For the model based on the single methyl alkanes having an even number of carbons, 31 of the 37 alkanes were used to develop the model while the remaining 6 were withheld to serve as a predictive data. Since this data does not oscillate, the same indices used in Model 4 were used to develop Model 5. The coefficient of determination for Model 5 is $R^2 = 0.988475$ and the standard deviation is 3.8 K.

Model 5.

$$f_{\text{Model 5}}(^1P, ^2P, ^3P, ^4P, \overline{W}) = -552 + 1731(^1P)^{0.2345} - 1090.5(^2P)^{0.2120} - 4.866(^3P)^{1.245} - 254.0(^4P)^{0.1865} + 0.006393\overline{W}^{4.092}$$

Table 9 gives the number and percentage of alkanes whose absolute melting point deviations determined by Model 5 are within the specified ranges. Table 10 illustrates the predictive ability of Model 5 for the given set of six withheld alkanes. The standard deviation of Model 5 was notably improved over that of Model 4 since four of the six predicted melting points were within 2 K of the experimental melting points.

As seen in Fig. 21, the melting points of single methyl alkanes having an odd number of alkanes oscillate as the methyl locant index varies. Therefore, we used both the number of carbon atoms and the methyl locant index with the cosine function

Table 9 Alkanes and Their Corresponding Absolute MP Deviations for Model 5

$\Delta T/K$	No. of Alkanes	Alkanes (%)
0–1	7	22.6
1–2	7	22.6
2–4	8	25.8
4–6	6	19.4
6–9	3	9.7

Table 10 Predictive Ability of Model 5

Alkane	T_{fus}/K	$T_{fus,cal}/K$	$\Delta T/K$
Dodecane	263.6	261.9	1.7
2-Methylundecane	227.4	225.3	2.0
4-Methylundecane	205.3	206.1	0.8
2-Methyltridecane	247.0	246.0	0.9
4-Methylpentadecane	250.9	244.8	6.1
Eicosane	311.0	315.1	4.0

as one of the terms in Model 6. Model 6 is fit to 27 of the 32 single methyl alkanes using the Wiener ^iP numbers ($1 \leq i \leq 4$), the mean Wiener index, the number of carbon atoms (n), and the methyl locant index. Five data points were randomly chosen to serve as predictive data and were therefore not used to form the model. Model 6 has a coefficient of determination of $R^2 = 0.99241$ and the standard deviation is 3.0 K.

Model 6.

$$\begin{aligned} f_{\text{Model 6}}(^1P, ^2P, ^3P, ^4P, \overline{W}, n, loc) = & -6695 + 2159(^1P)^{0.16275} - 240.5(^2P)^{0.3518} \\ & + 1076(^3P)^{-0.4243} - 2.537(^4P)^{1.407} \\ & + 3750\overline{W}^{0.03078} + 1.203n^{0.5095} \cos(loc\,\pi) \end{aligned}$$

Table 11 gives the number and percentage of alkanes whose absolute melting point deviations determined by Model 6 are within the specified ranges. Table 12 illustrates the predictive ability of Model 6 for the given set of five withheld alkanes.

Table 11 Alkanes and Their Corresponding Absolute MP Deviations for Model 6

$\Delta T/K$	No. of Alkanes	Alkanes (%)
0–1	14	51.9
1–2	4	14.8
2–4	6	22.2
4–6	2	7.4
6–9	1	3.7

Table 12 Predictive Ability of Model 6

Alkane	T_{fus}/K	$T_{fus,cal}$/K	ΔT/K
3-Methyldodecane	213.0	217.9	4.9
4-Methyltetradecane	241.5	238.3	3.1
4-Methylhexadecane	258.4	256.1	2.2
Nonadecane	305.1	306.9	1.8
2-Methyloctadecane	284.4	291.9	7.5

Table 13 Eleven Predicted Melting Points

Alkane	$T_{fus,cal}$/K
6-Methylhexadecane	253.9
7-Methylhexadecane	243.1
8-Methylhexadecane	253.0
6-Methylheptadecane	258.8
7-Methylheptadecane	258.6
8-Methylheptadecanee	258.4
6-Methylnonadecane	271.4
7-Methylnonadecane	271.1
8-Methylnonadecane	270.8
9-Methylnonadecane	270.7
10-Methylnonadecane	270.7

Given the improved standard deviation of both Models 5 and 6, these models were used to predict the melting points of the 11 single methyl alkanes for which no experimental data exists in the set of 80 single methyl alkanes having 10–20 carbon atoms. Model 5 was used to predict the melting point if the alkane had an even number of carbon atoms and Model 6 was used for the alkanes having an odd number of carbons. Table 13 gives the predicted melting points of these 11 single methyl alkanes. Fig. 22 gives the plot of the predicted melting points of the octadecanes versus their methyl locant indices. Existing melting point data for the octadecanes are also included. Fig. 23 shows the predicted melting points of the octadecanes with available existing experimental data and Fig. 24 shows the predicted melting points of the eicosanes with existing experimental data.

4 CONCLUSIONS

Chemical graph theory is a productive application of mathematics to the physical properties of chemical molecules. Chemists have used graph theory to model the molecules they studied long before graph theory was established as an area of mathematics. Chemists played a major role in the growth of chemical graph

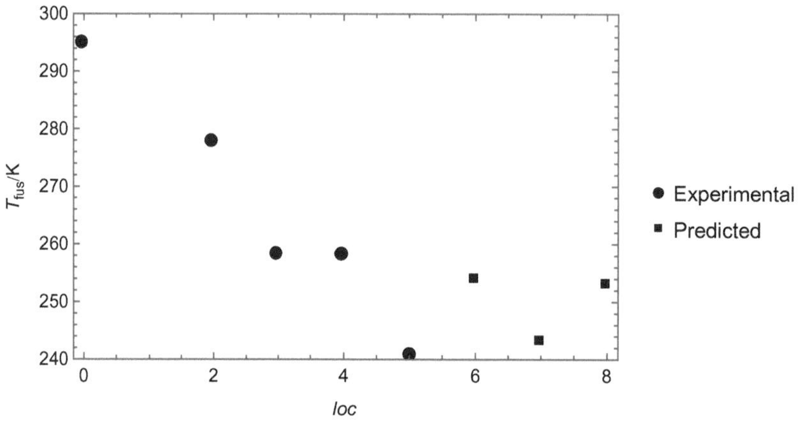

FIG. 22

Predicted melting points of the missing heptadecane methyl substituted alkanes.

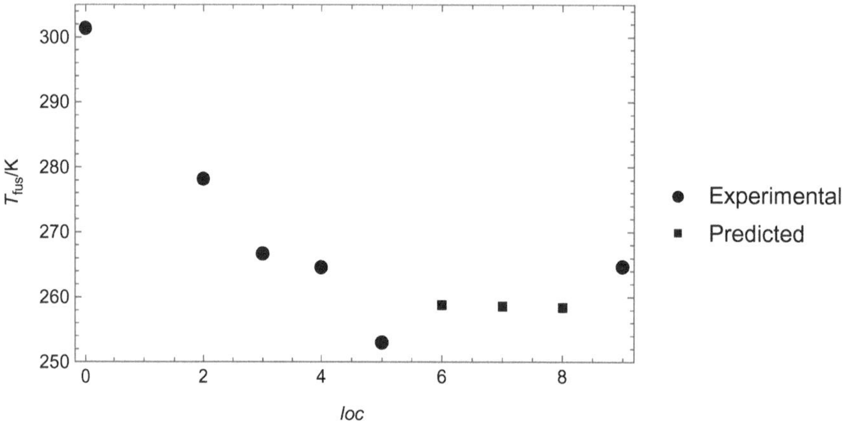

FIG. 23

Predicted melting points of the missing octadecane methyl substituted alkanes.

theory as they realized that mathematics was more than just a tool for crunching numerical data.

Using several topological indices to construct models of the physical properties of alkanes has proven successful as seen in this chapter. Model 3 does extremely well at predicting the boiling points of the additional 52 alkanes as seen in Table 5. The boiling point deviations for 24 of the 52 alkanes were less than 2°C. Given this is often an acceptable error for experimental measurement of physical properties, this model could be used to predict the boiling points of alkanes for which no experimental boiling point data exist. Although modeling the melting points of

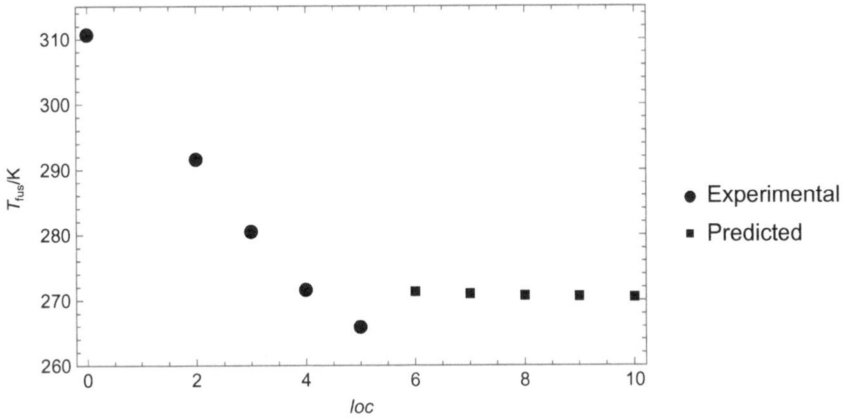

FIG. 24

Predicted melting points of the missing eicosane methyl substituted alkanes.

alkanes proved to be more difficult, splitting the data by the parity of the number of carbons resulted in greatly improved models. Model 4, which attempted to model the melting points of all the methyl substituted alkanes having between 10 and 20 carbon atoms, had 11.1 K as its worst predictive error. In contrast, Model 5, which modeled the methyl-substituted alkanes having an even number of carbons, had 6.1 K as its worst predictive error. The worst predictive error for Model 6, which modeled the alkanes having an odd number of carbons, was 7.5 K. Using models similar to those presented in this chapter would allow chemists to predict physical properties of molecules which may never be measurable in a traditional lab setting.

REFERENCES

[1] Todeschini R, Consonni V. Handbook of molecular descriptors. Weinheim: Wiley-VCH; 2000.

[2] Hosoya H. Topological index. A newly proposed quantity characterizing the topological nature of structural isomers of saturated hydrocarbons. Bull Chem Soc Jpn 1971;44(9): 2332–9.

[3] Gutman I, Polansky OE. Mathematical concepts in organic chemistry. Berlin: Springer-Verlag; 1986.

[4] Burch KJ, Wakefield DK, Whitehead EGJ. Boiling point models of alkanes. Match 2003; 47:25–52.

[5] Wiener H. Structural determination of paraffin boiling points. J Am Chem Soc 1947; 1(69):17–20.

[6] Rouvray D. The rich legacy of half a century of the Wiener index. In: Topology in chemistry: discrete mathematics of molecules. Horwood Publishing; 2002. p. 295–306.

[7] Gutman I, Miljković O. Molecules with smallest connectivity indices. Match 2000;41: 57–70.

[8] Randić M. On characterization of molecular branching. J Am Chem Soc 1975;97: 6609–15.
[9] Balaban AT. Highly discriminating distance-based topological index. Chem Phys Lett 1982;89(5):399–404.
[10] Plavšić D, Nikolić S, Trinajstić N, Mihalić Z. On the Harary index for the characterization of chemical graphs. J Math Chem 1993;12:235–50.
[11] Database BC; 2001. Available from: http://www.beilstein.com/products/xfire/.
[12] Frenkel MD. Thermodynamics Research Center, NIST Boulder Laboratories. In: Linstrom PJ, Mallard WG, editors. "Thermodynamics Source Database" in NIST Chemistry WebBook, NIST Standard Reference Database Number 69. Gaithersburg MD: National Institute of Standards and Technology; 2003. p. 20899. Available from: http://webbook.nist.gov [Accessed October 2001].
[13] Balaban AT. A personal view about topological indices for QSAR/QSPR. In: QSPR/QSAR studies by molecular descriptors. Huntington, NY: Nova Science Publishers; 2001. p. 1–30.
[14] Espinosa G, Yaffe D, Cohen Y, Arenas A, Giralt F. Neural network based quantitative structural property relations (QSPRs) for predicting boiling points of aliphatic hydrocarbons. J Chem Inf Comput Sci 2000;40:859–79.
[15] Fletcher R, Leyffer S. Nonlinear programming without a penalty function. Math Program 2002;91(2):239–69.
[16] Burch KJ, Whitehead EGJ. Melting point models of alkanes. J Chem Eng Data 2004;49: 858–63.
[17] Baeyer A. Ueber Regelmässigkeiten im Schmelzpunkt homologer Verbindungen. Ber Chem Ges 1877;10:1286.
[18] Mihalić Z, Trinajstić N. A graph-theoretical approach to structure-property relationships. J Chem Educ 1992;69(9):701–12.

CHAPTER

Singularity analysis in quantum chemistry

9

David Z. Goodson

Department of Chemistry and Biochemistry, University of Massachusetts Dartmouth, North Dartmouth, MA, United States

1 MATHEMATICAL BACKGROUND

Complex analysis is a branch of mathematics with widespread applications in mathematical physics [1,2]. One of its core strategies is to develop approximate descriptions of unknown functions by allowing the variable to range over the complex numbers, and then analyzing the behavior in the neighborhoods of special points in the complex plane, called *singularities*, where the functions become poorly defined. This chapter describes various applications of singularity analysis to the Schrödinger equation. The function in question is the energy of an atom or molecule and the variable is some parameter that affects the energy, such as a bond distance, or a nuclear charge, or an artificial parameter in a perturbation theory.

Let us begin with a simple example. The Schrödinger equation is typically solved within a finite basis-set approximation, with the wavefunction Ψ expressed as a linear combination

$$\Psi = \sum_{j=1}^{N} a_j \psi_j, \tag{1}$$

in terms of a set of N linearly independent functions ψ_j. This is conveniently formulated as a matrix eigenvalue equation,

$$\mathbf{H}\boldsymbol{a} = E\boldsymbol{a}, \tag{2}$$

where $\boldsymbol{a}$ is an N-dimensional column vector of the unknown coefficients a_j and $\mathbf{H}$ is an $N \times N$ matrix of the scalar values $H_{jk} = \langle\psi_j|H|\psi_k\rangle$ constructed from the Hamiltonian operator H for each of the N possible values of j and k. N, the size of the basis set, is referred to as the basis set's *dimension*. The eigenvalues E approach

Mathematical Physics in Theoretical Chemistry. https://doi.org/10.1016/B978-0-12-813651-5.00009-7

the true eigenvalues as N approaches infinity. We are interested here in how the eigenvalues depend on some parameter λ that appears explicitly in the Hamiltonian.

Eq. (2) is usually solved with some numerical approximation method, but it is informative to consider a simple case that can be solved analytically. Consider a Hamiltonian that can be written in the form

$$H(\lambda) = h + \lambda v, \tag{3}$$

where the operators h and v are independent of λ, with a basis of dimension $N = 2$. Let us consider in particular the case with, for $j \neq k$, matrix elements $h_{jk} = 0$ and $v_{jk} = v_{kj} = w$,

$$\left[\begin{pmatrix} h_{11} & 0 \\ 0 & h_{22} \end{pmatrix} + \lambda \begin{pmatrix} v_{11} & w \\ w & v_{22} \end{pmatrix}\right] \begin{pmatrix} a_1 \\ a_2 \end{pmatrix} = E \begin{pmatrix} a_1 \\ a_2 \end{pmatrix}. \tag{4}$$

This is a system of two simultaneous linear equations in the coefficients a_1 and a_2 that can be solved to give two solutions for the energy E as a function of λ,

$$E(\lambda) = \frac{h_{11} + h_{22}}{2} + \lambda \frac{v_{11} + v_{22}}{2} \pm \frac{1}{2} \left(\delta_v^2 + 4w^2\right)^{1/2} \sqrt{(\lambda_s - \lambda)(\lambda_s^* - \lambda)}, \tag{5}$$

where $\delta_v = v_{11} - v_{22}$ and

$$\lambda_s = \frac{h_{22} - h_{11}}{\delta_v^2 + 4w^2}(\delta_v + 2wi), \quad \lambda_s^* = \frac{h_{22} - h_{11}}{\delta_v^2 + 4w^2}(\delta_v - 2wi), \tag{6}$$

with $i = \sqrt{-1}$. One can see from Eq. (4) that the parameter w is responsible for the coupling between the equations for a_1 and a_2.

Fig. 1 shows the distinctive behavior of the two solutions for $E(\lambda)$. The left panel shows $E(\lambda)$ for values of λ along the real axis of the complex λ-plane. There is an avoided crossing with the closest approach occurring at the value of λ equal to $\operatorname{Re} \lambda_s$. The spacing between the curves at closest approach is

$$\Delta E_{\min} = \frac{2|(h_{22} - h_{11})w|}{\sqrt{\delta_v^2 + 4w^2}}. \tag{7}$$

This decreases with decreasing w. For very small w the curves would come so close that they would appear, under low resolution, to pass through each other. Physical properties correspond to *real* numbers. Nevertheless, this example demonstrates that the existence of singular points at nonphysical parameter values, in the complex plane, can affect the behavior of the real, physical, solution.

The right panel of Fig. 1 shows the real parts of the two solutions for $E(\lambda)$ along a diagonal path through the complex plane passing through the origin and the singular point in the upper half-plane. Note the kinks, at $\lambda = \lambda_s$, where the two eigenvalues become equal. The slopes at that point become infinite. The general definition of a *singular point* (or *singularity*, for short) is a point at which the function and/or its first derivative becomes infinite or undefined.

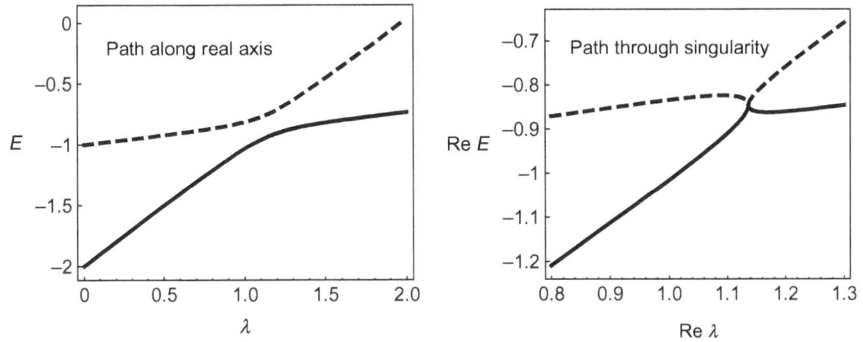

FIG. 1

Eigenvalues of Eq. (4), a 2 × 2 matrix eigenvalue equation with parameters $h_{11} = -2$, $h_{22} = -1$, $v_{11} = 1$, $v_{22} = 0.15$, and $w = 0.08$, which lead to branch points at $1.136 \pm 0.214i$. The eigenvalues are shown along two paths in the complex λ-plane; the *left panel* corresponds to a path along the real axis while the *right panel* corresponds to a path along the straight line that passes through $\lambda = 0$ and $\lambda = 1.136 + 0.214i$. The *solid curves* show the lower eigenvalue while the *dashed curves* show the upper eigenvalue.

The singularities of the matrix eigenvalue problem are an example of an important class of singularities called *branch points*. If one follows one of the two $E(\lambda)$ functions along a path in the complex λ-plane that completely encircles λ_s, then it will switch to the other function. Thus the two eigenvalue solutions, one with the "−" sign and the other with the "+" sign, can be thought of as two branches of a single function in the complex plane. To see how this happens in Eq. (5), consider a circular path of radius β centered at λ_s. This can be described in terms of an angle θ according to

$$\lambda = \lambda_s - \beta(\cos\theta + i\sin\theta) = \lambda_s - \beta e^{i\theta}. \tag{8}$$

(The second equality follows from Euler's relation, the definition of the exponential of a complex number [1].) As θ goes from 0 to 2π, the factor

$$(\lambda_s - \lambda)^{1/2} = \beta^{1/2} e^{i\theta/2} \tag{9}$$

in Eq. (5) goes from $\beta^{1/2}$ to $\beta^{1/2}e^{i\pi} = -\beta^{1/2}$, switching the square root's sign.

In this particular case, we have a *square-root* branch point; the singular behavior results from the square root in Eq. (5). The main focus of this chapter will be square-root branch points. The matrix eigenvalue equation for arbitrary finite dimension N was analyzed by Katz [3] in the context of nuclear physics. The exact analytical solution is not in general available at higher dimension, but Katz was able to prove that for nondegenerate eigenvalues $E_0(\lambda), E_1(\lambda), E_2(\lambda), \ldots, E_N(\lambda)$ for states of the same symmetry (i.e., the same values of angular momentum quantum numbers), there exists in general a complex-conjugate pair of square-root branch points λ_{jk} and λ_{jk}^* with nonzero imaginary parts that connect the eigenvalues $E_j(\lambda)$ and $E_k(\lambda)$ for any combination of j and k.

Branch points are not the only kind of singularity. For example, a function

$$f(\lambda) = \frac{c}{(\lambda_s - \lambda)^m}, \tag{10}$$

where m is a positive integer and c is a constant, is said to have a *pole* of order m at $\lambda = \lambda_s$. The function and its derivative are infinite at λ_s but there is only one branch.

Another kind of singularity is a *critical point*, a point at which the qualitative nature of the solution undergoes a sudden transition. These occur, for example, in thermodynamics when the system undergoes a phase transition. The singularities corresponding to phase transitions are branch points. Their explicit functional forms are usually unknown, but they are much more complicated than simple algebraic branch points. In contrast to the branch points of the matrix eigenvalue problem, the critical point lies *on* the real axis.

2 AVOIDED CROSSINGS OF MOLECULAR POTENTIAL ENERGY

2.1 BORN-OPPENHEIMER APPROXIMATION

The Schrödinger equation for a diatomic molecule can be formulated as a matrix eigenvalue equation as in Eq. (2), with λ corresponding to the distance x between the two atomic nuclei. Let N_e be the number of electrons. We will assume that the kinetic energy operator, in atomic units, can be expressed as

$$T = -\frac{1}{2}\sum_{j=1}^{N_e} \nabla_j^2, \tag{11}$$

where N_e is the number of electrons. This follows from the Born-Oppenheimer approximation, according to which the kinetic energy operators for nuclear motions are omitted because the nuclear masses are each much larger than the mass of an electron. The potential energy from electron-electron repulsion is

$$W = \sum_{j=1}^{N_e-1} \sum_{k=j+1}^{N_e} r_{jk}^{-1}, \tag{12}$$

where r_{jk} is the distance between electron j and electron k. The potential energy for nucleus-electron attraction for a diatomic molecule is

$$U(x) = -\sum_{j=1}^{N_e}\left(\frac{Z_1}{\rho_{1j}} + \frac{Z_2}{\rho_{2j}}\right), \tag{13}$$

where ρ_{1j} and ρ_{2j} are, respectively, the distance between electron j and nucleus 1 and the distance between electron j and nucleus 2. Z_1 and Z_2 are the atomic numbers of the nuclei (i.e., the nuclear charges in atomic units). The Hamiltonian, then, is

$$H(x) = T + U(x) + W, \tag{14}$$

and the dependence on x comes only from the expressions for $\rho_{1,j}$ and $\rho_{2,j}$, which depend on the relative positions of the nuclei.

We can write an electronic Schrödinger equation as

$$H(x)\Psi = \epsilon(x)\Psi. \tag{15}$$

The total energy of the molecule is obtained by adding to the electronic energy $\epsilon(x)$ the potential energy of internuclear repulsion,

$$E(x) = \epsilon(x) + Z_1 Z_2/x. \tag{16}$$

$E(x)$ can be treated as the potential energy for the dynamics of the nuclei.

Expressing Ψ in terms of some appropriate set of basis functions ψ_j according to Eq. (1) leads to a matrix eigenvalue equation for $\epsilon(x)$. If the basis set is constructed from Hartree-Fock atomic orbitals, then this is called the *full configuration interaction* (FCI) solution. This method reliably converges to the solution for ϵ with increasing basis-set size N, and what we obtain is a set of numerical values for $\epsilon(x)$ at an arbitrary selection of discrete values for internuclear separation x. The left panel of Fig. 2 shows benchmark FCI results for the hydrogen fluoride molecule for the

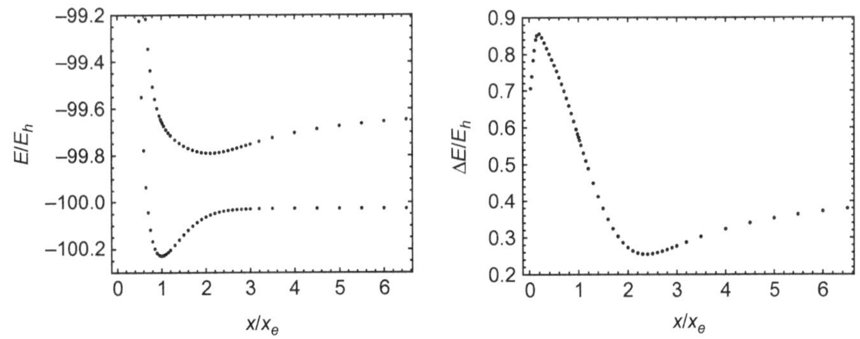

FIG. 2

Potential energy values for the hydrogen fluoride molecule, for the ground state and the first excited state of the same symmetry, from numerical solution of the matrix eigenvalue problem. (These results are from full configuration-interaction calculations within the cc-pVDZ basis set. See Ref. [4] for details.) The *left panel* shows the functions $E(x)$ and the *right panel* shows the difference between them. The internuclear distance is in units of the equilibrium bond length, $x_e = 0.92025$ Å and energy is in units of hartrees.

ground-state energy and for the first excited state of the same symmetry as the ground state, for discrete values of x over a wide range. The right panel of the figure shows the difference between the two eigenvalues. The closest approach is seen to occur at x/x_e between 2.3 and 2.4. It is only for the 2×2 matrix eigenvalues that closest approach will occur exactly at the real part of a branch point. However, we can expect that the real part of the branch point in this case will be somewhere in this vicinity.

2.2 INTERPOLATION WITH QUADRATIC APPROXIMANTS

Our goal now is use our knowledge of the nature of the singularity structure to construct from a discrete set of energy values a continuous analytical function that interpolates between them. Consider an arbitrary function $f(x)$ and suppose we have a dataset of K discrete points,

$$\{(x_1,f_1),(x_2,f_2),(x_3,f_3),\ldots,(x_K,f_K)\},$$

with $f_j = f(x_j)$. Let us define three polynomials,

$$P(x) = p_0 + p_1 x + p_2 x^2 + \cdots + p_{M_p} x^{M_p}, \tag{17}$$

$$Q(x) = 1 + q_1 x + q_2 x^2 + \cdots + q_{M_q} x^{M_q}, \tag{18}$$

$$R(x) = r_0 + r_1 x + r_2 x^2 + \cdots + r_{M_r} x^{M_r}, \tag{19}$$

such that

$$Q(x_j) f_j^2 - P(x_j) f_j + R(x_j) = 0 \tag{20}$$

for each point in the dataset. Dividing both sides of Eq. (20) by a constant has no effect on the solution. Therefore, one of the polynomial coefficients is arbitrary. It is traditional to account for this by setting $q_0 = 1$.

Eq. (20) is a set of K simultaneous equations that determine the values of the polynomial coefficients p_j, q_j, and r_j. We have $M_p + M_q + M_r + 2$ coefficients and K equations. Therefore, we must choose values M_p, M_q, M_r such that

$$M_p + M_q + M_r = K - 2. \tag{21}$$

If we solve Eq. (20) for arbitrary x_j we obtain

$$f_j = \frac{1}{2Q(x_j)}\left[P(x_j) \pm \sqrt{P(x_j)^2 - 4Q(x_j)R(x_j)}\right]. \tag{22}$$

This suggests that we model the function with

$$S(x) = \frac{1}{2Q(x)}\left[P(x) \pm \sqrt{P(x)^2 - 4Q(x)R(x)}\right]. \tag{23}$$

$S(x)$ is called a *quadratic approximant*.[1] It interpolates between the points in the dataset, because by construction $S(x_j) = f_j$, and, very importantly, it can model square-root branch points. $S(x)$ will generally have square-root branch points at x values in the complex plane that are roots of the discriminant polynomial; that is, x values for which

$$P(x)^2 - 4Q(x)R(x) = 0. \tag{24}$$

Quadratic approximants are a convenient tool for modeling functions that contain branch points. However, to take full advantage of their capabilities requires some thought. It is a good idea to incorporate into the approximant whatever we know in advance about the functional form of $\epsilon(x)$. Consider its behavior in the limit $x \to \infty$. This is the limit of dissociation into molecular fragments. For the hydrogen fluoride molecule this would be neutral isolated H and F atoms. Therefore, we want each branch of $S(x)$ to approach a different finite negative value. (Negative because the fragments are in bound electronic states.) For very large x we have

$$S(x) \sim \frac{1}{2q_{M_q}x^{M_q}}\left[p_{M_p}x^{M_p} \pm \left(p_{M_p}^2 x^{2M_p} - 4q_{M_q}r_{M_r}x^{M_q+M_r}\right)^{1/2}\right], \tag{25}$$

which will have the desired behavior only if both of the terms in the square brackets are proportional to x^{M_q} in the large-x limit. It follows that the only reasonable choice for the degrees of the three polynomials for this application is

$$M_p = M_q = M_r \equiv M. \tag{26}$$

Suppose that the sum of the energies of the dissociated fragments in the ground state is a known value $E_{d,0}$, and in an excited state is a known value $E_{d,1}$. (If we can calculate molecular energies, then it should be even easier to calculate the energies of the fragments, because each fragment has fewer electrons than does the original molecule.) Setting Eq. (25) equal to these values gives

$$p_M = (E_{d,0} + E_{d,1})q_M, \quad r_M = E_{d,0}E_{d,1}q_M. \tag{27}$$

A possible source of difficulty comes from the fact that $S(x)$ can have more branch points than needed. For the true $E(x)$, the branch point pair connecting the ground state with the excited state shown in Fig. 2, with real part $2.36\,x_e$, is the only one that strongly affects the behavior of the ground state in the regions of the potential that are of practical interest. There is a weak avoided crossing of the ground state with the next higher excited state at a much larger bond distance. There is also a complicated collection of closely spaced avoided crossings at internuclear distances less than $0.3\,x_e$, but because the potential energy is so high for such small internuclear

[1] Quadratic approximants were proposed in the 19th century but attracted little attention until they were rediscovered in the 1970s. Since then, a variety of applications has been found for them in different areas of physics. The idea of using them for molecular potential energies comes from Jordan [5].

distances, this region will not usually be of importance for chemical dynamics and will usually not be well enough represented in the dataset for the approximants to accurately model the corresponding branch points. Thus, at least in this case, we really only have one branch point pair concerning which the dataset provides much information. However, the number of branch points in the diagonal approximant is equal to the number of roots of the polynomial $P^2 - QR$. This is $2M$, which can be much larger than needed.

The approximants dispose of superfluous branch points in a few different ways. One way is to place them at distant points in the complex plane, far from the region of the real axis being modeled. These cause no problem. Another way is to have pairs of nearly coincident branch points. Suppose a point x_s is a double root. Then the square-root term in the approximant will be proportional to $\sqrt{(x - x_s)^2} = x - x_s$, which is in fact *not* singular. In practice the roots are not exactly coincident, which means that in their immediate vicinity $S(x)$ will have spurious singular behavior. If they are located near the real axis in the physically interesting region, this can be problematic. However, their positions typically shift with increasing M, which distinguishes them from physical branch points.

There is another kind of spurious singularity. Consider a point x_s on the positive real axis at which the denominator polynomial goes to zero; that is, $Q(x_s) = 0$. The lower branch of the approximant (with the minus sign) is not obviously affected. As x approaches x_s, $S(x)$ smoothly approaches $R(x_s)/P(x_s)$ with no singular behavior. However, for the upper branch (with the plus sign) in the neighborhood of x_s we have

$$S(x) \sim \frac{P(x)}{Q(x)} \sim \frac{c}{x_s - x}, \quad x \to x_s, \tag{28}$$

where c is a constant, which is a first-order pole. This singularity is completely spurious but it has a huge effect on the upper branch.

An easy way to prevent a pole on the positive real axis is to use

$$Q(x) = 1 + q_M x^M \tag{29}$$

as the denominator polynomial and then treat q_M as an arbitrary parameter. If q_M is chosen to be a positive real number, then $Q(x)$ will have no zeros on the positive real axis. Furthermore, one can take advantage of q_M as a free parameter by choosing it so as to optimize the approximant's branch point structure, moving spurious branch points away from the region of interest. The positions of spurious branch points will shift with changing q_M while physical branch points will remain stable.

We have $2M$ parameters still to be determined: $p_0, p_1, p_2, \ldots, p_{M-1}$ and $r_0, r_1, r_2, \ldots, r_{M-1}$. For a given M value we will need a dataset of $K = 2M$ points on the potential energy curve. The parameters are obtained from K simultaneous algebraic equations given by Eq. (20). The approximant $S(x)$ will give an interpolation between these points and, with its second branch, a prediction for the excited state.

Suppose that at some or all of the x_j the dataset includes the energy of the excited state as well as that of the ground state. It is easy to include this in the parameterization. If we add together the two branches of $S(x)$, given by Eq. (23), we get an approximant for the sum of the energies. It follows that

$$P(x_j) - (E_{j,0} + E_{j,1})Q(x_j) = 0. \tag{30}$$

These are additional linear equations, which we solve along with those for the ground-state energy from Eq. (20).

Fig. 3 shows that the quadratic approximants can give an accurate analytical expression for modeling the potential energy over a wide range of x. The upper panel shows a result from parameterization to a dataset including only ground-state energies. The dashed curves show the upper branch of the approximant for two different choices of q_M. The lower branch is virtually unchanged over a wide range of q_M, but the upper branch is strongly affected. However, the location of the dominant branch point is quite stable. Both cases in the figure have branch points at $2.0 \pm 0.9i$. The fact that the upper curve poorly models the excited state is due to the fact that $S(x)$ has only two branches, while the solution to the $N \times N$ matrix formulation of the Schrödinger equation has N branches. The upper branch of $S(x)$ models different excited states in different regions of x. In particular, at small x, where avoided crossings occur between the ground state and higher excited states, the upper branch will switch to a higher eigenstate. A relatively smaller value of q_M has the effect of giving greater weight to data points at smaller x. The lower panel of Fig. 3 demonstrates that the upper branch can be constrained to the first excited state by including some excited-state energies in the dataset.

The ability of quadratic approximants to describe the behavior in the vicinity of an avoided crossing nicely complements various computationally efficient but less accurate methods for solving the Schrödinger equation. For molecules with more than 10 or so electrons, the matrix eigenvalue problem with an appropriately large basis set becomes computationally intractable. More efficient computational methods, however, involve further approximations and these approximations can be particularly inaccurate near an avoided crossing.

The approximations typically are based on the assumption that the true wavefunction is described reasonably well by a single Hartree-Fock orbital configuration. This assumption fails as the bond is stretched into the avoided crossing region, where the wavefunction is more accurately described by a linear combination of two different orbitals, corresponding to the interacting eigenstates. Another problem arises from the treatment of spin. "Unrestricted" methods are based on an approximate wavefunction that is not an eigenfunction of the spin operator. In practice this fact does not significantly affect the accuracy of the energy if x is close to the equilibrium bond distance or if x is very large, approaching dissociation. However, at intermediate bond distances the approximate wavefunction will be a linear combination of orbitals with different spin eigenvalues and this will adversely affect the accuracy. It is possible to formulate approximate methods so as to restrict the wavefunction to the correct

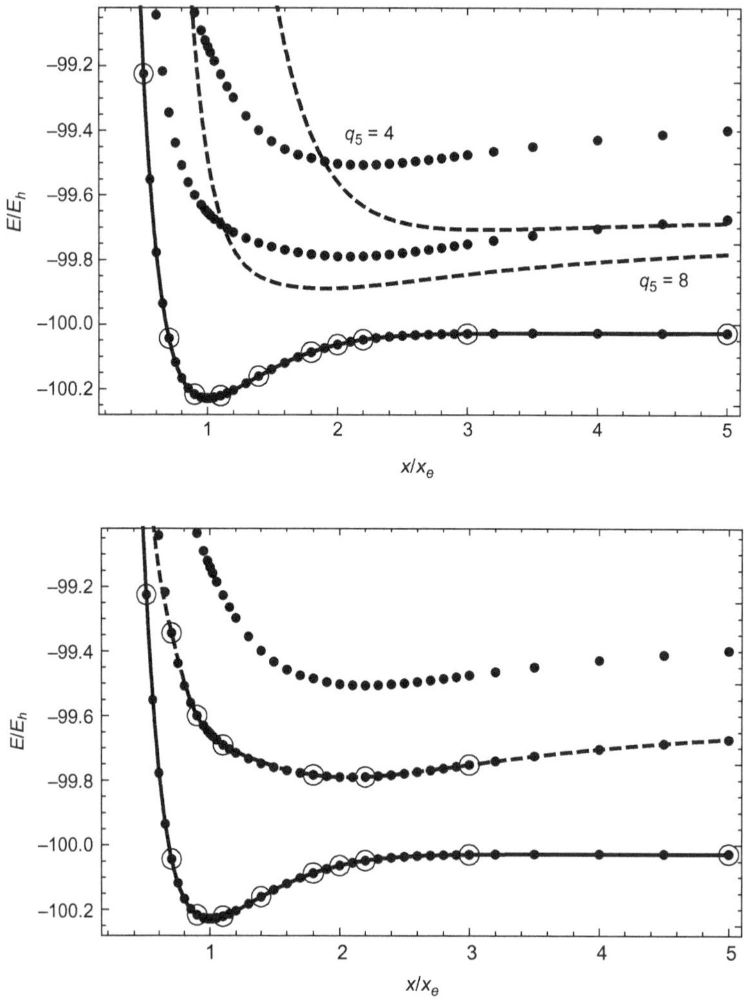

FIG. 3

Quadratic approximants for the potential energy curves of the ground state and the first excited state of the same symmetry for the hydrogen fluoride molecule, using two different datasets. The *dots* show FCI energies for the ground state and the first two excited states of the same symmetry as the ground state. *Circles* mark the points included in the dataset. The *solid curves* show the lower branch of the quadratic approximant while the *dashed curves* show the upper branch. The *upper panel* shows upper branches calculated with two different values of the parameter q_M.

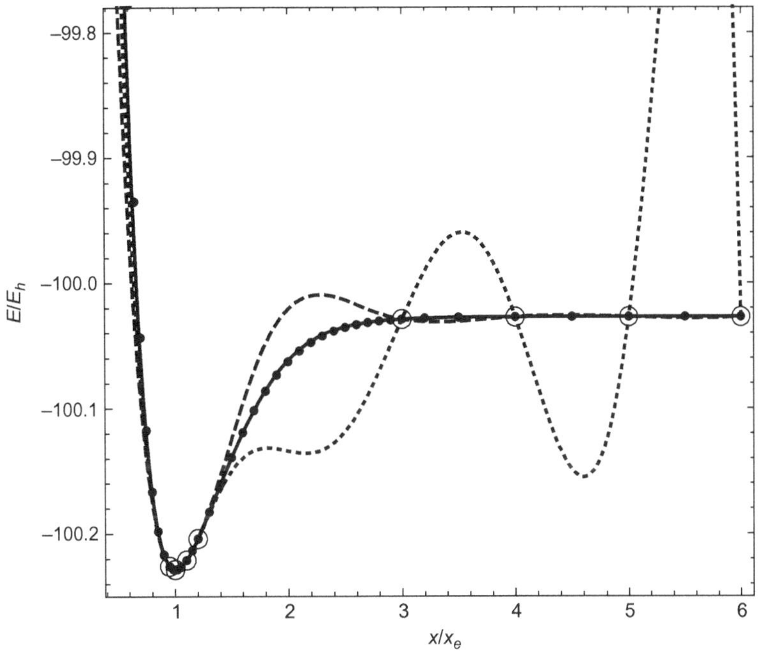

FIG. 4

Comparison of interpolations through the avoided crossing region using a dataset of points outside that region, for the ground-state potential energy curve for hydrogen fluoride. *Dots* indicate correct values. *Circled dots* indicate points used in the dataset for interpolation. The *solid curve* corresponds to the quadratic approximant, $S(x)$. The *dashed curve* shows the cubic spline interpolation while the *dotted curve* shows the interpolation from a seventh-degree polynomial.

spin eigenvalue. Spin restricted formulations of coupled-cluster methods have been found to give better accuracy than the unrestricted version for mildly stretched bonds [6]. Nevertheless, with further stretching restricted methods dissociate incorrectly. For the hydrogen fluoride molecule, for example, restricted methods dissociate into a fluoride anion and a hydrogen cation, rather than the correct limit of two neutral atoms.

Fig. 4 shows the result of interpolation through the avoided crossing region using a dataset that includes only points in the vicinity of the equilibrium bond distance and points at very large internuclear separation. This simulates a situation in which one is using an approximate method that can only give valid results in these two regions. The quadratic approximant predicts the intermediate points virtually exactly within the resolution of this plot. By contrast, the polynomial that interpolates between the eight data points gives a reasonable result only near the equilibrium bond length, and elsewhere it is quite useless. The cubic spline interpolation does fine in the parameterized regions but in the vicinity of the avoided crossing it is qualitatively incorrect. The polynomial and spline interpolations do not use any information other than the points in the dataset. By contrast, the quadratic approximant, by modeling the expected square-root branch point structure, uses in addition to the dataset, the known mathematical physics of the interaction between the ground state and the first excited state, which influences the shape of the potential energy curve.

3 CRITICAL POINTS IN ELECTRONIC STRUCTURE

3.1 IONIZATION AS A CRITICAL PHENOMENON

The Hamiltonian for an atom with atomic number Z and with N_e electrons can be written

$$H = T + ZU + W, \tag{31}$$

in terms of the kinetic energy operator T (Eq. 11), the potential energy for interelectron repulsion W (Eq. 12), and the potential energy ZU for nucleus-electron attraction, with

$$U = \sum_{j=1}^{N_e} r_j^{-1}, \tag{32}$$

where r_j is the distance between the nucleus and electron j. It is convenient to scale the units of distance by a factor of Z, substituting r_j/Z for r_j, r_{jk}/Z for r_{jk}, and $Z^2\nabla_j^2$ for ∇_j^2. Then the Schrödinger equation can be written

$$H(\lambda)\Psi = E(\lambda)\Psi, \tag{33}$$

where

$$H(\lambda) = T + U + \lambda W \tag{34}$$

and

$$\lambda = 1/Z. \tag{35}$$

The units of energy are now scaled by a factor of Z^2. The unscaled energy can be obtained at the end of the analysis by multiplying E by Z^2. The λ dependence in $H(\lambda)$ is now confined to the interelectron repulsion term λW.

Let us treat λ as a continuous variable. According to Eq. (34), the limit of $\lambda = 0$ represents the turning off of interelectron repulsion, giving a system that is more stable than the physical atom, while increasing λ has the effect of increasing the magnitude of interelectron repulsion. Clearly, there will be some value of λ on the positive real axis beyond which the effect of interelectron repulsion will overwhelm the potential energy of electron-nucleus attraction, and the system will dissociate, releasing an electron. This abrupt change in the character of the system in response to a continuous change in a parameter is analogous to a phase transition in equilibrium thermodynamics that occurs in response to a continuous change in a state variable, for example, a liquid dissociating into a gas in response to increasing temperature. The parameter value corresponding to the transition is called a *critical* value. Therefore, we will refer to the highest positive real value of λ that can support a fully bound state of the atom as λ_c, the critical λ.

The mathematical manifestation of this phenomenon will be a singularity in $E(\lambda)$ at λ_c, but in contrast to the branch points of the matrix eigenvalue problem, which must have nonzero imaginary parts, the critical point is a singularity *on* the real axis. A sophisticated and laborious analysis [7] of the simplest case, the two-electron atom, showed that $E(\lambda)$ has a branch point at $\lambda_c = 1.09766$ (consistent with the fact that the hydride ion, $\lambda = 1$, is stable in the gas phase) but found that the form of the singularity is more complicated than an algebraic function. The exact functional form of the singularity is not known. When approached from the direction of $\lambda < \lambda_c$ on the real axis, $E(\lambda) - E(\lambda_c)$ approaches 0 in proportion to $(\lambda_c - \lambda)$, but at λ_c the derivative becomes undefined and a path around λ_c causes the function to switch to a different branch.

The Schrödinger equation is usually solved within a basis-set approximation for the wavefunction. This converts Eq. (33) into a matrix eigenvalue equation, which, according to Katz's theorem, can only have square-root branch points, and none of those branch points can lie on the real axis. For an N-dimensional basis set the eigenvalue, will necessarily be nonsingular at λ_c, where we expect the true solution for the energy to be singular.

One might therefore presume that $E_N(\lambda)$ will be useless for studying ionization. In fact, this is not the case. The following section describes a method that uses the expected failure of the matrix formulation as a means to characterize the critical singularity.

3.2 FINITE-SIZE SCALING TO CALCULATE CRITICAL PARAMETERS

In principle, one might determine the value of λ_c by computing $E(\lambda)$ at different λ values and then identifying the critical point as the value of λ at which the energy of the neutral atom becomes equal to that of the +1 cation. This is difficult in practice. The true function $E(\lambda)$ is singular at the critical point while the matrix eigenvalue solution for E is nonsingular at that point. At λ values far from λ_c, the influence of the singularity will be small and a relatively small basis set will give high accuracy, but as λ approaches λ_c the size of the basis set will have to be increasingly large to give the same level of accuracy.

While this slowing of the basis-set convergence might be considered a problem, a study of the rate of convergence can be used to advantage to determine the value λ_c indirectly, as the value of λ where the convergence of the energy with increasing basis size begins to significantly slow. This is the rationale of the method of *finite size scaling*, an approach developed in the context of statistical thermodynamics to characterize critical points in thermodynamical state functions [8].

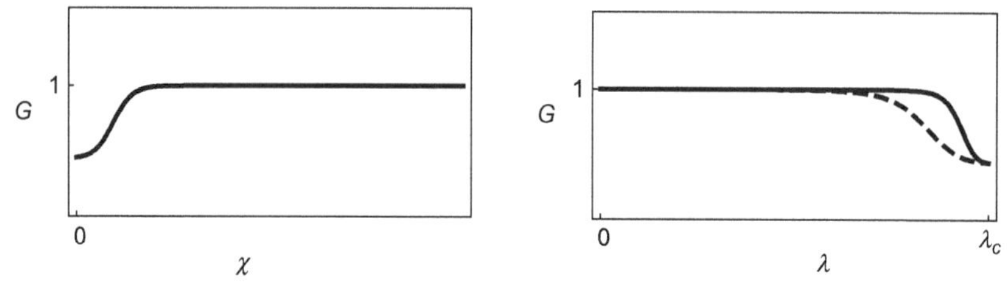

FIG. 5

Expected qualitative behavior of the correction factor $G(\chi)$ of Eq. (36) as a function of χ, and as a function of λ as implied by the scaling postulate, Eq. (37). The *dashed curve* shows the expected behavior for a relatively smaller value of N.

Let us assume that the basis functions do not themselves depend on λ. Then the basic idea is as follows: As the number N of basis functions increases, the neighborhood of λ_c in which the true $E(\lambda)$ is poorly described by the matrix eigenvalue $E_N(\lambda)$ shrinks. Let us introduce a multiplicative correction factor G to describe this situation, such that

$$E - E_c = (E_N - E_c)G, \tag{36}$$

where E_c is the true energy at the critical point, $E(\lambda_c)$. Our expectations for the behavior of G are as follows:

(i) G will be about equal to 1 except in the neighborhood of λ_c.
(ii) The neighborhood in which G significantly deviates from 1 will shrink with increasing N.

The method of finite size scaling is based on a postulate that ensures these expectations are met. Consider G as a function of some variable χ that depends on λ and N such that G rapidly goes to 1 with increasing χ. This is illustrated in the left panel of Fig. 5. As χ approaches 0, G drops below 1, because, according to the variational principle, E_N is an upper bound to E. The finite size scaling postulate is that χ depends on N and λ according to

$$\chi = N(\lambda_c - \lambda)^{\nu}. \tag{37}$$

ν is a parameter whose value we will determine later. $G(\chi)$, using Eq. (37) for χ with ν positive, satisfies our expectations. The first expectation is satisfied because the limit $\lambda \to \lambda_c$ is the same as the limit $\chi \to 0$. The second expectation is satisfied because for given value of λ, increasing N will increase χ, shifting it away from the region of small χ, where G deviates from 1. Thus, N can be thought of as a scale factor for the size of the region in which E_N is inaccurate. This is illustrated in the right panel of Fig. 5. The dashed curve corresponds to a relatively smaller value of N while the solid curve corresponds to a relatively larger value.

The scaling postulate has not been rigorously derived. It is a reasonable guess inspired by qualitative expectations and we will see that it will make it feasible to calculate the value of λ_c without having to solve the much harder problem of calculating G. It was proposed in the 1970s to describe the thermodynamics of magnetization in solids [9]. In that case the parameter N was the number of atomic layers in a thin film. That was a simple example of a more general problem: phase transitions are, strictly speaking, only a property of infinite systems. Thermodynamic functions can be expressed in terms of the partition function, which is an effectively infinite sum over the states of the system. However, if the infinite sum is truncated after a finite number of terms, then it gives a nonsingular function of temperature. This is a close analogy to the basis-set expansion in quantum mechanics. The finite expansion of the partition function can give an accurate description of a bulk system, but the accuracy drops significantly in the neighborhood of the critical temperature, where the system can spontaneously reorganize into a different phase and the true partition function is singular. Finite size scaling has been widely applied in condensed matter physics [8]. The idea of applying it to problems in quantum chemistry comes from Kais et al. [10]. They demonstrated that the scaling postulate gives accurate estimates of λ_c for various model problems in quantum mechanics and then they used the method to estimate critical nuclear charges for atoms and molecules.

In order to computationally study the dependence of E_N on the basis dimension, let us define

$$\Delta_E(\lambda; N_1, N_2) = \ln \frac{E_{N_1}(\lambda) - E_c}{E_{N_2}(\lambda) - E_c} \Big/ \ln \frac{N_2}{N_1} \tag{38}$$

and

$$\Delta'_E(\lambda; N_1, N_2) = \ln \frac{\partial E_{N_1}/\partial\lambda}{\partial E_{N_2}/\partial\lambda} \Big/ \ln \frac{N_2}{N_1}, \tag{39}$$

where N_1 and N_2 are the dimensions for two separate computations of E_N at various different values of λ, using basis functions that are independent of λ. Let us assume that $E(\lambda) - E_c$ is proportional to $(\lambda_c - \lambda)^\alpha$ as λ approaches λ_c from below. (The critical exponent α will be determined later.) In the immediate neighborhood of λ_c it is reasonable to describe E in terms of a series expansion valid as λ approaches λ_c from below, in the form

$$E(\lambda) - E_c \sim (\lambda_c - \lambda)^\alpha \left[\gamma_0 + \gamma_1(\lambda_c - \lambda) + \cdots\right], \quad \lambda_- \to \lambda_c. \tag{40}$$

Let us expand $G(\chi)$ about $\chi = 0$ and then substitute in the scaling postulate for χ. Eq. (36) implies that G will also have to be proportional to $(\lambda_c - \lambda)^\alpha$ in the limit $\lambda_- \to \lambda_c$. It follows that G can be expanded in the form

$$G\left(N(\lambda_c - \lambda)^\nu\right) \sim \left[N(\lambda_c - \lambda)^\nu\right]^{\alpha/\nu}\left[g_0 + g_1 N^{1/\nu}(\lambda_c - \lambda) + \cdots\right], \quad \lambda_- \to \lambda_c. \tag{41}$$

Note that the N and λ dependence appears only in terms of factors of $N(\lambda_c - \lambda)^\nu$. Using the Taylor series $(1+X)^{-1} \sim 1-X$, with $X = (g_1/g_0)N^{1/\nu}(\lambda_c - \lambda)$, we obtain from Eq. (36) the following expansion for E_N about the critical point:

$$E_N(\lambda) - E_c \sim \frac{\gamma_0}{g_0} N^{-\alpha/\nu} \left[1 + \left(\frac{\gamma_1}{\gamma_0} - \frac{g_1}{g_0} N^{1/\nu} \right) (\lambda_c - \lambda) + \cdots \right]. \tag{42}$$

We can now see why Δ_E was defined according to Eq. (38). Its leading-order behavior as λ approaches λ_c is

$$\Delta_E \sim \frac{\ln \left(\frac{N_1}{N_2} \right)^{-\alpha/\nu}}{\ln \left(\frac{N_2}{N_1} \right)} = \frac{\alpha}{\nu}. \tag{43}$$

The important feature to note is that Δ_E becomes independent of N in the limit $\lambda \to \lambda_c$. If the scaling postulate is valid, then plots of $\Delta_E(\lambda; N_1, N_2)$ versus λ for various different choices of N_1 and N_2 will all intersect when $\lambda = \lambda_c$, at the value α/ν. This is quite useful. It implies that λ_c, the location of the singularity, can be determined using computations of $E_N(\lambda)$, a function that is not singular there.

In order to estimate the value of α, let us define

$$\Gamma_E(\lambda; N_1, N_2) = \frac{\Delta_E}{\Delta_E - \Delta_E'}. \tag{44}$$

It follows from Eq. (42) that

$$\frac{\partial E_N}{\partial \lambda} \sim \frac{\gamma_0 g_1}{g_0^2} N^{(1-\alpha)/\nu} \left(1 - \frac{g_0 \gamma_1}{\gamma_0 g_1} N^{-1/\nu} \right), \quad \lambda_- \to \lambda_c. \tag{45}$$

If N is sufficiently large, specifically, if $N \gg (g_0\gamma_1/\gamma_0 g_1)^\nu$, then Γ_E becomes independent of basis dimension at the critical point,

$$\Gamma_E(\lambda; N_1, N_2) \sim \frac{\alpha/\nu}{\frac{\alpha}{\nu} + \frac{1-\alpha}{\nu}} = \alpha. \tag{46}$$

If N_1 and N_2 are sufficiently large and the scaling postulate is applicable, then plots of $\Gamma_E(\lambda; N_1, N_2)$ versus λ for various different choices of N_1 and N_2 will intersect at λ_c and their value at that point will be the critical exponent, α. To calculate Γ_E, we will need to calculate $\partial E_N/\partial\lambda$, but that is easily done by taking advantage of the Hellmann-Feynman theorem,

$$\frac{\partial E_N}{\partial \lambda} = \left\langle \frac{\partial H(\lambda)}{\partial \lambda} \right\rangle = \langle W \rangle. \tag{47}$$

Kais et al. [10] considered various simple model quantum mechanical problems and showed that the Δ_E and Γ_E plots had an intersection point, and that the resulting estimates for the critical parameters were reasonable. They also considered the two-electron atom, using a specialized basis set, and were able to accurately reproduce the expected results for λ_c and α. They then used the method with a simplified model Hamilton to predict critical charges for atoms throughout the periodic table [11]. They have more recently formulated the method for use with Gaussian basis sets, the basis sets used in most quantum chemistry software packages [12].

4 SUMMATION OF PERTURBATION SERIES

4.1 THE EFFECTS OF SINGULARITIES ON CONVERGENCE

Suppose that the Hamiltonian operator for a problem of interest is similar to that of a problem that can be solved more easily. Let H_0 be the Hamiltonian whose eigenfunctions and eigenvalues are known and then write the true Hamiltonian in the form

$$H = H_0 + \lambda v, \quad v = H - H_0, \tag{48}$$

where v is the *perturbation* and λ is the *perturbation parameter*, such that $\lambda = 0$ corresponds to the problem we can easily solve while $\lambda = 1$ corresponds to the physical problem. The eigenvalue $E(\lambda)$ of the Schrödinger equation

$$(H_0 + \lambda v)\Psi = E(\lambda)\Psi \tag{49}$$

can be treated as a continuous function of a variable λ in the complex plane. In contrast to the parameters considered previously in this chapter, which represented physical properties, this λ is an artificial construction that does not have to have any physical meaning. Nevertheless, the extrapolation from $E(0)$ to $E(1)$ will depend on the mathematical structure of the function $E(\lambda)$; in particular, on its singularities.

Eq. (49) is typically solved recursively. For example, if we substitute series expansions for the wavefunction and the eigenvalue,

$$\Psi \sim \sum_{k=0}^{\infty} \Psi_k \lambda^k, \quad E \sim \sum_{k=0}^{\infty} E_k \lambda^k, \tag{50}$$

and then collect terms according to power of λ, we obtain a set of equations that can be solved one after the other. At zeroth order we have only the unperturbed problem, $(H_0 - E_0)\Psi_0 = 0$, which by assumption we know how to solve. The equation obtained by collecting terms proportional to λ^1 can be solved in terms of the zeroth-order wavefunction Ψ_0, and so on to higher orders, with each subsequent equation solvable in terms of solutions from lower orders. However, for realistic problems the computational cost increases considerably with the level of recursion.

We have been using the symbol "$\sim$." Let us now define it. The power series for E is an example of an *asymptotic series*. The qualitative idea is that each additional term is a correction to the sum of the preceding terms, but it is worthwhile to describe this more precisely.[2] Let S_n be the *partial sum* of the series at arbitrary order n,

$$S_n(\lambda) = \sum_{k=0}^{n} E_k \lambda^k. \tag{51}$$

Then the relation "$\sim$," called *asymptotic equality*, is defined as follows: Consider a function $E(\lambda)$ and an infinite power series $\sum_{k=0}^{\infty} E_k \lambda^k$. If for any value of n,

$$\lim_{\lambda\to 0} \frac{E(\lambda) - S_n}{\lambda^n} = 0, \tag{52}$$

then $E(\lambda)$ is *asymptotic* to the infinite series,

$$E \sim \sum_{k=0}^{\infty} E_k \lambda^k, \quad \lambda \to 0. \tag{53}$$

Let us write

$$E - S_n \sim \mathcal{O}(\lambda^{n+1}) \tag{54}$$

to say that the error from truncating the infinite power series at order n goes to 0 at least as fast as λ^{n+1} in the limit $\lambda \to 0$. A theorem from complex analysis tells us that the asymptotic series of a function, in terms of a given variable about a given point, is unique. There exists only one series that will satisfy Eq. (52).

Asymptotic equality is a formal way of stating what is perhaps obvious: If we use S_n as an approximation for $E(\lambda)$, the error $\mathcal{O}(\lambda^{n+1})$ in the limit $\lambda \to 0$ goes to 0 at least as fast as λ^{n+1}. The implication is that for sufficiently small λ the error in S_2 will be less than the error in S_1, the error in S_3 will be less than that in S_2, and so on. It is important to emphasize, however, that this holds only for *sufficiently small* λ. How small is "sufficient" is not always obvious. For perturbation theories, as formulated in Eq. (48), the evaluation point is $\lambda = 1$, so the partial sums can converge only if the E_k decrease with k, which is not always the case in practice. For realistic physical problems, the computational cost of obtaining the coefficients of the series coefficients can increase very significantly with increasing order. It can be quite discouraging if that newly computed term at next higher order in fact *lowers* the accuracy!

Let us study this phenomenon with some simple model problems. If we have an explicit expression for $E(\lambda)$, we can calculate the series coefficients E_n using Taylor's theorem:

$$E_k = \left. \frac{1}{k!} \frac{d^k E}{d\lambda^k} \right|_{\lambda=0}. \tag{55}$$

[2]For a more detailed exposition of the theory of asymptotic series, see Baker and Graves-Morris [13].

Of course, in practical calculations we will go in the other direction—we will calculate the E_k in order to estimate $E(\lambda)$. However, Taylor's theorem is useful for obtaining asymptotic series of known functions for the purpose of method development and testing.

Our first example will be the 2×2 matrix eigenvalue problem of Eq. (4), but rather than solve this at various discrete values of λ and interpolate between them as we did for the molecular potential energy curve in Section 2, here we will treat this as a perturbation theory. The advantage of a perturbation theory is that it constructs a function of λ just from information at a single particular point where the analysis is simpler. Let us suppose that we know the eigenvalues and eigenvectors of the operators h_{11} and h_{22} in Eq. (4). We could solve for the asymptotic series for $E(\lambda)$ recursively using the standard, but laborious, methods of perturbation theory, but in this case, because we have the explicit exact solution for $E(\lambda)$ (Eq. 5), we can just use Taylor's theorem to calculate the series coefficients. Because the series is unique, the result will be the same.

We can apply quadratic approximants to an asymptotic series by parameterizing the polynomial coefficients to the series coefficients rather than to values of the function itself. Consider an arbitrary function $E(\lambda)$ with asymptotic series in the form of Eq. (53). As before, we define three polynomials,

$$P(\lambda) = p_0 + p_1\lambda + p_2\lambda^2 + \cdots + p_{M_p}\lambda^{M_p}, \tag{56}$$

$$Q(\lambda) = 1 + q_1\lambda + q_2\lambda^2 + \cdots + q_{M_q}\lambda^{M_q}, \tag{57}$$

$$R(\lambda) = r_0 + r_1\lambda + r_2\lambda^2 + \cdots + r_{M_r}\lambda^{M_r}. \tag{58}$$

If we know the series coefficients through order n, then we will have $n + 1$ simultaneous linear equations for the polynomial coefficients. This constrains the choice of polynomial degrees according to

$$M_p + M_q + M_r = n - 1, \tag{59}$$

so that we have the same number of equations as coefficients. We now construct a quadratic equation

$$Q(\lambda)E^2 - P(\lambda)E + R(\lambda) \sim \mathcal{O}(\lambda^{n+1}), \tag{60}$$

substituting for E the asymptotic series (Eq. 53). We multiply through by the polynomials, collect the terms into separate equations according to powers of λ, and ignore any terms proportional to λ raised to powers $n + 1$ or higher. This gives us $n + 1$ simultaneous linear equations for the coefficients. The approximants are then given by

$$S(\lambda) = \frac{1}{2Q(\lambda)}\left[P(\lambda) \pm \sqrt{P(\lambda)^2 - 4Q(\lambda)R(\lambda)}\right]. \tag{61}$$

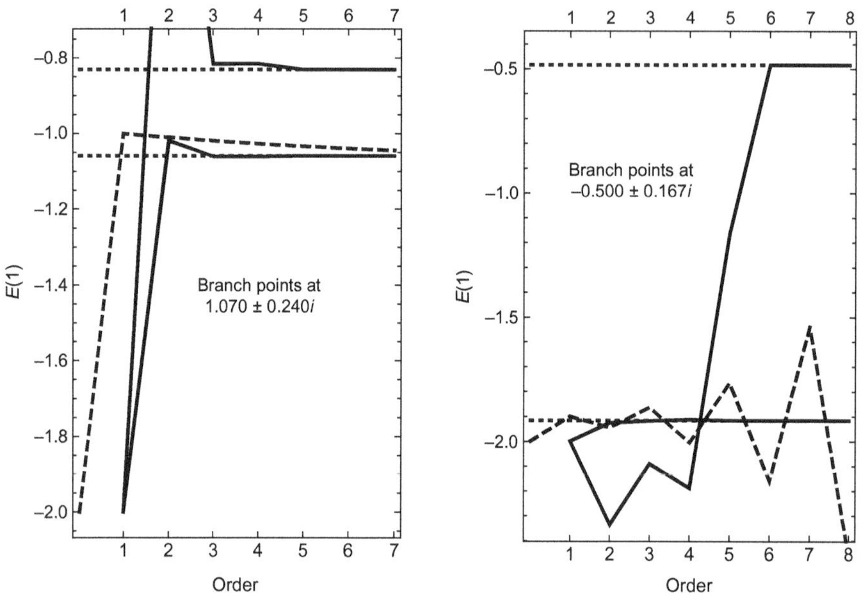

FIG. 6

Convergence of perturbation series for 2 × 2 matrix eigenvalue problems in the form of Eq. (4), summed at $\lambda = 1$. The parameters are $h_{11} = -2$, $h_{22} = -1$, $v_{11} = 1$, $v_{22} = 0.11$, $w = 0.1$ for the *left panel*, and $h_{11} = -2$, $h_{22} = -1.5$, $v_{11} = 0.1$, $v_{22} = 1.0$, $w = 0.15$ for the *right*. *Dashed line* segments are drawn between the results for the partial sums. *Solid line* segments are drawn between results for each branch of a sequence of quadratic approximants. The *dotted lines* show the exact values of the two eigenvalues.

We obtain a different approximant for each choice of the polynomial degrees, and for any $n > 1$ there will be more than one choice for given n, according to Eq. (59). Let us indicate the polynomial degrees with the symbol $[M_p/M_q, M_r]$. A common practice is to calculate approximants with increasing n in a *diagonal* sequence, with polynomial degrees increasing in concert, such as

$$[0/0,0],\ [1/0,0],\ [1/0,1],\ [1/1,1],\ [2/1,1],\ldots. \quad (62)$$

In general, the number of branch points in the $[M_p/M_q, M_r]$ approximant is

$$n_{\text{bp}} = \max(2M_p, M_q + M_r). \quad (63)$$

The diagonal sequence (62) ensures that the approximant will always have an even number of branch points, which is consistent with the Katz theorem for matrix eigenvalue problems.

Fig. 6 shows how singularity structure can affect the convergence of an asymptotic series. The dashed curves show the results from the partial sums S_n evaluated

Table 1 Locations of Branch Points of the Quadratic Approximants Used for the Left Panel of Fig. 6

Order	**2**	**3**	**4**	**5**	**6**	
Index	[1/0, 0]	[1/0, 1]	[1/1, 1]	[2/1, 1]	[2/1, 2]	Exact
Branch points	$1.920 \pm 0.560i$	$1.075 \pm 0.229i$	$1.076 \pm 0.230i$	$1.070 \pm 0.240i$	$1.070 \pm 0.240i$	$1.070 \pm 0.240i$
			$-7.1 \pm 0.001i$	$2 \times 10^{10} \pm 8 \times 10^{7}i$	-1×10^{9}	
					1×10^{9}	

at $\lambda = 1$ while the solid curves show the results for both branches from sequences of quadratic approximants. The left panel is for a case in which the partial sums converge steadily, but slowly, to the correct answer. A well-known theorem from complex analysis states that the partial sums of an asymptotic series will converge only if there are no singular points within the circular region of the complex plane centered at the expansion point ($\lambda = 0$ in this case) with radius equal to the distance to the evaluation point ($\lambda = 1$ in this case). The *circle of convergence* is the circular region centered at the expansion point with radius equal to the distance to the nearest singularity in the complex plane. For the left panel, the evaluation point is just barely inside this circle. The partial sums converge, but the convergence is slow. The lower branch of the quadratic approximant, in contrast, converges almost immediately, already giving essentially the exact result for the lower eigenvalue by third order. The upper branch of the quadratic approximant performs very poorly at first, but by third order begins a quick convergence to the upper eigenvalue.

The convergence pattern of the quadratic approximants can be explained by the convergence patterns for the branch point location. As shown in Table 1, there is a close correspondence between the two. At second order, there is no branch point in the vicinity of the correct location. As a result, the accuracy of the quadratic approximant in the left panel for the ground state at second order is no better than that of the partial sum. The series we are summing is the asymptotic expansion of the lower eigenvalue and it is only via the branch point that anything can be inferred from it about the upper eigenvalue. Therefore, it is not surprising that the second-order result for the upper eigenvalue is not accurate at all. By third order, the branch point is already being modeled accurately, and the results for both eigenvalues are quite good. The superfluous branch points, which first appear at fourth order, are placed far away from the vicinities of the expansion point and the evaluation point.

The right panel of the figure shows the convergence for a case in which the singular points are in the negative half-plane and within the circle of convergence. Because the singular points are far from the evaluation point, the first-order partial sum gives a very accurate estimate, but then it *diverges* with increasing order, as expected from the partial-sum convergence theorem. The bracketing behavior, alternating above and below the correct result, is due to the fact that the real part

Table 2 Locations of Branch Points of the Quadratic Approximants Used for the Right Panel of Fig. 6

Order	2	3	4	5	6	
Index	[1/0, 0]	[1/0, 1]	[1/1, 1]	[2/1, 1]	[2/1, 2]	Exact
Branch points	−0.589	−0.298	−0.352	$-0.502 \pm 0.170i$	$-0.500 \pm 0.167i$	$-0.5 \pm 0.167i$
	−679	−4.1	3.63	$24 \pm 2.7i$	$-3\times10^{12} \pm 7\times10^{11}i$	

of the singular points is negative. Despite the divergence of the partial sums, the sequence of quadratic approximants converges well. The lower branch is already very accurate by second order. The approximant's branch point structure, shown in Table 2, is not quite right below fifth order, with a single branch point on the real axis approximately between the two true singular points, but this has little effect on the accuracy of the lower branch because the evaluation point is far away. The upper branch of the approximant converges only when the branch point positions become very accurate, at sixth order.

Fig. 7 treats a more complicated situation. The function is

$$E(\lambda) = -[(\lambda_1 - \lambda)(\lambda_1^* - \lambda)]^{1/2} - [(\lambda_2 - \lambda)(\lambda_2^* - \lambda)]^{1/2},$$
$$\lambda_1 = 0.65 + 0.2i, \quad \lambda_2 = 1.3 + 0.2i. \tag{64}$$

The behavior of the partial sums of the asymptotic series is determined by the singular points at λ_1 and λ_1^*, which determine the circle of convergence. Because the evaluation point $\lambda = 1$ is outside that circle, the partial sums must diverge. However, in contrast to the previous case, in which the singularities were in the negative half-plane and the partial sums initially converged to the correct result before diverging, these singularities are in the positive half-plane, between the expansion point and the evaluation point. As shown in the figure, the partial sums in this situation are useless even at low order. The basic problem is that the partial summation approximant, $S_n(\lambda)$ of Eq. (51), is a polynomial. A polynomial can have no singularity at any finite point in the complex plane. It can accurately approximate singular functions in nonsingular regions, but not in regions affected by singularities. In this case, $S_n(\lambda)$ is attempting to extrapolate through a pair of branch points and fails.

This function has four branches, corresponding to the different possible choices for the signs of the two square roots. The quadratic approximant, by construction, has only two branches. Fig. 7 shows that it convergences to the two branches of the function connected by the branch points at λ_1 and λ_1^*. Again, the accuracy of the approximant corresponds to the accuracy of the branch point positions, listed in Table 3. Note in the figure the glitch in the convergence of the lower branch at fifth order. The lower-order approximants have only two branch points, and therefore focus their efforts on accurately describing the dominant singularities at λ_1 and λ_1^*.

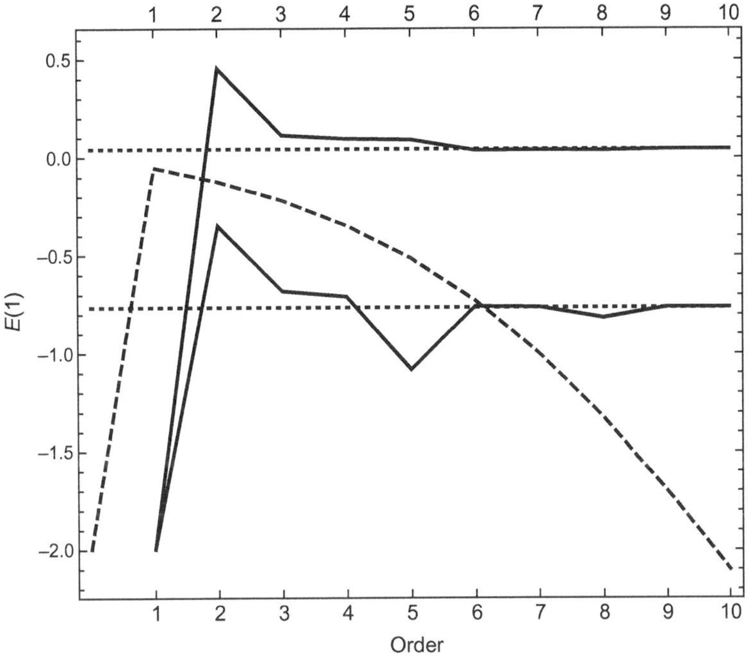

FIG. 7

Convergence of perturbation series summed at $\lambda = 1$ for the function $E(\lambda)$ of Eq. (64), which has two pairs of branch points in the positive half-plane. *Dashed line* segments are drawn between the results for the partial sums. *Solid line* segments are drawn between results for each branch of a sequence of quadratic approximants. The *dotted lines* show the exact values of the two eigenvalues connected by the innermost of the branch-point pairs.

The $[2/1, 1]$ approximant is the first in the sequence to have four branch points, because the discriminant $P^2 - 4QR$ is now a polynomial of fourth degree. The initial attempt in an approximant sequence to model a singularity is typically relatively inaccurate. $S_{[2/1,1]}$ in this case gives a poor description of the singularities at λ_2 and λ_2^* and a slightly less accurate description of the dominant singularities. This is remedied at sixth order. Beyond seventh order the results do not significantly improve, because after that, newly introduced singularities are superfluous. The true function has only four singularities.

The quadratic approximant is a special case of a more general construction called an *algebraic approximant*. Let us generalize Eq. (60) to arbitrary degree m,

$$P_{M_m}(\lambda)E^m + P_{M_{m-1}}(\lambda)E^{m-1} + P_{M_{m-2}}(\lambda)E^{m-2} + \cdots + P_{M_0}(\lambda) \sim \mathcal{O}(\lambda^{n+1}), \qquad (65)$$

where P_{M_j} is a polynomial of degree M_j. Of the $\sum_j(M_j + 1)$ coefficients, one is always arbitrary. (Dividing both sides of the equation by one of the coefficients will have no effect on the equations that determine the coefficients.) Usually, one sets

Table 3 Locations of Branch Points of the Quadratic Approximants of Fig. 7

Order	**2**	**3**	**4**	**5**	
Index	[1/0, 0]	[1/0, 1]	[1/1, 1]	[2/1, 1]	
Branch points	$0.948 \pm 0.393i$	$0.676 \pm 0.230i$	$0.660 \pm 0.218i$	$0.680 \pm 0.187i$	
				$1.811 \pm 0.115i$	
Order	**6**	**7**	**8**	**9**	
Index	[2/1, 2]	[2/2, 2]	[3/2, 2]	[3/2, 3]	Exact
Branch points	$0.649 \pm 0.200i$	$0.650 \pm 0.200i$	$0.650 \pm 0.201i$	$0.650 \pm 0.200i$	$0.65 \pm 0.2i$
	$1.293 \pm 0.024i$	$1.300 \pm 0.026i$	$1.291 \pm 0.053i$	$1.297 \pm 0.086i$	$1.3 \pm 0.2i$
			-3.282	$1.303 \pm 0.116i$	
			-3.286		

the zeroth-order coefficient of the polynomial multiplying E^m to 1. Substituting the asymptotic series for E and collecting terms according to power of λ yields, as before, a set of simultaneous linear equations that determine the values of the polynomial coefficients. The approximant is obtained as the numerical solution of

$$P_{M_m}(\lambda)S^m + P_{M_{m-1}}(\lambda)S^{m-1} + P_{M_{m-2}}(\lambda)S^{m-2} + \cdots + P_{M_0}(\lambda) = 0, \tag{66}$$

with λ set to the evaluation point. For high-order series, it is wise to use a recursive algorithm in order to control numerical instabilities from round-off error [14,15].

An algebraic approximant of degree m has m branches. This is appealing, because the functions that describe realistic physical problems often have many more than two branches. However, the rate of convergence of the approximants is highest for the principal branch (i.e., the branch for which the asymptotic series was calculated) and slows significantly for higher branches, especially for branches that interact only weakly with the principal branch. High-degree approximants can be useful if the series coefficients are known to high order. However, in practice, quadratic approximants are often adequate, especially if one is only interested in studying the principal branch of the function. They provide a simple explicit analytical expression for $S(\lambda)$. Furthermore, they are able to model the effects on the principal eigenvalue of many different branches with appropriately placed branch points. Of course, all the branch points of a quadratic approximant connect with a single second branch, rather than to different branches as in higher-degree algebraic approximants and in the true solution. The higher branch of the quadratic approximant will have to simultaneously model multiple branches, modeling different branches in different regions of λ.

4.2 SUMMATION METHODS FOR VARIOUS PROBLEMS

4.2.1 Molecular vibrations

Consider the vibrational spectrum of a diatomic molecule. For a first approximation, it is usually treated as a harmonic oscillator, with Hamiltonian

$$H_0 = -\frac{\hbar^2}{2\mu}\frac{d^2}{d\xi^2} + \frac{1}{2}k\xi^2, \tag{67}$$

where μ is the reduced mass of the molecule and the coordinate $\xi = x - x_e$ is the deviation from the equilibrium bond distance. This Hamiltonian gives one of the few exactly solvable Schrödinger equations. Its eigenfunctions are well known and are commonly used as a basis set for more complicated problems.

Of course, actual molecular potential curves are not symmetric in ξ (see Fig. 2). However, the harmonic oscillator is a reasonable approximation for low-lying eigenstates of tightly bound molecules, for which only the immediate vicinity of the bottom of the potential well has a significant probability distribution. For a more accurate description, a common practice is to add a polynomial perturbation, the simplest being a cubic term,

$$H = H_0 + \lambda\xi^3. \tag{68}$$

This slightly broadens the well in the direction of stretched bond lengths and makes it steeper for compressed bond lengths, which is physically reasonable. The eigenvalues and eigenfunctions of H can be calculated from a matrix eigenvalue problem in the harmonic oscillator basis. We can expect square-root branch points in the complex λ-plane, according to the Katz theorem.

Fig. 8 compares the qualitative shapes of the unperturbed and perturbed potential energy curves. Suppose that a wavefunction is initially localized in the well of the potential. There will be a nonzero probability of its tunneling out of the well, giving dissociated free atoms. There must be a critical point, corresponding to the largest λ at which the molecule is stable. The critical point in this case is at $\lambda = 0$, because for any nonzero λ the potential will eventually drop to negative infinity on one side of the well. This is just an artifact of the form chosen for the potential. However, it causes the radius of the circle of convergence for the perturbation theory in λ to be zero. There is a singularity at the origin of the complex λ-plane.

The functional form of this singularity is complicated—certainly more complicated than the branch points of an algebraic approximant. Nevertheless, algebraic approximants give convergent results for the perturbation theory in λ [15]. They manage to accomplish this using clusters of algebraic branch points near the origin.

For practical applications to molecular vibrations, the singularity at the origin is not usually a problem, because adequate accuracy can be obtained from moderately low orders off the perturbation theory. For vibrations of polyatomic molecules, square-root branch points of the real axis tend to be the more serious problem. Eigenstates that are degenerate in the zeroth-order approximation can be affected differently by the perturbation and have their degeneracy broken. This can result in

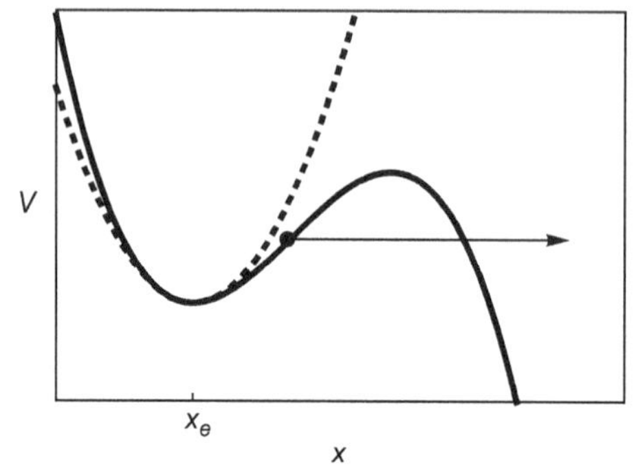

FIG. 8

Illustration of the instability of the cubic oscillator potential energy (*solid curve*) to tunneling, as opposed to the harmonic oscillator potential (*dotted curve*), which is stable.

branch points close to the origin that can seriously degrade the convergence of partial sums. Algebraic approximants have been applied with success to perturbation series for excited vibrational states of molecules with three or four atoms [16]. Fourth-degree approximants seem best for states with moderately high excitation, with no further improvement from going to higher degree. For very high excitation levels, there is a proliferation of branch points that in practice overwhelms the approximants' ability to distinguish between the effects of interactions with so many different states. In such cases, repartitioning the Hamiltonian, redefining H_0 so as to selectively shift the levels of the unperturbed eigenstates and thereby change the positions of branch points, can help [17].

4.2.2 Dimensional perturbation theory

As another example of a theory in which singularity analysis plays an important role, consider the electronic Schrödinger equation treated with dimensional perturbation theory, which, in effect, formulates the electronic structure calculation as a vibrational problem. The Schrödinger equation is generalized to a space of arbitrary dimension D and then expressed in appropriately scaled units such that D appears only as an effective mass multiplying the derivatives in the kinetic energy operator. In the limit $D \to \infty$ the electrons adopt a fixed configuration at the minimum of an effective potential energy surface. Within first-order perturbation theory in terms of $\delta = 1/D$, the electrons vibrate harmonically. Higher orders bring in anharmonic corrections.

One can prove that the ground-state energy of the Schrödinger equation in the dimension-scaled units has a second-order pole at $\delta = 1$, and writing the energy as an interpolation between the first-order series in δ at the $\delta \to 0$ limit and the pole at the $D \to 1$ limit results in a surprisingly reasonable estimate for the physical energy at $D = 3$ [18]. However, the attempt to systematically improve the accuracy simply by adding in higher powers of δ fails because the partial sums of the series diverge. This is another perturbation theory with a zero radius of convergence, with a complicated singular point at the origin of the complex δ-plane.

Convergent summation approximants for the δ series have been developed [19]. However, the high-order series coefficients are difficult to calculate, and in practice, the effect of the singularity at the origin on the behavior of the low-order coefficients is small. Furthermore, it results from generic behavior of the kinetic energy operator, not from the detailed form of the potential energy. This means that the large-order behavior of the δ series from an approximate model such as the Hartree-Fock approximation that simplifies the potential energy but not the kinetic energy should be similar to that of the exact series. Herschbach et al. [20] have taken advantage of this to construct a very simple yet quite accurate expression for the ground-state energy of the two-electron isoelectronic sequence (H^-, He, Li^+, etc.) as a function of atomic number Z. What they did is interpolate the correlation energy (the difference between the exact energy and the energy from the Hartree-Fock approximation) between the first-order δ series and the $\delta = 1$ pole, evaluate this at $D = 3$, and then add it to the accurately known Hartree-Fock energy. The effect of the singularity at the origin mostly canceled out.

4.2.3 Møller-Plesset perturbation theory

The classic perturbation theory for calculations of electronic structure is the Møller-Plesset (MP) perturbation theory (also called many-body perturbation theory). The Hamiltonian is partitioned so that the zeroth-order part H_0 is the sum of one-electron Fock operators. A perturbation parameter z is introduced into the full Hamiltonian according to

$$H(z) = H_0 + z(H_{\text{phys}} - H_0), \tag{69}$$

where H_{phys} is the actual physical Hamiltonian. Evaluation of the perturbation series at $z = 1$ yields an approximate solution of the physical problem. The zeroth-order approximation is qualitatively similar to the exact solution, and as a result, the eigenvalue $E(z)$ does not have a singularity at the origin. However, the zeroth-order Hamiltonian omits from the potential energy the terms that explicitly describe interelectron repulsion, replacing those terms with a one-electron mean field.

The Hartree-Fock approximation is solved within a basis-set expansion for the wavefunction. Hence, the perturbation theory yields a series in z that is asymptotic to the ground-state eigenvalue of the matrix representation of $H(z)$ in that basis set. This is an asymptotic series for the FCI solution as a function of the perturbation parameter. Let us call this solution $E_N(z)$, where N is the size of the basis set. Thus, the MP series gives an approximate solution for $E_N(z)$, which in turn is an approximate solution for $E(z)$, the exact ground-state eigenvalue of $H(z)$.

The first-order series in z gives the Hartree-Fock approximation for the energy. The second-order series (MP2) is quite popular, on account of its balance of computational efficiency and reasonable accuracy for predicting molecular geometries, but for quantitative calculations of energy differences, higher-order series are needed. MP3, MP4, and MP5 are available in standard quantum chemistry software packages. MP4, in particular, has long been of interest as the highest-order MP method for which the computational cost is not unreasonable.

Traditionally, the MP series was evaluated as a partial sum, simply adding up the terms of the series. This is now known to be inadequate [21] because it does not take into account the singularity structure of the energy as a function of z. The singularity structure of the exact MP energy function $E(z)$ is quite interesting [22]. To begin with, there are the expected Katz-type branch points. These are square-root branch points at discrete complex-conjugate points in the complex z plane, which connect pairs of eigenstates. Then there are critical points, of two different kinds.

The perturbation operator $z(H_{\text{phys}} - H_0)$ contains the term $z \sum_{j,k} r_{jk}^{-1}$, the potential energy for interelectron repulsion in H_{phys}, and the term $-zV_0$, where V_0 is the mean field repulsive potential for a given electron obtained by averaging over the exact potential with the Hartree-Fock orbitals. The interelectron term for increasing z on the positive real axis becomes increasingly repulsive. This is counteracted by the mean field term. V_0 itself is repulsive but the minus sign makes it an attractive potential in the perturbation operator for positive real z. If the repulsive term dominates, then the net effect of the perturbation operator will be destabilizing and there will be some value z_c beyond which the system will ionize. This will give a critical point on the positive real axis. Such singularities have been identified [22] but do not seem to be of significant importance in practice. For physically stable systems, this z_c will be quite distant from $z = 1$, while for less stable systems, a square-root branch point pair from an avoided crossing with an excited state can be expected that has smaller real part than z_c and a stronger effect on series convergence.

Consider, however, what can happen for *negative* real z. The interelectron repulsion is now multiplied by a negative number. This means that the electron-electron interaction is not repulsive, but attractive! Of course, this is completely nonphysical, but we know from complex analysis that singularities in nonphysical regions of the z-plane can affect the behavior of $E(z)$ in the physical region. The term $-zV_0$ is now a repulsive mean field, becoming increasingly destabilizing at increasingly negative real z. There will inevitably be a z_c beyond which the system will ionize, arranging itself into a bound cluster of mutually attractive electrons dissociated from the nuclei. This curious phenomenon was apparently first predicted in the context of nuclear physics [23], where a similar perturbation theory is used to calculate nuclear binding energies. This singularity in practice quite often has a significant effect on the MP series convergence.

Given enough terms in the z series, quadratic approximants can deal with all these singularities and give convergent summation. Although the functional form of a critical point in $E(z)$ is much more complicated than a square-root branch point, the series we are summing in fact corresponds to $E_N(z)$, which has only square-root branch points. In the neighborhood of z_c, $E_N(z)$ uses a clustering of branch points in complex-conjugates pairs close to the real axis to mimic the true singularity structure there.

The situation with MP4, however, is more complicated. The quadratic approximant at fourth order has only two branch points. If $E_N(z)$ has significant singularity structure in both the positive and negative half-planes, the approximant will have a tendency to place its branch points somewhere in between, thereby introducing singular behavior in a nonsingular region and failing to model any of the true singularities. An easy way to mitigate this effect is to use the quadratic approximant in combination with a bilinear conformal mapping of the complex plane [24]. Let us replace z with a new variable u according to

$$u = \frac{z}{1 - \lambda + \lambda z}, \quad z = \frac{(1 - \lambda)u}{1 - \lambda u}, \tag{70}$$

where λ is an arbitrary free parameter. The expression for z can be expanded in a Taylor series in u and then substituted into the perturbation series. If the coefficients of the series in z are E_k, then the coefficients of the new u series are

$$\tilde{E}_k(\lambda) = \sum_{j=1}^{k} \binom{k-1}{j-1} \lambda^{k-j}(1 - \lambda)^j E_j, \quad \binom{n}{m} = \frac{n!}{(n-m)!\,m!}. \tag{71}$$

This mapping has two fixed points, at $z = 0$ and $z = 1$, where z and u have the same value. Thus, the mapping leaves the zeroth-order solution and the physical solution unchanged. All other points in the complex plane are shifted, depending on the value chosen for λ. This is illustrated in Fig. 9, which shows the locations in the complex u-plane of singularities for the hydrogen fluoride molecule as a function of λ.

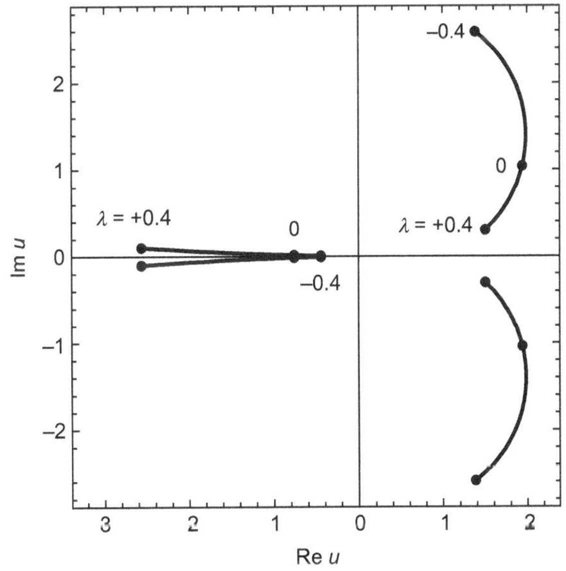

FIG. 9

Shift in singularity positions from a bilinear conformal mapping of the complex plane. The values of the parameter λ, of Eqs. (70), are indicated. The original singularity positions ($\lambda = 0$) correspond to the hydrogen fluoride molecule with the aug-cc-pVDZ basis.

A negative value of λ shifts the branch points in the positive half-plane away from the origin, and the critical point (modeled in $E_N(z)$ by a pair of square-root branch points very close to the real axis) is shifted toward the origin. The reverse happens with a positive value of λ. This usually makes it possible to force the MP4 approximant to accurately model singularity structure in one half-plane while ignoring that in the other half-plane. The MP4qλ method [21,25] is a reasonably systematic procedure for implementing this idea.

Because of the convergence problems with the partial sums, MP4 has been mostly supplanted in practice by the CCSD(T) method, a hybrid of coupled-cluster theory and perturbation theory that has only a slightly higher computational cost but is somewhat more reliable. The MP4qλ results seem to be comparable in accuracy to CCSD(T) for atoms and for molecules near their equilibrium geometries. Both methods fail as bonds become significantly stretched. MP4qλ fails somewhat sooner than CCSD(T), but at least has the advantage of a built-in diagnostic, in the position of its branch points. If the conformal mapping cannot significantly shift the branch points, this indicates that the branch points are too close to the fixed points, and in practice it means that the accuracy of the approximant will suffer.

REFERENCES

[1] Goodson DZ. Mathematical methods for physical and analytical chemistry. Hoboken, NJ: Wiley; 2011 [Chapters 6 and 7; For brief and relatively painless introduction to this subject].

[2] Arfken GA. Mathematical methods for physicists. Orlando, FL: Academic Press; 1985.

[3] Katz A. The analytic structure of many-body perturbation theory. Nucl Phys 1962;29:353.

[4] Goodson DZ. On the use of quadratic approximants to model diatomic potential energy curves. Mol Phys 2012;110:1681.

[5] Jordan KD. Applications of analytic continuation in the construction of potential energy curves. Int J Quant Chem Symp Ser 1975;9:325.

[6] Olsen J, Jørgensen P, Koch H, Balkova A, Bartlett RJ. Full configuration-interaction and state of the art correlation calculations on water in a valence double-zeta basis with polarization functions. J Chem Phys 1996;104:8007.

[7] Baker JD, Freund DE, Hill RN, Morgan JD, III. Radius of convergence and analytic behavior of the $1/Z$ expansion. Phys Rev A 1990;41:1247.

[8] Privman V. Radius of convergence and analytic behavior of the $1/Z$ expansion. Finite size scaling and numerical simulations of statistical systems. Singapore: World Scientific; 1990.

[9] Fisher ME, Barber MN. Scaling theory for finite-size effects in the critical region. Phys Rev Lett 1972;28:1516.

[10] Neirotto JP, Serra P, Kais S. Electronic structure critical parameters from finite-size scaling. Phys Rev Lett 1997;79:3142.

[11] Sergeev AV, Kais S. Critical nuclear charges for N-electron atoms. Int J Quant Chem 1999;75:533.

[12] Moy W, Kais S, Serra P. Finite size scaling with gaussian basis sets. Mol Phys 2008;106:203.
[13] Baker GAJ, Graves-Morris P. Padé approximants. Cambridge: Cambridge University Press; 1996.
[14] Sergeev AV. A recursive algorithm for Padé-Hermite approximations. USSR Comput Math Math Phys 1986;26:17.
[15] Sergeev AV, Goodson DZ. Summation of asymptotic expansions of multiple-valued functions using algebraic approximants: Application to anharmonic oscillators. J Phys A Math Gen 1998;31:4301.
[16] Duchko AN, Bykov AD. Resummation of divergent perturbation series: Application to the vibrational states of formaldehyde molecule. J Chem Phys 2015;143:154102.
[17] Surjan PR, Szabados A. Optimized partitioning in perturbation theory: Comparison to related approaches. J Chem Phys 2000;112:4438.
[18] Doren DJ, Herschbach DR. Accurate semiclassical electronic structure from dimensional singularities. Chem Phys Lett 1985;118:115. Loeser JG. Atomic energies from the large-dimension limit. J Chem Phys 1986;86:5635. Frantz DD, Herschbach DR. Lewis electronic structures as the large-dimension limit for H_2^+ and H_2 molecules. Chem Phys 1988;126:59.
[19] Elout MO, Goodson DZ, Elliston CD, Huang SW, Sergeev AV, Watson DK. Improving the convergence and estimating the accuracy of summation approximants of $1/D$ expansions for Coulombic systems. J Math Phys 1998;39:5112.
[20] Herschbach DR, Loeser JG, Virgo WL. Exploring unorthodox dimensions for two-electron atoms. J Phys Chem A 2017;121:6336.
[21] Goodson DZ. Resummation methods. WIREs Comput Mol Sci 2012;2:743.
[22] Sergeev AV, Goodson DZ, Wheeler SE, Allen WD. On the nature of the Møller-Plesset critical point. J Chem Phys 2005;123:064105.
[23] Baker GA, Jr. Singularity structure of the perturbation series for the ground-state energy of a many-fermion system. Rev Mod Phys 1971;43:479.
[24] Feenberg E. Invariance property of the Brillouin-Wigner perturbation series. Phys Rev 1956;103:1116.
[25] Goodson DZ. A summation procedure that improves the accuracy of the fourth-order Møller-Plesset perturbation theory. J Chem Phys 2000;113:6461.

CHAPTER

Diagrams in coupled-cluster theory: Algebraic derivation of a new diagrammatic method for closed shells

10

Devin A. Matthews*, John F. Stanton†

Institute for Computational Engineering and Sciences, The University of Texas at Austin, Austin, TX, United States Department of Chemistry, University of Florida, Gainesville, FL, United States†*

1 INTRODUCTION

The use of diagrammatic methods in chemical physics can trace its origin to a paper published by Montroll and Mayer in the early part of World War II [1]. In a paper entitled "The Statistical Mechanics of Imperfect Gases," they proposed an ingenious way to approach the enumeration and computation of multidimensional integrals over connected products of pair interactions required to calculate virial coefficients. While evaluation of the simplest of these "irreducible integrals" is a standard exercise in a first course in statistical mechanics, the higher-order terms that involve three or more interactions quickly become formidably complicated. Montroll and Mayer pointed out that these integrals "correspond to 'graphs'," the set of which are now called "cluster diagrams" and constitute part of the canonical pedagogy of statistical mechanics [2]. This pioneering work introduced concepts still used in diagrammatic methods, such as the notion of topological equivalence and connectivity.

During this time, Feynman was experimenting with his own "graphs" to calculate matrix elements appearing in quantum electrodynamics, the fundamentals of which were perhaps best described by Dyson in a paper [3] that united the approaches to QED of Tomonaga, Schwinger, and Feynman, who would go on to share the 1965 Nobel Prize in Physics. Feynman famously used them in a lecture that he gave at the Pocono Manor Conference in 1948 that is widely viewed as a springboard toward the subsequent development of a mature quantum theory of electrodynamics. However, he does not appear to have published a paper that contained any diagrams until the next year when two papers featuring them appeared in the Physical Review [4,5]. Now, Feynman diagrams are an essential part of modern quantum field theory,

Mathematical Physics in Theoretical Chemistry. https://doi.org/10.1016/B978-0-12-813651-5.00010-3

and widely appreciated to be a conceptually simple tool that leads to enormous simplification in doing calculations [6].

The development of diagrammatic methods to evaluate complicated integrals has continued in both chemistry and physics; one application that spans both areas is the problem of electron (and other fermionic) correlation. While the conceptually simple technique of configuration interaction [7] was the first to be applied to the electronic correlation problem, methods related to the perturbation treatments used in quantum field theory began to emerge. It was Goldstone [8] that first brought diagrams similar to those of Feynman to this area of application, based on a zeroth-order picture provided by the independent particle model (in which the electron-electron interaction is treated as a perturbation, to make the correspondence with the basic structure of QED). He significantly extended and formalized Brueckner's previous formulation of what is now called many-body perturbation theory (MBPT) [9] "using the technique of Feynman graphs" and proved diagrammatically that MBPT is manifestly size extensive at all orders, while Brueckner had only managed previously to show this to the lowest few orders. This landmark paper was the first demonstration of the power of diagrammatic methods for treating the problem of electron correlation, and the utility of the diagrammatic method was subsequently seized upon by others. Hugenholtz [10,11] developed a type of "skeleton diagram" to address the rapid proliferation at higher orders of what were already called "Goldstone diagrams." By enumerating the conceptually simpler skeleton diagrams, one could follow a well-defined protocol to translate these into the (many) associated Goldstone diagrams, the latter immediately giving the explicit matrix elements appearing in the equations to be summed.

Another type of diagram used for the many-body treatment of the electron-electron interaction was introduced by Brandow [12], and can be viewed as a hybrid of the Goldstone and Hugenholtz approaches, but which is also well suited for the use of spin-dependent Hamiltonians. The Brandow-type diagrams have been extensively used in the infinite-order generalization of MBPT known as coupled-cluster theory, particularly by the efforts of Bartlett et al. [13,14]. Bartlett et al. refer to Brandow diagrams as "antisymmetrized Goldstone diagrams," or simply ASGs. The use of these Brandow diagrams in CC theory facilitates the organization of "spin-orbital" equations used to solve the nonlinear system of equations that result from the coupled-cluster wavefunction ansatz and the projection approach that is characteristically used to solve them. For closed-shell systems, appropriate spin-integration can be performed to obtain the simpler equations that result by imposition of spin-symmetry. For the simple singles and doubles coupled-cluster model known as CCSD, optimal equations were determined quite early [15]. But for more complicated models—the simplest of these being CCSDT—progress came more slowly, with optimal equations apparently not being formulated for another decade, and these have only recently been published [16].

In the remainder of this contribution, we will show how a new diagrammatic method—specifically tailored to the derivation of efficient equations for closed-shell systems—may be obtained rigorously by algebraic manipulation of the existing spin-orbital (Brandow) diagrammatic theory. We emphasize the so-called "high-order" coupled-cluster methods (CCSDT, CCSDTQ, and beyond), and develop a new set of diagrams and diagrammatic rules that can be used to easily generate a (highly nontrivial) combination of fundamental spin-free (Goldstone) terms that may be exploited to gain maximum efficiency in solving the associated coupled-cluster equations.

2 THE COUPLED-CLUSTER METHOD AND ITS DIAGRAMMATIC REPRESENTATIONS

The diagrammatic methods that will be derived here are not strictly limited to coupled cluster. However, it is advantageous to discuss the basic theory of this treatment of fermionic correlation in order to better understand the nature of the diagrams studied here, as well as the origin of some of the diagrammatic rules pertaining to exponential expansions. In addition to the diagrammatic representation of coupled-cluster theory, the general Brandow diagrammatic technique will also be considered. After the coupled-cluster method has been revisited in the context of closed-shell systems, the Goldstone diagrammatic technique will also be reviewed in Section 3.

2.1 SPIN-ORBITAL PICTURE

A convenient basis for the derivation of many quantum chemical methods is the spin-orbital second-quantized representation. In this framework, operators are written as strings of annihilation and creation operators, each of which acts on a specific molecular spin-orbital. For example, in the coupled-cluster singles and doubles excitation (CCSD) model [15,17–20] the relevant operators are

$$\begin{aligned}\hat{H}_N &= \hat{F}_N + \hat{V}_N \\ &= \sum_{pq} f^p_q \left\{a^\dagger_p a_q\right\}_N + \frac{1}{4}\sum_{pqrs} v^{pq}_{rs}\left\{a^\dagger_p a^\dagger_q a_s a_r\right\}_N \end{aligned} \tag{1}$$

$$\hat{T}_1 = \sum_{ai} t^a_i a^\dagger_a a_i \tag{2}$$

$$\hat{T}_2 = \frac{1}{4}\sum_{abij} t^{ab}_{ij} a^\dagger_a a^\dagger_b a_j a_i \tag{3}$$

where $\{\ldots\}_N$ denotes normal ordering, $\hat{H}_N$ is the normal-ordered Hamiltonian operator with one-electron ($\hat{F}_N$) and two-electron ($\hat{V}_N$) components, and the cluster operators $\hat{T}_1$ and $\hat{T}_2$ produce single and double excitations from the reference determinant $|\Phi_0\rangle$, respectively:

$$\hat{T}_1|\Phi_0\rangle = \sum_{ai} t^a_i |\Phi^a_i\rangle \tag{4}$$

$$\hat{T}_2|\Phi_0\rangle = \frac{1}{4}\sum_{abij} t^{ab}_{ij} |\Phi^{ab}_{ij}\rangle \tag{5}$$

More generally, the cluster operator which produces k-fold excitations $|\Phi^{a_1 \dots a_k}_{i_1 \dots i_k}\rangle$ is given by

$$\hat{T}_k = \frac{1}{(k!)^2} \sum_{a_1 \dots a_k} \sum_{i_1 \dots i_k} t^{a_1 \dots a_k}_{i_1 \dots i_k} a^\dagger_{a_1} \dots a^\dagger_{a_k} a_{i_k} \dots a_{i_i} \tag{6}$$

Several different types of indices are used here: $pqrs$ for labeling either occupied or virtual (unoccupied) spin-orbitals, $ab \dots ef \dots$ for only virtual spin-orbitals, and $ij \dots mn \dots$ for only occupied ones.

2.2 SPIN-INTEGRATION

However, the indices in each case run over orbitals with *both* α *(spin-up) and* β *(spin-down) spin functions*. Due to the spin-selection rules of the spin-independent Hamiltonian, only matrix elements which independently conserve the number of α and β spin electrons are nonzero. This indicates that many elements of the tensors f^p_q and v^{pq}_{rs} are zero. Specifically, we may decompose these spin-orbital quantities into a set of unique *spin cases*, each of which represents a possible combination of α and β spin-orbitals:

$$f^p_q \rightarrow \left\{ f^p_q, f^{\bar{p}}_{\bar{q}} \right\} \tag{7}$$

$$v^{pq}_{rs} \rightarrow \left\{ v^{pq}_{rs}, v^{p\bar{q}}_{r\bar{s}}, v^{p\bar{q}}_{\bar{r}s}, v^{\bar{p}q}_{r\bar{s}}, v^{\bar{p}q}_{\bar{r}s}, v^{\bar{p}\bar{q}}_{\bar{r}\bar{s}} \right\} \tag{8}$$

where an overbar denotes β spin and no overbar α spin. Due to the antisymmetry of the tensors that arises from the anticommutation properties of the annihilation and creation operators (which ultimately derives from the Pauli principle), many of the spin cases for v^{pq}_{rs} are redundant, except when p and q (or r and s) are of different occupancies. Expanding all the possible occupancy as well as spin combinations, the following unique spin cases for $\hat{H}_N$ exist:

$$f^p_q \rightarrow \left\{ f^a_b, f^{\bar{a}}_{\bar{b}} \right\} \cup \left\{ f^a_i, f^{\bar{a}}_{\bar{i}} \right\} \cup \{ f^i_a, f^{\bar{i}}_{\bar{a}} \} \cup \left\{ f^i_j, f^{\bar{i}}_{\bar{j}} \right\} \tag{9}$$

$$\begin{aligned} v^{pq}_{rs} \rightarrow & \left(\left\{ v^{ab}_{ij}, v^{a\bar{b}}_{i\bar{j}}, v^{\bar{a}\bar{b}}_{\bar{i}\bar{j}} \right\} \cup \left\{ v^{ab}_{ci}, v^{a\bar{b}}_{c\bar{i}}, v^{\bar{a}b}_{\bar{c}i}, v^{\bar{a}\bar{b}}_{\bar{c}\bar{i}} \right\} \cup \left\{ v^{ij}_{ka}, v^{i\bar{j}}_{k\bar{a}}, v^{\bar{i}j}_{\bar{k}a}, v^{\bar{i}\bar{j}}_{\bar{k}\bar{a}} \right\} \cup h.c. \right) \\ & \cup \left\{ v^{ab}_{cd}, v^{a\bar{b}}_{c\bar{d}}, v^{\bar{a}\bar{b}}_{\bar{c}\bar{d}} \right\} \cup \left\{ v^{ij}_{kl}, v^{i\bar{j}}_{k\bar{l}}, v^{\bar{i}\bar{j}}_{\bar{k}\bar{l}} \right\} \cup \left\{ v^{ai}_{bj}, v^{a\bar{i}}_{b\bar{j}}, v^{\bar{a}i}_{\bar{b}j}, v^{a\bar{i}}_{j\bar{b}}, v^{\bar{a}i}_{\bar{j}b}, v^{\bar{a}\bar{i}}_{\bar{b}\bar{j}} \right\} \end{aligned} \tag{10}$$

where the Hermitian conjugate of, for example, $v^{a\bar{b}}_{c\bar{i}}$ is $v^{c\bar{i}}_{a\bar{b}}$ and all terms in parentheses are so conjugated.

The amplitudes (elements of $\hat{T}_k$) are also generally chosen to be spin-free, giving rise to the spin cases

$$t^a_i \rightarrow \left\{ t^a_i, t^{\bar{a}}_{\bar{i}} \right\} \tag{11}$$

$$t^{ab}_{ij} \rightarrow \left\{ t^{ab}_{ij}, t^{a\bar{b}}_{i\bar{j}}, t^{\bar{a}\bar{b}}_{\bar{i}\bar{j}} \right\} \tag{12}$$

$$t_{ijk}^{abc} \rightarrow \left\{ t_{ijk}^{abc}, t_{ij\bar{k}}^{ab\bar{c}}, t_{i\bar{j}\bar{k}}^{a\bar{b}\bar{c}}, t_{\bar{i}\bar{j}\bar{k}}^{\bar{a}\bar{b}\bar{c}} \right\} \tag{13}$$

$$\vdots$$

In other theories such as coupled-cluster equation-of-motion theory (EOM-CC) an excitation operator is used which may add or remove an electron (so that the spin-projection quantum number is $\pm\frac{1}{2}$) [21] or change the total spin of the wavefunction (i.e., a spin-projection quantum number of ± 1) [22]. These operators may have more complicated spin-selection rules, but are outside the focus of this work. The resolution of equations written in terms of spin-orbital quantities into (a set of) equations written in terms of distinct spin cases is known as *spin-integration* and is essential to arriving at efficient programmable equations. This process will be illustrated shortly.

2.3 COUPLED CLUSTER

In the coupled-cluster method, the exact wavefunction is expanded as

$$|\Psi\rangle = e^{\hat{T}}|\Phi_0\rangle, \quad \hat{T} = \sum_{k=1}^{N} \hat{T}_k \tag{14}$$

where N is the number of electrons. Inserting this wavefunction into the Schrödinger equation and projecting by $e^{-\hat{T}}$ gives

$$e^{-\hat{T}}\hat{H}_N e^{\hat{T}}|\Phi_0\rangle \equiv \bar{H}_N|\Phi_0\rangle = E_{CC}|\Phi_0\rangle \tag{15}$$

Because the Slater determinants are orthonormal, an equation for the energy and a set of coupled nonlinear equations for the amplitudes can be obtained by projection:

$$E_{CC} = \langle \Phi_0 | \bar{H}_N | \Phi_0 \rangle \tag{16}$$

$$0 = \left\langle \Phi_{i_1 \ldots i_k}^{a_1 \ldots a_k} | \bar{H}_N | \Phi_0 \right\rangle \tag{17}$$

The transformed Hamiltonian $\bar{H}_N$ can then be expanded by using the Baker-Campbell-Hausdorff expansion:

$$\begin{aligned} \bar{H}_N &= \hat{H}_N + \left[\hat{H}_N, \hat{T}\right] + \frac{1}{2}\left[\left[\hat{H}, \hat{T}\right], \hat{T}\right] + \cdots \\ &= \left(\hat{H}_N e^{\hat{T}}\right)_c \end{aligned} \tag{18}$$

where the subscript c stands for *connected*, the meaning of which will become clear shortly. Because of the structure of the bare Hamiltonian, only four commutators (or an expansion of $e^{\hat{T}}$ to $\hat{T}^4$) are necessary.

In practice, only a limited set of $\hat{T}_k$ operators will be included, for example, using $\hat{T}_1 + \hat{T}_2$ defines the CCSD method, adding $\hat{T}_3$ gives CCSDT, and so on.

When the number of electrons $N > 2$ for CCSD (or $N > 3$ for CCSDT, etc.) the method is no longer exact, but generally gives a very good approximation to the true wavefunction and energy. Algebraically, Eqs. (16), (17) may be evaluated by methods such as Wick's theorem. As an example, the energy equations become

$$E_{CC} = \sum_{em} t_m^e f_e^m + \frac{1}{4} \sum_{efmn} \left(t_{mn}^{ef} + 2t_m^e t_n^f \right) v_{ef}^{mn} \tag{19}$$

The amplitude equations similarly include products of amplitudes and integrals (Hamiltonian elements), except that not all indices will be summed over as they are in the energy expression. These remaining *external* indices must match those of one of the excited determinants $\langle \Phi_{i_1 \ldots i_k}^{a_1 \ldots a_k} |$. As an example, consider the following contribution to the quadruply excited determinants in CCSDTQ:

$$\begin{aligned} \left\langle \Phi_{ijkl}^{abcd} | \bar{H}_N | \Phi_0 \right\rangle &= z_{ijkl}^{abcd} \\ &\leftarrow \left\langle \Phi_{ijkl}^{abcd} \left| \frac{1}{2} [[\hat{H}_N, \hat{T}_2], \hat{T}_2] \right| \Phi_0 \right\rangle \\ &\leftarrow -P(ab/c/d)P(i/jk/l) \sum_{em} v_{el}^{md} t_{im}^{ab} t_{jk}^{ec} \end{aligned} \tag{20}$$

where "$\leftarrow$" denotes one of many contributions to the left-hand side. The permutation factors $P(\ldots)$ include antisymmetric exchange of labels between each group separated by a slash. For example, $P(ab/c/d) = (1+P(ad)+P(bd)+P(cd))(1+P(ac)+P(bc))$ where $P(pq)$ is a simple antisymmetric permutation (exchange) operator, e.g., $P(ac) v_{el}^{md} t_{im}^{ab} t_{jk}^{ec} = -v_{el}^{md} t_{im}^{cb} t_{jk}^{ea}$.

In order to obtain programmable equations for the general open-shell case, the fundamental spin-orbital equations must be spin-integrated. The spin-integrated energy equation is given by the not too much more complicated expression

$$\begin{aligned} E_{CC} = &\sum_{em} t_m^e f_e^m + \sum_{\bar{e}\bar{m}} t_{\bar{m}}^{\bar{e}} f_{\bar{e}}^{\bar{m}} + \frac{1}{4} \sum_{efmn} \left(t_{mn}^{ef} + 2t_m^e t_n^f \right) v_{ef}^{mn} \\ &+ \frac{1}{4} \sum_{\bar{e}\bar{f}\bar{m}\bar{n}} \left(t_{\bar{m}\bar{n}}^{\bar{e}\bar{f}} + 2t_{\bar{m}}^{\bar{e}} t_{\bar{n}}^{\bar{f}} \right) v_{\bar{e}\bar{f}}^{\bar{m}\bar{n}} + \sum_{e\bar{f}m\bar{n}} \left(t_{m\bar{n}}^{e\bar{f}} + t_m^e t_{\bar{n}}^{\bar{f}} \right) v_{e\bar{f}}^{m\bar{n}} \end{aligned} \tag{21}$$

However, contributions to the amplitude equations may become much more involved upon spin-integration. Taking the example previously, and considering only one of the five possible spin cases of z_{ijkl}^{abcd}, the result is

$$\begin{aligned} z_{ij\bar{k}\bar{l}}^{ab\bar{c}\bar{d}} \leftarrow &-P(a/b)P(c/d) \left\{ P(k/l) \sum_{em} v_{e\bar{l}}^{m\bar{d}} t_{\bar{k}m}^{\bar{c}b} t_{ji}^{ea} + P(i/j) \sum_{\bar{e}\bar{m}} v_{\bar{e}i}^{\bar{m}a} t_{j\bar{m}}^{b\bar{c}} t_{\bar{k}\bar{l}}^{\bar{e}\bar{d}} \right\} \\ &- P(i/j)P(k/l) \left\{ P(c/d) \sum_{em} v_{e\bar{l}}^{m\bar{d}} t_{mj}^{ab} t_{i\bar{k}}^{e\bar{c}} + P(a/b) \sum_{\bar{e}\bar{m}} v_{\bar{e}i}^{\bar{m}a} t_{\bar{k}\bar{m}}^{\bar{c}\bar{d}} t_{\bar{l}j}^{\bar{e}b} \right\} \end{aligned}$$

$$
\begin{aligned}
&- P(a/b)P(k/l)\sum_{e\bar{m}} v^{\bar{m}b}_{\bar{l}e} t^{\bar{c}\bar{d}}_{\bar{k}\bar{m}} t^{ea}_{ji} - P(c/d)P(i/j)\sum_{\bar{e}m} v^{m\bar{d}}_{j\bar{e}} t^{ab}_{im} t^{\bar{e}\bar{c}}_{\bar{l}\bar{k}} \\
&- P(a/b)P(c/d)P(i/j)P(k/l)\left\{\sum_{em} v^{ma}_{ei} t^{\bar{d}b}_{\bar{l}m} t^{e\bar{c}}_{j\bar{k}} + \sum_{\bar{e}\bar{m}} v^{\bar{m}\bar{d}}_{\bar{e}\bar{l}} t^{b\bar{c}}_{j\bar{m}} t^{a\bar{e}}_{i\bar{k}}\right. \\
&\left. + \sum_{e\bar{m}} v^{\bar{m}a}_{\bar{l}e} t^{b\bar{d}}_{j\bar{m}} t^{e\bar{c}}_{i\bar{k}} + \sum_{\bar{e}m} v^{m\bar{d}}_{j\bar{e}} t^{\bar{c}b}_{\bar{k}m} t^{a\bar{e}}_{i\bar{l}}\right\}
\end{aligned} \tag{22}
$$

While the diagrammatic method examined in the next section does not assist with spin-integration (only with derivation of the spin-orbital equations), the spin-integration process is very mechanical and can easily be automated. The examination of the closed-shell case in Section 3 will, however, demonstrate that considerable simplification of the spin-integrated equations may be achieved in that context, and ultimately we will show how this simplification can be done rigorously and diagrammatically.

2.4 BRANDOW DIAGRAMS

In order to aid in the derivation and analysis of quantum chemical equations such as those in coupled cluster, the various operators may be represented diagrammatically. In the Brandow representation, the tensor elements of spin-orbital operators are represented by *vertices*, depicted as horizontal lines of various types, with the same number of lines branching off above or below the vertex as there are indices in the algebraic tensor representation.

As an example, consider the representations of the various Hamiltonian elements shown in Fig. 1. The semivertical lines in the Brandow representation are explicitly recognized as either occupied (down arrows) or virtual (up arrows). Additionally, the position of the lines above or below the vertex classifies the associated annihilation or creation operator as a *q-creation* or *q-annihilation* operator, respectively. *q*-creation operators create particles or holes relative to the reference determinant ($a_a^\dagger$ or a_i) while *q*-annihilation operators destroy particles and fill holes (a_a or $a_i^\dagger$). For example, we can see from Fig. 2 that all operators in the cluster amplitudes are *q*-creation operators. The virtual *q*-creation and occupied *q*-annihilation lines are grouped as *out* lines (because their arrows point out from the vertex) while occupied *q*-creation and virtual *q*-annihilation operators are *in* lines. These will later be further broken down into the four basic categories, *particle-out*, *hole-out*, *particle-in*, and *hole-in*.

The crucial importance of diagrams is not that operators may be represented graphically, but that *interactions of operators* may be represented *and analyzed* graphically. A simple example can be seen in the diagrammatic representation of the coupled-cluster energy given in Fig. 3. In these diagrams, vertices are placed top to bottom as the corresponding operators appear left to right in the algebraic expression, and then lines from the cluster operator are joined to lines from the Hamiltonian. In this case the lines are called *contracted*, with the exact same meaning

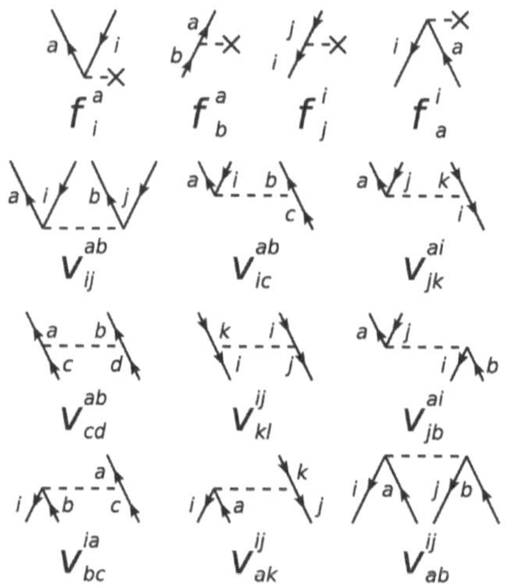

FIG. 1

The normal-ordered Hamiltonian in the Brandow diagrammatic representation.

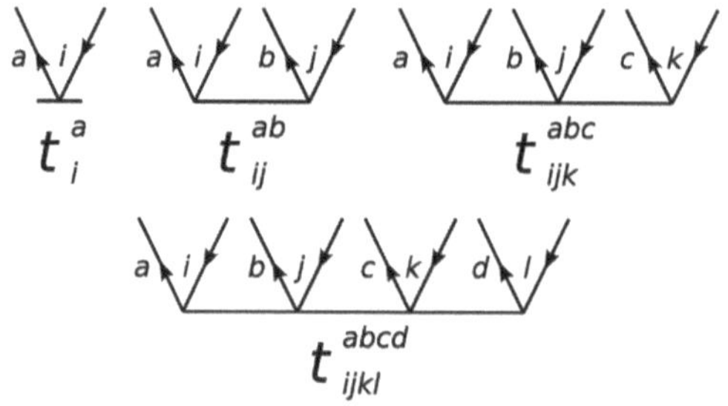

FIG. 2

The cluster operator in the Brandow diagrammatic representation.

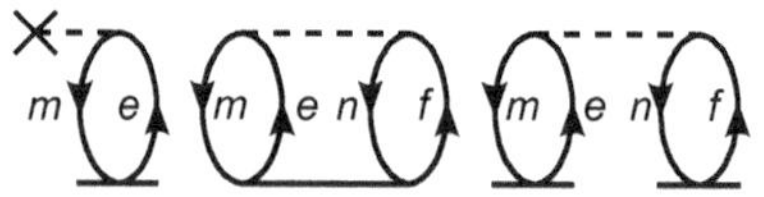

FIG. 3

Brandow diagrammatic representation of the coupled-cluster energy.

as in Wick's theorem. From the diagrammatic representation we can see that due to topological restrictions only q-creation lines may be contracted with q-annihilation lines of the same occupancy as required algebraically and that a fully contracted diagram has no uncontracted (external) lines, again as required by Wick's theorem. Where the diagrammatic method and Wick's theorem diverge, however, is that a

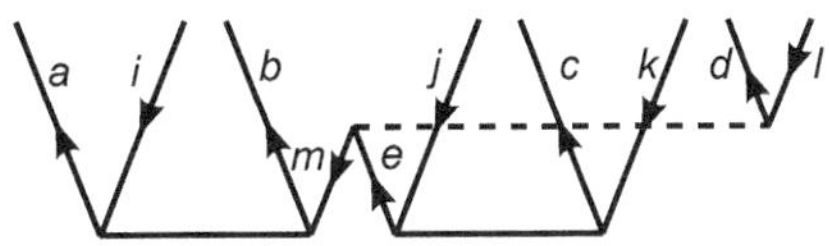

FIG. 4

Typical Brandow diagram in the CCSDTQ amplitude equations.

single diagram may stand for an arbitrarily large number of Wick contractions. For example, the middle diagram represents four (degenerate) Wick contractions. This lack of overcounting greatly simplifies the use of diagrams for deriving equations.

The amplitude equations may be represented similarly. The diagrammatic representation of the contribution to the quadruples amplitude equations given in Eq. (20) is depicted in Fig. 4. This diagram now has external lines which correspond to the excited Slater determinant and which must extend upwards (because the excited determinant is a bra state). As the formal rules for diagram interpretation stated below add, the permutation factors of Eq. (20) can also be read off of this single diagram, representing a huge number of distinct Wick contractions in the process. Because of the antisymmetry of the spin-orbital quantities, any diagram with the same *topological connectivity* (i.e., number of lines of each type contracted between each pair of vertices) are equivalent. Only a single topologically distinct Brandow diagram is then needed to determine each algebraic spin-orbital term. The energy diagrams as well as the amplitude diagram in Fig. 4 are *connected*, as there is a path from every vertex to every other. Connectedness of the diagrams is necessary in order to fulfill the commutator expansion in Eq. (18), but also, as Goldstone showed, in order to preserve size-extensivity of the method.

Finally, the formal rules for the Brandow diagrams are may be stated [12],

R1. Each vertex contributes a spin-orbital tensor element $\boldsymbol{x}^{\text{all-out-labels}}_{\text{all-in-labels}}$, where the identity of $\boldsymbol{x}$ is determined by the type of the horizontal line (e.g., $\boldsymbol{f}$, $\boldsymbol{v}$, or $\boldsymbol{t}$).

R2. Each set of n identical lines gives a factor of $\frac{1}{n!}$.

R3. The overall sign is equal to $(-1)^{l+h}$, where l is the number of "loops" (a path which either goes from one point along contracted lines returning to that same point, or from an external line to another external line on the same side) and h is the number of occupied (hole) lines.

R4. Each set of n topologically identical vertices from an exponential expansion (e.g., the coupled cluster amplitudes) gives a factor of $\frac{1}{n!}$.

R5. Contracted lines are summed over all spin-orbitals associated with the directionality of the line (e.g., all occupied orbitals are summed over for a downward-directed line).

R6. External lines of the same type ($\{hole, particle\} \otimes \{in, out\}$) which appear on different vertices are antisymmetrized.

From these rules we can easily see why, for example, Eq. (20) has a negative sign, why the $\hat{T}_2$ term in the energy expression (19) has a factor of 1/4 (and the $\hat{T}_1^2$ term a factor of 1/2), and so on. The reader is encouraged to fully verify Eqs. (19), (20) using these rules.

3 THE CLOSED-SHELL CASE

3.1 REDUCTION FROM SPIN TO SPATIAL ORBITALS

If the number of occupied α and β orbitals are the same, and the same spatial orbitals are used for both spin functions, then some interesting and useful relationships arise between the spin-integrated spin cases. The most useful relationship is spin-reversal symmetry, where the value of a tensor element is unchanged on interchange of α and β spin functions, for example

$$v_{ij}^{ab} = v_{\bar{i}\bar{j}}^{\bar{a}\bar{b}} \tag{23}$$

$$t_{ij\bar{k}}^{ab\bar{c}} = t_{\bar{i}\bar{j}k}^{\bar{a}\bar{b}c}, \ldots \tag{24}$$

This reduces the number of independent spin cases by roughly $\frac{1}{2}$, and similarly the complexity of the equations by a similar amount. However, it is possible to go further. Spin cases with all indices of the same spin can be formed by antisymmetrizing spin cases of mixed spin. For example,

$$v_{ij}^{ab} = v_{i\bar{j}}^{a\bar{b}} - v_{j\bar{i}}^{a\bar{b}} = v_{i\bar{j}}^{a\bar{b}} - v_{i\bar{j}}^{b\bar{a}} \tag{25}$$

$$\begin{aligned} t_{ijkl}^{abcd} &= t_{ijk\bar{l}}^{abc\bar{d}} - t_{ijl\bar{k}}^{abc\bar{d}} - t_{ilk\bar{j}}^{abc\bar{d}} - t_{ljk\bar{i}}^{abc\bar{d}} \\ &= t_{ij\bar{k}\bar{l}}^{ab\bar{c}\bar{d}} - t_{ik\bar{j}\bar{l}}^{ab\bar{c}\bar{d}} - t_{kj\bar{i}\bar{l}}^{ab\bar{c}\bar{d}} - t_{il\bar{k}\bar{j}}^{ab\bar{c}\bar{d}} - t_{lj\bar{k}\bar{i}}^{ab\bar{c}\bar{d}} + t_{kl\bar{i}\bar{j}}^{ab\bar{c}\bar{d}} \\ &= \ldots \end{aligned} \tag{26}$$

Using these types of relations, the number of spin cases for the Hamiltonian, $\hat{T}_1$, $\hat{T}_2$, and $\hat{T}_3$ can be reduced to one, and for $\hat{T}_4$ to two ($t_{ijk\bar{l}}^{abc\bar{d}}$ and $t_{ij\bar{k}\bar{l}}^{ab\bar{c}\bar{d}}$, which cannot be interrelated by simple antisymmetrization).

This is still not the end, however, as it is possible to use tensor elements which have no spin functions (labels) at all [17,23,24]. These quantities are called orbital tensor elements, also called spin-free and sometimes skeleton quantities, because for the Hamiltonian they represent integrals over purely spatial orbitals. These tensor elements are distinguished here by an inverted chevron (a hacek) over the quantity, for example, $\check{v}_{ij}^{ab}$. The spin-integrated tensor elements, even those of mixed spin, are formed from the orbital elements by antisymmetrization of the same-spin indices. For example,

$$v_{i\bar{j}}^{a\bar{b}} = \check{v}_{ij}^{ab} \tag{27}$$

$$v^{ab}_{ij} = \check{v}^{ab}_{ij} - \check{v}^{ab}_{ji} \tag{28}$$

$$t^{abc\bar{d}}_{ijk\bar{l}} = \check{t}^{abcd}_{ijkl} - \check{t}^{abcd}_{ijlk} - \check{t}^{abcd}_{ilkj} - \check{t}^{abcd}_{ljki} + \check{t}^{abcd}_{kijl} + \check{t}^{abcd}_{jkil} \tag{29}$$

$$t^{ab\bar{c}\bar{d}}_{ij\bar{k}\bar{l}} = \check{t}^{abcd}_{ijkl} - \check{t}^{abcd}_{jikl} - \check{t}^{abcd}_{ijlk} + \check{t}^{abcd}_{jilk}, \ldots \tag{30}$$

The relationship for the one-particle operators ($\hat{F}_N$ and $\hat{T}_1$) is trivial, as well as for the mixed-spin cases of the two-particle Hamiltonian and $\hat{T}_2$. Since CCSD only requires one- and two-particle quantities, which expand into usual spin cases (e.g., $t^{\bar{a}\bar{b}}_{\bar{i}\bar{j}} = \check{t}^{ab}_{ij} - \check{t}^{ab}_{ji} = t^{a\bar{b}}_{i\bar{j}} - t^{a\bar{b}}_{j\bar{i}}$), there are no additional considerations in using the orbital representation. For CCSDT and CCSDTQ, however, the orbital relationship is more complicated. The orbital quantities do not have antisymmetry of their labels, since anticommutation of the creation and annihilation operators is only fixed after readdition of the spin functions. They do have a different kind of symmetry, though, in that interchange of columns of indices (i.e., simultaneously interchanging labels in the same positions among the top and bottom labels) gives the same result with no change in sign. This column symmetry reflects the indistinguishability of Slater determinants, irrespective of spin. Thus, the integrals and amplitudes have symmetries like

$$\check{v}^{pq}_{rs} = \check{v}^{qp}_{sr} \tag{31}$$

$$\check{t}^{ab}_{ij} = \check{t}^{ba}_{ji} \tag{32}$$

$$\check{t}^{abc}_{ijk} = \check{t}^{acb}_{ikj} = \check{t}^{cba}_{kji} = \check{t}^{bac}_{jik} = \check{t}^{bca}_{jki} = \check{t}^{cab}_{kij} \tag{33}$$

$$\check{t}^{abcd}_{ijkl} = \check{t}^{abdc}_{ijlk} = \check{t}^{cbad}_{kjil} = \check{t}^{bdac}_{jlik} = \cdots \tag{34}$$

Because the indices of orbital amplitudes and integrals do not correspond to either α or β spin functions, there is no concept of a spin case for these operators. However, due to the lack of antisymmetry among the upper and (separately) among the lower indices, there are in general multiple ways to order the indices which produce a unique amplitude or integral. Sometimes a permutation results in a distinct numerical value but a quantity of the same general form; for example, $\check{v}^{ab}_{ij}$ and $\check{v}^{ab}_{ji}$ will in general have a different value, but the latter (as a tensor) can be obtained from the former by tensor transposition. On the other hand, the integrals $\check{v}^{ai}_{bj}$ and $\check{v}^{ai}_{jb}$, which also have distinct numerical values, are not simply the tensor transpose of one another. In this case, these two permutations are considered as different *orbital cases*. Two orbital cases are distinct if there is no permutation among indices of the same type (occupied or virtual) on either the top or the bottom which relates them, including column symmetry. So, while $\check{v}^{ai}_{bj}$ and $\check{v}^{ai}_{jb}$ are distinct orbital cases, $\check{v}^{ai}_{bj}$ and $\check{v}^{ia}_{jb}$ are not (they are related by column symmetry), nor are $\check{v}^{ai}_{bc}$ and $\check{v}^{ia}_{bc}$ (since the latter is equivalent by column symmetry to $\check{v}^{ai}_{cb}$ which is a permutation of $\check{v}^{ai}_{bc}$ of the same type). As shown in Section 4, orbital cases can be classified by the number of columns which have both an occupied and a virtual label.

Lastly, because the orbital amplitudes are specific to a closed-shell reference function, the final correlated wavefunction is (somewhat trivially) guaranteed to be a pure singlet state. For this reason, the orbital form is also known as the nonorthogonally spin-adapted formalism. The "nonorthogonal" qualifier refers to the fact that the excited single-determinental functions produced by applying the orbital amplitude operator are not orthogonal to one another. Additionally, the excited functions are actually linearly dependent for triple excitations and higher, so that the orbital amplitude elements for $\hat{T}_3$ and $\hat{T}_4$ are not unique or even well-defined [25]. Applying the antisymmetrization operations necessary to get the spin-integrated spin cases removes these indeterminacies, so that the spin-integrated (and of course spin-orbital) equations are numerically well-determined.

The use of the orbital quantities instead of the spin-integrated ones has the potential to produce much simpler and easier-to-implement equations for closed-shell systems. But, it is tedious to do a direct expansion of the spin-orbital or spin-integrated equations to the orbital ones, and from there it is necessary to factor the result (in a nonobvious way) to obtain the most compact representation. This "optimal" orbital form has been used for some time for CCD and CCSD [15,19,20], but for CCSDT and especially CCSDTQ, the techniques of the following sections provide insight that is not easily achieved by other means.

3.2 GOLDSTONE DIAGRAMS

Another related diagrammatic method is due to Goldstone and later Čížek. In this formulation, the spin-free nature of the Hamiltonian and cluster operators are taken into account, and all diagrams are interpreted in terms of orbital instead of spin-orbital tensor elements. One important consequence of this is that antisymmetry of tensor labels and hence lines in the diagram is lost. Thus, diagrams which were previously topologically equivalent are now distinct. An example of this can be seen in Fig. 5 where the coupled-cluster energy is represented in Goldstone diagrams. Two diagrams each are required for the $\hat{H}_N\hat{T}_2$ and $\frac{1}{2}\hat{H}_N\hat{T}_2^2$ terms, which differ only by a permutation of labels/lines on the Hamiltonian vertex.

A second important consequence of the use of orbital quantities is that not only are the orbitals represented by just their spatial part, but *summations over orbitals*

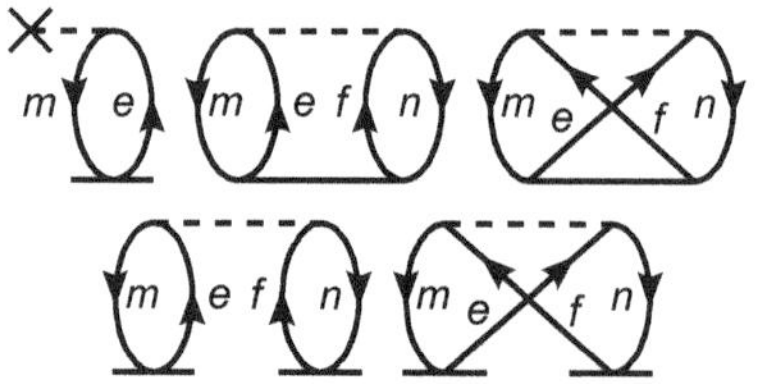

FIG. 5

Goldstone diagrammatic representation of the coupled-cluster energy.

run over only the spatial part as well. In the spin-orbital theory, sums run over both α and β spin, with possible spin combinations on two or more lines determined by the spin-integration rules. In short, Goldstone noted that for *closed loops*, the spin of all lines around the loop must be the same. Thus, a sum over lines in a closed loop has two parts: a sum over α spin-orbitals and a sum over β ones. In the closed-shell case, the spatial part of the α and β spin-orbitals is identical and so we should include a factor of 2 for a sum over each closed loop in the Goldstone case. Note that the number of closed loops is a topologically dependent factor, as can be seen from Fig. 5 where the second and fourth diagrams have two closed loops and hence a factor of 4, while the first, third, and fifth diagrams have only one closed loop and so a factor of 2. The sign factor (which is unchanged from the Brandow picture) is also topologically dependent.

Lastly, the symmetry due to indistinguishability of Slater determinants (the column symmetry) must also be taken into account. Goldstone showed that, for MBPT, this symmetry requires diagrams which have left-right mirror symmetry (like the last four diagrams in Fig. 5) to receive an additional factor of 1/2. Putting all of these considerations together, the rules for interpreting Goldstone diagrams (omitting MBPT energy denominators for simplicity) can be written as [26]:

R1′. Each vertex contributes an orbital tensor element: $\check{x}_{\text{all-in-labels}}^{\text{all-out-labels}}$, where $\check{x}$ is determined by the type of the horizontal line.
R2′. Each *closed* loop contributes an overall factor of 2.
R3′. The overall sign is equal to $(-1)^{l+h}$, where l is the number of loops and h is the number of hole lines.
R4′. If the diagram is symmetric with respect to left-right reflection, the overall factor is multiplied by $\frac{1}{2}$.
R5′. Each set of n topologically identical vertices from an exponential expansion (e.g., the coupled-cluster amplitudes) gives a factor of $\frac{1}{n!}$.
R6′. Internal lines are summed over all spatial orbitals of the type associated with the directionality of the line.

Applying these to the diagrams of Fig. 5 gives

$$\begin{aligned} E_{CC} &= 2\sum_{em} \check{t}_m^e \check{v}_e^m + 2\sum_{efmn} \check{t}_{mn}^{ef} \check{v}_{ef}^{mn} - \sum_{efmn} \check{t}_{mn}^{ef} \check{v}_{ef}^{nm} + 2\sum_{efmn} \check{t}_m^e \check{t}_n^f \check{v}_{ef}^{mn} - \sum_{efmn} \check{t}_m^e \check{t}_n^f \check{v}_{ef}^{nm} \\ &= 2\sum_{em} \check{t}_m^e \check{v}_c^m + \sum_{efmn} \left(\check{t}_{mn}^{ef} + \check{t}_m^e \check{t}_n^f\right)\left(2\check{v}_{cf}^{mn} - \check{v}_{cf}^{nm}\right) \end{aligned} \tag{35}$$

The reader is encouraged to derive this equation from Eq. (21) using the relationship of the spin-orbital and orbital quantities (27), (28) and similarly for $\check{t}_{ij}^{ab}$.

Applying these rules directly to coupled cluster, even to CCD or CCSD, is not as satisfactory as with the Brandow approach. One problem is that consideration of external lines and their symmetry is absent. For example, take the "ring" part of the $\hat{H}_N\hat{T}_2$ contributions to the doubles amplitude equations, whose Goldstone diagrams

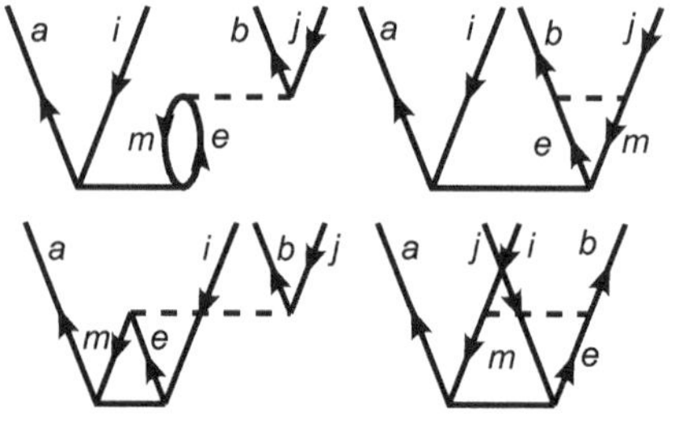

FIG. 6

Diagrams for the "ring" part of $\hat{Z}_2 \leftarrow \hat{H}_N \hat{T}_2$. The Goldstone diagrams which result from interchanging the *ai* and *bj* labels are not shown.

are given in Fig. 6. A straightforward interpretation of these diagrams (including the additional diagrams noted in the caption) gives

$$\begin{aligned} \check{z}_{ij}^{ab} \leftarrow & \sum_{em} \Big\{ 2\check{t}_{im}^{ae}\check{v}_{ej}^{mb} - \check{t}_{im}^{ae}\check{v}_{je}^{mb} - \check{t}_{mi}^{ae}\check{v}_{ej}^{mb} - \check{t}_{mi}^{be}\check{v}_{je}^{ma} \\ & + 2\check{t}_{jm}^{be}\check{v}_{ei}^{ma} - \check{t}_{jm}^{be}\check{v}_{ie}^{ma} - \check{t}_{mj}^{be}\check{v}_{ei}^{ma} - \check{t}_{mj}^{ae}\check{v}_{ie}^{mb} \Big\} \\ = & \left(1 + P_{bj}^{ai}\right) \sum_{em} \Big\{ 2\check{t}_{im}^{ae}\check{v}_{ej}^{mb} - \check{t}_{im}^{ae}\check{v}_{je}^{mb} - \check{t}_{mi}^{ae}\check{v}_{ej}^{mb} - \check{t}_{mi}^{be}\check{v}_{je}^{ma} \Big\} \end{aligned} \tag{36}$$

where P_q^p exchanges each label in group p with the label in the same position in group q simultaneously; for example, $P_{bj}^{ai}\check{t}_{im}^{ae}\check{v}_{ej}^{mb} = \check{t}_{jm}^{be}\check{v}_{ei}^{ma}$. The factor $(1 + P_{bj}^{ai})$ replaces the usual antisymmetrization operators in the Brandow representation, and gives the result the proper permutational symmetry. While such permutation and antisymmetrization factors are explicit in the Brandow approach, a strict application of Goldstone's rules requires the evaluation of additional permuted diagrams.

A second problem with the Goldstone approach is that the working equations derived from the diagrammatic form require additional analysis and reformulation in order to minimize the operation count. Taking the previous "ring" contribution, after some algebra, an optimally factorized expression can be obtained [15,19,20]:

$$\check{z}_{ij}^{ab} \leftarrow \left(1 + P_{bj}^{ai}\right) \sum_{em} \left\{ \frac{1}{2}\left(2\check{t}_{im}^{ae} - \check{t}_{mi}^{ae}\right)\left(2\check{v}_{ej}^{mb} - \check{v}_{je}^{mb}\right) - \left(\frac{1}{2} + P_b^a\right)\check{t}_{mi}^{ae}\check{v}_{je}^{mb} \right\} \tag{37}$$

While this form of the equation may seem bizarre, it in fact exemplifies the general structure of many factorized nonorthogonally spin-adapted equations. As we will see in the following sections, the optimal factored form of the equations as seen here can in fact be derived diagrammatically.

Lastly, and of critical importance to correlated theories beyond two-body operators, such as higher-order coupled cluster (CCSDT, CCSDTQ, etc.), the Goldstone rules do not capture the correct constant factors required, as rule R4′ is only valid

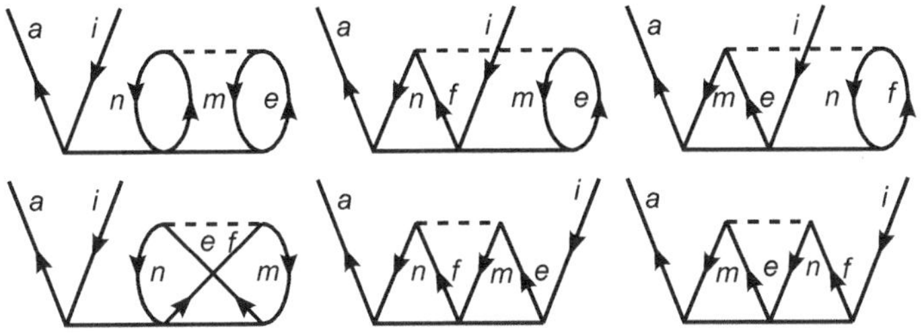

FIG. 7

Goldstone diagrams for $\hat{Z}_1 \leftarrow \hat{H}_N \hat{T}_3$.

for MBPT. An example can be seen in the $\hat{H}_N\hat{T}_3$ contribution to the CCSDT singles amplitude equations, whose Goldstone diagrams are given in Fig. 7. As none of these diagrams exhibits left-right mirror symmetry, the Goldstone rules give the following contribution

$$\check{z}_i^a \leftarrow \sum_{mnef} \check{v}^{mn}_{ef} \left(4\check{t}^{aef}_{imn} - 2\check{t}^{aef}_{nmi} - 2\check{t}^{aef}_{min} - 2\check{t}^{aef}_{inm} + \check{t}^{aef}_{nim} + \check{t}^{aef}_{mni}\right) \tag{38}$$

However, we may also interpret one of the diagrams in Fig. 7 as a Brandow diagram and follow the derivation starting from the spin-orbital representation in order to arrive at

$$z_i^a \leftarrow \frac{1}{4} \sum_{mnef} v^{mn}_{ef} t^{aef}_{imn} \tag{39}$$

$$z_i^a = z_{\bar{i}}^{\bar{a}} \leftarrow \frac{1}{4} \sum_{mnef} v^{mn}_{ef} t^{aef}_{imn} + \sum_{m\bar{n}e\bar{f}} v^{m\bar{n}}_{e\bar{f}} t^{ae\bar{f}}_{im\bar{n}} + \frac{1}{4} \sum_{\bar{m}\bar{n}\bar{e}\bar{f}} v^{\bar{m}\bar{n}}_{\bar{e}\bar{f}} t^{a\bar{e}\bar{f}}_{i\bar{m}\bar{n}} \tag{40}$$

$$\begin{aligned}
\check{z}_i^a \leftarrow & \frac{1}{4} \sum_{mnef} \left(\check{v}^{mn}_{ef} - \check{v}^{mn}_{fe}\right) \left(\check{t}^{aef}_{imn} - \check{t}^{aef}_{nmi} - \check{t}^{aef}_{min} - \check{t}^{aef}_{inm} + \check{t}^{aef}_{nim} + \check{t}^{aef}_{mni}\right) \\
& + \sum_{mnef} \check{v}^{mn}_{ef} \left(\check{t}^{aef}_{imn} - \check{t}^{aef}_{min}\right) + \frac{1}{4} \sum_{mnef} \left(\check{v}^{mn}_{ef} - \check{v}^{mn}_{fe}\right) \left(\check{t}^{aef}_{imn} - \check{t}^{aef}_{inm}\right) \\
= & \frac{1}{2} \sum_{mnef} \check{v}^{mn}_{ef} \left(4\check{t}^{aef}_{imn} - 2\check{t}^{aef}_{nmi} - 2\check{t}^{aef}_{min} - 2\check{t}^{aef}_{inm} + \check{t}^{aef}_{nim} + \check{t}^{aef}_{mni}\right)
\end{aligned} \tag{41}$$

which is the same as Eq. (38), but with an extra factor of 1/2. The derivation from the spin-orbital equation provides the "right answer," showing that some modification of Goldstone's rules is necessary for higher-order coupled cluster. While it is possible to formulate other variations of rule R4′ to handle cases such as this correctly, it is clear that the Goldstone formulation lacks the general applicability of Brandow diagrams, and is not on its own sufficient to derive highly efficient working equations for the closed-shell case. We will show that such a diagrammatic technique may be derived algebraically, starting from the Brandow diagrammatic picture and using the key components of the Goldstone approach which are necessary to relate the spin-orbital and orbital representations.

4 NONORTHOGONALLY SPIN-ADAPTED DIAGRAMS

As illustrated earlier, the Goldstone rules, while quite useful for MBPT, CCSD, and other two-body theories, are not sufficiently general for CCSDT and CCSDTQ diagrammatics. Furthermore, the Goldstone rules require the enumeration of all permutationally unique diagrams for each equivalent Brandow diagram, which leads to a "Goldstone explosion" in CCSDT and especially in CCSDTQ. The resulting equations must also be factorized by hand after applying the Goldstone rules to obtain a compact, efficient set of equations. So, it would be desirable to derive a diagrammatic interpretation which

1. requires only the unique Brandow diagrams or only a very few additional diagrams,
2. produces compact, *already-factored* orbital equations, and
3. can be unambiguously applied to any order of coupled-cluster and other many-body theories.

Rather than use the MBPT Goldstone rules as a starting point, the approach taken here is to start with the spin-orbital interpretation of the Brandow diagrams, and then to use the basic ideas of the Goldstone diagrams—that the spin-orbital quantities are related to the orbital ones by antisymmetrization and that summation over closed loops gives a factor of 2 due to summation over both spins—to derive "nonorthogonally spin-adapted diagrammatics" directly. While the previous sections cover well-known diagrammatic methods; the content of this section is new.

4.1 PERMUTATIONS AND ANTISYMMETRIZERS

Since the orbital quantities are related to the spin-orbital and spin-integrated ones by antisymmetry, this property will serve as the fundamental linking principle between the two diagrammatics. To formalize the issue, consider the set of antisymmetric permutations of indices on a vertex v, $\{P_v(pq)\}$, where p and q run over all indices out of either the top (*out*) or bottom (*in*) labels of v. Note that "top" and "bottom" here refer to the indices when written as a tensor and not as a diagram. For example, the integral $\check{v}^{ab}_{ic}$ has two top/*out* labels, a and b, while there are three indices (a, b, and c) on "top" of the diagram in Fig. 1. Each of the permutations $P_v(pq)$ exchanges the indicated labels p and q, but only on the given vertex v, and introduces a factor of -1. Lastly, we must restrict p and q to be either both *in* or both *out* labels, as the spin-orbital quantities only exhibit antisymmetry which each set. Thus for $\check{v}^{ab}_{ci}$ we have $P_v(ab)$ and $P_v(ci)$.

Of course pair-wise permutations are self-inverse:

$$P_v(pq)^2 = 1 \tag{42}$$

Additionally, each label p has, due to the column symmetry of the orbital quantities, a column-symmetric partner p^*. The column-symmetric partner is simply the label in the same column when all labels are written out, for example, $v^{\ldots p \ldots q \ldots}_{\ldots p^* \ldots q^* \ldots}$. Then, since $v^{\ldots p \ldots q \ldots}_{\ldots p^* \ldots q^* \ldots} = v^{\ldots q \ldots p \ldots}_{\ldots q^* \ldots p^* \ldots}$, we have

$$P_v(pq) = P_v(p^* q^*) \tag{43}$$

$$P_v(pq) P_v(p^* q^*) = 1 \tag{44}$$

The product of a permutation with its column-symmetric partner (or equivalently itself) can also be written as a symmetric double permutation:

$$P_v(pq) P_v(p^* q^*) \equiv S_v \begin{pmatrix} p & q \\ p^* & q^* \end{pmatrix} \tag{45}$$

where the symmetric permutation interchanges both sets of labels simultaneously. This type of permutation has no effect on a quantity which is already column symmetric, but *creates* column symmetry when it does not already exist (e.g., when applied to the product of a tensor contraction). This equivalence can also be used to represent the product of symmetry-equivalent antisymmetrization operations as a combined antisymmetrization-symmetrization operation. For example,

$$\begin{aligned}
(1 + P_v(pq))(1 + P_v(p^* q^*)) &= 1 + P_v(pq) + P_v(p^* q^*) + P_v(pq) P_v(p^* q^*) \\
&= 1 + P_v(pq) + P_v(pq)^2 P_v(p^* q^*) + P_v(pq) P_v(p^* q^*) \\
&= (1 + P_v(pq))(1 + P_v(pq) P_v(p^* q^*)) \\
&= (1 + P_v(pq)) \left(1 + S_v \begin{pmatrix} p & q \\ p^* & q^* \end{pmatrix}\right)
\end{aligned} \tag{46}$$

Given a spin-orbital diagram, all of the related orbital Goldstone diagrams correspond to an antisymmetrization operation built up from the set of column-symmetry-unique permutations applied to the spin-orbital diagram. This antisymmetrization operation, however, generally contains permutations which involve more than two indices. In the general case, the unique permutations of more than two out of n indices can be formed from the set of two-index permutations $P \equiv \{P(pq)\}^n_{p<q=1}$ by taking only ordered products:

$$\tilde{P}^k \equiv \{P(p_1 q_1) \ldots P(p_k q_k) \,|\, p_1 < \cdots < p_k \wedge p_l < q_l \forall l\}^n_{p_1 \ldots p_k, q_1 \ldots q_k = 1} \tag{47}$$

$$\mathscr{A}_P \equiv \bigcup_{k=0}^{n-1} \tilde{P}^k \tag{48}$$

The special case $\tilde{P}^0 = \{1\}$ even if P is the empty set. The permutations in $\tilde{P}^k$ are termed *ordered permutations*. All of the possible permutations of n objects $p_1 \ldots p_n$

are then given by $\mathscr{A}_P p_1 \ldots p_n$ (see Theorem 1). Taking the set of indices abc as an example, we have

$$P = \{P(ab), P(ac), P(bc)\} \tag{49}$$
$$\tilde{P}^0 = \{1\} \tag{50}$$
$$\tilde{P}^1 = P \tag{51}$$
$$\tilde{P}^2 = \{P(ab)P(bc), P(ac)P(bc)\} \tag{52}$$

From these, it can be seen that $\mathscr{A}_P$ gives the correct result:

$$\begin{aligned}\mathscr{A}_P abc &= \tilde{P}^0 abc + \tilde{P}^1 abc + \tilde{P}^2 abc \\ &= (abc) + (-bac - cba - acb) + (bca + cab)\end{aligned} \tag{53}$$

In manipulating sets of permutations and their antisymmetrizers, it is quite helpful to be able to separate the permutations of a set of indices into those of some disjoint partitioning. In particular, the total antisymmetrizer can be written as a product of subset antisymmetrizers and a special partial antisymmetrizer, which contains permutations between the subsets (see Theorem 2):

$$\begin{aligned}\mathscr{A}_P &= \mathscr{A}_{P_1}\mathscr{A}_{P_2}\tilde{\mathscr{A}}_{P_1-P_2} \\ &= \tilde{\mathscr{A}}_{P_1-P_2}\mathscr{A}_{P_1}\mathscr{A}_{P_2}\end{aligned} \tag{54}$$

where $\mathscr{A}_{P_1}$ is the antisymmetrizer for the indices in set 1 (of size n_1, and similarly for set 2 of size n_2), and P (the set of permutations for the concatenations of sets 1 and 2) is equal to $P_1 \cup P_2 \cup P_{1-2}$. The "special antisymmetrizer" $\tilde{\mathscr{A}}_{P_1-P_2}$ is given by

$$\begin{aligned}\tilde{\tilde{P}}^k_{1-2} \equiv \{&P_{1-2}(p_1q_1)\ldots P_{1-2}(p_kq_k)\,|\,p_1 < \cdots < p_k \wedge q_1 < \cdots < q_k \\ &\wedge p_l < q_l \forall\, l\}^{n_1,n_2}_{p_1\ldots p_k=1,q_1\ldots q_k=1}\end{aligned} \tag{55}$$

$$\tilde{\mathscr{A}}_{P_1-P_2} \equiv \bigcup_{k=0}^{\min\{n_1,n_2\}} \tilde{\tilde{P}}^k_{1-2} \tag{56}$$

where the set of permutations P_{1-2} interchanges indices p_i from set 1 with indices q_i from set 2 (i.e., $P_{1-2} = P \setminus (P_1 \cup P_2)$). The permutations in $\tilde{\tilde{P}}^k$ are *doubly ordered permutations.* To give another example, consider the index sets ab and cd. For these sets, the special antisymmetrizer is given by

$$P_1 = \{P(ab)\} \tag{57}$$
$$P_2 = \{P(cd)\} \tag{58}$$
$$P_{1-2} = \{P(ac), P(ad), P(bc), P(bd)\} \tag{59}$$
$$\tilde{\tilde{P}}^0_{1-2} = \{1\} \tag{60}$$
$$\tilde{\tilde{P}}^1_{1-2} = P_{1-2} \tag{61}$$

$$\tilde{\tilde{P}}^2_{1-2} = \{P(ac)P(bd)\} \tag{62}$$

from which one can easily see that

$$\begin{aligned}\tilde{\mathscr{A}}_{P_1-P_2}abcd &= \tilde{\tilde{P}}^0_{1-2}abcd + \tilde{\tilde{P}}^1_{1-2}abcd + \tilde{\tilde{P}}^2_{1-2}abcd \\ &= (abcd) + (-cbad - dbca - acbd - adcb) + (cdab)\end{aligned} \tag{63}$$

It is left as an exercise for the reader to verify that $\tilde{\mathscr{A}}_{P_1-P_2}\mathscr{A}_{P_1}\mathscr{A}_{P_2}abcd = \mathscr{A}_P abcd$ and $\mathscr{A}_{P_1}\mathscr{A}_{P_2}\tilde{\mathscr{A}}_{P_1-P_2}abcd = \mathscr{A}_P abcd$ (but *not* $\mathscr{A}_{P_1}\tilde{\mathscr{A}}_{P_1-P_2}\mathscr{A}_{P_2}abcd = \mathscr{A}_P abcd$).

4.2 DIAGRAMMATIC AND ALGEBRAIC FORMALISM

To simplify derivation of the new rules, it suffices to consider only diagrams that contain one or two vertices. This is not restrictive since the factorization of the target equations into unary and binary terms can always be performed at the spin-orbital diagram stage, and then the resulting diagrams in terms of intermediates transformed into orbital equations. Since there are only one or two vertices (of which one is presumably $\hat{H}_N$), then the exponential factor F_e does not come into play directly. Rather, this factor must be applied before factorization and will be ignored from here on.

Since diagrams of only one vertex are trivial (this corresponds to simply assigning or adding one quantity to another), we can consider only the general two-vertex diagram of Fig. 8. Each vertex v in this form has: $n_{a;v}$ *particle-out* (particle being synonymous with virtual or unoccupied and represented by an upward-oriented arrow) and $n_{i;v}$ *hole-in* (hole meaning occupied and downward-oriented) lines which extend upward (above the vertex) and $n_{b;v}$ *particle-in* and $n_{j;v}$ *hole-out* lines which extend downward (below the vertex), to give n_v total lines. The upper vertex is labeled A and the lower B. Usually, the operators will be number-conserving such that $n_{a;v} + n_{j;v} = n_{b;v} + n_{i;v}$, and it will simplify the discussion to make this assumption, although the results obtained here are applicable to nonnumber-conserving cases (i.e., those in EOM-CC) with only slight modification.

Of the labels on each vertex, some $n_e \leq \min(n_{b;A}, n_{a;B})$ of the *particle-in* vertices on A and the *particle-out* vertices on B are contracted together as are some $n_m \leq \min(n_{j;A}, n_{i;B})$ of the *hole-out* and *hole-in* vertices on A and B, respectively. The individual labels are conveniently called $a_{1;A} \ldots a_{n_{a;A};A}$, $e_1 \ldots e_{n_e}$, etc. The possible permutations of the labels relevant to antisymmetry are then

$$\begin{aligned}P_{o;A} \equiv{}& \{P_A(a_{k;A}a_{l;A})\}_{k<l=1}^{n_{a;A}} \cup \{P_A(j_{k;A}j_{l;A})\}_{k<l=1}^{n_{j;A}} \\ &\cup \{P_A(a_{k;A}j_{l;A})\}_{k,l=1}^{n_{a;A},n_{j;A}}\end{aligned} \tag{64}$$

$$\begin{aligned}P_{i;A} \equiv{}& \{P_A(b_{k;A}b_{l;A})\}_{k<l=1}^{n_{b;A}} \cup \{P_A(i_{k;A}i_{l;A})\}_{k<l=1}^{n_{i;A}} \\ &\cup \{P_A(b_{k;A}i_{l;A})\}_{k,l=1}^{n_{b;A},n_{i;A}}\end{aligned} \tag{65}$$

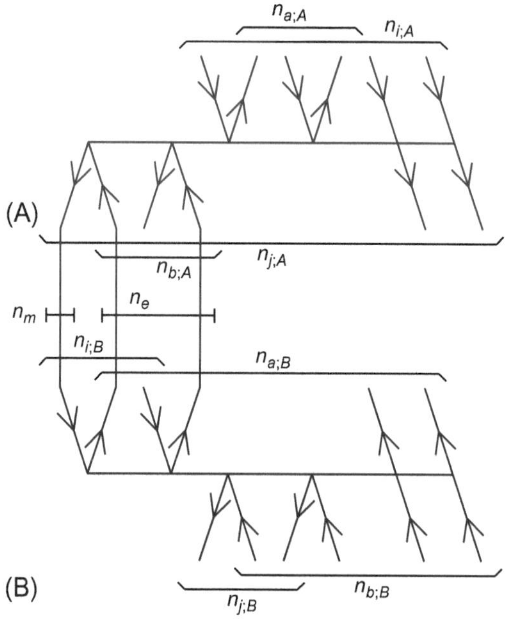

FIG. 8

General form of a two-vertex Brandow diagram and the grouping of lines into various sets. The "extra" lines on the right may be either hole or particle lines.

and similarly for B. Since there are an equal number of *out* and *in* vertices, the column symmetry of the orbital quantities means that each permutation in $P_{o;A}$ is related to a permutation in $P_{i;A}$ and vice versa, for example, $P_A(a_{1;A}a_{2;A}) = P_A(a^*_{1;A}a^*_{2;A}) = P_A(i_{1;A}b_{1;A})$ for $A^{a_1a_2}_{i_1b_1}$. Only one set is needed to produce distinct Goldstone diagrams and we can arbitrarily choose $P_A \equiv P_{i;A}$ (i.e., permutations among bottom/*in* indices). The final set of permutations P_A and its unique products form the relationship between the spin-orbital and orbital quantities (assuming the factor for conversion from spin-orbital summations to orbital ones is accounted for separately):

$$\begin{aligned} A^{pq\ldots}_{rs\ldots} &\Rightarrow \sum_{k=0}^{n_A-1} \tilde{P}^k_A \check{A}^{pq\ldots}_{rs\ldots} \\ &= \mathscr{A}_A \check{A}^{pq\ldots}_{rs\ldots} \end{aligned} \tag{66}$$

where the application of the set $\tilde{P}^k_A$ sums over each element applied to the object in turn.

Besides permutations of the two diagram vertices, we must also consider permutations of the external labels. To use the previously defined language, the external lines can be considered to belong also to a third (C) vertex, with sets of permutations defined similarly to A and B (except of course no lines are contracted). Rule 6 of the Brandow diagrammatics gives permutations (and products of permutations since multiple indices may be antisymmetrized) between labels of the same type on different vertices, conveniently denoted by

$$\begin{aligned} P_{Brandow} &\equiv P_{a;Brandow} \cup P_{b;Brandow} \cup P_{i;Brandow} \cup P_{j;Brandow} \\ &= \{P(a_{k;A}a_{l;B})\}_{k=1,l=1}^{n_{a;A},n_{a;B}-n_e} \cup \{P(j_{k;A}j_{l;B})\}_{k=1,l=1}^{n_{j;A},n_{j;B}-n_m} \\ &\cup \{P(i_{k;A}i_{l;B})\}_{k=1,l=1}^{n_{i;A}-m,n_{i;B}} \cup \{P(b_{k;A}b_{l;B})\}_{k=1,l=1}^{n_{b;A}-n_e,n_{b;B}} \end{aligned} \tag{67}$$

$$\begin{aligned} \tilde{\mathscr{A}}_{Brandow} \equiv &\left(\bigcup_{k=0}^{\min\{n_{a;A},n_{a;B}-n_e\}} \tilde{\tilde{P}}^{k}_{a;Brandow}\right) \otimes \left(\bigcup_{k=0}^{\min\{n_{j;A},n_{j;B}-n_m\}} \tilde{\tilde{P}}^{k}_{j;Brandow}\right) \\ &\otimes \left(\bigcup_{k=0}^{\min\{n_{i;A}-n_m,n_{i;B}\}} \tilde{\tilde{P}}^{k}_{i;Brandow}\right) \otimes \left(\bigcup_{k=0}^{\min\{n_{b;A}-n_e,n_{b;B}\}} \tilde{\tilde{P}}^{k}_{b;Brandow}\right) \\ = &\tilde{\mathscr{A}}_{a;Brandow}\tilde{\mathscr{A}}_{j;Brandow}\tilde{\mathscr{A}}_{i;Brandow}\tilde{\mathscr{A}}_{b;Brandow} \end{aligned} \tag{68}$$

where $a_{l;B}$, $j_{l;B}$, $i_{k;A}$, and $b_{k;A}$ run only over noncontracted labels. The permutations here are not labeled with a vertex subscript because they affect the labeling of all vertices (including the external "vertex" C).

Armed with the relationship between the spin-orbital ($\boldsymbol{A}$, $\boldsymbol{B}$, and $\boldsymbol{C}$) and orbital ($\check{A}$, $\check{B}$, and $\check{C}$) quantities and an explicit form for the Brandow antisymmetrization operator, we can now write a preliminary relationship between the two diagrammatics:

$$\boldsymbol{C} = \tilde{\mathscr{A}}_{Brandow}F_sF_hF_p\sum \boldsymbol{AB} \tag{69}$$

$$\Updownarrow$$

$$\mathscr{A}_C\check{C} = \tilde{\mathscr{A}}_{Brandow}F_sF_hF_pF_c\sum \mathscr{A}_A\mathscr{A}_B\check{A}\check{B} \tag{70}$$

F_h and F_p are the factors for identical contracted hole and particle lines, respectively, from rule R2. F_s is the sign factor from rule R3. The summation from rule R5 goes over spin-orbitals in the first equation, but over spatial orbitals in the second, with the additional factor F_c accounting for the factor of 2 for each closed loop. The indices on A, B, and C and on the summation symbol are implicit here and in the following derivation.

4.3 FROM SPIN-ORBITAL TO ORBITAL

The first logical thing to do is to somehow remove the antisymmetrization operation on $\check{C}$ so that an equation for a specific permutation of the indices on $\check{C}$ can be obtained (and more importantly, an expression for each unique orbital case can be derived). The antisymmetrizer $\mathscr{A}_C$ contains some permutations of labels on $\check{C}$ where the labels are of the same type. For example, given $C = W^{abm}_{ief}$, $\mathscr{A}_C$ contains (among others) $P(ab)$, $P(ef)$, etc., where the orderings induced by these permutations correspond to the same orbital case. However, $\mathscr{A}_C$ also contains permutations which mix labels of different types, for example, $P(am) = P(if)$, where now the different induced orderings are distinct orbital cases. Note that, for example, $P(bm)$ does not induce a separate orbital case since it is equivalent by column symmetry to $P(ef)$. The different orbital cases related to the same spin-orbital quantity can be classified by a number n_s which counts the number of pairs of hole labels and particle labels in the same column (this is also the number of external loops when C itself is viewed as a diagram). For example, $\check{v}^{ai}_{bj}$ has $n_s = 0$ while $\check{v}^{ai}_{jb}$ has $n_s = 2$.

Using Theorem 4, $\mathscr{A}_C$ can be split into several parts, each of which applies to a specific n_s value:

$$
\begin{aligned}
\mathscr{A}_C\check{C} &= \sum_{n_s} \mathscr{A}_{a;C}\mathscr{A}_{j;C}\tilde{\mathscr{A}}_{\substack{a\\i}-\substack{j\\i};C}\tilde{\mathscr{A}}_{\substack{a\\b}-\substack{j\\b};C}\check{C}_{n_s} \\
&= \sum_{n_s} \tilde{\mathscr{A}}_{Brandow}F_sF_hF_pF_c\sum\left(\mathscr{A}_A\mathscr{A}_B\check{A}\check{B}\right)_{n_s}
\end{aligned}
\tag{71}
$$

$$\Downarrow$$

$$
\mathscr{A}_{a;C}\mathscr{A}_{j;C}\tilde{\mathscr{A}}_{\substack{a\\i}-\substack{j\\i};C}\tilde{\mathscr{A}}_{\substack{a\\b}-\substack{j\\b};C}\check{C}_{n_s} = \tilde{\mathscr{A}}_{Brandow}F_sF_hF_pF_c\sum\left(\mathscr{A}_A\mathscr{A}_B\check{A}\check{B}\right)_{n_s}
\tag{72}
$$

where, for example, $\mathscr{A}_{a;C}$ antisymmetrizes the *particle-out* labels on $\check{C}$ and $\tilde{\mathscr{A}}_{\substack{a\\i}-\substack{j\\i};C}$ exchanges *hole-in* (i) labels which are partnered with *particle-in* (a) labels with any *hole-in* (i) labels which are partnered with *hole-out* (j) labels (on $\check{C}$). The notation $(\ldots)_{n_s}$ means that only permutations which give external labels consistent with the given value of n_s are retained ($\tilde{\mathscr{A}}_{Brandow}$ does not change n_s).

The antisymmetrization operators on the left-hand side of Eq. (71) can be cleaned up somewhat by inserting the identity as $(n_{\substack{a\\i};C}!n_{\substack{j\\i};C}!n_{\substack{a\\b};C}!n_{\substack{j\\b};C}!)^{-1}\mathscr{A}_{\substack{a\\i};C}\mathscr{A}_{\substack{j\\i};C}\mathscr{A}_{\substack{a\\b};C}\mathscr{A}_{\substack{j\\b};C}$, which is possible since, for example, $\mathscr{A}_{a;C}\mathscr{A}_{\substack{a\\i};C} = \mathscr{A}_{a;C}(n_{\substack{a\\i};C}!)$ due to the fact that all permutations in $\mathscr{A}_{\substack{a\\i};C}$ are also present in $\mathscr{A}_{a;C}$ and using Eq. (42). Now, we can simplify by using Theorem 2:

$$
\begin{aligned}
\mathscr{A}_{a;C}\mathscr{A}_{j;C}\tilde{\mathscr{A}}_{\substack{a\\i}-\substack{j\\i};C}\tilde{\mathscr{A}}_{\substack{a\\b}-\substack{j\\b};C}\check{C}_{n_s} &= \frac{\mathscr{A}_{a;C}\mathscr{A}_{j;C}\mathscr{A}_{\substack{a\\i};C}\mathscr{A}_{\substack{j\\i};C}\mathscr{A}_{\substack{a\\b};C}\mathscr{A}_{\substack{j\\b};C}\tilde{\mathscr{A}}_{\substack{a\\i}-\substack{j\\i};C}\tilde{\mathscr{A}}_{\substack{a\\b}-\substack{j\\b};C}}{n_{\substack{a\\i};C}!n_{\substack{j\\i};C}!n_{\substack{a\\b};C}!n_{\substack{j\\b};C}!}\check{C}_{n_s} \\
&= \frac{\mathscr{A}_{a;C}\mathscr{A}_{j;C}\left(\mathscr{A}_{\substack{a\\i};C}\mathscr{A}_{\substack{j\\i};C}\tilde{\mathscr{A}}_{\substack{a\\i}-\substack{j\\i};C}\right)\left(\mathscr{A}_{\substack{a\\b};C}\mathscr{A}_{\substack{j\\b};C}\tilde{\mathscr{A}}_{\substack{a\\b}-\substack{j\\b};C}\right)}{n_{\substack{a\\i};C}!n_{\substack{j\\i};C}!n_{\substack{a\\b};C}!n_{\substack{j\\b};C}!}\check{C}_{n_s} \\
&= \frac{\mathscr{A}_{a;C}\mathscr{A}_{j;C}\mathscr{A}_{i;C}\mathscr{A}_{b;C}}{n_{\substack{a\\i};C}!n_{\substack{j\\i};C}!n_{\substack{a\\b};C}!n_{\substack{j\\b};C}!}\check{C}_{n_s}
\end{aligned}
\tag{73}
$$

A similar trick can be applied to simplify the Brandow antisymmetrizer, by inserting additional antisymmetrizers so that it becomes

$$\begin{aligned}
\tilde{\mathscr{A}}_{Brandow} &= \mathscr{N}^{-1}\tilde{\mathscr{A}}_{Brandow}\mathscr{A}_{a;A}\mathscr{A}_{j;A\backslash m}\mathscr{A}_{i;A}\mathscr{A}_{b;A\backslash e}\mathscr{A}_{a;B\backslash e}\mathscr{A}_{j;B}\mathscr{A}_{i;B\backslash m}\mathscr{A}_{b;B} \\
&= \mathscr{N}^{-1}\left(\tilde{\mathscr{A}}_{a;Brandow}\mathscr{A}_{a;A}\mathscr{A}_{a;B\backslash e}\right)\left(\tilde{\mathscr{A}}_{j;Brandow}\mathscr{A}_{j;A\backslash m}\mathscr{A}_{j;B}\right) \\
&\quad\times\left(\tilde{\mathscr{A}}_{i;Brandow}\mathscr{A}_{i;A}\mathscr{A}_{i;B\backslash m}\right)\left(\tilde{\mathscr{A}}_{b;Brandow}\mathscr{A}_{b;A\backslash e}\mathscr{A}_{b;B}\right) \\
&= \mathscr{N}^{-1}\mathscr{A}_{a;C}\mathscr{A}_{j;C}\mathscr{A}_{i;C}\mathscr{A}_{b;C}
\end{aligned} \tag{74}$$

$$\mathscr{N} = n_{a;A}!(n_{j;A} - n_m)!n_{i;A}!(n_{b;A} - n_e)!(n_{a;B} - n_e)!n_{j;B}!(n_{i;B} - n_m)!n_{b;B}! \tag{75}$$

where, for example, $a;B \setminus e$ is the set of *uncontracted particle-out* labels on $\check{B}$. Both the $\check{C}$ and $\check{A}\check{B}$ portions of Eq. (71) now have the same set of antisymmetrizers applied, so that a single orbital case of $\check{C}$ can be extracted from both sides by removing this antisymmetrization step.

Of course, the equation obtained this way is not unique, and that is the origin of the indeterminacy of the orbital tensor elements. However, the equation is valid when the antisymmetrizers are reapplied, and is unique when the procedure here is followed, so that the orbital quantities obtained this way can be considered as a "canonical" form. After removing the antisymmetrizers, the orbital equation becomes

$$\check{C} = \frac{n_{{}^{a}_{i};C}!n_{{}^{j}_{i};C}!n_{{}^{a}_{b};C}!n_{{}^{j}_{b};C}!F_sF_hF_pF_c\sum\left(\mathscr{A}_A\mathscr{A}_B\check{A}\check{B}\right)_{\check{C}}}{n_{a;A}!(n_{j;A} - n_m)!n_{i;A}!(n_{b;A} - n_e)!(n_{a;B} - n_e)!n_{j;B}!(n_{i;B} - n_m)!n_{b;B}!} \tag{76}$$

where now $\check{C}$ is has a definite ordering of the labels, and the notation $(\ldots)_{\check{C}}$ retains only contributions which match this ordering.

4.4 FACTORIZING THE ORBITAL EQUATIONS

At this point, Eq. (76) is very similar to the usual Goldstone interpretation, in that the antisymmetrizers $\mathscr{A}_A\mathscr{A}_B$ produce the unique Goldstone diagrams (with some overcounting which cancels the factorials in the denominator). However, with the equation in a formal representation, the antisymmetrizers can be manipulated further to expose the desired compact, factorized result directly. In order to proceed we must do two things: first, while the antisymmetrizers produce all of the Goldstone diagrams from some initial diagram determined by the label ordering of $\check{A}\check{B}$, it will prove advantageous to choose this initial ordering carefully as the one (or ones, as noted below) which has the maximal number of closed loops, and second, starting from this ordering, we will subdivide the labels on each fragment into three groups: c, x, and u. The c labels are contracted labels whose partner label is also contracted on both vertices. The x labels are either contracted *and* have an uncontracted partner or vice versa on at least one vertex. The u labels are uncontracted and have uncontracted partners. Choosing a well-defined initial ordering is important since these classifications depend on that ordering, and for this ordering, any permutation *decreases* the number of closed loops. For a small number of Brandow diagrams, there will be more than one distinct Goldstone diagram with the maximal number of closed loops. In this case it will prove advantageous to forbid permutations which relate one "primary" diagram to another and to sum together the contributions from each such diagram.

The antisymmetrizers $\mathscr{A}_A$ and $\mathscr{A}_B$ can be decomposed into different partitions of these groups using Theorem 2. The first step is to remove the redundant permutations over c and x labels, since a permutation of these labels on $\check{A}$ is the same as permutation on $\check{B}$ followed by relabeling the contracted labels. This gives

$$\begin{aligned}\check{C} &= \mathscr{M} F_s F_h F_p F_c \sum \left(\mathscr{A}_A \mathscr{A}_B \check{A}\check{B}\right)_{\check{C}} \\ &= \mathscr{M} F_s F_h F_p F_c \sum \left(\mathscr{A}_A \left\{\mathscr{A}_{cx;B} \mathscr{A}_{u;B} \tilde{\mathscr{A}}_{cx-u;B}\right\} \check{A}\check{B}\right)_{\check{C}} \\ &= \mathscr{M} F_s \max\{F_h, F_p\} F_c \sum \left(\mathscr{A}_A \mathscr{A}_{u;B} \tilde{\mathscr{A}}_{cx-u;B} \check{A}\check{B}\right)_{\check{C}}\end{aligned} \tag{77}$$

$$\mathscr{M} = \frac{n_{{}^a_i;C}!\, n_{{}^{\;j}_i;C}!\, n_{{}^a_b;C}!\, n_{{}^{\;j}_b;C}!}{n_{a;A}!(n_{j;A} - n_m)!\, n_{i;A}!(n_{b;A} - n_e)!(n_{a;B} - n_e)!\, n_{j;B}!(n_{i;B} - n_m)!\, n_{b;B}!} \tag{78}$$

Since the number of $cx \equiv c \cup x$ labels is the greater of the number of contracted hole and particle lines, then $\mathscr{A}_{cx;B} = n_{cx}!$ (when applied with $\mathscr{A}_A$ which contains the same permutations) cancels either F_h or F_p. $\mathscr{A}_A$ can also be decomposed, but this time it is better to break it into c and xu groups:

$$\begin{aligned}\check{C} &= \mathscr{M} F_s \max\{F_h, F_p\} F_c \sum \left(\mathscr{A}_A \mathscr{A}_{u;B} \tilde{\mathscr{A}}_{cx-u;B} \check{A}\check{B}\right)_{\check{C}} \\ &= \mathscr{M} F_s \max\{F_h, F_p\} F_c \sum \left(\left\{\mathscr{A}_{xu;A} \mathscr{A}_{c;A} \tilde{\mathscr{A}}_{c-xu;A}\right\} \mathscr{A}_{u;B} \tilde{\mathscr{A}}_{cx-u;B} \check{A}\check{B}\right)_{\check{C}}\end{aligned} \tag{79}$$

The next step is to use the rather esoteric form of Theorem 3, although its purpose will become clear after some more manipulation:

$$\begin{aligned}\check{C} &= \mathscr{M} F_s \max\{F_h, F_p\} F_c \sum \left(\mathscr{A}_{xu;A} \mathscr{A}_{c;A} \tilde{\mathscr{A}}_{c-xu;A} \mathscr{A}_{u;B} \tilde{\mathscr{A}}_{cx-u;B} \check{A}\check{B}\right)_{\check{C}} \\ &= \mathscr{M} F_s \max\{F_h, F_p\} F_c \sum \left(\mathscr{A}_{xu;A} \mathscr{A}_{c;A} \tilde{\mathscr{A}}_{c-xu;A} \mathscr{A}_{xu;B} \frac{n_{u;B}!}{(n_x + n_{u;B})!} \tilde{\mathscr{A}}_{cx-u;B} \check{A}\check{B}\right)_{\check{C}} \\ &= \mathscr{M} F_s \max\{F_h, F_p\} F_c \\ &\quad \times \sum \left(\mathscr{A}_{xu;A} \mathscr{A}_{c;A} \tilde{\mathscr{A}}_{c-xu;A} \mathscr{A}_{xu;B} \sum_{k=0}^{\min\{n_c, n_{u;B}\}} \frac{k!}{(n_x + k)!} \tilde{P}^k_{c-u;B} \check{A}\check{B}\right)_{\check{C}}\end{aligned} \tag{80}$$

The second equality is obtained by inserting $\frac{n_{u;B}!}{(n_x+n_{u;B})!} \mathscr{A}_{x;B} \tilde{\mathscr{A}}_{x-u;B}$ and using Theorem 2.

Before attempting to clean up the factorials in the denominator of $\mathscr{M}$ which will cancel out eventually, we can use the factorials in the numerator to build a useful operator. Since this equation is only "really" valid after reapplication of the antisymmetrizers $\mathscr{A}_{a;C}\mathscr{A}_{j;C}\mathscr{A}_{i;C}\mathscr{A}_{b;C}$, then the fact that these contain all of the permutations in, for example, $\mathscr{A}_{\substack{a\\i};C}$ allows us to "reconstitute" these antisymmetrizers on the RHS:

$$\check{C} = \frac{\mathscr{A}_{\substack{a\\i};C}\mathscr{A}_{\substack{j\\i};C}\mathscr{A}_{\substack{a\\b};C}\mathscr{A}_{\substack{j\\b};C}F_s \max\{F_h, F_p\}F_c}{n_{a;A}!(n_{j;A}-n_m)!n_{i;A}!(n_{b;A}-n_e)!(n_{a;B}-n_e)!n_{j;B}!(n_{i;B}-n_m)!n_{b;B}!} \times \sum \left(\mathscr{A}_{xu;A}\mathscr{A}_{xu;B}\mathscr{A}_{c;A}\tilde{\mathscr{A}}_{c-xu;A} \sum_{k=0}^{\min\{n_c,n_{u;B}\}} \frac{k!}{(n_x+k)!}\tilde{\tilde{P}}^k_{c-u;B}\check{A}\check{B} \right)_{\check{C}} \tag{81}$$

But, since we are now applying the same permutations twice, we can use Eq. (46) to change this into a symmetrizer:

$$\check{C} = \frac{\mathscr{S}_C F_s \max\{F_h, F_p\}F_c}{n_{a;A}!(n_{j;A}-n_m)!n_{i;A}!(n_{b;A}-n_e)!(n_{a;B}-n_e)!n_{j;B}!(n_{i;B}-n_m)!n_{b;B}!} \times \sum \left(\mathscr{A}_{xu;A}\mathscr{A}_{xu;B}\mathscr{A}_{c;A}\tilde{\mathscr{A}}_{c-xu;A} \sum_{k=0}^{\min\{n_c,n_{u;B}\}} \frac{k!}{(n_x+k)!}\tilde{\tilde{P}}^k_{c-u;B}\check{A}\check{B} \right)_{\check{C}} \tag{82}$$

where the symmetrizer $\mathscr{S}_C$ is defined as

$$\begin{aligned} S_C &\equiv S_{\substack{a\\i};C} \cup S_{\substack{j\\i};C} \cup S_{\substack{a\\b};C} \cup S_{\substack{j\\b};C} \\ &= \left\{ S_C \begin{pmatrix} a_k a_l \\ a_k^* a_l^* \end{pmatrix} \middle| \; a_k^* \in i \wedge a_l^* \in i \right\}_{k<l=1}^{n_{a;C}} \\ &\cup \left\{ S_C \begin{pmatrix} j_k j_l \\ j_k^* j_l^* \end{pmatrix} \middle| \; j_k^* \in i \wedge j_l^* \in i \right\}_{k<l=1}^{n_{j;C}} \\ &\cup \left\{ S_C \begin{pmatrix} a_k a_l \\ a_k^* a_l^* \end{pmatrix} \middle| \; a_k^* \in b \wedge a_l^* \in b \right\}_{k<l=1}^{n_{a;C}} \\ &\cup \left\{ S_C \begin{pmatrix} j_k j_l \\ j_k^* j_l^* \end{pmatrix} \middle| \; j_k^* \in b \wedge j_l^* \in b \right\}_{k<l=1}^{n_{j;C}} \end{aligned} \tag{83}$$

$$\begin{aligned} \mathscr{S}_C \equiv &\left(\bigcup_{k=0}^{n_{\substack{a\\i};C}-1} \tilde{S}^k_{\substack{a\\i};C} \right) \otimes \left(\bigcup_{k=0}^{n_{\substack{j\\i};C}-1} \tilde{S}^k_{\substack{j\\i};C} \right) \\ &\otimes \left(\bigcup_{k=0}^{n_{\substack{a\\b};C}-1} \tilde{S}^k_{\substack{a\\b};C} \right) \otimes \left(\bigcup_{k=0}^{n_{\substack{j\\b};C}-1} \tilde{S}^k_{\substack{j\\b};C} \right) \end{aligned} \tag{84}$$

This symmetrization step is important as it ensures the column symmetry of $\check{C}$, which is vital to practical implementation of nonorthogonally spin-adapted methods (and of course is assumed in this diagrammatic derivation).

The last simplification to make is to deal with the factorials in the denominator and the antisymmetrizers $\mathscr{A}_{xu;A}$ and $\mathscr{A}_{xu;B}$. Since the permutations in these antisymmetrizers which relate external labels of different types (and whose partners are of different types) change the value of n_s, we may ignore them when computing the equations for a specific orbital case (and hence value of n_s). In some cases, though, one permutation (say from $\mathscr{A}_{xu;B}$) may change n_s while a later permutation (from $\mathscr{A}_{xu;A}$) may change it back to the "correct" value. However, permutations such as this relate one Brandow diagram with a particular number of closed loops to another with the same number of loops. So, as mentioned previously, it is prudent to forbid these types of permutations and simply sum the equations for all Brandow diagrams with the maximal number of loops for a given n_s together.

The remaining permutations in $\mathscr{A}_{xu;A}$ (and similarly for $\mathscr{A}_{xu;B}$) can be decomposed as

$$\mathscr{A}_{xu;A} = \mathscr{A}_{aj;A\backslash m} \mathscr{A}_{\substack{mm\\i\,b};A} \tilde{\mathscr{A}}_{aj-\substack{mm\\i\,b};A} \tag{85}$$

where the set $\substack{mm\\i\,b}$ is the set of indices in $x;A$ where one hole label is contracted. Furthermore, we can decompose

$$\begin{aligned} \mathscr{A}_{aj;A\backslash m} &= \mathscr{A}_{a;A} \mathscr{A}_{j;A\backslash m} \tilde{\mathscr{A}}_{a-j;A\backslash m} \\ &\overset{\delta n_s=0}{=} \mathscr{A}_{a;A} \mathscr{A}_{j;A\backslash m} \tilde{\mathscr{A}}_{\substack{a\\i}-\substack{j\\i};\substack{A\backslash m\\x}} \tilde{\mathscr{A}}_{\substack{a\\b}-\substack{j\\b};\substack{A\backslash m\\x}} \end{aligned} \tag{86}$$

where, for example, $\tilde{\mathscr{A}}_{\substack{a\\i}-\substack{j\\i};\substack{A\backslash m\\x}}$ interchanges labels which are of type a with those of type $j \setminus m$ on $\check{A}$, but only when their partners (through the column symmetry of $\check{C}$) are both of type i on $\check{A}$ or, if contracted, on $\check{B}$. The second equality uses the fact the we care only about permutations which conserve n_s. Similarly,

$$\begin{aligned} \mathscr{A}_{\substack{mm\\i\,b};A} \tilde{\mathscr{A}}_{aj-\substack{mm\\i\,b};A} &= \mathscr{A}_{\substack{m\\i};A} \mathscr{A}_{\substack{m\\b};A} \tilde{\mathscr{A}}_{\substack{m\\i}-\substack{m\\b};A} \tilde{\mathscr{A}}_{aj-\substack{mm\\i\,b};A} \\ &\overset{\delta n_s=0}{=} \mathscr{A}_{\substack{m\\i};A} \mathscr{A}_{\substack{m\\b};A} \tilde{\mathscr{A}}_{\substack{a\\i}-\substack{a\\b};\substack{B\\A}} \tilde{\mathscr{A}}_{\substack{j\\i}-\substack{j\\b};\substack{B\\A}} \tilde{\mathscr{A}}_{aj-\substack{mm\\i\,b};A} \end{aligned} \tag{87}$$

These antisymmetrizers have three types of permutations: (1) permutations of external labels (i.e., from sets x and u) of the same type (or which have partners of the same type), (2) permutations of external labels which are not of the same type, and therefore change n_s, and (3) permutations involving only partially contracted labels (labels from the set x) not already included elsewhere. Since the factor of $n_{a;A}!$ could be canceled by antisymmetrizer $\mathscr{A}_{a;A}$ and so on for each factor, we must examine which permutations from $\mathscr{A}_{xu;A}$ are "missing" to form $\mathscr{A}_{a;A} \mathscr{A}_{j;A} \mathscr{A}_{i;A/m} \mathscr{A}_{b;A/e}$ and similarly for B. Additionally, if a factor of $n_x!$ is multiplied and divided in the sum over k then, as careful analysis shows, it is actually simpler to relate the xu antisymmetrizers for both vertices into the daunting string of operators, $\mathscr{A}_{a;A} \mathscr{A}_{j;A} \mathscr{A}_{i;A/m} \mathscr{A}_{b;A/e} \mathscr{A}_{a;B/e} \mathscr{A}_{j;B/m} \mathscr{A}_{i;B} \mathscr{A}_{b;B} \mathscr{A}_{x}$, which can simultaneously cancel all of the remaining factorials. Since permutations of type (2) and permutations in (3) which, due to the column symmetry of $\check{C}$ also change n_s can be ignored (an exception to this is discussed later), the "only" remaining permutations needed to cancel the factors are $\mathscr{A}_{\substack{a\\i};A} \mathscr{A}_{\substack{a\\b};A} \mathscr{A}_{\substack{j\\i};A} \mathscr{A}_{\substack{j\\b};A} \mathscr{A}_{\substack{a\\i};A} \mathscr{A}_{\substack{j\\i};A} \mathscr{A}_{\substack{a\\b};A} \mathscr{A}_{\substack{j\\b};A} \mathscr{A}_{\substack{a\\i};x} \mathscr{A}_{\substack{j\\i};x} \mathscr{A}_{\substack{a\\b};x} \mathscr{A}_{\substack{j\\b};x}$ where the $\mathscr{A}_{\substack{p\\q};x}$ antisymmetrizers permute labels which are of the type p on $\check{A}$ and whose column-symmetric partners in $\check{C}$ are of the type q on $\check{B}$.

What appears to be a startlingly complicated set of antisymmetrizers is actually quite simple when examined diagrammatically, as it simply antisymmetrizes pairs of external labels of the same combined type ($\substack{a\\i}$, $\substack{j\\i}$, etc.) which are either on the same vertex or both partially contracted. In essence these are "identical" external pairs. Since these permutations are lacking in $\mathscr{A}_{xu;A}\mathscr{A}_{xu;B}$, they must be inserted as an identity with the corresponding factors $n_{\substack{a\\i};A}!$, etc. in the denominator (this can be done since all of these permutations are in $\mathscr{A}_{a;C}\mathscr{A}_{j;C}\mathscr{A}_{i;C}\mathscr{A}_{b;C}$). However, these need not remain in the denominator for long as they also exactly cancel the overcounting (i.e., symmetrization of already symmetric pairs of labels) in $\mathscr{S}_C$. These can then be removed, and a "unique" symmetrizer $\tilde{\mathscr{S}}_C$ inserted instead, which only symmetrizes pairs of external labels of the same type where either the *in* and/or the *out* labels are on different vertices. Thus, the final expression for the nonorthogonally spin-adapted tensor element is

$$
\begin{aligned}
\check{C} &= \tilde{\mathscr{S}}_C F_s \max\{F_h, F_p\} F_c \\
&\quad \times \sum \left(\left\{ \mathscr{A}_{C;A} \tilde{\mathscr{A}}_{C-xu;A} \right\} \left\{ \sum_{k=0}^{\min\{n_c, n_{u;B}\}} \binom{n_x + k}{k}^{-1} \tilde{\tilde{P}}^{k}_{C-u;B} \right\} \check{A}\check{B} \right)_{\check{C}} \\
&\equiv \tilde{\mathscr{S}}_C F_s \max\{F_h, F_p\} F_c \sum \left(SS_A \widetilde{SS}_B \check{A}\check{B} \right)_{\check{C}}
\end{aligned} \tag{88}
$$

The sign factor F_s is determined for the initial diagrammatic ordering of $\check{A}\check{B}$, as the individual permutations in the previous expression carry their own sign change factors. The factor of $\max\{F_h, F_p\} = 1/n_c!$ is also determined once for the initial ordering, as further permutations will change the number of loops (and hence the "instantaneous" value of n_c), but the overcounting factor which led to this term is fixed by the initial ordering. However, the additional closed-loop factor F_c *does* depend on the permutation of $\check{A}\check{B}$ and so will be affected by the previous permutations. The way in which this factor varies can be determined easily, though, since we choose the initial ordering of $\check{A}\check{B}$ with the maximal number of closed loops. Thus, each individual permutation in SS_A or $\widetilde{SS}_B$ necessarily reduces the number of loops by one, and hence F_c by a factor of 2. When there are multiple closed loops, SS_A or $\widetilde{SS}_B$ by itself will continue to reduce the number of loops with additional compound permutations, until all closed loops are gone and these operators run out of permutations.

4.5 SPIN-SUMMATION AND DEALING WITH CROSS-PRODUCTS

However, when permutations from both SS_A and $\widetilde{SS}_B$ are applied simultaneously, the result may only have one fewer closed loop than the initial ordering, although we would expect to have at least two fewer loops. This occurs when the permutations from both operators permute the same contracted label (or its partner).

Then, the first permutation breaks the loop, but the second then simply moves one end of the new broken loop to the other vertex. If the ends of the loop (the external labels) are of the same type, then this gives a relabeling of the result (with a change in sign!). If not, then this changes the result to a different value of n_s, and so these permutations are discarded. Note that in the first case, even though the ordering does not match $\check{C}$ due to the relabeling, we could simply have considered the initial $\check{A}\check{B}$ to be correspondingly relabeled so that the result still is the desired $\check{C}$, and so these contributions must be kept. Also, in the second case, even though the permutations are discarded, the orbital case they describe would have opposite sign from what the permutations give since the number of loops is not in fact decreased. These anomalous "cross-products" of the SS_A and $\widetilde{SS}_B$ operators means that we cannot simply replace $SS_A\widetilde{SS}_B\check{A}\check{B}$ by the very tempting expression $\left(SS_A\check{A}\right)\left(\widetilde{SS}_B\check{B}\right)$ without some correction term(s).

In order to determine the necessary correction terms, we proceed by forcing this convenient factorization and defining *spin-summed* quantities $\check{A}_{SS}$ and $\check{B}_{\widetilde{SS}}$:

$$\begin{aligned}
\check{A}_{SS} &= F_c SS_A \check{A} \\
&= \sum_{k=0}^{\min\{n_c, n_x+n_{u;A}\}} \sum_{l=0}^{n_c-1} F_c \tilde{\tilde{P}}^k_{c-xu;A} \tilde{P}^l_{c;A} \check{A} \\
&= \left\{ \prod_{k=1}^{n_c} \left(2 + \sum_{i=k+1}^{n_c} P(c_k c_i) + \sum_{i=1}^{n_x+n_{u;A}} P(c_k xu_i) \right) \right\} \check{A}
\end{aligned} \tag{89}$$

$$\begin{aligned}
\check{B}_{\widetilde{SS}} &= \frac{F_c}{2^{n_c}} \widetilde{SS}_B \check{B} \\
&= \left\{ \sum_{k=0}^{\min\{n_c, n_{u;B}\}} 2^{-k} \binom{n_x + k}{k}^{-1} \tilde{\tilde{P}}^k_{c-u;B} \right\} \check{B}
\end{aligned} \tag{90}$$

where c is the set of contracted labels (which is the same for $\check{A}$ and $\check{B}$) and xu is all other labels on $\check{A}$ (the same as $x \cup u; A$). Note that the factor of $F_c/2^{n_c}$ cancels out in the original ordering, and only introduces a factor of 1/2 for each closed loop broken. This is to avoid overcounting with $\check{A}_{SS}$. The quantity $\check{A}_{SS}$ is *fully spin-summed*, while $\check{B}_{\widetilde{SS}}$ is *partially spin-summed*. Of course, we could also have derived this expression with the roles of $\check{A}$ and $\check{B}$ reversed so that we finally obtained $\check{A}_{\widetilde{SS}}$ and $\check{B}_{SS}$, so that we consider full and partial spin-summation only by their action and not by which vertex they are applied to.

As an example of the permutations that can result from SS_A and $\widetilde{SS}_B$, consider the diagram in Fig. 9. Of course, such a diagram does not occur in normal coupled-cluster theories, but it is an interesting example of a topologically complicated diagram. The relevant sets of indices are (including column-symmetric partners as p/p^* for clarity) $c = \{e/m, f/n\}$, $x;A = \{a/g\}$, $x;B = \{g/i\}$, $u;A = \{b/j\}$, and $u;B = \{c/k\}$ (and of course $xu;A = \{a/g, b/j\}$). For SS_A we get

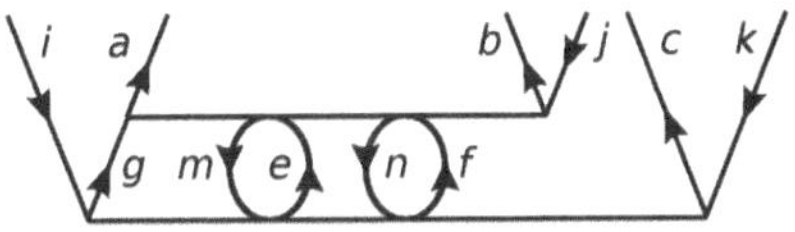

FIG. 9

Example diagram with nontrivial SS_A and $\widetilde{SS}_B$ permutations.

$$\begin{aligned}F_c SS_A \check{A}^{mnab}_{efgj} &= (2 + P(fg) + P(fj)) \\ &\quad \times (2 + P(ef) + P(eg) + P(ej))\, \check{A}^{mnab}_{efgj} \\ &= 4\check{A}^{mnab}_{efgj} - 2\check{A}^{mnab}_{fegj} - 2\check{A}^{mnab}_{gfej} - 2\check{A}^{mnab}_{jfge} \\ &\quad - 2\check{A}^{mnab}_{egfj} - 2\check{A}^{mnab}_{jegf} + \check{A}^{mnab}_{gefj} + \check{A}^{mnab}_{jegf} \\ &\quad + \check{A}^{mnab}_{fgej} + \check{A}^{mnab}_{gjef} + \check{A}^{mnab}_{jgfe} + \check{A}^{mnab}_{fjge}\end{aligned} \tag{91}$$

The result for $\widetilde{SS}_B$ is

$$\begin{aligned}\frac{F_c}{2^{n_c}} \widetilde{SS}_B \check{B}^{efgc}_{mnik} &= \left(\binom{1+0}{0}^{-1} \check{B}^{efgc}_{mnik} \right) \\ &\quad + \left(2^{-1} \binom{1+1}{1}^{-1} (P(mk) + P(nk))\, \check{B}^{efgc}_{mnik} \right) \\ &= \check{B}^{efgc}_{mnik} - \frac{1}{4}\check{B}^{efgc}_{knim} - \frac{1}{4}\check{B}^{efgc}_{mkin}\end{aligned} \tag{92}$$

From this one can see that 36 terms which would normally require individual Goldstone diagrams are accounted for here from a single diagram and in a naturally factored form.

The factorized expression is now

$$\check{C} = \tilde{\mathscr{S}} \frac{F_s}{n_c!} \sum \check{A}_{SS} \check{B}_{\widetilde{SS}} + \text{Real cross terms} - \text{Extraneous cross terms} \tag{93}$$

As we saw previously, each cross term results in (1) an anomalous number of loops (one more than expected for each contracted line in common between the permutations from SS_A and $\widetilde{SS}_B$), and (2) a possible permutation of externals labels of the same type with sign change. The "extraneous" cross terms introduced by forced factorization into $\check{A}_{SS}$ and $\check{B}_{\widetilde{SS}}$ will thus have an extra factor of 1/2 for each of n'_c shared contracted lines, and so may be canceled by a factor $2^{-n'_c}X$, where "X" is the as-yet undetermined cross-product term (without any permutation). Additionally, when the "real" cross term results in a permutation of external labels of the same type and concomitant sign change, then we may add back in this term as $P(pq)X = -P^p_q X$. Thus, assuming we can easily determine X, we must only add $-2^{-n'_c}X$ or $-\left(2^{-n'_c} + P^p_q\right)X$ as appropriate.

Carrying forth the remaining terms from SS_A and $\widetilde{SS}_B$ in combination with the overall factors of Eq. (88), it is easy to see that in fact each cross term X is proportional to Eq. (88) (or Eq. 93) applied to an appropriate orbital case with fewer closed loops. So, for each possible cross term, we must include a suitable contribution which is itself the result of interpreting a nonorthogonally spin-adapted diagram. This term requires two additional numerical factors: the $\binom{n_x+k}{k}^{-1}$ factor from $\widetilde{SS}_B$ which is not accounted for in the new diagram (note, however, that factors of 1/2 due to F_c and relative sign changes *are* included in the overall diagrammatic factors), and the (lack of) sign change expected from SS_A and $\widetilde{SS}_B$ as only one closed loop is broken. If we define the algebraic interpretation of this new diagram as $\check{C}'$ (where n_s' may be less than n_s), then we can see that $X = -\binom{n_x+k}{k}^{-1}\check{C}'$ and that the necessary correction term is $2^{-n_c'}\binom{n_x+k}{k}^{-1}\check{C}'$ or $\left(2^{-n_c'}+P_q^p\right)\binom{n_x+k}{k}^{-1}\check{C}'$. If $\check{C}'$ requires correction for cross-product terms itself, the argument continues inductively.

5 EXAMPLES I: SIMPLE CASES

Before going on to state the final nonorthogonally spin-adapted theory as a set of diagrammatic rules, let us look at some simpler cases pertinent to coupled-cluster theory, so that we may gain some experience with the algebraic techniques (especially spin-summation).

As a first example, let us return to the coupled-cluster energy represented in Figs. 3 and 5. Considering for a moment only the $\hat{T}_2$ contribution, the single Brandow diagram depicted is also the Goldstone diagram with the largest number of closed loops, and so will be our primary diagram. In this case, we have only contracted (c) labels, and if we choose, say, $\check{v}$ to be fully spin-summed, we must consider the following permutations:

$$\begin{aligned}\left\{\check{v}_{ef}^{mn}\right\}_{SS} &= \left\{\prod_{k=1}^{n_c}\left(2+\sum_{i=k+1}^{n_c}P(c_kc_i)+\sum_{i=1}^{n_x+n_{u;A}}P(c_kxu_i)\right)\right\}\check{v}_{ef}^{mn}\\ &= 4\check{v}_{ef}^{mn}-2\check{v}_{fe}^{nm}\end{aligned} \tag{94}$$

$$\begin{aligned}\left\{\check{t}_{mn}^{ef}\right\}_{\widetilde{SS}} &= \left\{\sum_{k=0}^{\min\{n_c,n_{u;B}\}}2^{-k}\binom{n_x+k}{k}^{-1}\tilde{\tilde{P}}_{c-u;B}^{k}\right\}\check{t}_{mn}^{ef}\\ &= \check{t}_{mn}^{ef}\end{aligned} \tag{95}$$

Additionally, we have the factors $F_s = 1$ and $1/n_c! = \frac{1}{2}$ and a sum over *efmn* spatial orbitals, while $\tilde{\mathscr{S}}$ has no effect (because there are no external lines). Putting this together gives

$$E_{CC}(\hat{T}_2) = \sum_{efmn} \check{t}^{ef}_{mn} \left(2\check{v}^{mn}_{ef} - \check{v}^{nm}_{ef} \right) \tag{96}$$

as expected. The $\hat{T}_1$ part is trivial, and for the $\hat{T}_1^2$ part we may break the diagram down into binary terms by defining the intermediate $F^m_e = \frac{1}{2}\hat{H}_N\hat{T}_1$. We encourage the reader to follow the derivation of this intermediate and its contribution to the energy using the methods developed here.

Next, let us examine the "ring" part of the doubles amplitudes in CCSD again, except we now consider the $\frac{1}{2}\hat{H}_N\hat{T}_2^2$ ring term, which must also be split into an intermediate as illustrated in Fig. 10. In this case we have defined an intermediate $\tilde{\boldsymbol{W}}^{am}_{ie} = \frac{1}{2}\boldsymbol{t}^{af}_{in}\boldsymbol{v}^{mn}_{ef}$, whose diagrams are the first one in both subfigures of Fig. 10. The contribution of this intermediate to the amplitude residual $\boldsymbol{z}^{ab}_{ij}$ is given by the second diagram of each subfigure. Unlike the residual $\boldsymbol{z}^{ab}_{ij}$ which has only one orbital case $\check{z}^{ab}_{ij}$, the intermediate has two orbital cases, $\check{\tilde{W}}^{am}_{ie}$ and $\check{\tilde{W}}^{am}_{ei}$, for which $n_s = 2$ and 0, respectively. The two diagrams in the subfigures correspond to the primary diagram for each orbital case, while the diagram in the second subfigure is also a secondary diagram (i.e., a diagram which must be recursively considered) for the $n_s = 2$ orbital case.

Since it is used twice, consider $\check{\tilde{W}}^{am}_{ei}$ first. Because this diagram has no c labels, we have trivially

$$\left\{ \check{t}^{af}_{ni} \right\} ss = \check{t}^{af}_{ni} \tag{97}$$

$$\left\{ \check{v}^{ef}_{nm} \right\} \tilde{s}\tilde{s} = \check{v}^{ef}_{nm} \tag{98}$$

Additionally, we have $F_s = -1$ and $1/n_c! = 1$, a sum over fn, and no effect from $\tilde{\mathscr{P}}$. Thus, remembering the factor of 1/2 from the exponential expansion, we have

$$\check{\tilde{W}}^{am}_{ei} = -\frac{1}{2} \sum_{fn} \check{t}^{af}_{ni} \check{v}^{ef}_{nm} \tag{99}$$

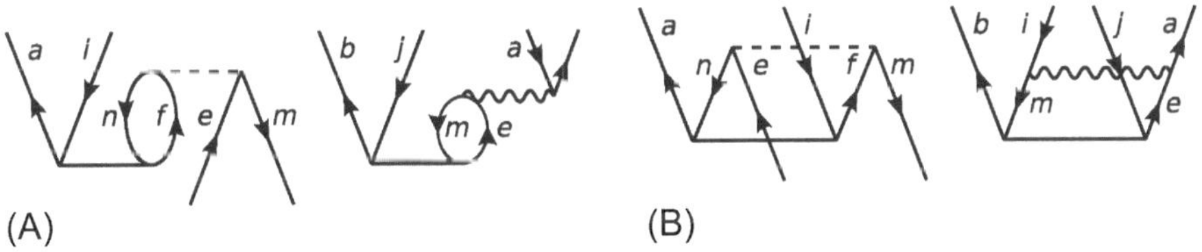

FIG. 10

Factorized diagrams for the ring part of $\hat{Z}_2 \leftarrow \frac{1}{2}\hat{H}_N\hat{T}_2^2$. (A) Primary orbital cases; (B) secondary orbital cases.

For $\check{\tilde{W}}_{ie}^{am}$ we now have c labels as well as u labels on both vertices. Spin-summation then gives

$$\left\{\check{t}_{in}^{af}\right\}_{SS} = 2\check{t}_{in}^{af} - \check{t}_{ni}^{af} \tag{100}$$

$$\left\{\check{v}_{mn}^{ef}\right\}_{\widetilde{SS}} = \check{v}_{mn}^{ef} - \frac{1}{2}\check{v}_{nm}^{ef} \tag{101}$$

The scalar factors are $F_s = 1$ and $1/n_c! = 1$, and there is a sum over fn. Again $\tilde{\mathscr{S}}$ has no effect since the external pairs are of different types. If we look at the permutations from SS_A and $\widetilde{SS}_B$, though, we see that they are $P(af)$ and $P(ef)$ (or equivalently $P(in)$ and $P(mn)$). Since these permute the same contracted line, we have to recursively consider the diagram formed by applying both simultaneously. Of course, this is just the diagram for the $n_s = 0$ orbital case, $\check{\tilde{W}}_{ei}^{am}$. Since we have already determined its algebraic contribution, we multiply it by a factor of 1/2 (since $n_c = 1$ and $n_x = 0$) and add it to the total expression for $\check{\tilde{W}}_{ie}^{am}$:

$$\check{\tilde{W}}_{ie}^{am} = \frac{1}{2}\sum_{fn}\left(2\check{t}_{in}^{af} - \check{t}_{ni}^{af}\right)\left(\check{v}_{mn}^{ef} - \frac{1}{2}\check{v}_{nm}^{ef}\right) - \frac{1}{4}\sum_{fn}\check{t}_{ni}^{af}\check{v}_{nm}^{ef} \tag{102}$$

Note that we have obtained an already-factored form of the expression similar to Eq. (37).

Performing a similar analysis for the contribution of the intermediate to the residual, we can see that

$$\check{z}_{ij}^{ab} \leftarrow \left(1 + P_{bj}^{ai}\right)\left\{\sum_{em}\left(2\check{t}_{mj}^{eb} - \check{t}_{jm}^{eb}\right)\left(\check{\tilde{W}}_{ie}^{am} - \frac{1}{2}\check{\tilde{W}}_{ei}^{am}\right) - \left(\frac{1}{2} + P_j^i\right)\sum_{fn}\check{t}_{jm}^{eb}\check{\tilde{W}}_{ei}^{am}\right\} \tag{103}$$

The analysis here is very similar to that for $\check{\tilde{W}}_{ei}^{am}$ except that (1) the symmetrization factor $\tilde{\mathscr{S}}$ now produces the permutation factor $\left(1 + P_{bj}^{ai}\right)$, and (2) the cross-product term, which comes from the permutations $P(mj)$ and $P(am)$, gives a diagram which is the $i \leftrightarrow j$ permutation of the original ordering. Thus we must include the full $\left(\frac{1}{2} + P_j^i\right)$ factor on the second term. Comparing to Eq. (37) we can see that the full, factored, and symmetrized expression is unambiguously obtained.

Let us also revisit the $\hat{H}_N\hat{T}_3$ contribution to the CCSDT singles equations, whose Goldstone diagrams are given in Fig. 7. From the primary diagram (the first in the figure), we can deduce that the effect of spin-summation is

$$\left\{\check{t}_{imn}^{aef}\right\}_{SS} = 4\check{t}_{imn}^{aef} - 2\check{t}_{inm}^{aef} - 2\check{t}_{nmi}^{aef} - 2\check{t}_{min}^{aef} + \check{t}_{mni}^{aef} + \check{t}_{nim}^{aef} \tag{104}$$

$$\left\{\check{v}_{ef}^{mn}\right\}_{\widetilde{SS}} = \check{v}_{ef}^{mn} \tag{105}$$

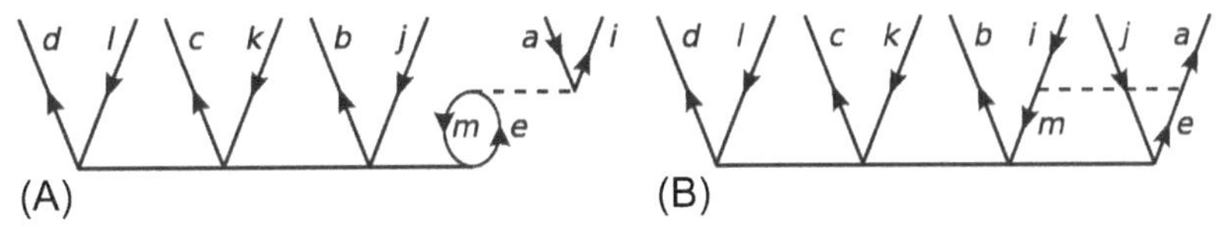

FIG. 11

Diagrams for the ring part of $\hat{Z}_4 \leftarrow \hat{H}_N \hat{T}_4$. (A) Primary diagram; (B) example secondary diagram.

We can also see that $F_s = 1$, $1/n_c! = \frac{1}{2}$, there is a sum over *efmn*, and $\tilde{\mathscr{S}}$ has no effect. In total,

$$\check{z}_i^a \leftarrow \frac{1}{2} \sum_{efmn} \left(4\check{t}_{imn}^{aef} - 2\check{t}_{inm}^{aef} - 2\check{t}_{nmi}^{aef} - 2\check{t}_{min}^{aef} + \check{t}_{mni}^{aef} + \check{t}_{nim}^{aef} \right) \check{v}_{ef}^{mn} \tag{106}$$

Using our newly derived method, the factor of 1/2 which was not captured in the original Goldstone interpretation occurs naturally, as does the summation over all six of the Goldstone diagrams in Fig. 7.

Last, let us examine two examples which show that the new technique is not limited to cases which are also easily derived using Goldstone diagrammatics. First, let us extend the ring diagram from previously to the $\hat{H}_N \hat{T}_4$ contribution in the CCSDTQ quadruples amplitudes (Fig. 11). The analysis of this diagram is quite similar to that in the case of the doubles amplitudes, and we can see that spin-summation gives

$$\left\{ \check{t}_{mjkl}^{ebcd} \right\}_{SS} = 2\check{t}_{mjkl}^{ebcd} - \check{t}_{jmkl}^{ebcd} - \check{t}_{kjml}^{ebcd} - \check{t}_{ljkm}^{ebcd} \tag{107}$$

$$\left\{ \check{v}_{ie}^{am} \right\}_{\tilde{SS}} = \check{v}_{ie}^{am} - \frac{1}{2} \check{v}_{ei}^{am} \tag{108}$$

and identical scalar factors and summation labels. The symmetrizer is expanded now as $\tilde{\mathscr{S}} = \left(1 + P_{bj}^{ai} + P_{ck}^{ai} + P_{dl}^{ai} \right)$, since there are four external pairs of lines. Correction of the cross terms is also the same as in the doubles case, except that now there are three similar cross terms, giving a final result:

$$\begin{aligned} \check{z}_{ijkl}^{abcd} \leftarrow & \left(1 + P_{bj}^{ai} + P_{ck}^{ai} + P_{dl}^{ai} \right) \sum_{em} \Bigg\{ \left(2\check{t}_{mjkl}^{ebcd} - \check{t}_{jmkl}^{ebcd} - \check{t}_{kjml}^{ebcd} - \check{t}_{ljkm}^{ebcd} \right) \left(\check{v}_{ie}^{am} - \frac{1}{2} \check{v}_{ei}^{am} \right) \\ & - \left(1 + P_{ck}^{bj} + P_{dl}^{bj} \right) \left(\frac{1}{2} + P_j^i \right) \check{t}_{jmkl}^{ebcd} \check{v}_{ei}^{am} - \left(1 + P_{ck}^{ai} + P_{dl}^{ck} \right) \left(\frac{1}{2} + P_k^i \right) \check{t}_{kjml}^{ebcd} \check{v}_{ei}^{am} \\ & - \left(1 + P_{dl}^{ai} + P_{dl}^{bj} \right) \left(\frac{1}{2} + P_l^i \right) \check{t}_{ljkm}^{ebcd} \check{v}_{ei}^{am} \Bigg\} \end{aligned} \tag{109}$$

Note that the cross terms include additional symmetrization factors as there are fewer equivalent external pairs after the permutations. As will be seen in the

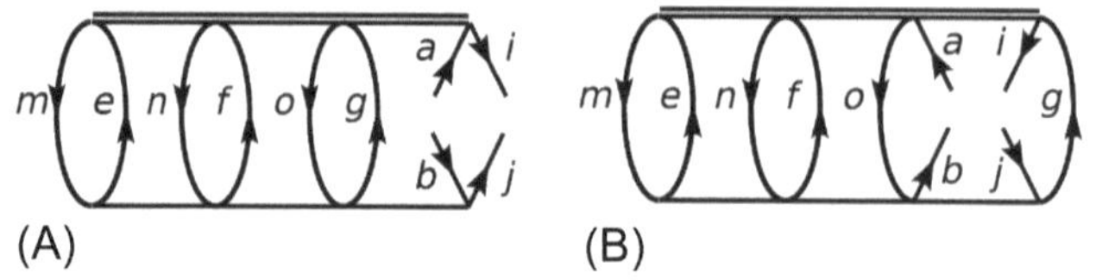

FIG. 12

Diagrams for the ring part of $\Gamma \leftarrow \hat{\Lambda}_4 \hat{T}_4$. (A) Primary diagram; (B) secondary diagram.

following sections, this and other expressions can also be written using the total antisymmetrizer $\mathscr{S}$, which simplifies the expression somewhat, as well as reducing the number of cross terms as only topologically unique terms must be considered:

$$\check{z}^{abcd}_{ijkl} \leftarrow \left(1+P^{ai}_{bj}+P^{ai}_{ck}+P^{ai}_{dl}\right)\left(1+P^{bj}_{ck}+P^{bj}_{dl}\right)\left(1+P^{ck}_{dl}\right)\sum_{em}\left\{\frac{1}{6}\left(2\check{t}^{ebcd}_{mjkl} - \check{t}^{ebcd}_{jmkl} - \check{t}^{ebcd}_{kjml} - \check{t}^{ebcd}_{ljkm}\right)\left(\check{v}^{am}_{ie} - \frac{1}{2}\check{v}^{am}_{ei}\right) - \frac{1}{2}\left(\frac{1}{2}+P^{i}_{j}\right)\check{t}^{ebcd}_{jmkl}\check{v}^{am}_{ei}\right\} \tag{110}$$

From this we can see that the new method is immediately applicable to higher-order coupled cluster, and can produce efficient factorized forms for CCSDT and CCSDTQ as easily as for CCSD.

As our last example, let us consider the $\boldsymbol{\Gamma}^{ai}_{bj}$ two-particle density matrix elements encountered in CCSDTQ analytic gradient theory. In this example, the CCSDTQ quadruples amplitudes, $\hat{T}_4$, are contracted with the quadruples amplitude response vector $\hat{\Lambda}_4$ to form a contribution to the two-particle density. The relevant diagrams for the orbital case $\check{\Gamma}^{bi}_{ja}$ are given in Fig. 12. Looking at the primary diagram, we see that spin-summation gives

$$\begin{aligned}\left\{\check{\lambda}^{mnoi}_{efga}\right\}_{SS} &= 8\check{\lambda}^{mnoi}_{efga} - 4\check{\lambda}^{mnoi}_{efag} - 4\check{\lambda}^{mnoi}_{eagf} - 4\check{\lambda}^{mnoi}_{afge} - 4\check{\lambda}^{mnoi}_{egfa} - 4\check{\lambda}^{mnoi}_{gfea} - 4\check{\lambda}^{mnoi}_{fega} \\ &+ 2\check{\lambda}^{mnoi}_{egaf} + 2\check{\lambda}^{mnoi}_{eafg} + 2\check{\lambda}^{mnoi}_{agfe} + 2\check{\lambda}^{mnoi}_{gfae} + 2\check{\lambda}^{mnoi}_{gaef} + 2\check{\lambda}^{mnoi}_{afeg} + 2\check{\lambda}^{mnoi}_{feag} + 2\check{\lambda}^{mnoi}_{fage} \\ &+ 2\check{\lambda}^{mnoi}_{aegf} + 2\check{\lambda}^{mnoi}_{fgea} + 2\check{\lambda}^{mnoi}_{gefa} - \check{\lambda}^{mnoi}_{fgae} - \check{\lambda}^{mnoi}_{faeg} - \check{\lambda}^{mnoi}_{agef} - \check{\lambda}^{mnoi}_{geaf} \\ &- \check{\lambda}^{mnoi}_{gafe} - \check{\lambda}^{mnoi}_{aefg} \equiv \check{\lambda}^{\breve{m}\check{n}\breve{o}i}_{\check{e}f\breve{g}a}\end{aligned} \tag{111}$$

$$\begin{aligned}\left\{\check{t}^{efgb}_{mnoj}\right\}_{\widetilde{SS}} &= \check{t}^{efgb}_{mnoj} - \frac{1}{2}\check{t}^{efgb}_{mnjo} - \frac{1}{2}\check{t}^{efgb}_{mjon} - \frac{1}{2}\check{t}^{efgb}_{jnom} \\ &\equiv \frac{1}{2}\check{t}^{\overparen{efgb}}_{\underparen{mnoj}}\end{aligned} \tag{112}$$

where we have introduced a notation which will be defined fully in the next section. Combining this with the factors $F_s = 1$, $1/n_c! = \frac{1}{6}$, a sum over *efgmno*, and the symmetrization factor $\tilde{\mathscr{S}} = \left(1+P^{ai}_{bj}\right)$ gives

$$\check{\Gamma}_{ja}^{bi} \leftarrow \frac{1}{6}\left(1 + P_{bj}^{ai}\right) \sum_{efgmno} \left\{ \frac{1}{2} \check{\lambda}_{\check{e}\check{f}\check{g}a}^{\check{m}\check{n}\check{o}i} \check{t}_{\widetilde{mnoj}}^{\widetilde{efgb}} - \frac{3}{2} \check{\lambda}_{\check{e}\check{f}ag}^{\check{m}\check{n}oi} \check{t}_{mnjo}^{efgb} \right\} \tag{113}$$

where the spin-summation in the cross term is given by

$$\begin{aligned} \left\{ \check{\lambda}_{efag}^{mnoi} \right\}_{SS} &= 4\check{\lambda}_{efag}^{mnoi} - 4\check{\lambda}_{egaf}^{mnoi} - 4\check{\lambda}_{gfae}^{mnoi} - 4\check{\lambda}_{feag}^{mnoi} + \check{\lambda}_{geaf}^{mnoi} + \check{\lambda}_{fgae}^{mnoi} \\ &\equiv \check{\lambda}_{\check{e}\check{f}ag}^{\check{m}\check{n}oi} \end{aligned} \tag{114}$$

$$\left\{ \check{t}_{mnjo}^{efgb} \right\}_{\widetilde{SS}} = \check{t}_{mnjo}^{efgb} \tag{115}$$

Unlike the analogous case in the CCSDTQ amplitudes, the three cross terms which correspond to the $n_s = 0$ orbital case are all identical through relabeling of the contracted labels (although this is similar to the fully symmetrized form). Also, in the secondary ($n_s = 0$) orbital case, there is still some spin-summation of $\hat{\Lambda}$. But, since the vertices in the cross term do not have any remaining uncontracted pairs of labels, it is not necessary to recursively generate any additional cross terms. From the number of distinct factors, we can see that there are contributions from 78 distinct Goldstone diagrams ($78 = 24 \times 4 - 3 \times 6$, as the cross terms cancel diagrams which would not normally contribute). The derivation of the $\check{\Gamma}_{aj}^{bi}$ orbital case is given already by the previous secondary diagram as seen in the cross terms of the $\check{\Gamma}_{bj}^{ai}\ \check{\Gamma}_{ja}^{bi}$ orbital case.

6 NEW DIAGRAMMATIC RULES

The final procedure for forming the compact, factored nonorthogonally spin-adapted equations can now be summarized in the following rules:

- For each primary diagram (Goldstone diagram with the maximal number of closed loops):
  - **R1″.** Each vertex contributes an orbital tensor element: $\check{x}_{\text{all-in-labels}}^{\text{all-out-labels}}$, where $\check{x}$ is determined by the type of the horizontal line.
  - **R2″.** One vertex is *fully spin-summed* and the other is *partially spin-summed.*
  - **R3″.** The overall sign is equal to $(-1)^{l+h}$, where l is the number of loops and h is the number of hole lines.
  - **R4″.** Apply a factor of $\frac{1}{c!}$ where c is the number of *closed* loops.
  - **R5″.** Internal lines are summed over all spatial orbitals.
  - **R6″.** Pairs of external lines of the same type, which either go *in* or *out* on different vertices, are symmetrized.
  - **R7″.** For each pair of permutations from the spin-summation of the two vertices which involve the same set of k contracted lines, apply R1″–R7″ recursively to the resulting diagram and multiply by $2^{-k}\binom{n_x+k}{k}^{-1}$ (or

$(2^{-k} + P^p_q)\binom{n_x+k}{k}^{-1}$ if the permutations exchange sets of labels p and q of the same type), where n_x is the number of contracted lines with uncontracted partners.

As we saw with the case of the ring contribution of $\hat{H}_N\hat{T}_4$ to the CCSDTQ quadruples amplitudes, it may be beneficial in many cases to apply the total symmetrizer $\mathscr{S}$, instead of the partial symmetrized $\tilde{\mathscr{S}}$, in order to collapse the cross terms into only those that are represented by topologically distinct secondary diagrams. A somewhat modified set of diagrammatic rules may also be derived in that case by considering how the total antisymmetrizer overcounts in the case that there are equivalent pairs of external lines:

- For each primary diagram (Goldstone diagram with the maximal number of closed loops):

R1‴. Each vertex contributes an orbital tensor element: $\check{x}^{\text{all-out-labels}}_{\text{all-in-labels}}$, where $\check{x}$ is determined by the type of the horizontal line.
R2‴. One vertex is *fully spin-summed* and the other is *partially spin-summed.*
R3‴. The overall sign is equal to $(-1)^{l+h}$, where l is the number of loops and h is the number of hole lines.
R4‴. Apply a factor of $\frac{1}{c!}$ for each set of c equivalent *closed* or *external* loops (equivalent external loops go *in* and separately *out* on the same vertex).
R5‴. Internal lines are summed over all spatial orbitals.
R6‴. All pairs of external lines of the same type are symmetrized.
R7‴. For each *topologically unique* pair of permutations from the spin-summation of the two vertices which involve the same set of k contracted lines, apply R1‴–R7‴ recursively to the resulting diagram and multiply by $2^{-k}\binom{n_x+k}{k}^{-1}$ (or $(2^{-k} + P^p_q)\binom{n_x+k}{k}^{-1}$ if the permutations exchange sets of labels p and q of the same type), where n_x is the number of contracted lines with uncontracted partners.

As we can see in this case, use of the total antisymmetrizer in rule R6‴ requires modification of the overcounting factor in rule R4‴. The counting of topologically equivalent pairs of permutations encountered in rule R7‴ is also achieved by the change in factor determined in rule R4‴ before and after the permutations. Lastly we note that in the vast majority of cases (including all of the examples in the previous section), there is never both more than one closed loop *and* more than one external pair of lines on either vertex (this means that $k \leq 1$ in rule R7‴). Additionally, in factorizations of nonlinear theories such as coupled cluster, it is almost always the case when an optimal factorization is used that *both* vertices *do not* have a closed loop, an external pair of lines, *and* semicontracted lines (i.e., $n_x = 0$). In this case the vertex with all three types of lines (c, x, and u) may be chosen to be fully spin-summed and the other partially spin-summed. When both conditions are met, the numerical factors in rule R7‴ (R7″) simplify to 1/2 or $\left(\frac{1}{2} + P^p_q\right)$.

7 SPIN-SUMMATION REVISITED

In order to tackle some more complicated examples, let us first develop some new notation to simplify writing down spin-summed quantities. Here, a combined notation for $\check{A}_{SS}$ and $\check{B}_{\widetilde{SS}}$ is adopted which shows directly how to compute the spin-summed quantity from decorations on sets of index labels, using a combination of haceks ($\check{}$) and chevrons ($\hat{}$).

First, let us consider full spin-summation. We place a hacek over each individual c label and its column-symmetric partner. We then place a single chevron over all other (xu) indices. For the case of full spin-summation, we will usually omit the chevron and instead assume that all unhaceked indices get a chevron if none is specified. Then, we can see schematically that

$$\check{A}_{SS} = \check{A}^{\check{c}_1\ \check{c}_2\ \ldots x_1\, \widehat{x_2 \ldots u_1}\, u_2 \ldots}_{\check{c}_1{}^*\check{c}_2{}^*\ldots x_1^*\widehat{x_2^* \ldots u_1^*}u_2^*\ldots} \equiv \check{A}^{\check{c}_1\ \check{c}_2\ \ldots x_1\, x_2 \ldots u_1\, u_2 \ldots}_{\check{c}_1{}^*\check{c}_2{}^*\ldots x_1^* x_2^* \ldots u_1^* u_2^*\ldots} \tag{116}$$

For partial spin-summation, we will place a single hacek over all c labels, and a single chevron over all u indices. Of course, if there are no u indices then we should leave off all decorations so that an implicit chevron is not assumed. Then, we can see that

$$\check{B}_{\widetilde{SS}} = \check{B}^{\widecheck{c_1\, c_2}\ldots x_1\, x_2 \ldots \widehat{u_1\, u_2}\ldots}_{\widecheck{c_1^* c_2^*}\ldots x_1^* x_2^* \ldots \widehat{u_1^* u_2^*}\ldots} \tag{117}$$

The procedure for spin-summation is then:

1. Define the sets $h_0 \ldots h_{n-1}$, to be the sets of labels topped with a **h**acek and $h \equiv \bigcup_{i=0}^{n-1} h_i$. Also define r as the set of labels topped with a chev**r**on and x as the set of labels with no decoration.
2. For $i = 0 \ldots n-1$:
   (a) Update $h \leftarrow h \backslash h_i$.
   (b) Apply the set of permutations:

$$\sum_{k=0}^{\min\{n_{h_i}, n_h + n_r\}} 2^{1-k} \binom{n_x + k}{k}^{-1} \tilde{\tilde{P}}^k_{h_i - hr} \tag{118}$$

Note that for $\widetilde{SS}_B$, this procedure actually introduces a spurious factor of 2, which is otherwise necessary to reproduce the F_c factor in SS_A. Thus, we must always remember to include an additional factor of 1/2 in the overall expression when using this notation and when both vertices require spin-summation. The need for this caveat is an unfortunate side-effect of treating the two types of spin-summation with the same notation.

As we noted in the case of simplifying the numerical factors in rule R7‴, in the vast majority of cases there are (1) not both more than one contracted loop and more than one external pair on each vertex, and (2) no semicontracted (x) lines when partial spin-summation is required. We note that in satisfying the second condition it generally suffices to fully spin-sum the vertex of higher order (larger number of lines).

The first condition is also almost always satisfied in conventional many-body electronic theories for both energy and gradient expression because of the two-body nature of the Hamiltonian and the density matrix. So, in this case let us consider the partially spin-summed vertex. There are two cases:

$$\check{B}_{\widetilde{SS}} = \frac{1}{2}\check{B}^{\overbrace{c_1\,c_2}\ldots\hat{u}}_{\underbrace{c_1^*c_2^*}\ldots\hat{u}^*}$$

$$= \frac{1}{2}\check{B}^{\widehat{c_1\,c_2}\ldots\check{u}}_{\widehat{c_1^*c_2^*}\ldots\check{u}^*} \equiv \frac{1}{2}\check{B}^{c_1\,c_2\ldots\check{u}}_{c_1^*c_2^*\ldots\check{u}^*} \tag{119}$$

$$\check{B}_{\widetilde{SS}} = \frac{1}{2}\check{B}^{\check{c}\ \ldots\widehat{u_1\,u_2}\ldots}_{\check{c}^*\ldots\widehat{u_1^*u_2^*}\ldots} \equiv \frac{1}{2}\check{B}^{\check{c}\ \ldots u_1\,u_2\ldots}_{\check{c}^*\ldots u_1^*u_2^*\ldots} \tag{120}$$

where the need for the factor of 1/2 is explained previously. In both cases, the resulting spin-summation is functionally identical to full spin-summation of one closed loop, and may be represented using the simplified spin-summation notation.

Considering the rather complex procedure earlier for evaluation the spin-summation, we can also see that in the full spin-summation and this simplified partial spin-summation case, the procedure simplifies to applying a factor of 2 and subtracting permutations of labels recursively for each "haceked" label:

$$x^{p\,q\,r\,s}_{p^*q^*r^*s^*} \xrightarrow{2-P^p_q-P^p_r-P^p_s} x^{\check{p}\,q\,r\,s}_{\check{p}^*q^*r^*s^*} \xrightarrow{2-P^q_r-P^q_s} x^{\check{p}\,\check{q}\,r\,s}_{\check{p}^*\check{q}^*r^*s^*} \xrightarrow{2-P^r_s} x^{\check{p}\,\check{q}\,\check{r}\,s}_{\check{p}^*\check{q}^*\check{r}^*s^*} \xrightarrow{2} x^{\check{p}\,\check{q}\,\check{r}\,\check{s}}_{\check{p}^*\check{q}^*\check{r}^*\check{s}^*} \tag{121}$$

This simpler form of spin-summation suffices for most applications, for example, we have used it to derive nonorthogonally spin-adapted expressions for CCSDTQ amplitude equations, amplitude response equations, one- and two-particle density matrices, and equation-of-motion coupled-cluster (EOM-CC) excitation energies.

8 EXAMPLES II: COMPLEX CASES

Of course, ring-type diagrams are not the only complicated cases made easier by the nonorthogonally spin-adapted rules. Consider the case of Fig. 13, which now has c, x, and u labels on both vertices (which we will simply label as $\check{A}$ and $\check{B}$ as in the previous section). The rules now call for a nontrivial partial spin-summation of $\check{B}$ as well as two distinct secondary diagrams to be added to cancel spurious cross terms:

$$-\sum_{emn}\left\{\frac{1}{2}\check{A}^{\check{m}nj}_{\check{e}db}\check{B}^{\check{e}c\hat{a}}_{\check{m}n\hat{i}} - \frac{1}{2}\left(\frac{1}{2}+P^a_c\right)\check{A}^{mnj}_{deb}\check{B}^{eca}_{inm} - \frac{1}{4}\check{A}^{mnj}_{bde}\check{B}^{eca}_{inm}\right\} \tag{122}$$

The second cross term is computed from the $n_s = 0$ orbital case of the result, so that the only two additional tensor contractions are needed to complete this ($n_s = 2$) orbital case. Note that the presence of x labels in $\check{B}_{\widetilde{SS}}$ necessitates the additional factors of $\frac{1}{2}$ on the added orbital cases. Also note that while the $n_s = 0$ orbital case has a pair of external loops of the same type ($a \to b$ and $c \to d$), they are equivalent and so do not affect the calculation of the numerical or symmetrization prefactors.

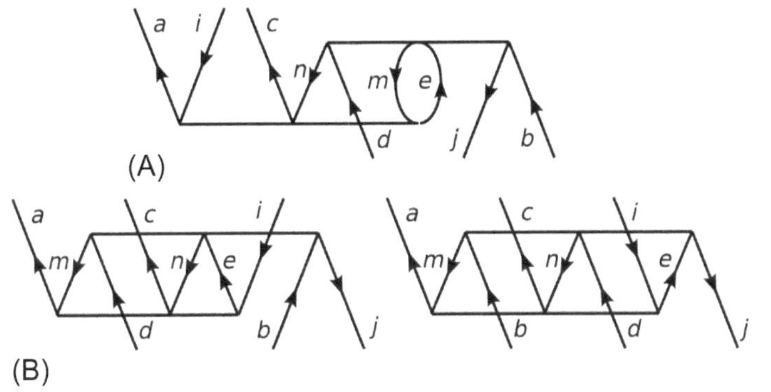

FIG. 13

More complicated diagrams. (A) Primary diagram; (B) secondary diagrams.

Despite the nuances introduced by having all three types of labels on each vertex, the overall structure and complexity of this expression is very similar to the others already presented.

A closely related diagram is given in Fig. 14. The only change is to take the bj labels on $\check{A}$ and make them upward-pointing, but now we find that there are two primary diagrams and four secondary diagrams. Both of these primary diagrams must be considered, and the rules must be applied to each independently. The total result is

$$\begin{aligned}&-\left(1+P_{bj}^{ai}\right)\sum_{emn}\left\{\frac{1}{2}\check{A}_{\check{e}dj}^{\check{m}nb}\check{B}_{\check{m}n\hat{i}}^{\check{e}c\hat{a}}-\frac{1}{2}\left(\frac{1}{2}+P_{c}^{a}\right)\check{A}_{dej}^{mnb}\check{B}_{inm}^{eca}-\frac{1}{2}\left(\frac{1}{2}+P_{b}^{a}\right)\check{A}_{jde}^{mnb}\check{B}_{inm}^{eca}\right\}\\&-\left(1+P_{bj}^{ai}\right)\sum_{emn}\left\{\frac{1}{2}\check{A}_{\check{e}jd}^{\check{m}nc}\check{B}_{\check{m}n\hat{i}}^{\check{e}b\hat{a}}-\frac{1}{2}\left(\frac{1}{2}+P_{b}^{a}\right)\check{A}_{jed}^{mnc}\check{B}_{inm}^{eba}-\frac{1}{2}\left(\frac{1}{2}+P_{c}^{a}\right)\check{A}_{dje}^{mnc}\check{B}_{inm}^{eba}\right\}\end{aligned} \tag{123}$$

There is only this one orbital case ($\check{C}_{ijd}^{abc}$, $n_s = 2$) to which all six of these contractions contribute. While it may seem uncommon, this diagram would in fact be encountered for a certain suboptimal factorization of CCSDT, where a three-body intermediate with the shape of $\check{A}$ is used. However, factorizations of this type are inefficient, not only because of the large number of contractions needed, but because the three downward lines of the intermediate must come from the two-particle Hamiltonian $\hat{V}_N$. Then, an alternate factorization would be to split the whole diagram on the other side of the $\hat{V}_N$ vertex to give an intermediate with one downward index. This alternate factorization will always have a leading-order asymptotic cost that is a factor of $\mathcal{O}(n^2)$ less. These contributions have also been evaluated using the Goldstone approach as an additional check, and the reduction in complexity, even though the final, factorized result is itself not particularly simple, is enormous. The form given here is also not at all obvious from the Goldstone equations, so that it would be

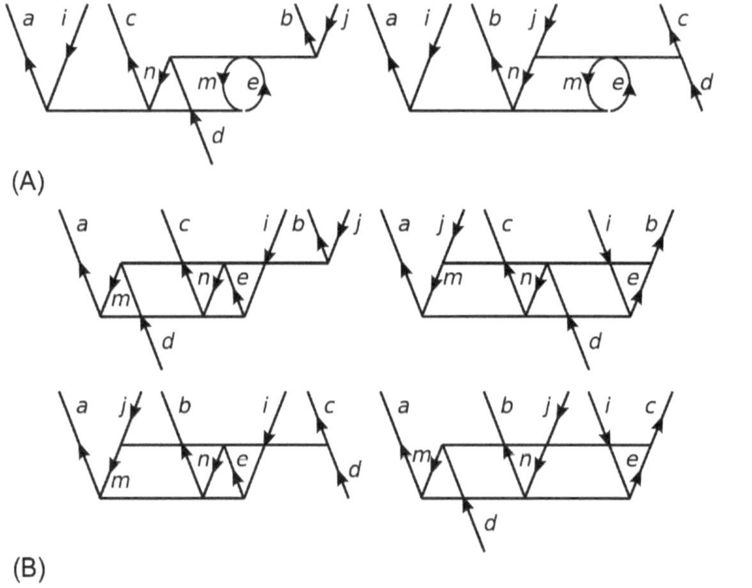

FIG. 14

Even more complicated diagrams. (A) Primary diagrams; (B) secondary diagrams.

quite difficult to arrive at the same result without the rules derived earlier or a deep understanding of these types of equations.

As a last example, let us consider the deceptively simple-looking diagram in Fig. 15, for which $n_s = 4$. The complexity of this diagram comes from the fact that the secondary diagrams, which must be added to correct the cross terms, themselves have a very complicated structure which involves yet another layer of tertiary diagrams (and additional factors from the presence of x labels). However, meticulously following the rules, recursively applying them to the "lower" orbital cases, one arrives at

$$
\begin{aligned}
\sum_{efmn} \Bigg\{ & \frac{1}{4}\check{A}^{\check{m}\check{n}kl}_{\check{e}\check{f}cd}\check{B}^{\overset{\vee}{ef}ab}_{\overset{\smile}{mn}ij} \\
& + \frac{1}{2}\left[-\frac{1}{2}\check{A}^{\check{m}nkl}_{\check{e}cfd}\check{B}^{\check{e}fa\hat{b}}_{\check{m}in\hat{j}} + \frac{1}{3}\left(\frac{1}{2}+P^{a}_{b}\right)\left(\check{A}^{mnkl}_{cefd}\check{B}^{efab}_{jinm} + \check{A}^{mnkl}_{fced}\check{B}^{efab}_{jinm}\right) + \frac{1}{6}\check{A}^{mnkl}_{dcfe}\check{B}^{efab}_{jinm}\right] \\
& + \frac{1}{2}\left[-\frac{1}{2}\check{A}^{\check{m}nkl}_{\check{e}dcf}\check{B}^{\check{e}fa\hat{b}}_{\check{m}in\hat{j}} + \frac{1}{3}\left(\frac{1}{2}+P^{a}_{b}\right)\left(\check{A}^{mnkl}_{decf}\check{B}^{efab}_{jinm} + \check{A}^{mnkl}_{fdce}\check{B}^{efab}_{jinm}\right) + \frac{1}{6}\check{A}^{mnkl}_{cdef}\check{B}^{efab}_{jinm}\right] \\
& + \frac{1}{2}\left[-\frac{1}{2}\check{A}^{\check{m}nkl}_{\check{e}cfd}\check{B}^{\check{e}f\hat{a}b}_{\check{m}j\hat{i}n} + \frac{1}{3}\left(\frac{1}{2}+P^{a}_{b}\right)\left(\check{A}^{mnkl}_{cefd}\check{B}^{efab}_{ijmn} + \check{A}^{mnkl}_{fced}\check{B}^{efab}_{ijmn}\right) + \frac{1}{6}\check{A}^{mnkl}_{dcfe}\check{B}^{efab}_{ijmn}\right] \\
& + \frac{1}{2}\left[-\frac{1}{2}\check{A}^{\check{m}nkl}_{\check{e}dcf}\check{B}^{\check{e}f\hat{a}b}_{\check{m}j\hat{i}n} + \frac{1}{3}\left(\frac{1}{2}+P^{a}_{b}\right)\left(\check{A}^{mnkl}_{decf}\check{B}^{efab}_{ijmn} + \check{A}^{mnkl}_{fdce}\check{B}^{efab}_{ijmn}\right) + \frac{1}{6}\check{A}^{mnkl}_{cdef}\check{B}^{efab}_{ijmn}\right]
\end{aligned}
$$

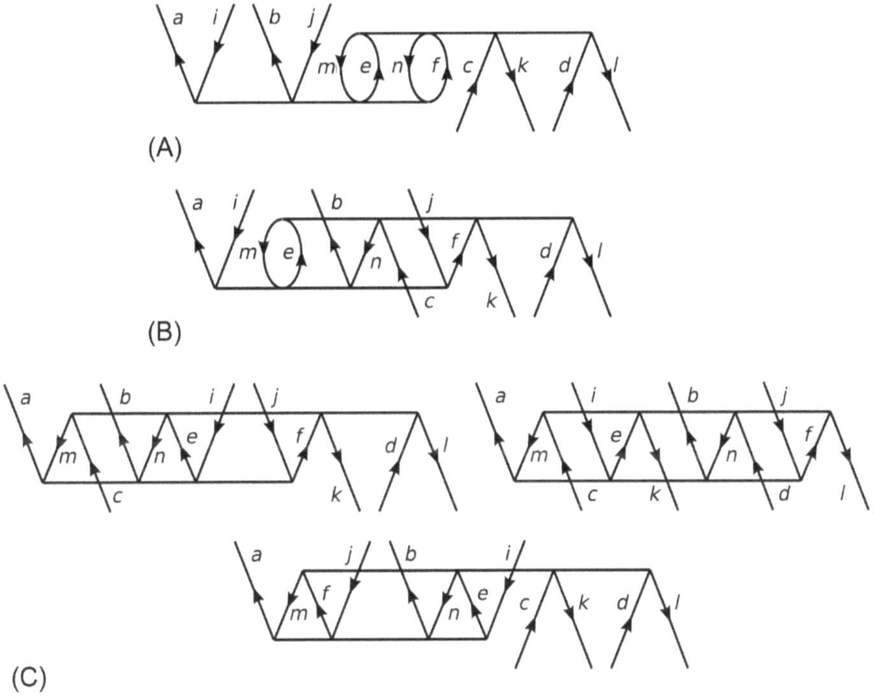

FIG. 15

Very complicated diagrams. (A) Primary diagram; (B) example secondary diagram; (C) tertiary diagrams.

$$
\begin{aligned}
&-\frac{1}{4}\left(\check{A}^{mnkl}_{fced}\check{B}^{efab}_{ijmn}+\check{A}^{mnkl}_{cefd}\check{B}^{efab}_{ijmn}+\check{A}^{mnkl}_{decf}\check{B}^{efab}_{ijmn}+\check{A}^{mnkl}_{fdce}\check{B}^{efab}_{ijmn}\right)\\
&-\frac{1}{4}\check{A}^{mnkl}_{cdef}\check{B}^{efab}_{ijmn}-\frac{1}{4}\check{A}^{mnkl}_{dcfe}\check{B}^{efab}_{ijmn}+\left(\frac{1}{2}+P^{a}_{b}\right)\check{A}^{mnkl}_{fecd}\check{B}^{efab}_{ijmn}\Big\}
\end{aligned}
\tag{124}
$$

Some terms in this expression are identical after relabeling of the contracted labels. Also, the parts in square brackets correspond to the total equation for one of the secondary orbital cases. Lastly, many terms are equivalent under permutations of pairs of external labels, so that by applying the total symmetrizer instead of $\tilde{\mathscr{S}}_C$, relabeling the contracted labels, and inserting "lower" orbital cases without expansion, we get the much simpler expression

$$
\begin{aligned}
&\left(1+P^{ai}_{bj}\right)\left(1+P^{ck}_{dl}\right)\sum_{efmn}\Big\{\frac{1}{16}\check{A}^{\check{m}\check{n}kl}_{\check{e}\check{f}cd}\check{B}^{\check{ef}ab}_{\widetilde{mn}ij}+\frac{1}{2}(\check{A}\check{B})^{abkl}_{cjid}\\
&-\frac{1}{8}\check{A}^{mnkl}_{fced}\check{B}^{efab}_{ijmn}+\frac{1}{8}(\check{A}\check{B})^{abkl}_{cdij}+\frac{1}{4}\left(\frac{1}{4}+P^{a}_{b}\right)\check{A}^{mnkl}_{fecd}\check{B}^{efab}_{ijmn}\Big\}
\end{aligned}
\tag{125}
$$

The orbital cases such as $(\check{A}\check{B})^{abkl}_{cjid}$ and $(\check{A}\check{B})^{abkl}_{cdij}$ are of course understood to be the diagram(s) of $\check{A}$ and $\check{B}$ with $n_s = 2$ and $n_s = 0$, respectively. This example is an extreme example of a complex diagram, whether in the Brandow or nonorthogonally spin-adapted diagrammatic formalism. However, it illustrates the generality and power of the new diagrammatic rules. It is significant to note that the authors, both of whom having significant experience and background in many-body theory, were, despite vigorous effort, unable to derive the factorized form of the last example without the aid of the new diagrammatics. To us, this underscores the practical merits of the approach presented here, despite its rather advanced level of abstraction and rather tedious (albeit wholly unambiguous) application.

9 CONCLUSIONS

Using the properties of simple pair-wise permutations, the rules for interpreting coupled-cluster diagrams in terms of spin-orbital quantities can be transformed into rules for generating compact, efficient equations in terms of orbital quantities. While the new rules are not as simple to understand or apply as the spin-orbital ones, they are vastly more expedient than direct translation of the spin-orbital equations into the closed-shell orbital framework, or the derivation of the orbital equations by using Goldstone diagrams (both of which require nonobvious factorizations of the result).

Fortunately, most of the diagrams encountered in the coupled-cluster amplitude equations (as well as the $\hat{\Lambda}$ and density equations, and in configuration interaction and excited state methods such as equation-of-motion coupled cluster), do not require some of the trickier aspects of the approach, and many of the complexities noted in the previous sections are not encountered in such applications. In fact, the rules given earlier can be simplified considerably for this use, as long as certain (necessarily suboptimal) factorizations are avoided. However, the simplification of the rules only enhances their benefit, as the translation from spin-orbital or factorization of Goldstone equations is still quite complex for CCSDT and CCSDTQ. Additionally, it is hoped that by deriving the rules in the most general case, that perhaps the theory of nonorthogonal spin-adaptation can be applied in new areas of chemical physics.

ACKNOWLEDGMENTS

Devin A. Matthews is supported by an Arnold O. Beckman Postdoctoral Fellowship from the Arnold and Mabel Beckman Foundation. This work was also supported by grants from the Welch Foundation (number F-1283) and the US National Science Foundation (number CHE-1748821 and CHE-1664325).

10 APPENDIX

Lemma 1. *Elements of $\tilde{P}^k$ are unique.*

Proof. The proof is trivial for $k = 0$ and $k = 1$. For $k = 2$, note that there are three distinct types of products: $P(ab)P(cd)$, $P(ab)P(bc)$, and $P(ac)P(bc)$. The permutations in the first case, since they share no common labels, commute, and so the product is equivalent only to itself and hence unique. The second type is equivalent by inspection to $P(bc)P(ac)$ and $P(ac)P(ab)$. However, since $a < b$ and a may not be repeated as the "from" label, neither of these is a permissible element of $\tilde{P}^2$ and so products of this type are unique. Similarly for the third type, $P(ac)P(bc)$ is equivalent to $P(bc)P(ab)$ and $P(ab)P(ac)$ and so these too are unique. For $k > 2$ we can proceed inductively, assuming that elements of $\tilde{P}^{k-1}$ are unique. From the relations for $\tilde{P}^2$, it can be seen that when going to any different related permutation, the set of "from" labels can never contain any labels which are not "from" labels in the original $\tilde{P}^k$ element (although one can exchange a larger "from" label for two of a smaller one). So, to relate this permutation to one in $\tilde{P}^k$ where the "from" labels must be strictly ordered, the last and largest "from" element must always be the same. Additionally, it can be seen from the analysis of $\tilde{P}^2$ that the "to" label associated with the larger "from" label always remains in the correct position such that when eventually the permutation is related to one with the maximal "from" label in the last position, the "to" label will also be the same as in the original permutation. Thus, any $\tilde{P}^k$ element related to the original is built from a $\tilde{P}^{k-1}$ element in conjunction with the same final simple permutation. However, since elements of $\tilde{P}^{k-1}$ are unique, any related $\tilde{P}^k$ element is in fact the original one and so it is also unique. Starting the induction with $\tilde{P}^2$ completes the proof. □

Lemma 2. $\tilde{P}^k \cap \tilde{P}^l = \emptyset \, \forall \, k \neq l$.

Proof. Assume for simplicity that $k < l$ and that the converse is true. For $\tilde{P}^k$ and $\tilde{P}^l$ to share any elements, some elements from $\tilde{P}^l$ would need to be removed due to the product of two permutations being the identity. Since two permutations only produce the identity when they are equal, then $\tilde{P}^l$, possibly in some equivalent form, would need to have two identical permutations next to each other. Taking this hypothetical permutation as $P(ab)$, we can use the observations from the proof of the previous lemma to see that another $P(ab)$ would have to be produced from $P(ac)P(cb)$ or $P(ac)P(bc)$ for some c. Since this would require two permutations with the same "from" index and violate the ordering restriction for $\tilde{P}^l$, then these permutations would in turn have to be produced from yet more permutations. However, the label a remains as the "from" label in some permutation for any decomposition of the permutations, and so it is never possible to create the necessary pair of $P(ab)$ permutations from a valid element of $\tilde{P}^l$. So, each permutation in $\tilde{P}^l$ cannot be equivalent to one with fewer individual permutations, in particular k of them. □

Lemma 3. *$\tilde{P}^k$ contains all k-fold permutations which are not reducible.*

Proof. For each successive pair of permutations in each k-fold permutation, the labels will be either (1) in an ordering consistent with $\tilde{P}^k$, (2) disjoint, such that the permutations commute, or (3) in one of three nonconsistent orderings: $P(bc)P(ac)$,

$P(bc)P(ab)$, or $P(ab)P(ac)$, where $a < b$ in all cases ($P(ab)P(ab)$ is not permissible since the total permutation is not reducible). However, the first is equivalent to $P(ab)P(bc)$ and the second and third to $P(ac)P(bc)$. Therefore, going from left to right, successive pairs or permutations can be placed into the proper order by (1) doing nothing, (2) interchanging the permutations if they are in the wrong order since they commute, or (3) switching to an equivalent pair of permutations which have the proper ordering restriction. In this way, any k-fold permutation can be related to an element of $\tilde{P}^k$. □

Theorem 1. *$\mathscr{A}_P = \{\tilde{P}^k\}_{k=0}^{n-1}$ contains all possible permutations of n labels exactly once.*

Proof. From the lemmas, we see that the elements of each $\tilde{P}^k$ are unique, and that the sets are also unique among each other. Since also each $\tilde{P}^k$ contains all unique products of k individual permutations, and $\tilde{P}^n$ and higher products are empty (since there are not enough labels to satisfy the ordering restriction), then the collection of all $\tilde{P}^k$, for which $0 \leq k < n$ are nonempty, gives all possible unique permutations of the n labels, denoted as $\mathscr{A}_P$. □

Theorem 2. *Given a set of n labels and their two-index permutations P, define on any disjoint partitioning (into partitions of size m and n − m) of the labels the sets of permutations P_1, P_2, and P_x which permute labels within in the first partition, within the second partition, and between the partitions, respectively. Define antisymmetrizers $\mathscr{A}_{P_1}$ and $\mathscr{A}_{P_2}$ for the sets of permutations P_1 and P_2 and the partial antisymmetrizer $\tilde{\mathscr{A}}_x = \{\tilde{P}_x^k\}_{k=0}^{min\{m,n-m\}}$. Then, the total antisymmetrizer $\mathscr{A}_P = \mathscr{A}_{P_1}\mathscr{A}_{P_2}\tilde{\mathscr{A}}_x = \tilde{\mathscr{A}}_x\mathscr{A}_{P_1}\mathscr{A}_{P_2}$ (and by extension $\mathscr{A}_{P_2}\mathscr{A}_{P_1}\tilde{\mathscr{A}}_x$ and $\tilde{\mathscr{A}}_x\mathscr{A}_{P_2}\mathscr{A}_{P_1}$ since $\mathscr{A}_{P_1}$ and $\mathscr{A}_{P_2}$ commute).*

Proof. From Theorem 1 we know that $\mathscr{A}_{P_1}$ and $\mathscr{A}_{P_2}$ have $m!$ and $(n-m)!$ unique permutations, respectively. For $\tilde{\mathscr{A}}_x$, the number of elements can be easily enumerated as the "from" and "to" labels come from distinct sets. The number of elements in $\tilde{P}_x^0$ is of course 1, the number in $\tilde{P}_x^1$ is $m(n-m)$, and in general the number in $\tilde{P}_x^k$ is $\binom{m}{k}\binom{n-m}{k}$ where $\binom{a}{b}$ are the binomial coefficients, since you must choose k unique labels from each set. Using a special case of Vandermonde's identity, the total number in $\tilde{\mathscr{A}}_x$ is then,

$$\begin{aligned} \left|\tilde{\mathscr{A}}_x\right| &= \sum_{k=0}^{\min\{m,n-m\}} \binom{m}{k}\binom{n-m}{k} \\ &= \binom{n}{m} \\ &= \frac{n!}{m!(n-m)!} \end{aligned} \tag{126}$$

The products $\mathscr{A}_{P_1}\mathscr{A}_{P_2}\tilde{\mathscr{A}}_x$ and $\tilde{\mathscr{A}}_x\mathscr{A}_{P_1}\mathscr{A}_{P_2}$ then have at most (since some elements may be equivalent) $m!(n-m)!\left|\tilde{\mathscr{A}}_x\right| = n!$ elements each.

By construction, one can also show that any total permutation can be represented as an element of both $\mathscr{A}_{P_1}\mathscr{A}_{P_2}\tilde{\mathscr{A}}_x$ and $\tilde{\mathscr{A}}_x\mathscr{A}_{P_1}\mathscr{A}_{P_2}$. In both cases, consider the positions of the labels from the two disjoint sets. After any permutation, some set of labels from set 1, which we will call set 1′ of size k, have moved to positions formerly occupied by set 2, and a set of the same size from set 2, called set 2′, are now in set 1 spots. For the first case ($\mathscr{A}_{P_1}\mathscr{A}_{P_2}\tilde{\mathscr{A}}_x$), the construction is as follows: (1) exchange the labels at the positions occupied in the final permutation by sets 1′ and 2′, which is an element of $\tilde{\mathscr{A}}_x$ (specifically $\tilde{\tilde{P}}^k_x$), (2) permute the labels of set 2 to the locations and order of the final permutation. Since the permutation from $\tilde{\mathscr{A}}_x$ puts *some* labels from set 2 (but not necessarily set 2′) in the proper set 1 spots, this is accomplished by permuting only set 2 labels which is an element of $\mathscr{A}_{P_2}$. (3) Similarly permute the set 1 labels, which is an element of $\mathscr{A}_{P_1}$. For the second case, the construction is similar: (1) permute the labels of set 2, which is of course an element of $\mathscr{A}_{P_2}$, so that set 2′ occupies the positions which will be taken by set 1′ in the final permutation, with the set 2′ labels in the order (based on magnitude) of the desired final set 1′ labels. This is necessary since we wish to later apply an element of $\tilde{\mathscr{A}}_x$, which has ordering restrictions on both "from" and "to" labels. (2) Similarly permute the labels of set 1, an element of $\mathscr{A}_{P_1}$, and (3) exchange the labels in sets 1′ and 2′ in order of magnitude, an element of $\tilde{\mathscr{A}}_x$. Thus, any total permutation of n labels from $\mathscr{A}_P$ can be written in either of these ways. Since there are $n!$ unique permutations in $\mathscr{A}_P$ and at most $n!$ permutations in each of $\mathscr{A}_{P_1}\mathscr{A}_{P_2}\tilde{\mathscr{A}}_x$ and $\tilde{\mathscr{A}}_x\mathscr{A}_{P_1}\mathscr{A}_{P_2}$, there must be $n!$ unique elements in each set and they must then be equal to $\mathscr{A}_P$. □

Theorem 3. *Given a set of n labels partitioned into three disjoint sets a, b, and c, the following relationship holds*

$$\mathscr{A}_{bc}\tilde{\mathscr{A}}_{ab-c} = \mathscr{A}_{bc} \sum_{k=0}^{\min\{n_a,n_c\}} \tilde{\tilde{P}}^k_{a-c} \frac{(n_b+n_c)!k!}{(n_b+k)!n_c!} \tag{127}$$

where $\mathscr{A}_{bc}$ is the antisymmetrizer for $b \cup c$, $\tilde{\mathscr{A}}_{ab-c}$ is the partial antisymmetrizer connecting $a \cup b$ and c, and finally $\tilde{\tilde{P}}^k_{a-c}$ are the k-fold double-ordered permutations connecting a and c.

Proof. Using the definition of the partial antisymmetrizers, we can write

$$\mathscr{A}_{bc}\tilde{\mathscr{A}}_{ab-c} = \mathscr{A}_{bc} \sum_{l=0}^{\min\{n_a+n_b,n_c\}} \tilde{\tilde{P}}^l_{ab-c} \tag{128}$$

For each element of $\tilde{\tilde{P}}^l_{ab-c}$, we can classify it by the number m of "from" labels which belong to the set b. This also gives the number of two-index permutations which belong to P_{b-c}, so that each element of $\tilde{\tilde{P}}^l_{ab-c}$ can be written as a product of elements from $\tilde{\tilde{P}}^m_{b-c}$ and $\tilde{\tilde{P}}^{l-m}_{a-c}$. Since $\mathscr{A}_{bc}$ already contains all antisymmetrizers from $\tilde{\tilde{P}}^m_{b-c}$, these elements reduce to unity. Additionally, since $\mathscr{A}_{bc}$ contains $\mathscr{A}_c$, an element of $\tilde{\tilde{P}}^{l}_{ab-c}$ with one set of "to" labels in c may be related to any other set of "to" labels (although the specific "from" labels must be the same). Specifically, we may "redistribute" elements of $\tilde{\tilde{P}}^{l-m}_{a-c}$ (which may all be related this way) such that each one receives an equal numerical factor. The number of elements in $\tilde{\tilde{P}}^{l}_{ab-c}$ for a fixed set of a "from" labels is $\binom{n_c}{l}\binom{n_b}{m}$ since some l indices from c must be selected as "to" labels and m from b as "from" labels. After redistribution, we are dividing the result among $\binom{n_c}{l-m}$ elements of $\tilde{\tilde{P}}^{l-m}_{a-c}$ for the same fixed a labels, so that each one receives the same factor and the total set $\tilde{\tilde{P}}^{l}_{ab-c}$ can be written (with the implicit understanding that $\mathscr{A}_{bc}$ is to be applied later),

$$\tilde{\tilde{P}}^{l}_{ab-c} = \sum_{m=\max\{0,l-n_a\}}^{\min\{l,n_b\}} \tilde{\tilde{P}}^{l-m}_{a-c}\binom{n_c}{l}\binom{n_b}{m}\binom{n_c}{l-m}^{-1} \tag{129}$$

The original equation of interest can then be written and refactored as,

$$\begin{aligned}
\mathscr{A}_{bc}\tilde{\mathscr{A}}_{ab-c} &= \mathscr{A}_{bc}\sum_{l=0}^{\min\{n_a+n_b,n_c\}}\ \sum_{m=\max\{0,l-n_a\}}^{\min\{l,n_b\}} \tilde{\tilde{P}}^{l-m}_{a-c}\binom{n_c}{l}\binom{n_b}{m}\binom{n_c}{l-m}^{-1} \\
&= \mathscr{A}_{bc}\sum_{k=0}^{\min\{n_a,n_c\}} \tilde{\tilde{P}}^{k}_{a-c}\sum_{m=0}^{\min\{n_b,n_c-k\}}\binom{n_c}{k+m}\binom{n_b}{m}\binom{n_c}{k}^{-1} \\
&= \mathscr{A}_{bc}\sum_{k=0}^{\min\{n_a,n_c\}} \tilde{\tilde{P}}^{k}_{a-c}\sum_{m=0}^{\min\{n_b,n_c-k\}}\binom{n_c}{(n_c-k)-m}\binom{n_b}{m}\binom{n_c}{k}^{-1} \\
&= \mathscr{A}_{bc}\sum_{k=0}^{\min\{n_a,n_c\}} \tilde{\tilde{P}}^{k}_{a-c}\binom{n_b+n_c}{n_c-k}\binom{n_c}{k}^{-1} \\
&= \mathscr{A}_{bc}\sum_{k=0}^{\min\{n_a,n_c\}} \tilde{\tilde{P}}^{k}_{a-c}\frac{(n_b+n_c)!k!}{(n_b+k)!n_c!}
\end{aligned} \tag{130}$$

where $k = l - m$ and the fourth equality is obtained using Vandermonde's identity. □

Theorem 4. *Given a set of n out-labels partitioned into two sets a and j of size n_a and n_j, and n in-labels partitioned into two sets b and i of size n_b and n_i, and a bijective mapping $* : a \cup j \mapsto b \cup i$ (where the mapping is altered by permutations of the labels), assume that the symmetry $P(pq) = P(p^*q^*)\ \forall\, p, q \in a \cup j$ exists. For a certain mapping $*$, define $n_s = n_{a;s} + n_{j;s}$ where $n_{a;s} = |a^* \cap i|$ and $n_{j;s} = |j^* \cap b|$. Then, define $\mathscr{A}_a$ and $\mathscr{A}_j$ as the total antisymmetrizers of the a and j labels, and $\tilde{\mathscr{A}}_{\frac{a}{i}-\frac{j}{i}}$ and $\tilde{\mathscr{A}}_{jb-ab}$ as,*

$$\tilde{\mathscr{A}}_{\substack{a\_j\\ i\_i}} = \bigcup_{k=0}^{\min(n_j - n_{j;x}, n_{a;x})} \tilde{\tilde{P}}^k_{ai-ji} \tag{131}$$

$$P_{\substack{a\_j\\ i\_i}} = \{P(a_k j_l) \mid a_k^*, j_l^* \in i\}_{k,l=1}^{n_a, n_j} \tag{132}$$

$$\tilde{\mathscr{A}}_{\substack{j\_a\\ b\_b}} = \bigcup_{k=0}^{\min(n_a - n_{a;x}, n_{j;x})} \tilde{\tilde{P}}^k_{jb-ab} \tag{133}$$

$$P_{\substack{j\_a\\ b\_b}} = \{P(j_k a_l) \mid j_k^*, a_l^* \in b\}_{k,l=1}^{n_j, n_a} \tag{134}$$

Lastly, define a set of permutations $P_s = \{P(a_k j_l) \mid a_k^* \in b \wedge j_l^* \in i\}_{k,l=1}^{n_a, n_j}$. *Then, the antisymmetrizer of all in or by symmetry of all out labels,* $\mathscr{A}$, *relative to some permutation with a mapping* $*$ *with* $n_s = |n_a - n_b|$ *is equal to* $\{\mathscr{A}_a \mathscr{A}_i \tilde{\mathscr{A}}_{\substack{a\_j\\ i\_i}} \tilde{\mathscr{A}}_{\substack{j\_a\\ b\_b}} (\tilde{\tilde{P}}^k_s)_0\}_{k=0}^{(n-|n_a-n_i|-|n_a-n_b|)/2}$, *where, for example,* $(S)_0$ *is an arbitrary element from the set S.*

Proof. All possible permutations of the *in* or *out* labels can be classified by their value of n_s. The minimum possible value is $|n_a - n_b|$, as at most $\min\{n_a, n_b\}$ labels may be paired as $a \leftrightarrow b$ and $\min\{n_i, n_j\}$ labels as $j \leftrightarrow i$, leaving $\frac{|n_a-n_b|+|n_i-n_j|}{2} = |n_a - n_b|$ pairs as $a \leftrightarrow i$ or $j \leftrightarrow b$. The maximum possible value is $n - |n_a - n_i|$, as similarly at most $\min\{n_a, n_i\}$ labels may be paired as $a \leftrightarrow i$ and $\min\{n_j, n_b\}$ labels as $j \leftrightarrow b$, leaving $\frac{|n_a-n_i|+|n_b-n_j|}{2} = |n_a - n_i|$ out of n pairs of labels as $a \leftrightarrow b$ or $j \leftrightarrow i$. Also, the value of n_s must vary between these values in steps of two since, for example, a permutation which results in ${}^{a}_{j} \leftrightarrow {}^{b}_{i} \Rightarrow {}^{a}_{j} \leftrightarrow {}^{i}_{b}$ (i.e., an element of P_s) changes two pairs of labels at a time. Thus, n_s may have $\frac{n-|n_a-n_i|-|n_a-n_b|}{2} + 1$ distinct values, which are generated by applying successive distinct elements of P_s, namely the elements $(\tilde{\tilde{P}}^k_s)_0$. For each value of n_s, the number of permutations in $\mathscr{A}_a \mathscr{A}_i \tilde{\mathscr{A}}_{\substack{a\_j\\ i\_i}} \tilde{\mathscr{A}}_{\substack{j\_a\\ b\_b}}$ is equal to $n_a! n_i! \binom{n_i}{n_{a;s}} \binom{n_b}{n_{j;s}}$ from previous analysis. Defining $n_{a;s}$ and $n_{j;s}$ in terms of a more convenient variable $k = \frac{n_s - |n_a - n_b|}{2}$ as $n_{a;s} = \max\{n_a - n_b, 0\} + k$ and $n_{j;s} = \max\{n_j - n_i, 0\} + k$, we can write the total number of permutations as,

$$N_{total} = n_a! n_j! \sum_{k=0}^{(n-|n_a-n_i|-|n_a-n_b|)/2} \binom{n_i}{\max\{n_a - n_b\} + k} \binom{n_b}{\max\{n_j - n_i, 0\} + k}$$
$$- \begin{cases} n_a! n_j! \sum_{k=0}^{(n-|n_a-n_i|-|n_a-n_b|)/2} \binom{n_i}{n_j - k} \binom{n_b}{k} \\ \quad \text{if } n_a > n_b \text{ and } n_i > n_j \\ n_a! n_j! \sum_{k=0}^{(n-|n_a-n_i|-|n_a-n_b|)/2} \binom{n_i}{k} \binom{n_b}{n_a - k} \\ \quad \text{if } n_b > n_a \text{ and } n_j > n_i \end{cases} \tag{135}$$

Using Vandermonde's identity again, these two cases are equal to $n_a! n_j! \binom{n}{n_j}$ and $n_a! n_j! \binom{n}{n_a}$, respectively, which are both equal to $n!$ upon expansion of the binomial coefficient. To show that these permutations are unique and hence equal to $\mathscr{A}$, we can show that all possible permutations which conserve n_s are related to elements of $\mathscr{A}_a\mathscr{A}_i\tilde{\mathscr{A}}_{\substack{a\\i}-\substack{j\\i}}\tilde{\mathscr{A}}_{\substack{j\\b}-\substack{a\\b}}$. The permutations which conserve n_s can be written nonuniquely as

$$\mathscr{A}_{\delta n_s=0} = \mathscr{A}_a\mathscr{A}_j\mathscr{A}_b\mathscr{A}_i \tag{136}$$

since these are all permutations which involve either "in" or "out" labels of the same type. However, as shown in Theorem 2, we can rewrite this as,

$$\mathscr{A}_{\delta n_s=0} = \tilde{\mathscr{A}}_{\substack{a\\i}-\substack{a\\b}}\tilde{\mathscr{A}}_{\substack{j\\b}-\substack{j\\i}}\mathscr{A}_{\substack{a\\b}}\mathscr{A}_{\substack{j\\i}}\mathscr{A}_{\substack{a\\i}}\mathscr{A}_{\substack{j\\b}}\mathscr{A}_{\substack{a\\b}}\mathscr{A}_{\substack{j\\i}}\mathscr{A}_{\substack{a\\i}}\mathscr{A}_{\substack{j\\b}}\tilde{\mathscr{A}}_{\substack{a\\i}-\substack{j\\i}}\tilde{\mathscr{A}}_{\substack{j\\b}-\substack{a\\b}} \tag{137}$$

where $\mathscr{A}_{\substack{p\\q}}$ is the antisymmetrizer for labels p which map as $p \leftrightarrow q$, etc., and $\tilde{\mathscr{A}}_{\substack{a\\i}-\substack{a\\b}}$ and $\tilde{\mathscr{A}}_{\substack{j\\b}-\substack{j\\i}}$ are defined with similar notation to $\tilde{\mathscr{A}}_{\substack{a\\i}-\substack{j\\i}}$ and $\tilde{\mathscr{A}}_{\substack{j\\b}-\substack{a\\b}}$. Since $\mathscr{A}_{\substack{p\\q}}^2 = K\mathscr{A}_{\substack{p\\q}}$ for some integer constant K and different $\mathscr{A}_{\substack{p\\q}}$ commute, we can remove some antisymmetrizers, recombine the pieces of $\mathscr{A}_a$ and $\mathscr{A}_j$, and arrive at

$$\mathscr{A}_{\delta n_s=0} = K\mathscr{A}_a\mathscr{A}_j\tilde{\mathscr{A}}_{\substack{a\\i}-\substack{j\\i}}\tilde{\mathscr{A}}_{\substack{j\\b}-\substack{a\\b}} \tag{138}$$

for some K. Thus, any permutation for a fixed n_s is relatable to $\mathscr{A}_a\mathscr{A}_j\tilde{\mathscr{A}}_{\substack{a\\i}-\substack{j\\i}}\tilde{\mathscr{A}}_{\substack{j\\b}-\substack{a\\b}}$. The elements are unique since the set of all unique $\mathscr{A}_{\delta n_s=0}$ for all n_s is the same size as the set of all $\mathscr{A}_a\mathscr{A}_j\tilde{\mathscr{A}}_{\substack{a\\i}-\substack{j\\i}}\tilde{\mathscr{A}}_{\substack{j\\b}-\substack{a\\b}}$, and so we have $n!$ unique permutations which must be equal to the total antisymmetrizer $\mathscr{A}$. □

REFERENCES

[1] Montroll EW, Mayer JE. Statistical mechanics of imperfect gases. J Chem Phys 1941;9(8):626–37.

[2] McQuarrie DA. Statistical mechanics. Suasalito, CA: University Science Books; 2000 (see, for example, Figs. 12-2 and 12-3 and the discussion on p. 230–3).

[3] Dyson FJ. The radiation theories of Tomonaga, Schwinger, and Feynman. Phys Rev 1949;75:486–502.

[4] Feynman RP. The theory of positrons. Phys Rev 1949;76(6):749–59.

[5] Feynman RP. Space-time approach to quantum electrodynamics. Phys Rev 1949;76:769–89.

[6] Mattuck RD. A guide to Feynman diagrams in the many-body problem. New York: McGraw-Hill; 1976.

[7] Shavitt I. The history and evolution of configuration interaction. Mol Phys 1998;94(1):3–17.

[8] Goldstone J. Derivation of the Brueckner many-body theory. Proc R Soc Lond A 1957;239(1217):267–79.

[9] Brueckner KA. Many-body problem for strongly interacting particles. II. Linked cluster expansion. Phys Rev 1955;100(1):36–45.

[10] Hugenholtz NM. Perturbation theory of large quantum systems. Physica 1957;23(1–5): 481–532.

[11] Paldus J, Čížek J. Time-independent diagrammatic approach to perturbation theory of fermion systems. Advances in quantum chemistry, vol. 9. London: Academic Press; 1975. p. 105–97.

[12] Brandow BH. Linked-cluster expansions for the nuclear many-body problem. Rev Mod Phys 1967;39(4):771–828.

[13] Bartlett RJ, Musiał M. Coupled-cluster theory in quantum chemistry. Rev Mod Phys 2007;79(1):291–352.

[14] Crawford TD, Schaefer III HF. An introduction to coupled cluster theory for computational chemists. Rev Comp Chem 2000;14:33.

[15] Scuseria GE, Janssen CL, Schaefer III HF. An efficient reformulation of the closed-shell coupled cluster single and double excitation (CCSD) equations. J Chem Phys 1988;89(12):7382–7.

[16] Matthews DA, Gauss J, Stanton JF. Revisitation of nonorthogonal spin adaptation in coupled cluster theory. J Chem Theory Comput 2013;9(6):2567–72.

[17] Čížek J. On the correlation problem in atomic and molecular systems. Calculation of wavefunction components in Ursell-Type expansion using quantum-field theoretical methods. J Chem Phys 1966;45(11):4256–66.

[18] Purvis GD, Bartlett RJ. A full coupled-cluster singles and doubles model: the inclusion of disconnected triples. J Chem Phys 1982;76(4):1910–8.

[19] Scuseria GE, Scheiner AC, Lee TJ, Rice JE, Schaefer III HF. The closed-shell coupled cluster single and double excitation (CCSD) model for the description of electron correlation. A comparison with configuration interaction (CISD) results. J Chem Phys 1987;86(5):2881–90.

[20] Lee TJ, Rice JE. An efficient closed-shell singles and doubles coupled-cluster method. Chem Phys Lett 1988;150(6):406–15.

[21] Stanton JF, Bartlett RJ. The equation of motion coupled-cluster method. A systematic biorthogonal approach to molecular excitation energies, transition probabilities, and excited state properties. J Chem Phys 1993;98(9):7029.

[22] Krylov AI. Spin-flip equation-of-motion coupled-cluster electronic structure method for a description of excited states, bond breaking, diradicals, and triradicals. Acc Chem Res 2006;39(2):83–91.

[23] Čížek J. On the use of the cluster expansion and the technique of diagrams in calculations of correlation effects in atoms and molecules. In: LeFebvre R, Moser C, editors. Advances in Chemical Physics. New York, NY: John Wiley & Sons, Inc.; 1969. p. 35–89.

[24] Paldus J, Čížek J, Shavitt I. Correlation problems in atomic and molecular systems. IV. Extended coupled-pair many-electron theory and its application to the BH_3 molecule. Phys Rev A 1972;5(1):50–67.

[25] Adams BG, Paldus J. Orthogonally spin-adapted coupled-cluster theory for closed-shell systems including triexcited clusters. Phys Rev A 1979;20(1):1.

[26] Shavitt I, Bartlett RJ. Many-body methods in chemistry and physics: MBPT and coupled-cluster theory. 1st ed. Cambridge; New York: Cambridge University Press; 2009.

CHAPTER

Quantum chemistry on a quantum computer

11

Guido Fano*, **S.M. Blinder**†

University of Bologna, Bologna, Italy University of Michigan, Ann Arbor, MI, United States*†

The preceding chapters in this monograph have dealt largely with applications of classical (conventional) computers to the problems of quantum chemistry. While this has been, in our modest opinion, a remarkable success story, there remains an insurmountable barrier limiting the extension of computational quantum chemistry to molecules of increasing size and to basis sets of increasing dimension. This is associated with the exponential growth in the size of the Hilbert spaces representing these quantum systems. Since about the 1980, a new paradigm of computation, based on direct analogy with quantum phenomena, has been proposed. One hope for circumventing the limitations of classical digital computation is the development of *quantum computers*. Although these are still at an early stage of practicable development, the theory of quantum information and quantum computation is a very active area of current research (see, e.g., [1]). Seminal proposals on the possibilities of quantum computation are due to Feynman [2] and Deutsch [3].

The prospect of applying quantum computers to atomic and molecular problems is highly attractive in view of the relative computational complexities of classical and quantum algorithms. Complexity is a general measure of how computer time and memory resources scale with increasing size. For classical computers, applications to atoms and molecules scale *exponentially* with system size, roughly of the order of 2^n with increasing number of electrons n. However, using quantum-computer algorithms, *in principle*, the increase is only *polynomial*, meaning something like n^3, n^4, etc. Problems, which can be solved using polynomially increasing computational resources, are commonly called *tractable*, while those that require exponentially growing resources remain *intractable*. In this chapter, we will focus entirely on the "software" aspects of quantum computation and not consider the "hardware" involved in the manipulation of qubits for quantum computers (for a review of quantum computing hardware, see, e.g., [4]).

The model of quantum computing we will consider is a generalization of the classical circuit model, based on classical bits and logic gates. A quantum circuit carries out transformations on qubits, using gates which correspond to unitary operations. It will suffice for our purposes to limit the unitary gates to those involving just one or two qubits. In classical computing, the NAND gate is universal, meaning

Mathematical Physics in Theoretical Chemistry. https://doi.org/10.1016/B978-0-12-813651-5.00011-5

than any logic circuit can, in principle, be constructed using NAND gates only. An analogous set of universal gates for a quantum computer consists of the CNOT gate to entangle qubits, plus a small number of single-qubit gates, for example, the Hadamard gate H and the S and T gates, defined here.

1 QUBITS

The fundamental idea of quantum computing is the replacement of the elementary unit of information in digital computers, the bit, with the *qubit* (this section follows the treatment in [5], Chapter 6). The qubit can be defined as a normalized quantum state vector belonging to a complex two-dimensional Hilbert space $\mathbb{C}^2$. The basis vectors are conventionally designated, using Dirac notation,

$$|0\rangle = \begin{pmatrix} 1 \\ 0 \end{pmatrix}, \quad |1\rangle = \begin{pmatrix} 0 \\ 1 \end{pmatrix}, \tag{1}$$

a generic element of $\mathbb{C}^2$ can be written as a complex linear combination of two classical states:

$$|\Psi\rangle = \alpha|0\rangle + \beta|1\rangle, \tag{2}$$

where α and β are complex numbers. But a qubit is an entirely different type of linear combination, which does not correspond to any well-defined classical entity. The norm $|\Psi|$ is assumed equal to 1; therefore α, β must obey the normalization condition:

$$|\alpha|^2 + |\beta|^2 = 1. \tag{3}$$

The *Bloch sphere*, shown in Fig. 1, provides a useful geometrical representation of a qubit. The angles θ and ϕ, analogous to latitude and longitude, describe the state of a qubit. Antipodal points represent *orthogonal states*, rather than just negatives of the state vector. On the Bloch sphere, $|0\rangle$ is mapped onto the north pole ($\theta = 0$) and $|1\rangle$ onto the south pole ($\theta = \pi$). The equator, with $\theta = \pi/2$, contains four additional *cardinal points*, with $\phi = 0, \pi/2, \pi, 3\pi/2$, corresponding to $(|0\rangle + |1\rangle)/\sqrt{2}$, $(|0\rangle + i|1\rangle)/\sqrt{2}$, $(|0\rangle - |1\rangle)/\sqrt{2}$, $(|0\rangle - i|1\rangle)/\sqrt{2}$, respectively. There is a one-to-one correspondence between qubit states and points on the Bloch sphere. A convenient parametrization of a qubit, consistent with the earlier, is

$$|\Psi\rangle = \cos\frac{\theta}{2}|0\rangle + \sin\frac{\theta}{2}\, e^{i\phi}|1\rangle, \tag{4}$$

noting the characteristic spinor half-angles.

Different physical systems can serve as realizations of a qubit: $|0\rangle$ and $|1\rangle$ can represent the two polarization states of a photon, the $\pm\frac{1}{2}$ spin states of an electron or nuclear spin, the direction of a vortex in a superconductor or two different energy levels of an atom or molecule. Since θ and ϕ are real numbers over a continuous

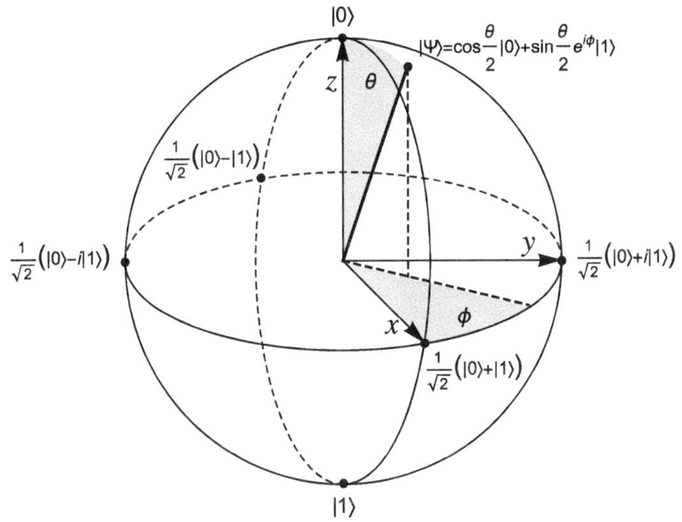

FIG. 1

Bloch sphere. The qubit $|\Psi\rangle$ and the six cardinal points are marked with *red dots* (in the color version).

range, a qubit can contain a vast amount of information. Quoting Nielsen and Chuang, "Paradoxically, there are an infinite number of points on the unit sphere, so that, in principle, one could store an entire text of Shakespeare in the infinite binary expansion of θ." However, an actual *measurement* of the qubit will give only one of two discrete results, ± 1, which collapses the qubit to either $|0\rangle$ or $|1\rangle$.

2 QUANTUM GATES AND CIRCUITS

Quantum gates are the analogs, for qubits, of classical logic gates. There is only one single-bit logic gate, namely, the NOT gate, but several other single-qubit quantum gates are possible. A single-qubit quantum gate transforms the qubit $\begin{pmatrix}\alpha \\ \beta\end{pmatrix}$ into the qubit $\begin{pmatrix}\alpha' \\ \beta'\end{pmatrix}$. To satisfy Eq. (3), we must have:

$$|\alpha'|^2 + |\beta'|^2 = |\alpha|^2 + |\beta|^2 = 1, \tag{5}$$

so that the *norm* of the state $\begin{pmatrix}\alpha \\ \beta\end{pmatrix}$ is invariant. Since we want to preserve linearity, there is only one class of operators that can serve as quantum gates, namely *unitary* operators. Unitary operators U leave the norm invariant, so that $|U\Psi| = |\Psi|$ for all Ψ. The most important single-qubit gates are shown in Fig. 2, along with the 2×2

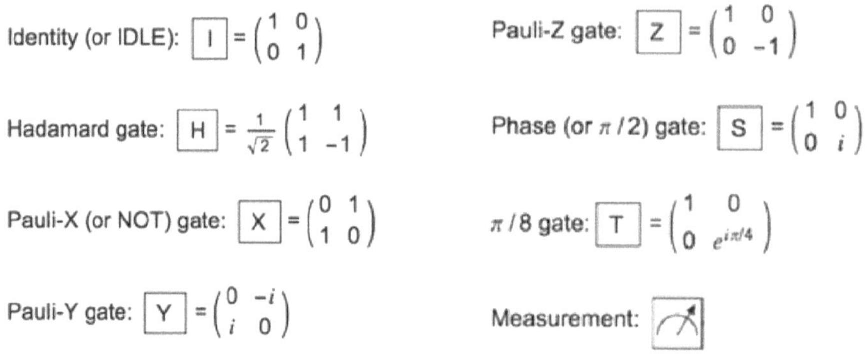

FIG. 2

Single-qubit gates.

unitary transformations they produce. For example, application of the Hadamard gate to each of the basis states gives

$$H|0\rangle = \frac{1}{\sqrt{2}}(|0\rangle + |1\rangle), \quad H|1\rangle = \frac{1}{\sqrt{2}}(|0\rangle - |1\rangle). \tag{6}$$

The Pauli-X, -Y, and -Z gates are named after the Pauli spin operators, σ_x, σ_y, and σ_z, which have the same matrix representations. The Pauli-X gate is equivalent to the NOT gate, which interchanges basis states

$$X|0\rangle = |1\rangle, \quad X|1\rangle = |0\rangle, \tag{7}$$

just as the classical NOT gate ▷o.

Operators representing rotation of a qubit on the Bloch sphere, about the x-, y-, and z-axes, can be constructed by exponentiation of the Pauli matrices, as the following:

$$\begin{aligned}
R_x(\alpha) &= e^{-i\alpha X/2} = \cos\frac{\alpha}{2}I - i\sin\frac{\alpha}{2}X = \begin{pmatrix} \cos\frac{\alpha}{2} & -i\sin\frac{\alpha}{2} \\ -i\sin\frac{\alpha}{2} & \cos\frac{\alpha}{2} \end{pmatrix}, \\
R_y(\beta) &= e^{-i\beta Y/2} = \cos\frac{\beta}{2}I - i\sin\frac{\beta}{2}Y = \begin{pmatrix} \cos\frac{\beta}{2} & -\sin\frac{\beta}{2} \\ \sin\frac{\beta}{2} & \cos\frac{\beta}{2} \end{pmatrix}, \\
R_z(\gamma) &= e^{-i\gamma Z/2} = \cos\frac{\gamma}{2}I - i\sin\frac{\gamma}{2}Z = \begin{pmatrix} e^{-i\gamma/2} & 0 \\ 0 & e^{i\gamma/2} \end{pmatrix}.
\end{aligned} \tag{8}$$

A generalization of Eq. (8), a rotation by θ about a unit vector to the Bloch sphere, $\mathbf{n} = (n_1, n_3, n_3)$ is represented by

$$R_{\mathbf{n}}(\theta) = e^{-i\theta\,\mathbf{n}\cdot\boldsymbol{\Sigma}/2} = \cos\frac{\theta}{2}I - i\sin\frac{\theta}{2}\,\mathbf{n}\cdot\boldsymbol{\Sigma}, \tag{9}$$

where $\mathbf{\Sigma} = (X, Y, Z)$. An arbitrary rotation on the Bloch sphere can be described by sequential rotations about three *Euler angles*. From a multitude of possible conventions, we will choose the sequence: (1) rotation about the z-axis by an angle α; (2) rotation about the new x-axis by an angle β; (3) rotation about the new z-axis by an angle γ. This sequence can be represented by the unitary operator

$$U = e^{i\delta} R_z(\gamma) R_x(\beta) R_z(\alpha), \tag{10}$$

where the global phase factor $e^{i\delta}$ enables some resulting matrices to be expressed in standard form. Any unitary operation on a single qubit can be written as a combination of rotations in the form (10), often in alternative ways. For example, you can show that U becomes the Hadamard gate H with the angles $\alpha = \beta = \gamma = \delta = \pi/2$. Another combination gives $H = iR_x(\pi)R_y(\pi/2)$. As we shall see later, rotation gates are if central importance in the simulation of quantum systems.

Multiqubit states and operations can be represented as *tensor products* and their matrices constructed as *Kronecker products*. For example, let $A = \begin{pmatrix} a_{11} & a_{12} \\ a_{21} & a_{22} \end{pmatrix}$ and $B = \begin{pmatrix} b_{11} & b_{12} \\ b_{21} & b_{22} \end{pmatrix}$ represent matrices in $\mathbb{C}^2$. Then the matrix for $A \otimes B$ in $\mathbb{C}^2 \otimes \mathbb{C}^2$ is given by

$$A \otimes B = \begin{pmatrix} a_{11}B & a_{12}B \\ a_{21}B & a_{22}B \end{pmatrix} = \begin{pmatrix} a_{11}b_{11} & a_{11}b_{12} & a_{12}b_{11} & a_{12}b_{12} \\ a_{11}b_{21} & a_{11}b_{22} & a_{12}b_{21} & a_{12}b_{22} \\ a_{21}b_{11} & a_{21}b_{12} & a_{22}b_{11} & a_{22}b_{12} \\ a_{21}b_{21} & a_{21}b_{22} & a_{22}b_{21} & a_{22}b_{22} \end{pmatrix}. \tag{11}$$

The two-qubit state with $\Psi_1 = \begin{pmatrix} \alpha_1 \\ \beta_1 \end{pmatrix}$, $\Psi_2 = \begin{pmatrix} \alpha_2 \\ \beta_2 \end{pmatrix}$ is given by

$$\Psi = \Psi_1 \otimes \Psi_2 = \begin{pmatrix} \alpha_1\alpha_2 \\ \alpha_1\beta_2 \\ \beta_1\alpha_2 \\ \beta_1\beta_2 \end{pmatrix}. \tag{12}$$

Next we consider two-qubit gates. Two-qubit gates operate in the $\mathbb{C}^2 \otimes \mathbb{C}^2$ space. Thus they can be represented by 4×4 unitary matrices. Since the inverse of the unitary matrix U is the matrix $U^\dagger$, all quantum gates are *reversible*, in contrast to classical logic gates. For a quantum gate, the input can be deduced from the output. But for a classical XOR gate $x \oplus y$, for the output 1, we do not know whether the input was $|10\rangle$ or $|01\rangle$. The output does not identify the input. The simplest two-qubit gate is the CNOT (controlled-NOT) gate. We denote the two input qubits by x, y. The first qubit x is called the *control qubit*, the second y the *target* qubit. This target qubit flips if and only if $x = 1$. If $x = 0$ the second qubit remains unchanged. A representation of the CNOT gate is shown in Fig. 3. The black dot represents the action of the control qubit. The unitary matrix corresponding to the CNOT gate is given by

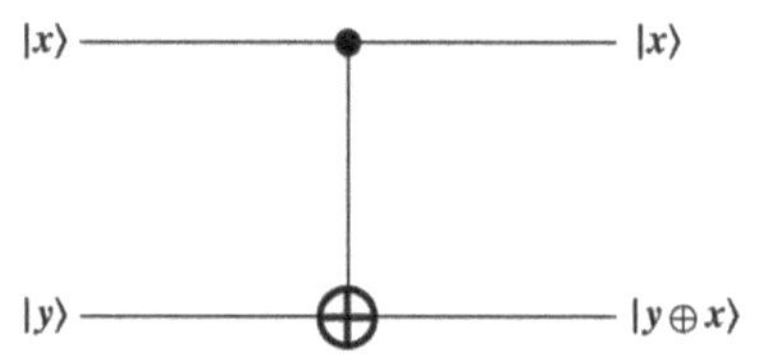

FIG. 3

Controlled-NOT (CNOT) gate.

$$U_{\mathrm{CNOT}} = \begin{pmatrix} 1 & 0 & 0 & 0 \\ 0 & 1 & 0 & 0 \\ 0 & 0 & 0 & 1 \\ 0 & 0 & 1 & 0 \end{pmatrix}. \tag{13}$$

In effect, U_{CNOT} leaves the states $|00\rangle, |01\rangle$ invariant, but swaps the states $|10\rangle$ and $|11\rangle$. The CNOT gate is the fundamental creator of entangled qubits, performing, in a sense, the "quantization" of a classical gate to a quantum gate.

A generalization of the CNOT gate is a controlled gates acting on two qubits, with the control qubit determining some unitary operation V on the target qubit. If the single-qubit gate V is represented by the 2×2 matrix:

$$V = \begin{pmatrix} V_{11} & V_{12} \\ V_{21} & V_{22} \end{pmatrix}. \tag{14}$$

The corresponding two-qubit matrix is

$$U_{\mathrm{CV}} = \begin{pmatrix} 1 & 0 & 0 & 0 \\ 0 & 1 & 0 & 0 \\ 0 & 0 & V_{11} & V_{12} \\ 0 & 0 & V_{21} & V_{22} \end{pmatrix}. \tag{15}$$

The symbol for the controlled-V gate is shown in Fig. 4.

Exchanging the control and target registers gives an *inverted controlled gate*. Fig. 5 shows an inverted controlled-V gate, which is represented by the matrix:

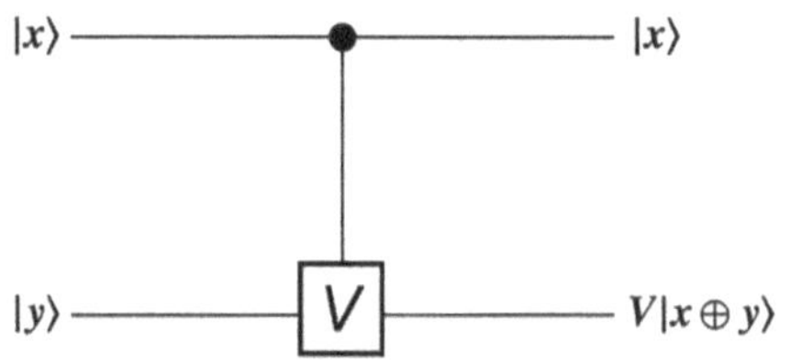

FIG. 4

Controlled-V gate.

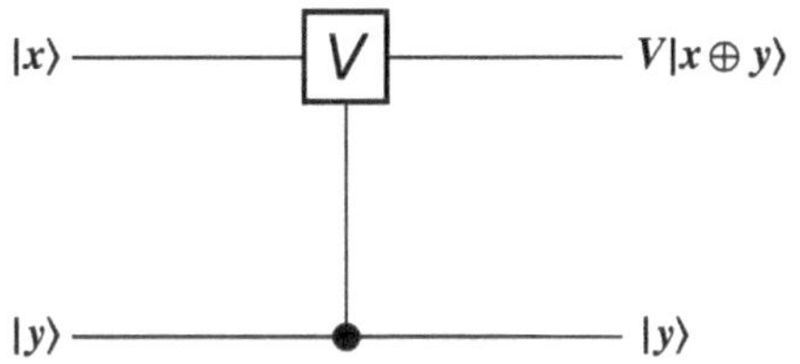

FIG. 5

Inverted controlled-V gate.

$$U_{\text{ICV}} = \begin{pmatrix} 1 & 0 & 0 & 0 \\ 0 & V_{11} & 0 & V_{12} \\ 0 & 0 & 1 & 0 \\ 0 & V_{21} & 0 & V_{22} \end{pmatrix}. \tag{16}$$

If V is a phase gate, such as S or T, in general $R_n = \begin{pmatrix} 1 & 0 \\ 0 & e^{i\pi/2^{n-1}} \end{pmatrix}$, then the corresponding controlled and inverted-controlled gates are identical:

$$U_{\text{CR}} = U_{\text{ICR}} = \begin{pmatrix} 1 & 0 & 0 & 0 \\ 0 & 1 & 0 & 0 \\ 0 & 0 & 1 & 0 \\ 0 & 0 & 0 & e^{i\pi/2^{n-1}} \end{pmatrix}. \tag{17}$$

Another important two-qubit gate is the SWAP gate, which swaps the states of two input qubits, such that

$$U_{\text{SWAP}} = \begin{pmatrix} 1 & 0 & 0 & 0 \\ 0 & 0 & 1 & 0 \\ 0 & 1 & 0 & 0 \\ 0 & 0 & 0 & 1 \end{pmatrix}. \tag{18}$$

This gate is shown in Fig. 6. It is easy to verify the equivalence of the SWAP gate with the sequence of CNOT operations is shown in Fig. 7.

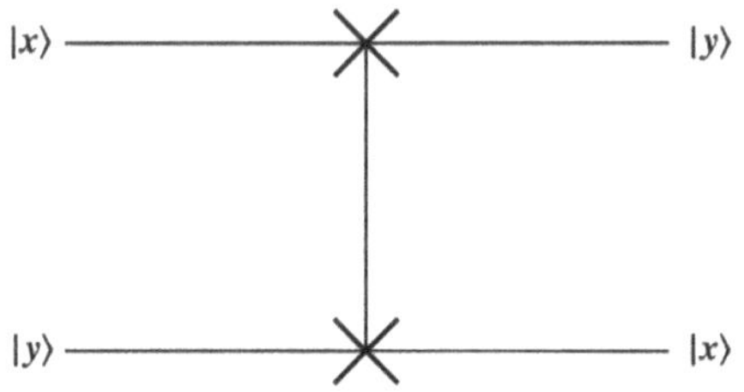

FIG. 6

SWAP gate.

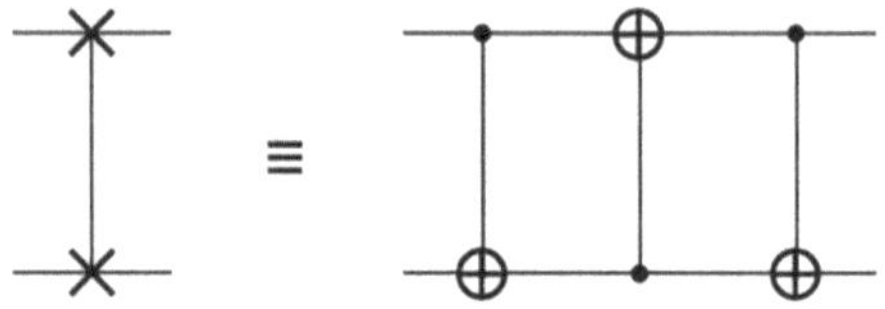

FIG. 7

Sequence of CNOT operations equivalent to SWAP gate.

In a quantum computer circuit, the path of each qubit is represented by a single horizontal line—which usually connects a sequence of gates. The qubit is understood to move along the "quantum wire" from left to right. A wire carrying a classical bit, 0 or 1, usually after a measurement, is indicated by a double line = =. As an example, let us show a quantum circuit which can produce *Bell states*, which are the superpositions $|\Phi^{\pm}\rangle = \frac{1}{\sqrt{2}}\big(|00\rangle \pm |11\rangle\big)$ and $|\Psi^{\pm}\rangle = \frac{1}{\sqrt{2}}\big(|01\rangle \pm |10\rangle\big)$. These are maximally entangled states of two qubits. Bell states can be produced by the circuit, as shown in Fig. 8. The first qubit is passed through a Hadamard gate and then the two qubits are entangled by a CNOT gate. If the input to the system is $|0\rangle \otimes |0\rangle = |00\rangle$, then the Hadamard gate changes the state to

$$\frac{1}{\sqrt{2}}\Big(|0\rangle + |1\rangle\Big) \otimes |0\rangle = \frac{1}{\sqrt{2}}\Big(|00\rangle + |10\rangle\Big),$$

and the CNOT gate converts this to $\frac{1}{\sqrt{2}}\Big(|00\rangle + |11\rangle\Big) = |\Phi^{+}\rangle$. For the four possible inputs, we find

$$\begin{aligned} |00\rangle &\to \frac{1}{\sqrt{2}}\Big(|00\rangle + |11\rangle\Big) = |\Phi^{+}\rangle, \quad |01\rangle \to \frac{1}{\sqrt{2}}\Big(|01\rangle + |10\rangle\Big) = |\Psi^{+}\rangle, \\ |10\rangle &\to \frac{1}{\sqrt{2}}\Big(|00\rangle - |11\rangle\Big) = |\Phi^{-}\rangle, \quad |11\rangle \to \frac{1}{\sqrt{2}}\Big(|01\rangle - |10\rangle\Big) = |\Psi^{-}\rangle. \end{aligned} \tag{19}$$

The quantum algorithms we are considering in this chapter are based on the *quantum circuit model*. This uses one- and two-qubit gates sequentially connected by wires. We begin with an initial register of qubits to be converted by the action of the circuit into a resultant register of classical bits, on which measurements can be made.

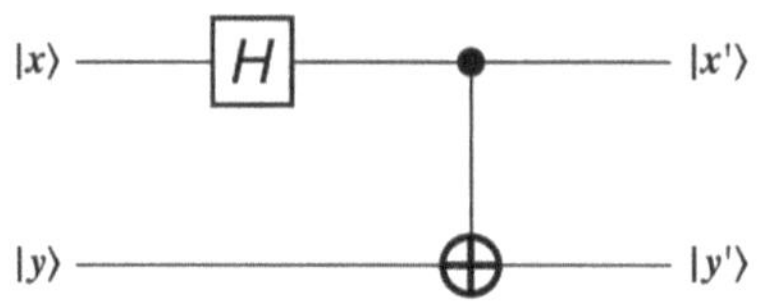

FIG. 8

Circuit producing a Bell state.

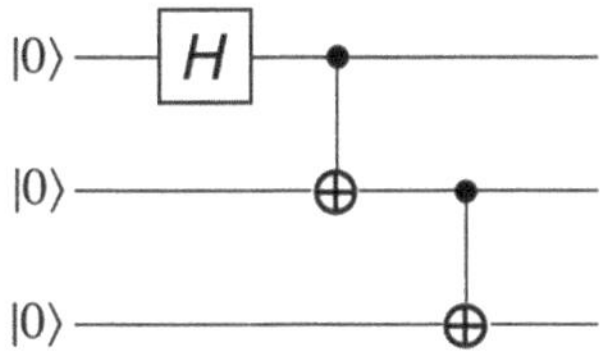

FIG. 9

Circuit producing GHZ state $\frac{1}{\sqrt{2}}(|000\rangle + |111\rangle)$.

Another interesting example is a quantum circuit, shown in Fig. 9, which can produce a *GHZ state*, which is an entangled superposition of *three* qubits:

$$|\Psi\rangle_{\mathrm{GHZ}} = \frac{1}{\sqrt{2}}\big(|000\rangle + |111\rangle\big). \tag{20}$$

GHZ refers to Greenberger, Horne, and Zeilinger, who first produced entangled sets of three (and four) photons. Such experiments enabled demonstration of the validity of Bell's theorem in a single run, rather than from statistical analysis of a multitude of runs. Bell's theorem states that the results of quantum mechanics are incompatible with any picture based on local realism.

3 QUANTUM FOURIER TRANSFORM

The quantum Fourier transform (QFT) is closely analogous to the well-known *discrete Fourier transform* (DFT), whereby set of N complex numbers x_j ($j = 0, 1, \ldots, N-1$) can be transformed into another set of N complex numbers y_k ($k = 0, 1, \ldots, N-1$) by the relations:

$$y_k = \frac{1}{\sqrt{N}} \sum_{j=0}^{N-1} x_j\, (\omega_N)^{kj}, \quad \omega_N = e^{2\pi i/N}. \tag{21}$$

The inverse DFT is then given by $x_j = \frac{1}{\sqrt{N}} \sum_{k=0}^{N-1} y_k\, (\omega_N^*)^{kj}$. A QFT, which we denote by F, is equivalent to a DFT on the *amplitudes* of a quantum state,

$$\Psi = \sum_{j=0}^{N-1} \alpha_j |j\rangle, \quad \text{with} \quad \sum_{j=0}^{N-1} |\alpha_j|^2 = 1, \tag{22}$$

such that

$$F\Psi = \sum_{k=0}^{N-1} \beta_k |k\rangle, \quad \text{with} \quad \beta_k = \frac{1}{\sqrt{N}} \sum_{j=0}^{N-1} \alpha_j (\omega_N)^{kj}. \tag{23}$$

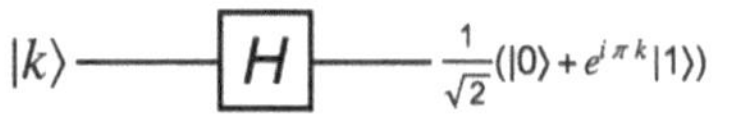

FIG. 10

Circuit with Hadamard gate; $k = 0$ or 1.

The unitary operator corresponding to a QFT is given by (with ω written for ω_N):

$$F = \frac{1}{\sqrt{N}} \begin{pmatrix} 1 & 1 & 1 & \cdots & 1 \\ 1 & \omega & \omega^2 & \cdots & \omega^{N-1} \\ 1 & \omega^2 & \omega^4 & \cdots & \omega^{2(N-1)} \\ \vdots & \vdots & \vdots & & \vdots \\ 1 & \omega^{N-1} & \omega^{2(N-1)} & \cdots & \omega^{(N-1)^2} \end{pmatrix}. \tag{24}$$

Most often, the QFT is applied to individual basis functions $|k\rangle$, such that

$$F|k\rangle = \frac{1}{\sqrt{N}} \sum_{j=0}^{N-1} (\omega_N)^{kj} |j\rangle. \tag{25}$$

The action of a Hadamard gate is the simplest example of a QFT, with $N = 2$, $\omega_2 = -1$, as shown in Eq. (6) and Fig. 10.

A quantum state represented by n-qubits has $N = 2^n$ basis functions. For example, a two-qubit state is spanned by four basis functions, which can be designated as follows

$$\begin{aligned} |\mathbf{0}\rangle &= |0\rangle \otimes |0\rangle = |00\rangle, \quad |\mathbf{1}\rangle = |0\rangle \otimes |1\rangle = |01\rangle, \\ |\mathbf{2}\rangle &= |1\rangle \otimes |0\rangle = |10\rangle, \quad |\mathbf{3}\rangle = |1\rangle \otimes |1\rangle = |11\rangle. \end{aligned} \tag{26}$$

To summarize:

$$|k\rangle = |k_1\rangle \otimes |k_2\rangle = |k_1 k_2\rangle, \tag{27}$$

where k is expressed as a binary number. With $N = 4$, $\omega = \omega_4 = e^{i\pi/2} = i$, the QFT of a basis function is given by

$$F|k\rangle = \frac{1}{2} \sum_{j=0}^{3} \omega^{kj} |j\rangle. \tag{28}$$

Noting that $|j\rangle = |j_1\rangle \otimes |j_2\rangle$,

$$\begin{aligned} F|k\rangle &= \frac{1}{2} \sum_{j_1=0}^{1} \sum_{j_2=0}^{1} \omega^{2kj_1} \omega^{kj_2} |j_1\rangle \otimes |j_2\rangle = \frac{1}{2} \left(|0\rangle + \omega^{2k}|1\rangle\right) \otimes \left(|0\rangle + \omega^{k}|1\rangle\right) \\ &= \frac{1}{\sqrt{2}} \left(|0\rangle + e^{i\pi k_2}|1\rangle\right) \otimes \frac{1}{\sqrt{2}} \left(|0\rangle + e^{i\pi\left(k_1 + \frac{k_2}{2}\right)}|1\rangle\right). \end{aligned} \tag{29}$$

It has also been noted that $e^{2\pi i k_1} = 1$.

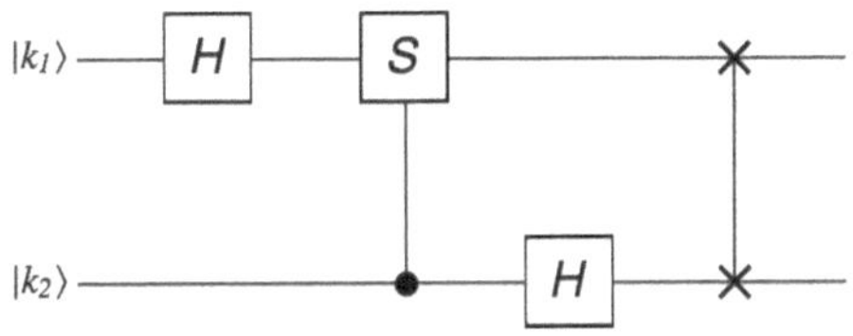

FIG. 11

Circuit for two-qubit quantum Fourier transform.

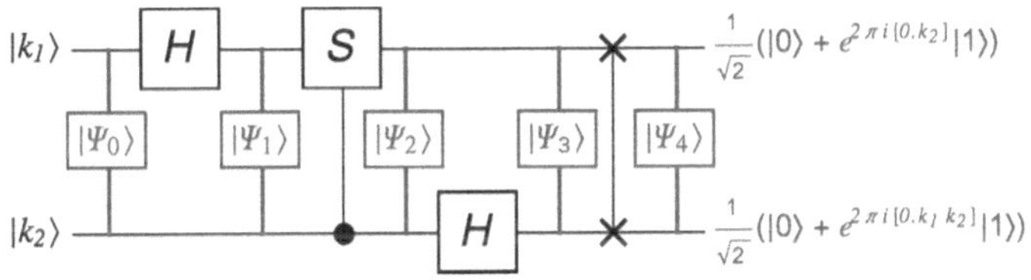

FIG. 12

Details for two-qubit quantum Fourier transform.

Fig. 11 shows a circuit, which implements the two-qubit QFT. The details of its operation are represented in Fig. 12. We begin with the state $|\Psi_0\rangle = |k_1\rangle \otimes |k_2\rangle$. The Hadamard gate on the first register produces the state

$$|\Psi_1\rangle = \frac{1}{\sqrt{2}}\left(|0\rangle + e^{i\pi k_1}|1\rangle\right) \otimes |k_2\rangle. \tag{30}$$

The subsequent action of the inverted controlled-S gate uses the unitary operator

$$U_{\text{ICS}} = \begin{pmatrix} 1 & 0 & 0 & 0 \\ 0 & 1 & 0 & 0 \\ 0 & 0 & 1 & 0 \\ 0 & 0 & 0 & e^{i\pi/2} \end{pmatrix}, \tag{31}$$

which gives

$$|\Psi_2\rangle = U_{\text{ICS}}|\Psi_1\rangle = \frac{1}{\sqrt{2}}\left(|0\rangle + e^{i\pi k_1}e^{i\pi\frac{k_2}{2}}|1\rangle\right) \otimes |k_2\rangle. \tag{32}$$

A Hadamard gate in the second register then results in

$$|\Psi_3\rangle = \frac{1}{\sqrt{2}}\left(|0\rangle + e^{i\pi\left(k_1+\frac{k_2}{2}\right)}|1\rangle\right) \otimes \frac{1}{\sqrt{2}}\left(|0\rangle + e^{i\pi k_2}|1\rangle\right). \tag{33}$$

The final result is produced by an SWAP gate:

$$|\Psi_4\rangle = \text{SWAP}|\Psi_3\rangle = \frac{1}{\sqrt{2}}\left(|0\rangle + e^{i\pi k_2}|1\rangle\right) \otimes \frac{1}{\sqrt{2}}\left(|0\rangle + e^{i\pi\left(k_1+\frac{k_2}{2}\right)}|1\rangle\right). \tag{34}$$

By analogy with the earlier procedure, we can derive the three-qubit QFT. Eq. (29) generalizes to

$$\begin{aligned} F|k\rangle &= \frac{1}{\sqrt{2}}\left(|0\rangle + e^{i\pi k_3}|1\rangle\right) \otimes \frac{1}{\sqrt{2}}\left(|0\rangle + e^{i\pi\left(k_2+\frac{k_3}{2}\right)}|1\rangle\right) \\ &\otimes \frac{1}{\sqrt{2}}\left(|0\rangle + e^{i\pi\left(k_1+\frac{k_2}{2}+\frac{k_3}{4}\right)}|1\rangle\right) \\ &= \frac{1}{\sqrt{2}}\left(|0\rangle + e^{2\pi i[0.k_3]}|1\rangle\right) \\ &\otimes \frac{1}{\sqrt{2}}\left(|0\rangle + e^{2\pi i[0.k_2k_3]}|1\rangle\right) \otimes \frac{1}{\sqrt{2}}\left(|0\rangle + e^{2\pi i[0.k_1k_2k_3]}|1\rangle\right). \end{aligned} \tag{35}$$

In the last line, we have expressed k in binary fractional notation, such that

$$k = k_1 + \frac{k_2}{2} + \frac{k_3}{4} + \cdots = 2[0 \cdot k_1k_2k_3 \ldots]. \tag{36}$$

The corresponding circuit is shown in Fig. 13. In all of the previous examples, a QFT applied to a single basis qubit transforms it into a equally weighted linear combination of all N basis qubits, with complex phase factors of unit magnitude.

The inverted controlled-T gate used in the circuit is represented by the unitary matrix

$$U_{\text{ICT}} = \begin{pmatrix} 1 & 0 & 0 & 0 \\ 0 & 1 & 0 & 0 \\ 0 & 0 & 1 & 0 \\ 0 & 0 & 0 & e^{i\pi/4} \end{pmatrix}. \tag{37}$$

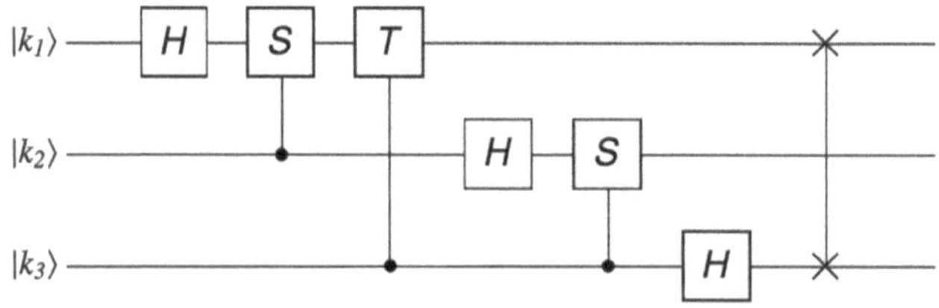

FIG. 13

Circuit for three-qubit quantum Fourier transform.

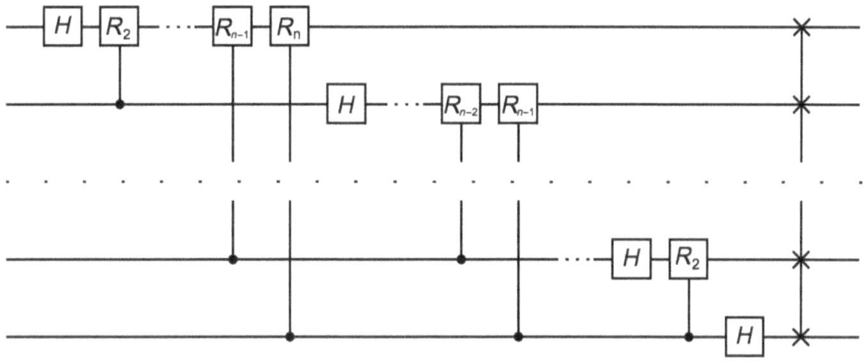

FIG. 14

Circuit for *n*-qubit quantum Fourier transform.

We can define a generalized *controlled-R_n gate* such that

$$R_n = \begin{pmatrix} 1 & 0 & 0 & 0 \\ 0 & 1 & 0 & 0 \\ 0 & 0 & 1 & 0 \\ 0 & 0 & 0 & e^{i\pi/2^{n-1}} \end{pmatrix}. \quad (38)$$

The general n-qubit QFT circuit is shown in Fig 14.

4 PHASE ESTIMATION ALGORITHM

An important application of the QFT is phase estimation. In applications to quantum chemistry, this enables energy eigenvalues to be calculated after the action of the evolution operator e^{-iHt}. Let U be a known $2^n \times 2^n$ unitary matrix and $|\Psi\rangle \in \mathbb{C}^{2^n}$, one of its eigenvectors. We can then write

$$U|\Psi\rangle = e^{i\varphi}|\Psi\rangle, \quad (39)$$

since the eigenvalues of a unitary matrix are complex numbers of unit magnitude (sometimes called *eigenphases*). The action of the operator U is performed by some otherwise unspecified "black box," also known as a *oracle*. This can be symbolically represented by $|\Psi\rangle$—$\boxed{U}$—$e^{i\varphi}|\Psi\rangle$. The basic maneuver leading to the phase-estimation algorithm is known as *phase kickback*. Consider the little circuit in Fig. 15, involving a single-qubit function $|\psi\rangle$. The initial input to the two registers can be written $|0\rangle \otimes |\psi\rangle$. The Hadamard gate transforms this to $\frac{1}{\sqrt{2}}(|0\rangle \otimes |\psi\rangle + |1\rangle \otimes |\psi\rangle)$. The controlled-$U$ gate then acts only for the $|1\rangle$ component of the control register. The result is

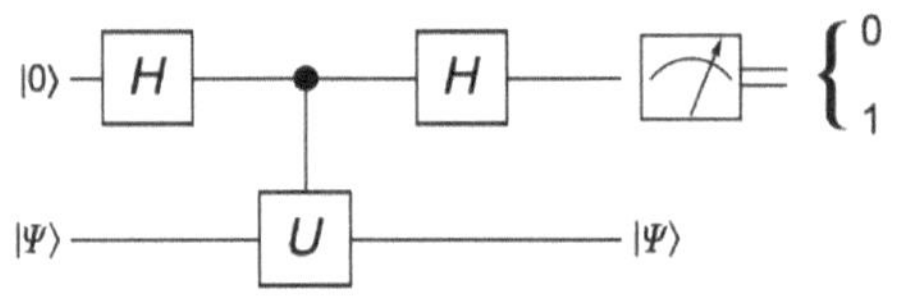

FIG. 15

Phase kickback.

$$\frac{1}{\sqrt{2}}\left(|0\rangle \otimes |\psi\rangle + |1\rangle \otimes e^{i\varphi}|\psi\rangle\right) = \frac{1}{\sqrt{2}}\left(|0\rangle + e^{i\varphi}|1\rangle\right) \otimes |\psi\rangle, \tag{40}$$

with the target register $|\psi\rangle$ emerging unchanged. After action of another Hadamard gate, the control register becomes

$$H\frac{1}{\sqrt{2}}\left(|0\rangle + e^{i\varphi}|1\rangle\right) = \frac{1}{2}\left(|0\rangle + |1\rangle\right) + \frac{1}{2}e^{i\varphi}\left(|0\rangle - |1\rangle\right) = \left(\frac{1+e^{i\varphi}}{2}\right)|0\rangle + \left(\frac{1-e^{i\varphi}}{2}\right)|1\rangle. \tag{41}$$

Finally, a measurement on the control register results in a bit 0 or 1, with probabilities:

$$p(0) = \left|\frac{1+e^{i\varphi}}{2}\right|^2 = \cos^2\frac{\varphi}{2}, \quad p(1) = \left|\frac{1-e^{i\varphi}}{2}\right|^2 = \sin^2\frac{\varphi}{2}. \tag{42}$$

A statistical analysis then enables the phase φ to be determined.

For a multiqubit function $|\Psi\rangle$, suppose it is desired to estimate the phase φ to an accuracy of n bits. We then need to subject the first n-qubits of the eigenvector $|\Psi\rangle$ to a series of controlled operators, involving powers of U, followed by an inverse QFT. The circuit implementing the phase estimation is shown in Fig. 16. The input of the first register consists of m-qubits, all prepared in the state $|0\rangle$. This is an example of an *ancilla*, which is an entangled component of a quantum computer intended to assist in carrying out an algorithm. The second register contains the input of

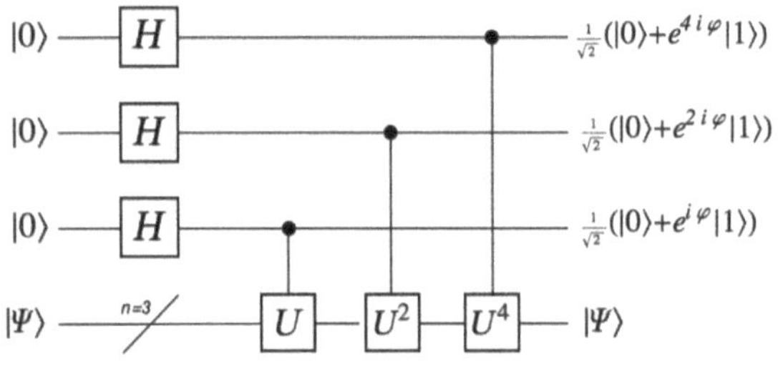

FIG. 16

Circuit implementing phase estimation.

n-qubits, which represent the state $|\Psi\rangle$. For concreteness, we consider the particular case $m = 3, n = 3$. A H gate, followed by a controlled U^{2n} (for $n = 0, 1, 2$), maps the qubit $|0\rangle$ into $\frac{1}{\sqrt{2}}\left(|0\rangle + e^{i2^n\varphi}|1\rangle\right)$. In this way the output of the top register is the product state:

$$|\text{OUT}\rangle = \frac{1}{\sqrt{2}}\left(|0\rangle + e^{4i\varphi}|1\rangle\right) \otimes \frac{1}{\sqrt{2}}\left(|0\rangle + e^{2i\varphi}|1\rangle\right) \otimes \frac{1}{\sqrt{2}}\left(|0\rangle + e^{i\varphi}|1\rangle\right). \tag{43}$$

Comparing this with the last line of Eq. (35), $|\text{OUT}\rangle$ is seen to represent the Fourier transform of a vector $|k\rangle = |k_1\rangle \otimes |k_2\rangle \otimes |k_3\rangle$. Accordingly, $|k\rangle$ can be determined by the inverse Fourier transform:

$$|k\rangle = F^{-1}|\text{OUT}\rangle. \tag{44}$$

The magnitude of k $(0 \le k \le 1)$ is given by $k = [0 \cdot k_1k_2k_3]$. This represents an approximation to the phase φ, to three-bit accuracy:

$$\varphi = 2\pi[0 \cdot k_1k_2k_3 \ldots]. \tag{45}$$

Therefore, the circuit of Fig. 16 can determine the phase φ to this level of accuracy. The error in this value is bounded by the first neglected bit, numbered $n+1$. Thus the error in the measured phase, say $\delta\varphi$, is of the order of $2\pi/2^{n+1}$. A plot showing a series of repeated determinations of φ might resemble a Gaussian centered at $2\pi k$ with $\sigma \approx \delta\varphi$. Clearly, increasing the number of qubits in $|k\rangle$ improves the accuracy of the computed phase.

5 MANY-ELECTRON SYSTEMS

Their representation requires systems of multiple qubits. The Hilbert space of D qubits, $\mathbb{C}^{2D}$, is the *tensor product* of D copies of $\mathbb{C}^2$, namely, $\mathbb{C}^2 \otimes \mathbb{C}^2 \cdots \otimes \mathbb{C}^2$, D times. An orthonormal in each space $\mathbb{C}^2$ is assumed. The extended basis then consists of 2^D product states:

$$|\delta_1, \delta_2, \ldots, \delta_D\rangle = |\delta_1\rangle \otimes |\delta_2\rangle \otimes \cdots \otimes |\delta_D\rangle. \tag{46}$$

We use the symbols δ_k, rather than n_k, for the occupation numbers to emphasize that they can equal only 0 or 1 (inspired by the Kronecker delta δ_{ij}). The multidimensional systems of particular interest are many-electron atoms or molecules, as represented by Fock states with D spin orbitals, occupied by N electrons in all combinations of occupation numbers. The D-dimensional Fock states will include N occupied spin-orbitals and $D - N$ unoccupied (or *virtual*) spin-orbitals. Most often, the virtual spin-orbitals have positive energies $\epsilon_k > 0$. On a classical computer, enumeration of all the possible Fock states for a D-dimensional system would require 2^D memory locations. But in a quantum computer, each qubit can be in a superposition of states, containing the same quantity of information. Moreover, computations can, in principle, be run simultaneously on the entire assemblage, exhibiting what is known as *quantum parallelism*. The classical algorithm has exponential complexity w.r.t. increasing system size, while the quantum algorithm is approximately polynomial. As an idea of the scales involved, a 30-qubit quantum computer could simultaneously process $10^{30} \approx 1$ billion states.

Fermion *creation operators* $a_k^\dagger$ and *annihilation operators* a_k can add or remove electrons from the Fock state. These obey the anticommutation relations

$$\{a_j, a_k^\dagger\} = \delta_{jk}, \quad \{a_j, a_k\} = \{a_j^\dagger, a_k^\dagger\} = 0. \tag{47}$$

The *vacuum state* $|\mathrm{vac}\rangle$, the formal starting point for manipulations in Fock space, is defined by

$$|\mathrm{vac}\rangle = |0_1, 0_2, \ldots, 0_D\rangle. \tag{48}$$

An electron in a spin-orbital k is added to the system using the creation operator $a_k^\dagger$:

$$a_k^\dagger|\mathrm{vac}\rangle = |0_1, 0_2, \ldots, 1_k, \ldots\rangle. \tag{49}$$

Adding a second electron in the spin-orbital j gives

$$a_j^\dagger a_k^\dagger|\mathrm{vac}\rangle = |0_1, 0_2, \ldots, 1_j, \ldots, 1_k, \ldots\rangle, \tag{50}$$

if j comes before k in the Fock space ordering, but

$$a_j^\dagger a_k^\dagger|\mathrm{vac}\rangle = -|0_1, 0_2, \ldots, 1_k, \ldots, 1_j, \ldots\rangle, \tag{51}$$

if k comes before j. The sign change is a consequence of the anticommutation $a_j^\dagger a_k^\dagger = -a_k^\dagger a_j^\dagger$. This accords with the antisymmetry principle for many-electron systems. Clearly, also, $a_k^\dagger a_k^\dagger = 0$, which is a statement of the Pauli exclusion principle. For the action of a creation operator on an arbitrary Fock state

$$a_k^\dagger|\delta_1, \delta_2, \ldots, 0_k, \ldots\rangle = (-1)^{\sum_{j=1}^{k-1}\delta_j}|\delta_1, \delta_2, \ldots, 1_k, \ldots\rangle, \tag{52}$$

where the sum counts the number of occupied spin-orbitals $j < k$. Of course, $a_k^\dagger|\delta_1, \delta_2, \ldots, 1_k, \ldots\rangle = 0$, if the spin-orbital k is already occupied.

Analogous relations hold for the annihilation operators a_k. In particular,

$$a_k|\delta_1, \delta_2, \ldots, 1_k, \ldots\rangle = (-1)^{\sum_{j=1}^{k-1}\delta_j}|\delta_1, \delta_2, \ldots, 0_k, \ldots\rangle, \tag{53}$$

and $a_k|\delta_1, \delta_2, \ldots, 0_k, \ldots\rangle = 0$.

6 ATOMIC AND MOLECULAR HAMILTONIANS

An atomic or molecular N-electrons system can be represented by the Hamiltonian

$$H = \sum_{i=1}^{N} \left\{ -\frac{1}{2}\nabla_i^2 - \sum_A \frac{Z_A}{|\mathbf{r}_i - \mathbf{R}_A|} \right\} + \sum_{i>j=1}^{N} \frac{1}{r_{ij}}, \tag{54}$$

in atomic units and with the usual assumption of the Born-Oppenheimer approximation and the neglect of relativistic effects. The Hamiltonian contains only one-body and two-body interactions, which makes it feasible to simulate them with quantum-computer circuits. In second-quantized notation, the Hamiltonian can be expressed as

$$H = \sum_{j,k} h_{j,k} a_j^\dagger a_k + \frac{1}{2} \sum_{j \neq k, p \neq q} v_{j,k,p,q} a_j^\dagger a_k^\dagger a_p a_q. \tag{55}$$

The coefficients $h_{j,k}$ and $v_{j,k,p,q}$ need to be evaluated in advance on a classical computer. In terms of a selected set of basis functions $\{\phi_k(x)\}$, we have

$$h_{j,k} = \left\langle \phi_j(x) \middle| \left\{ -\frac{1}{2}\nabla^2 - \sum_A \frac{Z_A}{|\mathbf{r} - \mathbf{R}_A|} \right\} \middle| \phi_k(x) \right\rangle, \tag{56}$$

and

$$v_{j,k,p,q} = \left\langle \phi_j(x)\phi_k(x') \middle| \frac{1}{|\mathbf{r} - \mathbf{r}'|} \middle| \phi_p(x)\phi_q(x') \right\rangle. \tag{57}$$

The next step is to assemble a set of quantum gates, which will simulate the actions of the creation and annihilation operators in this Hamiltonian. Qubits are often physically realized with the spin-$\frac{1}{2}$ states of electrons or nuclei. The spin-up state is written $|1\rangle$ (or sometimes $|1/2\rangle$), the spin-down state, $|-1\rangle$ (or $|-1/2\rangle$). The analogous qubits are designated $|0\rangle$ and $|1\rangle$, respectively. In Pauli spin algebra, raising and lowering operators are defined by

$$\sigma^{\pm} = \frac{1}{2}\left(\sigma_x \pm i\sigma_y\right), \tag{58}$$

such that

$$\sigma^{+}|-1\rangle = |1\rangle, \quad \sigma^{+}|1\rangle = 0, \quad \sigma^{-}|1\rangle = |-1\rangle, \quad \sigma^{-}|-1\rangle = 0. \tag{59}$$

These relations suggest the analogous quantum gates

$$Q_k^\dagger = \frac{1}{2}\left(X_k + iY_k\right), \quad Q_k = \frac{1}{2}\left(X_k - iY_k\right), \tag{60}$$

such that

$$Q_k^\dagger|1_k\rangle = |0_k\rangle, \quad Q_k^\dagger|0_k\rangle = 0, \quad Q_k|0_k\rangle = |1_k\rangle, \quad Q_k|1_k\rangle = 0. \tag{61}$$

It is useful to review the multiplicative relations involving Pauli gates on the same qubit:

$$X^2 = Y^2 = Z^2 = I, \quad XY = -YX = iZ, \quad YZ = -ZY = iX, \quad ZX = -XZ = iY,$$
$$QZ = -ZQ = Q, \quad Q^\dagger Z = -ZQ^\dagger = -Q^\dagger. \tag{62}$$

The correspondence between creation/annihilation operators and Pauli gates works very well for actions on a single qubit. However, a problem arises for multiple qubits: operators on different spins commute, but operators on multifermion states need to obey the appropriate anticommutation relations. This can be fixed by means of a *Jordan-Wigner transformation*. Essentially, this determines the proper $\pm$ signs, just like the $(-1)^\delta$ factors in Eqs. (52), (53). Recall the action of the Pauli-Z gate

$$Z_k|0_k\rangle = +|0_k\rangle, \quad Z_k|1_k\rangle = -|1_k\rangle. \tag{63}$$

This suggests a form for the quantum gate corresponding to the creation operator $a_k^\dagger$, which we designate $A_k{}^\dagger$:

$$A_k{}^\dagger = \otimes_{j=1}^{k-1} Z_j \otimes Q_k^\dagger \otimes_{j=k+1}^{D} I_j. \tag{64}$$

Analogously,

$$A_k = \otimes_{j=1}^{k-1} Z_j \otimes Q_k \otimes_{j=k+1}^{D} I_j. \tag{65}$$

Note that the gates $A_k{}^\dagger$ and A_k are now *nonlocal*, since their action depends on the other qubits in the Fock state. For example, for $D = 4, k = 3$, we have

$$A_3{}^\dagger = Z_1 \otimes Z_2 \otimes Q_3^\dagger \otimes I_4, \quad A_3 = Z_1 \otimes Z_2 \otimes Q_3 \otimes I_4. \tag{66}$$

Next, we show how to construct quantum-computer circuits to represent parts of the molecular Hamiltonian. For the diagonal terms in the one-particle sum in Eq. (55), the operator products $a_k{}^\dagger a_k = N_k$, which represent number operators. The corresponding Jordan-Wigner operators are given by

$$A_k{}^\dagger A_k = \left(\otimes_{j=1}^{k-1} Z_j \otimes Q_k^\dagger \otimes_{j=k+1}^{D} I_j\right)\left(\otimes_{j=1}^{k-1} Z_j \otimes Q_k \otimes_{j=k+1}^{D} I_j\right) = Q_k^\dagger Q_k$$
$$= \frac{1}{2}\left(X_k + iY_k\right)\frac{1}{2}\left(X_k - iY_k\right) = \frac{1}{4}\left(2I_k - i[X_k, Y_k]\right) = \frac{1}{2}\left(I_k + Z_k\right), \tag{67}$$

noting that $Z_j^2 = I_j$, $X_k^2 = Y_k^2 = I_k$, and $[X_k, Y_k] = 2iZ_k$.

For off-diagonal contributions to the one-particle sum, such as $a_j{}^\dagger a_k$, consider the case $j, k = 2, 3$:

$$A_3{}^\dagger A_2 = \left(Z_1 \otimes Z_2 \otimes Q_3^\dagger \otimes I_4\right)\left(Z_1 \otimes Q_2 \otimes I_3 \otimes I_4\right)$$
$$= Z_1^2 \otimes Z_2 Q_2 \otimes Q_3^\dagger I_3 \otimes I_4 = -I_1 \otimes Q_2 \otimes Q_3^\dagger \otimes I_4, \tag{68}$$

while

$$A_2{}^\dagger A_3 = \left(Z_1 \otimes Q_2^\dagger \otimes I_3 \otimes I_4\right)\left(Z_1 \otimes Z_2 \otimes Q_3 \otimes I_4\right)$$
$$= Z_1^2 \otimes Q_2^\dagger Z_2 \otimes I_3 Q_3 \otimes I_4 = -I_1 \otimes Q_2^\dagger \otimes Q_3 \otimes I_4. \tag{69}$$

Thus

$$A_2{}^\dagger A_3 + A_3{}^\dagger A_2 = -I_1 \otimes \left(Q_2^\dagger \otimes Q_3 + Q_2 \otimes Q_3^\dagger\right) \otimes I_4 = -\frac{1}{2} I_1 \otimes \left(X_2 \otimes X_3 + Y_2 \otimes Y_3\right) \otimes I_4. \tag{70}$$

If the indices j, k are not consecutive, then the intervening indices m will contribute factors of Z_m interposed between the j and k factors. Thus

$$A_1{}^\dagger A_3 + A_3{}^\dagger A_1 = -\frac{1}{2}\left(X_1 \otimes Z_2 \otimes X_3 + Y_1 \otimes Z_2 \otimes Y_3\right) \otimes I_4,$$
$$A_1{}^\dagger A_4 + A_4{}^\dagger A_1 = -\frac{1}{2}\left(X_1 \otimes Z_2 \otimes Z_3 \otimes X_4 + Y_1 \otimes Z_2 \otimes Z_3 \otimes Y_4\right), \tag{71}$$

and so forth.

Proceeding analogously for the two-particle terms in the Hamiltonian Eq. (55), we need to evaluate products of the form $A_j{}^\dagger A_k{}^\dagger A_p A_q$ using Eqs. (64), (65). For the sum of two-electron Coulomb plus exchange terms, we find

$$A_j{}^\dagger A_k{}^\dagger A_j A_k - A_j{}^\dagger A_k{}^\dagger A_k A_j = \frac{1}{2}\left(I_j + Z_j\right) \otimes \left(I_k + Z_k\right). \tag{72}$$

For the case $j \neq k \neq p \neq q$,

$$\begin{aligned} A_j{}^\dagger A_k{}^\dagger A_p A_q + A_q^\dagger A_p{}^\dagger A_k A_j = -\frac{1}{2}\Big(&X_j \otimes X_k \otimes X_p \otimes X_q - X_p \otimes X_q \otimes Y_j \otimes Y_k \\ &+ X_k \otimes X_q \otimes Y_j \otimes Y_p + X_j \otimes X_q \otimes Y_k \otimes Y_p + X_k \otimes X_p \otimes Y_j \otimes Y_q \\ &+ X_j \otimes X_p \otimes Y_k \otimes Y_q - X_j \otimes X_k \otimes Y_p \otimes Y_q + Y_j \otimes Y_k \otimes Y_p \otimes Y_q\Big). \end{aligned} \tag{73}$$

7 TIME-EVOLUTION OF A QUANTUM SYSTEM

The principal goal of the quantum-computer program we are outlining is to find accurate values of energy eigenvalues, most often just the ground-state energy. The method exploits the fact that the time-evolution of a quantum system exhibits its eigenvalues, in the form of a Fourier superposition of its eigenstates. Let us begin with the time-dependent Schrödinger equation for an arbitrary quantum state $\Psi(x_1, x_2, \ldots, x_N, t)$, abbreviated $\Psi(x, t)$:

$$i\frac{\partial \Psi(x,t)}{\partial t} = H\Psi(x,t). \tag{74}$$

The formal solution, with an initial state $\Psi(x,0)$, can be written

$$\Psi(x,t) = e^{-iHt}\Psi(x,0), \tag{75}$$

where the exponential, known as the *evolution operator* or *propagator*, is explicitly a representation of the power series

$$e^{-iHt} = \sum_{n=0}^{\infty} \frac{(-iHt)^n}{n!} = 1 + (-it)H + \frac{(-it)^2}{2!}H^2 + \frac{(-it)^3}{3!}H^3 + \cdots. \tag{76}$$

The initial state $\Psi(x,0)$ is formally represented by a superposition of eigenstates of the Hamiltonian H:

$$\Psi(x,0) = \sum_m c_m \psi_m(x) = c_0\psi_0(x) + c_1\psi_1(x) + c_2\psi_2(x) + \cdots + \text{continuum}, \tag{77}$$

where $H\psi_m(x) = E_m\psi_m(x) = \omega_m\psi_m(x)$. It is useful to express the energies in frequency units $E_m = \hbar\omega_m = \omega_m$. We label the ground state as $m = 0$. Applying the evolution operator, we find

$$\begin{aligned}\Psi(x,t) &= e^{-iHt}\Psi(x,0) = \sum_m c_m e^{-i\omega_m t}\psi_m(x)\\ &= c_0 e^{-i\omega_0 t}\psi_0(x) + c_1 e^{-i\omega_1 t}\psi_1(x) + \cdots + \text{continuum}.\end{aligned} \tag{78}$$

The scalar product of $\Psi(x,t)$ with $\Psi(x,0)$ is an instance of an *autocorrelation function*:

$$F(t) = \langle\Psi(x,0)|\Psi(x,t)\rangle = \langle\Psi(x,0)|e^{-iHt}|\Psi(x,0)\rangle = \sum_m |c_m|^2 e^{-i\omega_m t} = |c_0|^2 e^{-i\omega_0 t} + \cdots. \tag{79}$$

The Fourier transform of the autocorrelation function exhibits the spectrum of eigenvalues

$$G(\omega) = \frac{1}{2\pi}\sum_m |c_m|^2 \delta(\omega - \omega_m) + \text{continuum}. \tag{80}$$

In practice, the peaks of $G(\omega)$ will be finite, because of approximations and inaccuracies, something like the spectrum shown in Fig. 17.

The objective pursued in this chapter is computation of the quantized energy levels of a molecular system. This will be realized by simulation of the dynamics of the system, generated by the time-evolution operator e^{-iHt}. To do this on a quantum computer, we must construct a set of one- and two-qubit gates that implement this exponential transformation on a set of qubits. Approximations to the energy eigenvalues then appear in the phases of qubit states, which are determined using the phase-estimation algorithm.

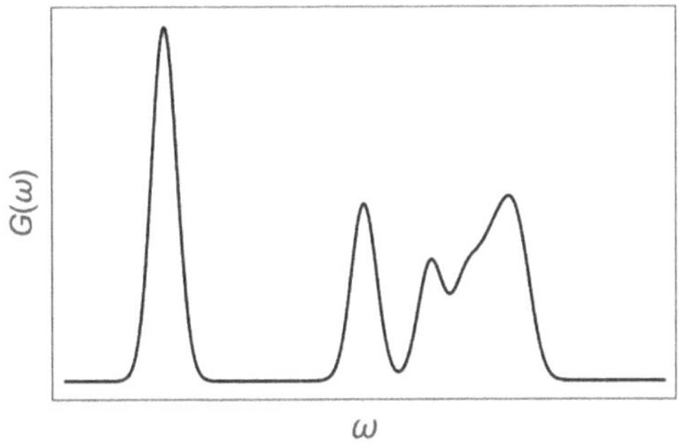

FIG. 17

Simulation of computed spectrum in quantum computation.

8 TROTTER EXPANSIONS

The time-evolution operator e^{-iHt} is constructed by exponentiation of the second-quantized representation of the molecular Hamiltonian. However, the terms the Hamiltonian, $H = h_1 + h_2 + \cdots$, do not generally commute among themselves. Therefore, approximate expansions involving noncommuting exponential operators are necessary.

Consider an exponential operator of the form $e^{(A+B)t}$ where A and B do not, in general, commute.[1] If the time interval t is broken up into n subintervals $\Delta t = t/n$, the *Lie-Trotter product formula* states that[2]

$$e^{(A+B)t} = \lim_{\Delta t \to 0} \left(e^{A\Delta t} e^{B\Delta t} \right)^n. \tag{81}$$

For a finite value of n, this is approximated by

$$e^{(A+B)t} \approx \left(e^{A\Delta t} e^{B\Delta t} \right)^n + O(\Delta t^{n+1}). \tag{82}$$

The right-hand side can be rearranged to

$$\left(e^{A\Delta t/2} e^{B\Delta t/2} e^{A\Delta t/2} e^{B\Delta t/2} \right)^n = \left(e^{A\Delta t/2} e^{B\Delta t} e^{A\Delta t/2} \right)^n \text{ or } \left(e^{B\Delta t/2} e^{A\Delta t} e^{B\Delta t/2} \right)^n. \tag{83}$$

This is the form generally used in applications of Trotter approximations for exponential operators, with values of n anywhere from 1 to 10, as appropriate. The

[1] Recall the Baker-Hausdorff formula $e^{(A+B)t} = e^{At} e^{Bt} e^{-\frac{1}{2}[A,B]t^2} + O(t^3)$.

[2] A simple proof:

$$e^{(A+B)t} = \lim_{n\to\infty} \left(1 + \frac{(A+B)t}{n}\right)^n = \lim_{n\to\infty} \left(1 + \frac{At}{n}\right)^n \left(1 + \frac{Bt}{n}\right)^n = \lim_{n\to\infty} \left(e^{At/n}\right)^n \left(e^{Bt/n}\right)^n.$$

propagator e^{-iHt} should act over sufficiently long time to resolve the leading Fourier frequencies to a desired level of precision.

As shown in Eq. (8), exponential operators involving Pauli matrices can be evaluated in closed form.

9 SIMULATIONS OF MOLECULAR STRUCTURE

By applying the Jordan-Wigner transformation and Trotter decomposition, it becomes possible to approximate the energy levels of a fermionic many-body Hamiltonian system on a quantum computer [6]. The energy levels of a quantum system can, in principle, be determined by simulating its time evolution, by action of the operator $U = e^{-iHt}$. This might be done on a quantum computer using a set of one- and two-qubit gates that implement the time evolution on a set of qubits. Recent work by Aspuru-Guzik and coworkers has been able to create a quantum circuit to compute the energy of a molecular system with fixed nuclear geometry using a recursive quantum phase estimation algorithm to simulate a full configuration interaction computation [7–10].

Let us develop the method beginning with a simple example. A single-particle term in the Hamiltonian, under a Jordan-Wigner transformation (see Eq. 67), has the form:

$$H = ha^\dagger a \rightarrow hA^\dagger A = \frac{h}{2}(I + Z). \tag{84}$$

If this were the only term in the Hamiltonian, the time-evolution operator would be given by

$$U = e^{-i\frac{h}{2}(I+Z)t} = e^{-iht/2}e^{-ihtZ/2}. \tag{85}$$

The exponential operator reduces to separate factors here since I and Z commute. The second factor has the form of a "rotated" Pauli operator (see Eq. 8). Therefore,

$$e^{-ihZt/2} = \cos\left(\frac{ht}{2}\right)I - i\sin\left(\frac{ht}{2}\right)Z = \begin{pmatrix} e^{-iht/2} & 0 \\ 0 & e^{iht/2} \end{pmatrix} \tag{86}$$

and

$$U = \begin{pmatrix} e^{-iht} & 0 \\ 0 & 1 \end{pmatrix}. \tag{87}$$

Taking $\Psi(0) = \begin{pmatrix} 1 \\ 0 \end{pmatrix}$, the autocorrelation function is given by

$$F(t) = \Psi^\dagger(0)U(t)\Psi(0) = e^{-iht}, \tag{88}$$

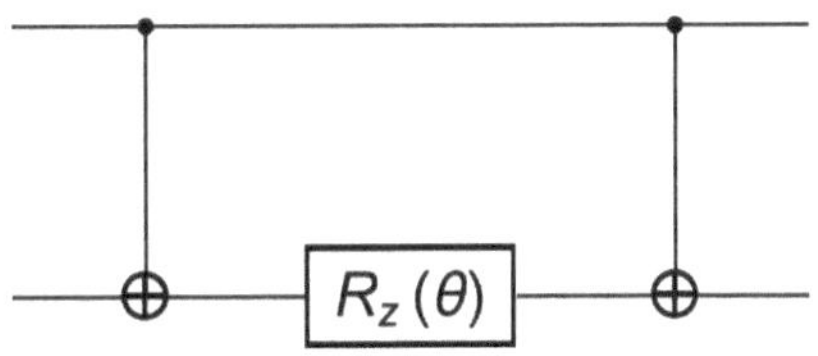

FIG. 18

Gate for two-particle terms.

which implies that the system has a single eigenvalue: $E_0 = h$.[3]

Consider next a two-particle contribution to the Hamiltonian, representing the Coulomb plus exchange integrals for spin-orbitals 1 and 2. Using Eq. (72) we have

$$v(a_1^\dagger a_2^\dagger a_2 a_1 - a_1^\dagger a_2^\dagger a_1 a_2) \rightarrow v(A_1^\dagger A_2^\dagger A_2 A_1 - A_1^\dagger A_2^\dagger A_1 A_2) = \frac{v}{2}(I+Z)\otimes(I+Z)\,. \quad (89)$$

The propagator corresponding to this Hamiltonian is given by

$$U_{12} = e^{-i\frac{vt}{2}(I+Z)\otimes(I+Z)}. \quad (90)$$

The unitary gate that can carry out operations of the form $\exp\left(-i\frac{\theta}{2}Z\otimes Z\right)$ is shown in Fig. 18. The CNOT gate first entangles the two qubits. Then the rotation $R_z(\theta) = e^{-i\frac{\theta}{2}Z}$ is applied, followed by a second CNOT gate.

Fig. 19 is a highly schematic diagram of a circuit, which might be applicable to a simplified two-electron system, such as a helium atom or hydrogen molecule, in which there are two single-particle terms and one interaction term in the Hamiltonian. A minimum basis set of qubits is designated by $|\phi_1\rangle, |\phi_2\rangle$. The input parameters are $h_{11} = h_{22} = h$ and $v_{1221} = v$. The gate $T(\theta)$ is represented by the unitary matrix $\begin{pmatrix} 1 & 0 \\ 0 & e^{-i\theta} \end{pmatrix}$. PEA represents the gates for the phase-estimation algorithm, which are shown fully in Fig. 16.

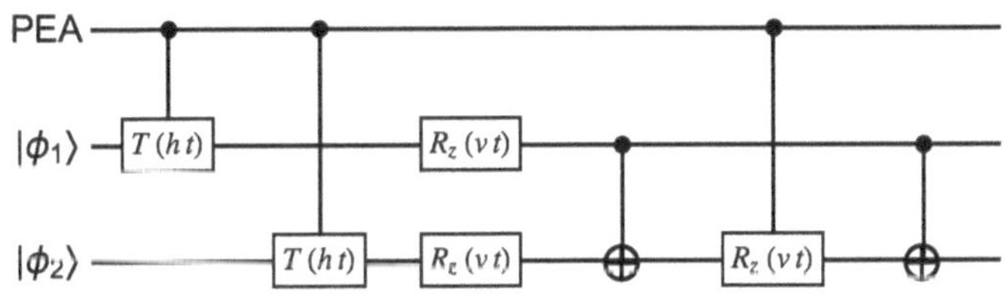

FIG. 19

Simplified circuit for two-particle Hamiltonian.

[3] If this were applied to the harmonic oscillator, with $H = a^\dagger a + \frac{1}{2}$, the result would give $E_0 = \frac{1}{2}$.

In summary, we outline the general procedure for a simulation of an atomic or molecular system by quantum computer circuit: (1) The Hamiltonian is expressed as a sum of products of Pauli operators, using the Jordan-Wigner transformation, in conjunction with a set of molecular integrals calculated by conventional computation. (2) Each of these operators is converted into unitary gates such that their sequential execution on a quantum computer can approximate the action of the propagator e^{-iHt}, with application of the Trotter decomposition. (3) The phase estimation algorithm is applied to approximate the eigenvalue of an eigenstate produced by the QFT of the time domain propagation. The output results can serve as an input for an improved approximation, and the procedure can be further iterated, as desired. The rather intricate and extensive details of the circuits, which carry out these operations are described more fully in the references.

REFERENCES

[1] Nielsen MA, Chuang IL. Quantum computation and quantum information. Cambridge: Cambridge University Press; 2011.

[2] Feynman RP. Simulating physics with computers. Int J Theor Phys 1982;21:467–88.

[3] Deutsch D. Quantum theory, the church-Turing principle and the universal quantum computer. Proc R Soc Lond A 1985;400:97–117.

[4] Ladd TD, Jelezko F, Laflamme R, Nakamura Y, Monroe C, O'Brien JL. Quantum computers. Nature 2010;464:45–53.

[5] Fano G, Blinder SM. Twenty-first century quantum mechanics: Hilbert space to quantum computers. New York, NY: Springer; 2017.

[6] Ovrum E, Hjorth-Jensen M. Quantum computation algorithm for many-body studies. Available from: https://arxiv.org/abs/0705.1928v1.

[7] Aspuru-Guzik A, Dutoi A, Love P, Head-Gordon M. Simulated quantum computation of molecular energies. Science 2005;309:1704.

[8] Whitfield JD, Biamonte J, Aspuru-Guzik A. Simulation of electronic structure Hamiltonians using quantum computers. Mol Phys 2011;109:735–50.

[9] Kassal I, Whitfield JD, Perdomo-Ortiz A, Yung MH, Aspuru-Guzik A. Simulating chemistry using quantum computers. Ann Rev Phys Chem 2011;62:185–207.

[10] O'Malley PJJ, et al. Scalable quantum simulation of molecular energies. Phys Rev X 2016;6:031007.

Index

Note: Page number followed by *f* indicate figure, *t* indicate tables and *np* indicate footnote.

D

T

U

V

W

Z

Kolibri 360 GmbH
Ferdinand-Jühlke-Straße 7
99095 Erfurt, Deutschland
produktsicherheit@kolibri360.de